LASERS, OPTICS, AND PHOTONICS

Lasers, Optics, and Photonics

Mário F. S. Ferreira

Professor of Physics, University of Aveiro, Portugal

OXFORD
UNIVERSITY PRESS

OXFORD
UNIVERSITY PRESS

Great Clarendon Street, Oxford, OX2 6DP,
United Kingdom

Oxford University Press is a department of the University of Oxford.
It furthers the University's objective of excellence in research, scholarship, and education by publishing worldwide. Oxford is a registered trade mark of Oxford University Press in the UK and in certain other countries.

Published in the United States of America by Oxford University Press
198 Madison Avenue, New York, NY 10016, United States of America

British Library Cataloguing in Publication Data
Data available

Library of Congress Control Number: 2026936020

ISBN 9780198952770

ISBN 9780198952787 (pbk.)

DOI: 10.1093/oso/9780198952770.001.0001

Printed and bound by
CPI Group (UK) Ltd., Croydon, CR0 4YY

The manufacturer's authorized representative in the EU for product safety is Oxford University Press España S.A. of Parque Empresarial San Fernando de Henares, Avenida de Castilla, 2 – 28830 Madrid (www.oup.es/en or product.safety@oup.com). OUP España S.A. also acts as importer into Spain of products made by the manufacturer.

MIX
Paper | Supporting responsible forestry
FSC® C013604
FSC www.fsc.org

To the memory of my parents
To Sara Maria, Sara Raquel, Mário Alberto, Maria do Céu, and Miguel Fernando

Preface

During the last 60 years, there has been a truly revolutionary change in the field of optics. In fact, optics has been assuming an increasingly protagonism role, not only in the area of physics but also in other areas of science, engineering, medicine, industry and, in general, our day-to-day life.

The driving force behind this revolution was the discovery and development of various types of lasers, and their subsequent use in cutting-edge scientific research and far-reaching technological applications, with a direct impact on broad segments of the population. Today, it is easy to find in our homes various applications of optics which did not exist, or which were the subject of research, only 40 years ago. Among these applications, we can mention CD players, laser printers, optical surveillance systems, high-definition televisions, liquid crystal screens, etc. We also often find holograms on bank cards, bar-code readers in stores and supermarkets, optical systems for monitoring traffic in cities, optical sensors for water quality control, etc. In the area of medicine and health care, eye surgery with lasers, non-contact infrared thermometers, optical tomography, etc., are already common.

Despite this impressive horizon of applications of optics, it is, however, in the manipulation of information that light currently has a greater impact on our society. This is basically due to the ability to transmit information at the speed of light and to the large bandwidth offered by the electromagnetic spectrum. Indeed, the last four decades have seen a profound transformation in the field of information transmission. The so-called *optical communications* have become generalized and have radically modified the conceptions of the generation, manipulation and distribution of information. Compared to the optical information flows with which we worked in the mid-seventies of the last century, the ability to transmit information by optical means has since increased more than a million times. For this reason, the end of the 20th century was characterized as the historical moment in which the cost of transporting information became negligible and practically independent of distance, thus laying the foundations for the century in which we now live.

To sustain the expansion of optical telecommunications, several research methods and techniques have been developed, which have led to a new area of physics called photonics. Photonics is defined as *the science that studies the generation, manipulation, transport, and harnessing of light and other forms of radiant energy, whose quantum unit is the photon.* The most notable applications of photonics are currently in the fields of communications and information processing. These applications assume the existence of a wide range of components: lasers, light-emitting diodes, optical fibres, photodetectors, optical gates, optical amplifiers, optical modulators, etc. The development of this area has been such that, especially in the field of long-distance information transmission, both transoceanic and terrestrial, photonics has almost completely replaced electronics. In fact, it turns out that photonics currently represents one of the most dynamic and lucrative markets in industrialized countries.

This book is intended to be used from the third year at undergraduate level. The prerequisites are a basic knowledge of electromagnetism, physics of waves, geometry and differential equations. However, it will also be useful at the graduate level, particularly considering Chapters 15–25. In fact, the book grew out of the notes prepared by the author in the scope of several courses taught during the last three decades in the areas of optics, photonics and lasers at the University of Aveiro. Finding an appropriate textbook for some of these courses has always been a difficult task because the existing books in these areas are aimed at either too high a level or too low a level, or they cover only a portion of the topics that are needed. In teaching these classes, I came to rely mostly on my lecture notes as the reading material for the course.

The need for a truly introductory-level book, covering a wide range of topics in optics, photonics and lasers, was my motivation for writing this book. Including those topics in a single book is justified, since there is a natural synergy between them. For example, understanding the operating principle of certain lasers requires knowledge of photonics concepts such as waveguiding, and analysing the stability conditions for the laser resonant cavity requires the use of matrix theory, developed within the geometrical optics framework. On the other hand, some knowledge of lasers is necessary to understand the principles of optical communications, which is actually a main area of photonics.

Since the area of influence of optics, photonics and lasers has been going beyond the limits of physics and progressively extending to other areas of knowledge, namely chemistry, biology, engineering and medicine, we think that this book may be useful as a reference text in some disciplines and training programmes developed in these latter areas. It can also be a useful tool for professionals with diverse backgrounds who wish to start the study of this field. In order to facilitate the understanding of the various subjects presented in this book, emphasise is placed on the basic concepts involved in each of them. For the same reason, some examples of application are discussed throughout the text, and a set of problems to be solved by the students is presented at the end of each chapter.

The chapters are grouped together to provide a progressive presentation of the concepts. The first seven chapters correspond to the fundamental concepts relating to geometric optics, optical instruments, transport and measurement of the energy contained in optical radiation, waves, Fourier theory, electromagnetic theory, and polarization of light. Chapters 8 to 12 cover both the fundamentals and some applications of the phenomena of interference and diffraction of light. Chapter 13 is devoted to holography, while Chapter 14 describes the main characteristics of anisotropic media. Chapter 15 is dedicated to the basic properties of both conventional and microstructured optical fibres, including their main applications, whereas Chapter 16 discusses, in particular, their dispersive characteristics. Lasers and the main properties of laser light, as well as an introduction to the physics of semiconductors, are presented in Chapters 17 to 19. The non-linear properties of a generic medium in which light propagates are discussed in Chapter 20, whereas Chapter 21 is concerned with non-linear effects in optical fibres. Chapter 22 describes the main characteristics of fibre amplifiers and lasers, whereas Chapter 23 is dedicated to the application of non-linear effects for all-optical signal processing. Chapter 24 discusses the properties of a particularly interesting non-linear pulse: the optical soliton. Finally, Chapter 25 provides an overview of the basic features and main applications of one of the most impressive non-linear phenomena that can be observed in optical fibres: supercontinuum generation.

I am deeply grateful to the many students and colleagues with whom I have interacted over the years. All of them contributed to this book, either directly or indirectly. Last but not least, I thank my family for understanding why I needed to continue working during many of our weekends and holidays.

Mário F. S. Ferreira
Aveiro, Portugal
August 2025

Contents

About the Author

Mário F. S. Ferreira is a professor at the physics department of Aveiro University, Portugal. Between 1990 and 1991 he was at the University of Essex, UK, performing experimental work on external cavity semiconductor lasers and non-linear optical fibre amplifiers. His research interests have been concerned with the modelling and characterization of multi-section semiconductor lasers for coherent systems, quantum well lasers, optical fibre amplifiers and lasers, optical sensors, soliton propagation, polarization and non-linear effects in optical fibers. He has been the leader of the Optics and Optoelectronics Group of the *I3N—Institute of Nanostructures, Nanomodelling and Nanofabrication*. He has written about 400 scientific journal and conference publications and several books, namely: *Optics and Photonics* (Lidel, 2003, in Portuguese), *Topics of Mathematical Physics* (Editora Ciência Moderna, 2018, Brazil, in Portuguese), *Optical Fibers: Technology, Communications and Recent Advances* (Ed., NOVA Science Publishers, 2017), *Advances in Optoelectronic Technology and Industry Development* (CRC Press, 2019), *Nonlinear Effects in Optical Fibers* (John Wiley & Sons, OSA, 2011), *Optical Signal Processing in Highly Nonlinear Fibers* (CRC Press, 2020), *Optical Fiber Technology and Applications—Recent Advances* (IOP Publishing, 2021), *Solitons in Optical Fiber Systems* (John Wiley & Sons, 2022), *Dissipative Optical Solitons* (Springer, 2022), *Photonic Crystal Fibers—Recent Advances* (NOVA Science Publishers, 2023), *Specialty Optical Fibers—Materials, Fabrication Technology and Applications* (Elsevier, 2024), *Optical Fiber Lasers—Recent Advances and Applications* (Springer Nature, 2025). He was the guest editor of four special issues of *Fiber and Integrated Optics* (Taylor & Francis): *Fiber Optics in Portugal* (2005), *Nonlinear Fiber Optics* (2015), *Optical Fiber Sources and Amplifiers* (2020) and *Quantum Communications* (2020), a joint special issue of *Optics Express* and *Applied Optics* (OSA) on *Optical Sensors and Sensing 2019*, one special issue of the *Journal of Nonlinear Mathematical Physics* (Springer) on *Complex Ginzburg-Landau Equation*, and two special issues of *Fibers* (MDPI) on *Optical Fiber Communications* and *Specialty Optical Fibers—Material, Fabrication and Applications* (2022), and two special issues of *Fibers* (MDPI) on *Optical Fiber Communications* (2020) and *Characterization and Applications of Specialty Optical Fibers* (2025).

He is a fellow and a travelling lecturer of both OPTICA (Optical Society of America) and SPIE—The International Society for Optics and Photonics, and a member of IEEE. He served in various committees of OSA/OPTICA and of SPIE, as well as in the organizing and scientific committees of various international conferences. He also serves as an associate editor or as an advisor board member of several international journals in the area of optics and photonics.

List of Abbreviations

AM	Amplitude modulation
ASE	Amplified spontaneous emission
ASK	Amplitude-shift keying
BER	Bit error rate
BFL	Brillouin fibre laser
CGLE	Complex Ginzburg–Landau equation
CP	Composite pulse
CW	Continuous wave
DBR	Distributed Bragg reflector
DC	Directional coupler
DCF	Dispersion compensating fibre
DFB	Distributed feedback
DFT	Dispersive Fourier transform
DGD	Differential group delay
DM	Dispersion management
DSF	Dispersion-shifted fibre
DSR	Dissipative soliton resonance
ECL	External cavity laser
EDF	Erbium-doped fibre
EDFA	Erbium-doped fibre amplifier
ESA	Excited state absorption
FBG	Fiber Bragg grating
FM	Frequency modulation
FOG	Fiber optic gyroscope
FOPA	Fiber optical parametric amplifier
FOPO	Fiber optic parametric oscillator
FP	Fabry–Perot
FRA	Fiber Raman amplifier
FSK	Frequency-shift keying
FWHM	Full width at half maximum

FWM	Four-wave mixing
GNLSE	Generalized non-linear Schrödinger equation
GVD	Group velocity dispersion
HNLF	Highly non-linear fibre
HOE	Higher-order effect
HWP	Half-wave plate
IRS	Intrapulse Raman scattering
KdV	Korteweg–de Vries
LAN	Local area network
LD	Laser diode
LED	Light-emitting diode
MAN	Metropolitan area network
MDDV	Minimum distance of distinct vision
MDM	Mode-division multiplexing
MI	Modulation instability
MIR	Mid-infrared
MOF	Microstructured optical fibre
MSR	Mode-suppression ratio
MZI	Mach–Zehnder interferometer
NALM	Non-linear amplifying loop mirror
NF	Noise figure
NIR	Near infrared
NLSE	Non-linear Schrödinger equation
NOLM	Non-linear optical loop mirror
NPR	Non-linear polarization rotation
NSR	Nonsoliton radiation
NZDF	Nonzero dispersion fibre
OOK	On-off keying
OPC	Optical phase conjugation
OTDM	Optical time-division multiplexing
PBG	Photonic bandgaps
PCF	Photonic crystal fibers
PCM	Phase conjugation mirror
PDM	Polarization-division multiplexing
PM	Phase modulation

PMD	Polarization-mode dispersion
PPLN	Periodically poled lithium niobate
PSK	Phase-shift keying
PSP	Principal states of polarization
QWP	Quarter-wave plate
RZ	Return to zero
SA	Saturable absorber
SBS	Stimulated Brillouin scattering
SCG	Supercontinuum generation
SHG	Second harmonic generation
SL	Semiconductor laser
SLM	Single longitudinal mode
SMF	Single mode fibre
SNR	Signal to noise ratio
SOA	Semiconductor optical amplifier
SOP	State of polarization
SPM	Self-phase modulation
SRS	Stimulated Raman scattering
SSFS	Soliton self-frequency shift
SST	Self-steepening
SWIR	short-wave infrared
TOD	Third-order dispersion
UV	Ultraviolet
VHA	Very high amplitude
XPM	Cross-phase modulation
WAN	Wide area network
WDM	Wavelength-division multiplexing
WSC	Wavelength selective coupler
ZDW	Zero-dispersion wavelength

Dixitque Deus: 'Fiat lux'. Et facta est lux. (Genesis 1:3)

The Nature of Light

1

Light plays a fundamental role in the spectacle that nature offers to us. Some scenes of this spectacle are briefly described in the appendix to this book. Furthermore, light is one of the basic ingredients of life on the earth, and understanding its true nature becomes really important. Consider, for example, the importance of sunlight for plants, which convert it into chemical energy through the photosynthesis process. Moreover, light is one of the main means through which we can transmit and receive information from beings around us. Actually, visual, and hence optical, communication is older than language. Hand signals, waving of the arms, and fire and smoke signals are basic means of communication.

The *visible spectrum* refers to the portion of the electromagnetic spectrum perceived as different colours by the human eye, ranging from the violet ($\sim$ 380 nm) to the red ($\sim$ 780 nm) colours. Actually, the visible spectrum is properly referred to as '*light*' and corresponds—quite naturally—to the frequencies of electromagnetic radiation that predominate in the output of the sun. The *optical spectrum* contains the visible region of the electromagnetic spectrum and is bounded by the invisible *ultraviolet* and *infrared* regions.

1.1 A Brief Historical Perspective

In this section we will provide a brief historical perspective regarding the evolution of our understanding of the physical nature of light. Actually, light's true nature has eluded us for centuries, and the suggestion of successive theories aiming at its description forms one of the most fascinating accounts in the history of science.

Already in Ancient Greece, several philosophers, such as Pythagoras, Democritus, Empedocles, Plato, and Aristotle, developed diverse theories about the nature of light. Aristotle's theory, in particular, was very similar to the theory of ether that would be developed in the 19th century. The geometer Euclid (c. 325 BC–265 BC) was the first known author of a treatise on geometrical optics. Following Plato's teaching, he assumed that the luminous 'rays' had their origin in the observer's eyes and were directed to the contemplated objects.

Hero of Alexandria (c. 10–70 AD) extended the principles of geometrical optics and argued that light, when travelling between two points, always follows the shortest path. In the case of a homogeneous medium, this path corresponds to a rectilinear trajectory. Using this principle, Hero was able to geometrically demonstrate the law of reflection. The principle of the shortest path turned out to be basically correct, although its author also admitted that light rays proceeded at great speed from the observer's eye to the object seen.

The refraction of light was studied by Cleomedes (50 AD) and later by Claudius Ptolemy (100–170 AD), from Alexandria, whose observations were recorded in his book *Optics*. He carried out a series of very strict measurements of the angles of incidence (θ_i) and refraction (θ_t) for various media. From these measurements, it came to an empirical relationship between the two angles, given by $\theta_t = a\theta_i - b\theta_i^2$, where a and b are constants that depend on the media that form the interface. It is interesting to note that although this expression is valid only for very small angles of incidence, it remained unchanged for about 1500 years!

Lasers, Optics, and Photonics. Mário F. S. Ferreira, Oxford University Press. © Mário F. S. Ferreira (2026).
DOI: 10.1093/oso/9780198952770.003.0001

The fall of the Western Roman Empire in 475 AD marks the beginning of a period during which scientific activity in Europe virtually stagnated. Meanwhile, the expansion of Islam shifted the centre of intellectual activity to the Arab world. It is in this context that the figure of Alhazen (965–1039) emerged in Baghdad; he wrote a collection of seven books on optics. Among other subjects, Alhazen elaborated the law of reflection, studied the characteristics of spherical and parabolic mirrors, and described in detail the functioning of the human eye. In his conception, light rays did not originate in the eye but in the sources that illuminate the objects, from which they are directed to the eye. Alhazen also argued that light spreads at a finite, albeit very high, speed and that this speed should be lower in an optically denser medium.

Alhazen's work was translated into Latin and eventually had a major influence on optical studies in Europe. Among others, special mention must be made of the English Bishop Robert Grosseteste (1175–1253), who was one of the most prominent figures in 13th-century English intellectual life, and the Polish mathematician Vitelo (c.1230–1280), who produced a massive presentation of the subject of optics entitled *Perspectiva.* Both also contributed to the recovery of the works of Aristotle. These studies were continued by the English Franciscan Roger Bacon (1215–1294), who was the first to propose the use of lenses to improve vision and to combine multiple lenses to form a telescope. Another English Franciscan, John Pecham (c. 1230–1292), who also became Archbishop of Canterbury, built on the work of Bacon, Grosseteste, and a range of earlier writers to produce what became the most widely used textbook on optics of the Middle Ages, the *Perspectiva Communis.* In the late 13th and early 14th centuries, Theodoric of Freiberg (1250–1310) was among the first in Europe to provide the correct scientific explanation for the rainbow phenomenon.

A mention has to be made of the contribution of Leonardo da Vinci (1452–1519), who described the operation of the so-called *camera obscura* (pinhole camera) in his manuscript *Codex Atlanticus.* It should be noted, however, that the principle of operation of the *camera obscura* had already been mentioned in antiquity by Aristotle and discussed in the 11th century by Alhazen.

Using a telescope made by himself, Galileo Galilei (1564–1642) discovered Jupiter's moons and many other wonders of celestial space. After checking Galileo's findings, Johannes Kepler (1571–1630) published his book *Dioptrice* in 1609, in which many of the results obtained so far were summarized. Among other subjects, this book presents the theory of lens combinations and describes the phenomenon of refraction using a small-angle approximation. Another main book by Kepler is *Astronomiae Pars Optica,* published in 1604, in which he describes the inverse-square law governing the intensity of light, reflection by flat and curved mirrors, and the principles of pinhole cameras, among other subjects. This book is generally recognized as the foundation of modern optics.

The law of refraction was definitively discovered in 1621 by Willebrord Snell (1591–1626) in Leyden. However, Snell's work was essentially empirical in nature and was not published immediately. René Descartes (1596–1650), who was apparently unaware of Snell's work, published the work *La Dioptrique* in 1637, presenting for the first time the law of refraction as we know it today. This was a big step forward in designing and fabricating new optical instruments.

Pierre de Fermat (1601–1665) demonstrated in 1657 the law of reflection on the basis of his *principle of minimum time.* According to this principle, which is an evolution of the idea of the shortest path advocated by Hero, light propagates from one point to another following a path that minimizes the propagation time, even if to do so it has to deviate from the straight line that passes through these points. On the basis of the same principle, and introducing the concept of optical resistance of the medium, Fermat also obtained the law of refraction.

Both Descartes and Fermat admitted that light consisted of a stream of particles. However, the main architect of this corpuscular theory was Isaac Newton (1642–1727), who explained on this basis the main characteristics of light observed experimentally, namely the laws of reflection and refraction. In his *Hypothesis of Light* of 1675, Newton posited the existence of the ether to transmit forces between particles. In 1704 he published *Opticks*, which became for a long time a reference work for the description of the characteristics of light. In this work, Newton described a series of experiments related to refraction and the domain currently called interference. One of these experiments illustrated the phenomenon of the dispersion of sunlight in a prism. Newton found that the phenomenon of colour was an intrinsic feature of light and that white light from the sun was the result of the combination of rays of all colours. These rays were understood as streams of particles that would move through the omnipresent ether. From this work he concluded that any refracting telescope would suffer from the dispersion of light. In order to avoid this problem, he proposed in 1668 a reflecting telescope, today known as a Newtonian telescope, which uses mirrors instead of lenses.

As opposed to the corpuscular theory, there were also those who advocated, already in the 17th century, wave-like theories of light. Among the proponents of these theories was Robert Hooke (1635–1703), who came to confront Newton directly at the Royal Society of London. Among other works, Hooke described the interference patterns with various colours that occur in thin dielectric films. Based on the hypothesis of the ethereal medium, he proposed an analogy between light waves and waves that form in water.

The Dutch physicist Christian Huygens (1629–1695) can be considered the main architect of the wave model of light, on the basis of which he explained the laws of reflection and refraction. This wave model was explained in his book *Traité de la Lumiére*, published in 1690, in which it was enunciated for the first time that what would come to be known as *the Huygens principle*: *each point of a primary wave front constitutes a source of elementary spherical waves, the overlap of which allows one to obtain the primary wave front at a later instant.* Adopting his wave theory, Huygens was able to derive the *laws of reflection* and *refraction* and to explain *double refraction* in calcite as well.

For various reasons, the wave theory of light turned out not to receive a great welcome in the scientific community of that time. In fact, all kinds of waves then known (sound waves, waves in water, etc.) propagated through some kind of material medium, while light coming from the stars, for example, should reach us through empty space. On the other hand, the wave model seemed unlikely to many scientists also because of the difficulty in explaining the rectilinear propagation of light. These scientists argued that if light was indeed a wave phenomenon, then it should spread in all directions and even be able to bypass obstacles. This phenomenon—known *as diffraction*—is not easily observed with light due to its small wavelength.

In fact, Francesco Grimaldi (1618–1663) came to observe experimentally the phenomenon of diffraction of light. However, Newton's great reputation in the scientific community caused the corpuscular theory to prevail clearly over the wave theory for more than a century. Among the few scientists of the 18th century who accepted the wave theory and rejected Newton's corpuscular theory, we can mention Leonard Euler (1707–1783) and Benjamin Franklin (1706–1790).

The first experimental demonstration of the wave nature of light was carried out in 1801 by Thomas Young (1773–1829). He performed the so-called *double-slit experiment*, in which an opaque screen with two small, closely spaced slits was illuminated by monochromatic light. The observed transmitted light formed a complex pattern like those produced by water waves. Young roughly calculated the wavelength of light and introduced the concept of interference, which can be considered as a principle of linear wave superposition. Combining this principle with Huygens' theory, Augustin Fresnel (1788–1827) studied the diffraction of light by various objects. At that time, it was thought that light was made up of longitudinal waves. However, the experimental results obtained by Fresnel and Dominique Arago (176–1853) on the effect of polarization on interference required that light be a transverse wave. On this basis, double refraction in calcite could be understood as a phenomenon involving *polarized light*. For each of the two components of polarized light, Fresnel developed the so-called *Fresnel equations*, which give the amplitudes of light reflected and transmitted at the interface between two different optical media.

The first person to prove that light's speed was finite, and to determine it numerically, was the Danish astronomer Ole Roemer (1644–1710). Roemer made it around the year 1675, when he was working at the Royal Observatory in Paris, by timing the eclipses of Jupiter's moon Io. He estimated the speed of light to be approximately 2×10^8 m/s, which is not very far from modern measurements. In 1850, Jean Foucault (1791–1868) experimentally demonstrated that the speed of light in liquids was lower than in the air. This was another important victory for the wave theory of light, since the corpuscular theory predicted an opposite result.

The most important development on the theoretical side was due to the work of James Clerk Maxwell (1831–1879), who showed, in 1873, that light was an electromagnetic phenomenon. Optical frequencies occupy a band of the electromagnetic spectrum that extends from the infrared through the visible to the ultraviolet, as shown in Fig. 1.1. The *visible region* of the spectrum corresponds to a narrow range of electromagnetic waves, ranging from approximately 380 to 770 nm. Electromagnetic waves with these wavelengths predominate in the output of the sun and are capable of producing a visual sensation in the human eye, being properly referred to as 'light'. The range of wavelengths that is generally considered to define the optical domain extends from 10 nm to 300 μm, comprising the visible spectrum and the invisible *ultraviolet* and *infrared* regions. Maxwell's theory predicted that electromagnetic waves would propagate in space at a speed of about 3×10^8 m/s, a result that was similar to that obtained experimentally by Foucault. The German physicist Heinrich Hertz (1857–1894) eventually confirmed Maxwell's theory in 1888.

The next task was to determine the properties of the medium that would support electromagnetic waves and, in particular, light. In this regard, several difficulties soon appeared, since the ether, which had been postulated for this purpose, should present some rather strange properties. On the one hand, it had to be very tenuous, since the celestial bodies would cross it without being even minimally affected. On the other hand, it should have very intense restoring forces in order to produce extremely high frequencies $(\sim 10^{15}\ \text{Hz})$.

Efforts to measure the earth's velocity relative to the ether culminated in the experiment of Albert Michelson (1852–1931) and Edward Morley (1838–1923) in 1887. In this experiment, Michelson and Morley did not detect any motion of the earth with respect to the ether. Actually, in order to explain the phenomenon of stellar aberration, known

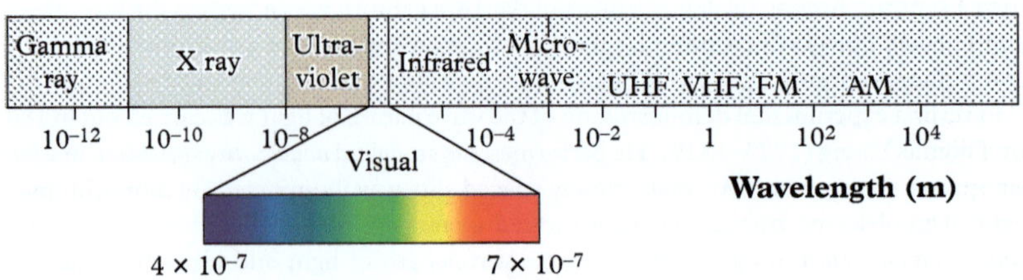

Figure 1.1 *The electromagnetic spectrum.*

since the time of James Bradley (1693–1762), it was necessary to admit such a relative movement. The solution to these difficulties was given by the German physicist Albert Einstein (1879–1955), when he enunciated in 1905 the principle of restricted relativity, which considered the existence of ether superfluous.

Although Maxwell's classical theory of electromagnetism was able to describe most of the characteristics of light, it could not explain some experimental results obtained as early as the end of the 19th century. In particular, experimental observations revealed that the kinetic energy of electrons emitted by a metal does not depend on the intensity of the light incident on its surface, which contradicts the predictions based on wave theory. The explanation of this phenomenon—known as the *photoelectric effect*—was given by Einstein in 1905, who used the quantization concept developed a few years before by the German physicist Max Planck (1858–1947). Actually, Planck presented his quantum theory in 1900, suggesting that black-body radiation is emitted in discrete energy packets called *quanta*. With this supposition, he was able to solve the ultraviolet catastrophe problem that had been puzzling physicists for many years.

Light quanta were called *photons* for the first time in 1926 by the American physical chemist Gilbert Lewis (1875–1946). The energy of a photon is proportional to the frequency, v, of the electromagnetic wave:

$$E = hv \tag{1.1}$$

where $h = 6.63 \times 10^{-34}$ J.s is *Planck's constant.* It is important to note that this theory combines some of the basic features of both the wave theory and the corpuscular theory of light. In fact, the photoelectric effect is the result of the transfer of energy between a given photon and a metal electron. That is, the electron interacts with a photon as if it were a particle. The model of light quanta was also used in 1922 by Arthur Compton (1892–1962), who explained the scattering of X-rays by electrons as particle-like collisions between photons and electrons, in which both energy and momentum were conserved.

Quantum theory developed rapidly between the years 1925 and 1930 with the wave mechanics of Erwin Schrödinger (1887–1961) and the matrix mechanics of Werner Heisenberg (1901–1976). The equivalence between the two theories was demonstrated by John von Neumann (1903–1957). As a result of this theory, it is considered today that any particle exhibits wave properties equally. According to Louis de Broglie (1892–1987), a particle with *momentum p* has an associated a wavelength λ, given by

$$\lambda = \frac{h}{p} \tag{1.2}$$

The experimental confirmation of the hypothesis suggested by Louis de Broglie occurred during the years 1927–1928, when Clinton Davisson (1881–1958) and Lester Germer (1896–1971) in the United States and Sir George Thomson (1892–1975) in England observed the phenomenon of diffraction with a beam of electrons.

The fact that both photons and electrons behave like particles or waves seemed, at an early stage, to be a contradiction, given that these two aspects are apparently irreconcilable. However, the so-called *principle of complementarity*, elaborated by Niels Bohr (1885–1962), highlighted that both photons and electrons are neither particles nor waves, but something more complex than the idea suggested by these models. Actually, *light behaves like waves in its propagation and in the phenomena of interference and diffraction; however, it exhibits particle-like behaviour when exchanging energy with matter, as in the Compton and photoelectric effects.* Similarly, electrons often behave like particles, as observed in the point-like scintillations of a phosphor exposed to a beam of electrons; in other cases, they are found to behave like waves, as in the diffraction produced by an electron microscope.

Quantum mechanics, together with the theory of relativity, shows that the momentum, p, and velocity, v, of both material particles and photons are given by the same expressions:

$$p = \frac{\sqrt{E^2 - m^2c^4}}{c} \tag{1.3}$$

$$v = \frac{pc^2}{E} \tag{1.4}$$

while the wavelength, λ, is given, in both cases, by Eq. (1.2). In Eqs. (1.3) and (1.4), m is the *mass at rest* and E is the total energy of the particle, given by the sum of the energy corresponding to the mass at rest, mc^2, and its kinetic energy.

An essential difference between electrons and photons is that the latter have no mass at rest. Thus, Eqs. (1.3) and (1.4) take on a simpler form for photons:

$$p = \frac{E}{c} \tag{1.5}$$

$$v = \frac{pc^2}{E} = c \tag{1.6}$$

Thus, it is verified that while particles with non-zero mass at rest have a velocity below c, particles such as photons have a constant velocity c in vacuum. The energy of a photon is not a function of its speed, but of its frequency, as indicated by Eq. (1.1).

Another significant difference between electrons and photons is that the former obey *Fermi–Dirac statistics*, whereas photons obey *Bose–Einstein statistics*. Fermi–Dirac statistics does not allow the existence, in a given system, of two electrons in the same state, whereas Bose–Einstein statistics do not impose such a prohibition on photons. In this way, one can have a large number of photons with exactly the same characteristics, which allows a given beam of light to be represented by a continuous electromagnetic wave.

In spite of the fact that quantum theory has developed since 1900, it is generally accepted that the modern subject of *quantum optics* was effectively born in 1956 with the work of Robert Hanbury Brown (1916–2002) and Richard Twiss (1920–2005), leading to the creation of intensity interferometers. They performed a series of experiments on correlations between the starlight intensities recorded on two separate detectors of the interferometer. In recent years, the subject of quantum optics has expanded to include the associated disciplines of quantum information processing and controlled light–matter interactions. The work of the French physicist Alain Aspect (1947–) and co-workers, starting from 1981 onwards, may perhaps be conceived as a landmark in this respect.

1.2 Models to Describe Light

Following the description in the previous section, *ray optics* can be considered as the simplest theory of light. In this model, light is described by rays that travel in different optical media in accordance with a set of geometrical rules. Ray optics is therefore also called *geometrical optics*. It provides an approximate description of the optical phenomenon, which is valid when the dimensions of the components of the optical system (mirrors, lenses, apertures, etc.) are much larger than the wavelength of light. When the optical system consists of several homogeneous media, the path of light is represented by a sequence of straight segments. Whenever light is reflected or refracted, there are discontinuities in the direction of these segments. The behaviour of the optical ray in these situations can be described on the basis of two laws: the *law of reflection* and the *law of refraction*.

If the dimensions of the optical components are not much bigger than the wavelength of light, the wave nature of light cannot be ignored, and we must use the concepts of the so-called *wave optics*. In such cases and to explain optical phenomena like interference and diffraction, one can use a *scalar* wave theory in which light is described by a single scalar wavefunction. Geometrical optics can be considered as a limiting case of wave optics when the wavelength is infinitesimally small.

As recognized by Maxwell, light is an electromagnetic phenomenon, governed by the same theoretical principles that govern all other forms of electromagnetic radiation. This conception of light is called *electromagnetic optics*. Electromagnetic radiation propagates in the form of two mutually coupled *vector waves*: an electric-field wave and a magnetic-field wave. Taking into account the vectorial nature of the optical field becomes necessary in many cases. Wave optics corresponds to a scalar simplification of electromagnetic optics.

Although electromagnetic optics provides the most complete treatment of light within the confines of classical optics, certain optical phenomena are characteristically quantum mechanical in nature and cannot be explained classically. For example, absorption and emission processes cannot be described using just Maxwell's equations. These nonclassical phenomena are described by a quantum version of electromagnetic theory known as *quantum electrodynamics*. For optical phenomena, this theory is also referred to as *quantum optics*.

Historically, the theories of optics were developed roughly in the order in which they were mentioned earlier: first, *geometrical optics*; second, *wave optics*; third, *electromagnetic optics*; and fourth, *quantum optics*. These models are progressively more complex and sophisticated and were developed successively to provide explanations for the outcomes of increasingly subtle optical experiments. Thus, the theory of quantum optics provides an explanation for virtually all optical phenomena. Electromagnetic optics provides the most complete treatment of light within the confines of classical optics. Wave optics is a scalar approximation of electromagnetic optics. Geometrical optics can be considered as the limit of wave optics when the wavelength is very short.

1.3 The Laser and the Emergence of Photonics

The invention of the *laser* in 1960 by Theodore Maiman (1927–2007) represented a major step forward in the recent expansion of the optics field. Actually, the laser can be considered as the most important quantum optical instrument, since it makes use of three basic physical processes that cannot be described classically: absorption, stimulated emission, and spontaneous emission of photons. A laser converts pump energy (which may be electrical or optical in nature) into intense, highly directional, and coherent electromagnetic wave radiation, spanning from the infrared to the ultraviolet regions.

The availability of high-power and coherent light sources has led to the discovery of new optical effects and the development of new methods and devices. Actually, they have a great influence on our daily life: scanning labels in supermarkets, performing surgery in hospitals, reading videodiscs in living rooms, cutting steel in factories, etc. Lasers are also used in diverse military applications, from spy satellites to infrared gadgets that see in the dark. The new technique known as *holography*, which provides the possibility of recording and reconstructing three-dimensional images, has found numerous applications and has significantly impacted the general public.

The invention of the laser also instigated the development of optical communications, namely via glass *optical fibres*, following the suggestion made in 1966 by Charles Kao (1933–2018) and George Hockham (1938–2013). Actually, fibre optic communications is the most important example illustrating the new area of *photonics*. This term will be taken, for the purpose of this book, to mean the *science that studies the generation, manipulation, transport and use of light and other forms of radiant energy, whose quantum unit is the photon*. The field of optical communications and information processing involves a wide range of components: lasers, light-emitting diodes, optical fibres, photodetectors, optical gates, optical amplifiers, optical modulators, etc. Actually, the development of this area is such that, especially in the field of long-distance transmission of information, photonics has almost completely replaced electronics. As such, photonics currently represents one of the most dynamic and profitable markets in industrialized countries.

1.4 Further Reading

Ditchburn, R. W., *Light*, Academic Press, London, 1976.

Hall, A. R., *All Was Light—An Introduction to Newton's Opticks*, Clarendon Press, Oxford, 1993.

Optics and Photonics News, 'The Nature of Light: What Is a Photon?' Vol. 3, No. 1, October 2003.

Ronchi, V., *The Nature of Light*, Harvard University Press, 1971.

Sabra, A. I., *Theories of Light from Descartes to Newton*, Osborn, 1967.

Simmons, J., and M. Guttmann, *States, Waves, and Photons: A Modern Introduction to Light*, Addison-Wesley, 1970.

1.5 Problems

1.1 Calculate the frequency, energy, momentum, and mass of a photon with wavelength $\lambda = 600\,\text{nm}$.

1.2 Calculate the wavelength of (a) a golf ball with a mass of 60 g moving at a speed of 50 m/s and (b) an electron with a kinetic energy of 10 eV.

1.3 Derive a relation between the wavelength of a photon, measured in angstroms, and its energy, measured in electron volts.

1.4 Consider a light beam with a wavelength $\lambda = 0.2\,\mu$ m and an optical power of 1 nW illuminating a given surface. Determine (i) the energy of a photon and (ii) the number of photons/s arriving at the surface.

1.5 A beam of green light ($\lambda = 500\,\text{nm}$) strikes a fully reflective mirror normally. Determine the number of photons that arrive at the mirror each second, considering that the force exerted on it by the light is 1N.

Geometrical Optics

2

Geometrical optics is an approximate description of optical phenomenon, which is valid when the dimensions of the components of the optical system (mirrors, lenses, apertures, etc.) are much larger than the wavelength of light. When this condition is not verified, and the wave nature of light cannot be ignored, we must use the concepts of so-called *wave optics*. In fact, geometrical optics can be considered as a limiting case of wave optics, corresponding to the case of an infinitesimally small wavelength.

The basic element in the description of geometrical optics is the concept of an *optical ray*, which corresponds to a straight line along which light energy is transmitted from one point to another in an optical system. When the optical system consists of several homogeneous media, the path of light is represented by a sequence of straight segments. Whenever light is reflected or refracted, there are discontinuities in the direction of these segments. The behaviour of the optical ray in these situations can be described by two basic laws: the *law of reflection* and the *law of refraction*.

2.1 Reflection in a Plane Mirror

Fig. 2.1 illustrates the reflection of a light ray in a plane mirror. Based on the geometry of this figure and Fermat's minimum time principle, referred to in Chapter 1, we can derive the law of reflection, which is one of the fundamental laws of geometrical optics.

Given that time is given by the ratio distance/speed and that the rays travel at constant speed in a uniform medium, the problem of minimizing the travel time can be replaced by the problem of minimizing the distance $L = \overline{AP} + \overline{PA'}$. As the exact position of the point P is not known, we use $\overline{OP} = x$ and $\overline{PO'} = l - x$, where $l = \overline{OO'}$. Considering the geometry of Fig. 2.1 and using the Pythagorean theorem, the following expression is obtained for the distance L:

$$L = \sqrt{d^2 + x^2} + \sqrt{d'^2 + (l - x)^2} \tag{2.1}$$

The minimum distance travelled by the light ray is obtained by calculating the derivative of L in order to x and equating the resulting expression to zero:

$$\frac{dL}{dx} = \frac{x}{\sqrt{d^2 + x^2}} - \frac{(l - x)}{\sqrt{d'^2 + (l - x)^2}} = 0 \tag{2.2}$$

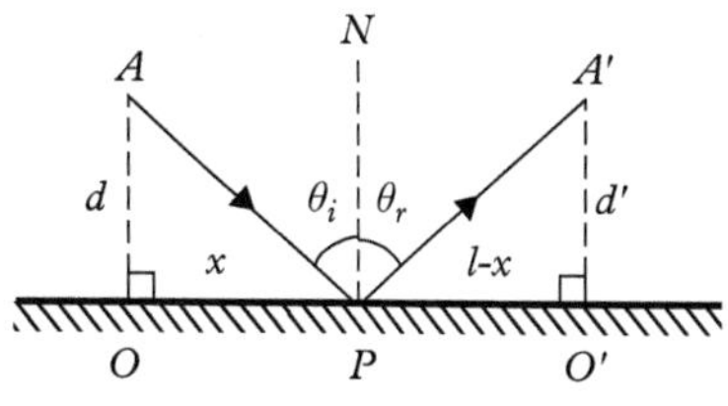

Figure 2.1 *Geometry used to derive the law of reflection in a plane mirror.*

Considering the triangles OAP and $O'A'P$, as well as the angles θ_i and θ_r in Fig. 2.1, Eq. (2.2) allows us to obtain the result:

$$\sin \theta_i = \sin \theta_r$$

Lasers, Optics, and Photonics. Mário F. S. Ferreira, Oxford University Press. © Mário F. S. Ferreira (2026).
DOI: 10.1093/oso/9780198952770.003.0002

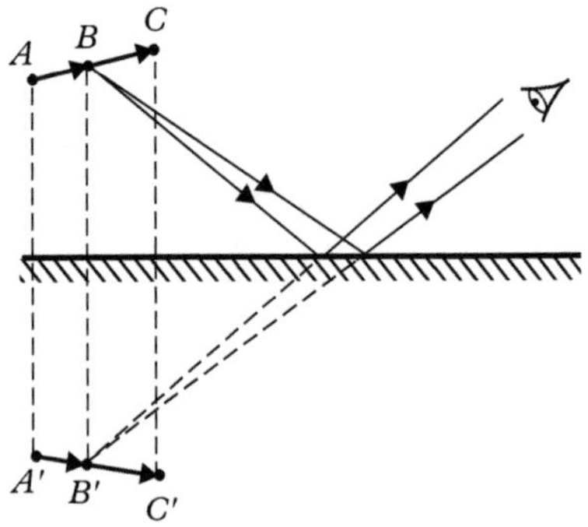

Figure 2.2 *Image formation in a plane mirror.*

Thus, we can conclude that

$$\theta_i = \theta_r \tag{2.3}$$

Eq. (2.3) corresponds to the *first law of reflection*, according to which *the angles made by the incident and reflected rays in relation to the normal at the point of incidence are equal*. The *second law of reflection* establishes *that the incident and reflected rays, as well as the normal to the mirror at the point of incidence, are located in the same plane*, called the *plane of incidence*. Usually, the set of the two previous laws is simply called the *law of reflection*.

Fig. 2.2 shows the formation of the image in a plane mirror. For an observer, the rays coming from point B of the object appear to come from the image point B' after being reflected by the mirror. Point B' *is* at a distance from the mirror equal to the distance between the mirror and point B. All points of an extended object have images that can be obtained in a similar way. Thus, it can be concluded that the position of the image does not depend on the position of the observer and that its size is equal to the size of the object, that is, the *magnification* is unitary. On the other hand, it turns out that the left side of the object appears as the right side in the image and vice versa.

2.2 Spherical Mirrors

A spherical mirror is part of a spherical surface, which has a given radius R and a centre of curvature C. Fig. 2.3 illustrates the cases of (a) a concave spherical mirror and (b) a convex spherical mirror. The *optical axis* is the line that passes through point V (called the *vertex*) and the centre of curvature C. The rays are reflected in the mirror according to the law of reflection ($\theta_i = \theta_r$). In the case of a concave mirror, when the incident rays are parallel to the optical axis, the reflected rays pass through the focal point F. In the case of a convex mirror, all reflected rays appear to come from the focal point, which is behind the mirror.

Fig. 2.4 illustrates the process of locating the image of a linear object formed by a concave mirror. Three rays are considered for this purpose: one (a) that strikes the mirror parallel to the optical axis, a second one (b) that strikes the mirror at the *vertex* V, and a third one (c) that passes through the centre of curvature of the mirror. In practice, the location of the image, given by the intersection of the reflected rays, can be obtained by considering only two of these rays.

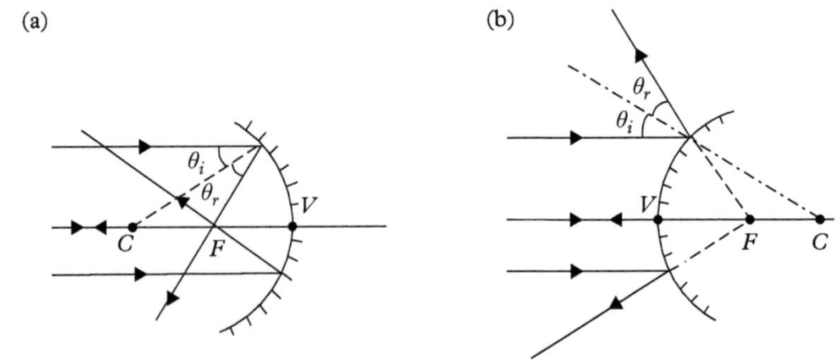

Figure 2.3 *Representation of (a) a concave spherical mirror and (b) a convex spherical mirror.*

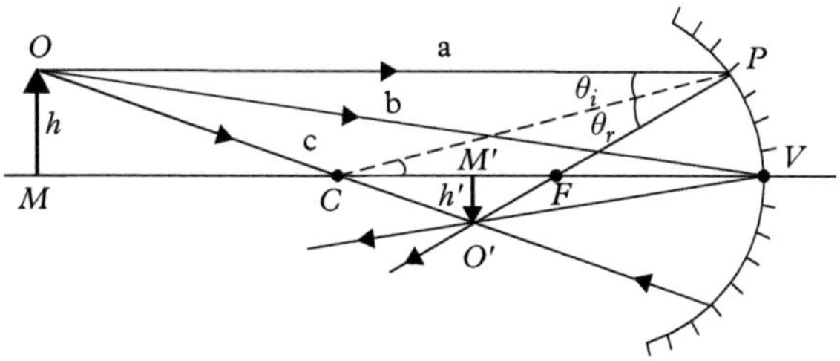

Figure 2.4 *Geometry for locating the image produced by a concave mirror.*

In the analysis of Fig. 2.4, we will consider the so-called *paraxial approximation*, in which the angle θ between the light ray and the optical axis is assumed to be sufficiently small. In these circumstances, we have

$$\cos\theta \approx 1$$

and

$$\tan\theta \approx \sin\theta \approx \theta$$

Based on the geometry of Fig. 2.4 and using this approximation, one has:

$$\tan(P\hat{C}V) = \theta_i = \frac{h}{|R|} \tag{2.4a}$$

and

$$\tan(P\hat{F}V) = 2\theta_i = \frac{h}{f} \tag{2.4b}$$

where $|R| = \overline{CP}$ is the radius and $f = \overline{FV}$ is the focal length of the mirror.

By convention, we will consider that R is negative when the centre of curvature C is to the left of V (i.e. the mirror is concave). On the other hand, R is positive when C is to the right *of* V (i.e. the mirror is convex). In the case of Fig. 2.4, we have $|R| = -R$ and, from Eqs. (2.4a, b), we find that

$$f = -\frac{R}{2} \tag{2.5}$$

Considering the OMC and $O'M'C$ triangles and making $s = \overline{MV}$, $s' = \overline{M'V}$, $f = \overline{FV}$ e $|R| = \overline{CV}$, one has:

$$\tan(O\hat{C}M) = \frac{h}{s - |R|} \tag{2.6a}$$

$$\tan(O'\hat{C}M') = \frac{h'}{|R| - s'} \tag{2.6b}$$

Since $\tan(O\hat{C}M) = -\tan(O'\hat{C}M')$, we find that

$$\frac{h'}{h} = -\frac{|R| - s'}{s - |R|} = -\frac{2f - s'}{s - 2f} \tag{2.7}$$

On the other hand, considering the *VPF and FM′O′* triangles, we have

$$\tan(V\hat{F}P) = \frac{h}{f} \tag{2.8a}$$

$$\tan(O'\hat{F}M') = \frac{h'}{s' - f} \tag{2.8b}$$

Since $\tan(V\hat{F}P) = -\tan(O'\hat{F}M')$, one has

$$\frac{h'}{h} = -\frac{s' - f}{f} \tag{2.9}$$

From Eqs. (2.7) and (2.9), one obtains the result

$$\frac{1}{s} + \frac{1}{s'} = \frac{1}{f} = -\frac{2}{R} \tag{2.10}$$

From the triangles OMV and $O'M'V$, one has

$$\tan(O\hat{V}M) = \frac{h}{s} \tag{2.11a}$$

$$\tan(O'\hat{V}M') = \frac{h'}{s'} \tag{2.11b}$$

As $\tan(O\hat{V}M) = -\tan(O'\hat{V}M')$, it is also concluded that

$$\frac{h'}{h} \equiv A_l = -\frac{s'}{s} \tag{2.12}$$

The quantity A_l given by Eq. (2.12) is the so-called *lateral magnification* of the image. If $A_l < 1$, the image is smaller than the object, whereas for $A_l > 1$, we have an enlarged image. On the other hand, $A_l > 0$ corresponds to an image that has the same orientation as the object, whereas $A_l < 0$ corresponds to an image that is inverted relative to the object.

In solving problems related to the determination of the image formed by plane or spherical mirrors, it is important to establish some conventions regarding the signs of the quantities involved. Let us assume the following conventions:

i) The distances s and s' are positive if the object and the image, respectively, are in front of the mirror. If the object (or the image) is behind the mirror, the corresponding distance is negative. In the first case, the object and image are *real*, whereas in the second case, they are *virtual*.
ii) The focal length f is positive for a concave mirror and negative for a convex mirror.
iii) The height, h, measured above the optical axis is positive, while below this axis it is negative.

Example 2.1

An object 2 cm high is situated 15 cm in front of (a) a concave and (b) a convex spherical mirror with a radius $|R| = 10$ cm. Determine the position and nature of the image in each case.

Solution

(a) Concave mirror: $f = \frac{|R|}{2} = 5\text{cm}$ and $s = +15$ cm

The position of the image is obtained from the relation

$$\frac{1}{s} + \frac{1}{s'} = \frac{1}{f}$$

from which we obtain

$$s' = \frac{fs}{s-f} = \frac{(5)(15)}{15-5} = +7.5\text{ cm}$$

The lateral magnification of the image is given by

$$A_l = -\frac{s'}{s} = -\frac{7.5}{15} = -\frac{1}{2}$$

The image is real (because s' is positive), located in front of the mirror and at 7.5 cm from its vertex, inverted (because A_l is negative) and half the size of the object (1 cm).

(b) Convex mirror: $f = -\frac{|R|}{2} = -5$ cm, $s = +15$ cm.

In this case, it is

$$s' = \frac{fs}{s-f} = \frac{(-5)(15)}{15-(-5)} = -3.75\text{ cm}$$

$$A_l = -\frac{s'}{s} = -\frac{(-3.75)}{15} = +\frac{1}{4}$$

The image is virtual (because s' is negative), located at 3.75 cm from the mirror vertex, erect (because A_l is positive) and a quarter of the size of the object (0.5 cm).

2.3 Snell's Law for Refraction

The optical density of a given medium is characterized by its refractive index, n, given by the ratio between the speed of light in a vacuum and in that medium, $n = c/v$. Fig. 2.5 illustrates the refraction of a light beam at the interface between two homogeneous and transparent media with different refractive indices, n_1 and n_2, respectively.

The travel time between points A and A' in Fig. 2.5 is given by

$$t = \frac{n_1}{c}\overline{AP} + \frac{n_2}{c}\overline{PA'} \tag{2.13}$$

Considering the geometry of Fig. 2.5 and using the Pythagorean theorem, one obtains

$$t = \frac{n_1}{c}\sqrt{d^2 + x^2} + \frac{n_2}{c}\sqrt{d'^2 + (l-x)^2} \tag{2.14}$$

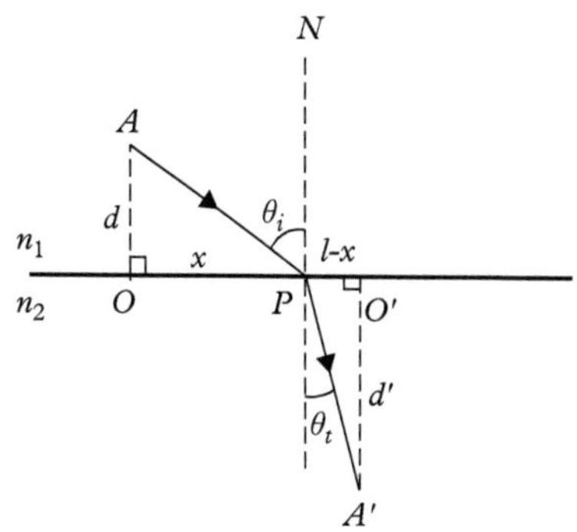

Figure 2.5 *Geometry used to demonstrate the law of refraction.*

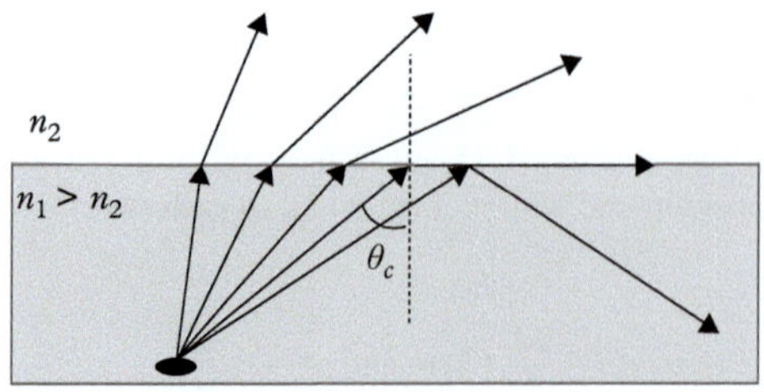

Figure 2.6 *Illustration of the phenomenon of total internal reflection, which occurs if $n_1 > n_2$ for $\theta_i > \theta_c$.*

where $l = \overline{OO'}$. Following Fermat's principle of minimum time, and in order to minimize $t(x)$, we impose the condition $dt/dx = 0$:

$$\frac{dt}{dx} = \frac{n_1}{c}\frac{x}{\sqrt{d^2 + x^2}} - \frac{n_2}{c}\frac{(l-x)}{\sqrt{d'^2 + (l-x)^2}} = 0 \tag{2.15}$$

Based on the previous condition and considering the geometry of Fig. 2.5, we obtain the following result:

$$n_1 \sin\theta_i = n_2 \sin\theta_t \tag{2.16}$$

Eq. (2.16) corresponds to *Snell's law for refraction*, which also states that *the incident and transmitted rays are in the plane of incidence.*

It should be noted that, if $n_1 < n_2$, we obtain from Eq. (2.16) that $\theta_t < \theta_i$. However, when the medium of incidence is optically denser than the transmission medium ($n_1 > n_2$), one has $\theta_t > \theta_i$. In this case, there is a critical angle of incidence, $\theta_i = \theta_c$, for which the transmission angle is $\theta_t = 90°$. Using Eq. (2.16), we obtain

$$\theta_c = \sin^{-1}\left(\frac{n_2}{n_1}\right) \tag{2.17}$$

For an angle of incidence $\theta_i > \theta_c$, there is no transmitted ray, and the incident ray undergoes *total internal reflection*. This phenomenon is illustrated in Fig. 2.6 and is essential in the field of guided optics, as will be seen in Chapter 15.

2.4 Principle of Reversibility

Let us consider Figs. 2.1 and 2.5 again, but change the roles of points A and A'. That is, let us suppose point A' is now the source of the rays. Fermat's principle of minimum time allows us to conclude that the light will, in such a case, follow the same trajectory obtained in the original situation but in the opposite direction. This result can be generalized, making it possible to state that, in an arbitrary optical system, a ray of light travels along the same trajectory when its direction of propagation is reversed. This statement is known as the *principle of reversibility* of ray propagation, and its application becomes very useful in a number of situations.

2.5 Refraction on a Spherical Surface

Fig. 2.7 illustrates the trajectory of a ray incident on a point P belonging to the spherical surface which is the boundary between two homogeneous media, with refractive indices n_1 and n_2, respectively. Snell's law for refraction at point P gives:

$$n_1 \sin\theta_i = n_2 \sin\theta_t \tag{2.18}$$

In the paraxial approximation, this equation can be written in the form:

$$n_1 \theta_i = n_2 \theta_t \tag{2.19}$$

From Fig. 2.7, and using the exterior angle theorem, one obtains:

$$\theta_i = \alpha + \gamma \tag{2.20}$$

and

$$\theta_t = \gamma - \beta \tag{2.21}$$

On the other hand, the paraxial approximation allows us to write

$$\alpha = \frac{h}{s}, \quad \beta = \frac{h}{s'}, \quad \gamma = \frac{h}{R} \tag{2.22}$$

Substituting Eqs. (2.20)–(2.22) into Eq. (2.19) provides the result

$$\frac{n_1}{s} + \frac{n_2}{s'} = \frac{n_2 - n_1}{R} = D \tag{2.23}$$

where D is the *power of the refracting surface*, which has the unit of *dioptres* if R is given in metres.

By convention, the distance s is considered positive (negative) if the light comes from a point on the left (right) of the refractive surface, whereas the distance s' is positive (negative) if the light is directed to a point on the right (left) of that surface. On the other hand, the radius R is considered positive (negative) if the centre of curvature of the spherical surface is located on the right (left) of vertex V.

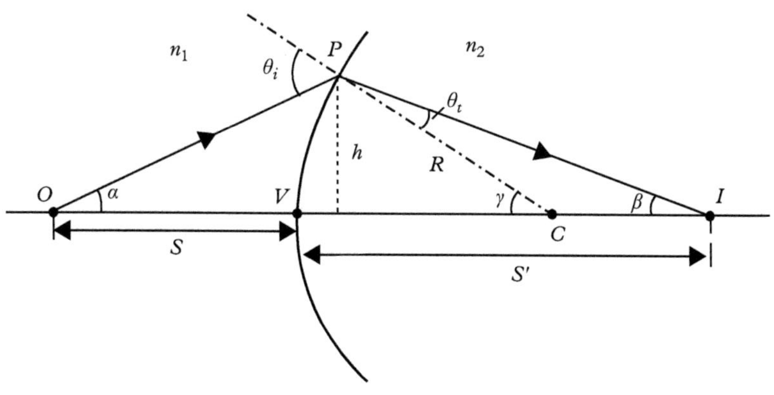

Figure 2.7 *Geometry for the study of light refraction at a spherical surface.*

2.6 The Thin Lens

A lens is a refractive system limited by two interfaces, at least one of which is curved. The case of a thin spherical lens, whose thickness is considered negligible, can be analysed based on Eq. (2.23). Suppose that the lens consists of a glass with a refractive index n, limited by two spherical surfaces with radii R_1 and R_2, located in the air. Refraction at the first surface, with radius R_1, is described by the following equation:

$$\frac{1}{s}+\frac{n}{s''}=\frac{n-1}{R_1} \tag{2.24}$$

where s'' is the distance of the image relative to that surface. This image acts as an object for the second refraction, located at a distance $-s''$ from the second surface, since the lens thickness is neglected. The equation for the refraction at the second surface will then be

$$\frac{n}{(-s'')}+\frac{1}{s'}=\frac{1-n}{R_2} \tag{2.25}$$

Adding Eqs. (2.24) and (2.25), we obtain:

$$\frac{1}{s}+\frac{1}{s'}=(n-1)\left(\frac{1}{R_1}-\frac{1}{R_2}\right) \tag{2.26}$$

which is the so-called *thin lens equation*, also known as the *lens manufacturers' equation*. Assuming $s\to\infty$ or $s'\to\infty$ in Eq. (2.26), we have $s'=f$ or $s=f$, respectively, where

$$f=\left[(n-1)\left(\frac{1}{R_1}-\frac{1}{R_2}\right)\right]^{-1} \tag{2.27}$$

is called the *focal length* of the lens. The inverse of the focal length is known as the *power of the lens*, given by the sum of the powers of the two refracting surfaces limiting this lens. If $f>0$, the lens is said to be *positive* or *converging*. On the other hand, if $f<0$, the lens is said to be *negative* or *diverging*.

Fig. 2.8 illustrates the formation of the image of a given object by a thin biconvex lens. Three rays are considered for this purpose: one ray leaves the object parallel to the optical axis and, after crossing the lens, passes through the secondary focus (F_i); another ray passes through the centre of the lens (O) and is not deviated; and a third ray passes through

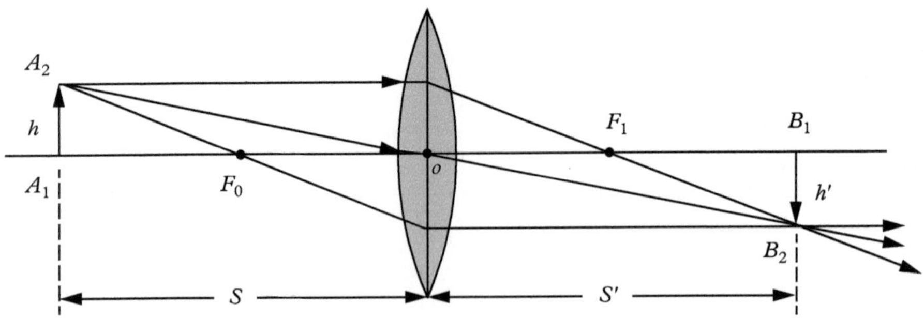

Figure 2.8 *Image formation by a thin biconvex lens.*

the primary focus (F_o) and emerges from the lens parallel to the optical axis. In fact, it is sufficient to use only two of these rays to find the point image of a given point of the object.

Considering the similarity of triangles A_1A_2O and B_1B_2O in Fig. 2.8, we find that the lateral magnification provided by the lens is given by

$$A_l \equiv \frac{h'}{h} = -\frac{s'}{s} \tag{2.28}$$

It can be verified from Eqs. (2.26)–(2.28) that, in the case of a real object located at a distance s from the lens, such that $s>f$, the corresponding image is also real ($s'>0$) but inverted ($A_l<0$). However, if the object is at a distance $s<f$, the corresponding image is virtual ($s'<0$) and upright ($A_l>0$).

Example 2.2

An object with a height of 1 cm is located 15 cm to the left of a thin lens, which has a focal length $f = 10$ cm. Find the position and characteristics of the corresponding image.

Solution

Using the thin lens equation

$$\frac{1}{s} + \frac{1}{s'} = \frac{1}{f}$$

we have

$$\frac{1}{15} + \frac{1}{s'} = \frac{1}{10}$$

From which $s' = 30$ cm.

The image is real (s' is positive) and is located 30 cm to the right of the lens.

The lateral magnification of the image is given by

$$A_l \equiv \frac{h'}{h} = -\frac{s'}{s} = -\frac{30}{15} = -2$$

The image has a length of 2 cm and is inverted (A_l is negative).

Example 2.3

An object with a height of 2 cm is placed at a distance of 30 cm from a biconvex lens with a focal length $f = 20$ cm. On the other side of the lens, at a distance of 100 cm from it, there is a concave mirror with radius $|R| = 60$ cm.

a) Determine the position of the image provided by the lens.

b) Determine the position and characterize the final image formed after reflection in the mirror.

continued

Example 2.3 continued

Solution

a) Using the thin lens equation, we obtain the distance between the lens and the image:

$$\frac{1}{s_1}+\frac{1}{s_1'}=\frac{1}{f}\rightarrow s_1'=\frac{s_1 f}{s_1-f}=\frac{(30)(20)}{30-20}=60\,\text{cm}$$

b) The image provided by the lens becomes the object for the mirror, which is at a distance:

$$s_2=d-s_1'=100-60=40\,\text{cm}$$

On the other hand, using the equation for mirror image formation, we obtain the distance from the mirror at which the final image is formed:

$$\frac{1}{s_2}+\frac{1}{s_2'}=\frac{1}{f_2}=\frac{2}{|R|}\rightarrow s_2'=\frac{s_2 f_2}{s_2-f_2}=\frac{(40)(30)}{40-30}=120\,\text{cm}$$

Given that $s_2'>0$, the image produced by the mirror is real. On the other hand, the total magnification is given by the product of the magnifications due to the lens and the mirror:

$$A_l=\left(-\frac{s_1'}{s_1}\right)\left(-\frac{s_2'}{s_2}\right)=(-2))(-3)=6>0$$

Therefore, the final image formed after transmission through the lens and reflection in the mirror has the same orientation ($A_l>0$) and is six times larger than the original object.

2.7 The Matrix Method

Geometrical optics can be used to trace rays through an optical system in order to study its performance. For this purpose, it is convenient to write the equations in matrix form, since matrix algebra makes it possible to treat with relative simplicity the problem of light propagation through any optical system. This section describes the main aspects of the matrix method, assuming the paraxial approximation already used in previous sections.

Fig. 2.9 represents the propagation of a light ray through a thick lens. It is assumed that the lens surfaces have a spherical shape, with radii of curvature R_1 (left surface) and R_2 (right surface). The points V_1 and V_2 are the vertices of the refracting surfaces and correspond to the intersections of these surfaces with the optical axis.

2.7.1 Refraction matrix

Applying Snell's law at point A in Fig. 2.9 and using the paraxial approximation, one obtains

$$n_1\theta_i=n_2\theta_t \tag{2.29}$$

Eq. (2.29) can be rewritten in the following form:

$$n_1(\gamma+\phi)=n_2(\gamma''+\phi) \tag{2.30}$$

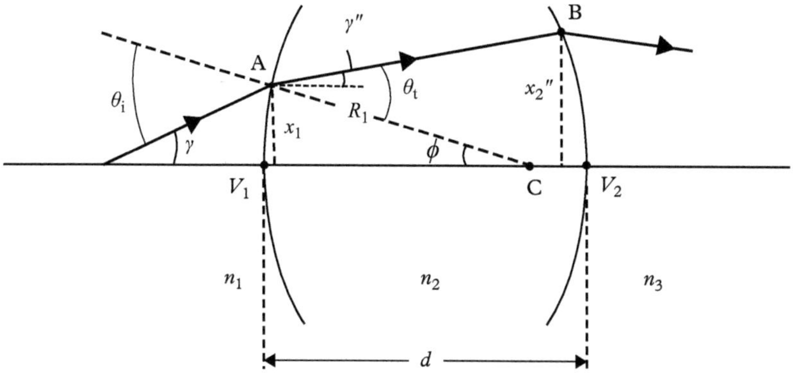

Figure 2.9 *Geometry for ray tracing in a thick lens.*

where γ and γ'' represent the angles between the incident and transmitted rays and the optical axis, whereas ϕ is the angle between this axis and the normal to the refractive surface at point A. Given the geometry of Fig. 2.9, we have $\phi = x_1/R_1$, and Eq. (2.30) becomes:

$$n_1\left(\gamma + \frac{x_1}{R_1}\right) = n_2\left(\gamma'' + \frac{x_1}{R_1}\right) \tag{2.31}$$

Solving this equation in order to γ'', we obtain:

$$\gamma'' = \frac{n_1}{n_2}\gamma + \frac{x_1(n_1 - n_2)}{n_2 R_1} \tag{2.32}$$

Eq. (2.32) can be written in the form:

$$\gamma'' = \frac{n_1}{n_2}\gamma - \frac{D_1}{n_2}x_1 \tag{2.33}$$

where

$$D_1 = \frac{n_2 - n_1}{R_1} \tag{2.34}$$

is the power of the refracting surface.

Since the distance between the ray and the optical axis does not vary during the refraction at point A, we have:

$$x_1 = x_1'' \tag{2.35}$$

where the primes indicate that one is near the interface but in the transmission medium, whose refractive index is n_2.

The system of Eqs. (2.32) and (2.34) can be written using matrices as follows:

$$\begin{bmatrix} x_1'' \\ \gamma'' \end{bmatrix} = \begin{bmatrix} 1 & 0 \\ -\frac{D_1}{n_2} & \frac{n_1}{n_2} \end{bmatrix} \begin{bmatrix} x_1 \\ \gamma \end{bmatrix} \tag{2.36}$$

The matrix

$$\mathbf{R}_1 = \begin{bmatrix} 1 & 0 \\ -\frac{D_1}{n_2} & \frac{n_1}{n_2} \end{bmatrix} \tag{2.37}$$

describes the refraction at point A of the spherical surface and is therefore called the *refraction matrix*. Its determinant is

$$|\mathbf{R}_1| = \frac{n_1}{n_2} \tag{2.38}$$

2.7.2 Transfer matrix

When the optical ray travels from point A to point B in Fig. 2.9, the angle that it makes with the optical axis does not vary, so

$$\gamma'' = \gamma_2'' \tag{2.39}$$

On the other hand, the height of the ray relative to the optical axis varies from x_1'' to x_2'', according to the expression:

$$x_2'' = x_1'' + d\gamma'' \tag{2.40}$$

where d is the thickness of the lens. Eqs. (2.39) and (2.40) can be written in matrix form as follows:

$$\begin{bmatrix} x_2'' \\ \gamma_2'' \end{bmatrix} = \begin{bmatrix} 1 & d \\ 0 & 1 \end{bmatrix} \begin{bmatrix} x_1'' \\ \gamma'' \end{bmatrix} \tag{2.41}$$

The matrix

$$\mathbf{T} = \begin{bmatrix} 1 & d \\ 0 & 1 \end{bmatrix} \tag{2.42}$$

describes the propagation of the optical ray between the two surfaces of the lens and is called the *transfer matrix*. Its determinant is

$$|\mathbf{T}| = 1 \tag{2.43}$$

2.7.3 System matrix for a thick lens

Ray refraction at point B of the second surface in Fig. 2.9 can be described using the same method that led to Eq. (2.37). In this case, the following matrix equation is obtained:

$$\begin{bmatrix} x_2' \\ \gamma' \end{bmatrix} = \begin{bmatrix} 1 & 0 \\ -\frac{D_2}{n_3} & \frac{n_2}{n_3} \end{bmatrix} \begin{bmatrix} x_2'' \\ \gamma_2'' \end{bmatrix} \tag{2.44}$$

where x_2' and γ' represent, respectively, the height and angle of the transmitted ray relative to the optical axis, whereas

$$D_2 = \frac{n_3 - n_2}{R_2} \tag{2.45}$$

is the power of the second refracting surface. The refraction matrix at point B on this surface is then

$$\mathbf{R}_2 = \begin{bmatrix} 1 & 0 \\ -\frac{D_2}{n_3} & \frac{n_2}{n_3} \end{bmatrix} \tag{2.46}$$

The product of the three matrices Eqs. (2.37), (2.42), and (2.46) gives the so-called *system matrix* **L** for the lens:

$$\mathbf{L} \equiv \begin{bmatrix} a_{11} & a_{12} \\ a_{21} & a_{22} \end{bmatrix} = \mathbf{R}_2\mathbf{T}\mathbf{R}_1 \tag{2.47}$$

The four elements of the matrix **L** are given by the following expressions:

$$a_{11} = 1 - \frac{dD_1}{n_2} \tag{2.48a}$$

$$a_{12} = \frac{dn_1}{n_2} \tag{2.48b}$$

$$a_{21} = \frac{1}{n_3}\left[\frac{dD_1D_2}{n_2} - D_1 - D_2\right], \tag{2.48c}$$

$$a_{22} = \frac{n_1}{n_3}\left[1 - \frac{dD_2}{n_2}\right] \tag{2.48d}$$

The determinant of the system matrix is

$$|\mathbf{L}| = |\mathbf{R}_2||\mathbf{T}||\mathbf{R}_1| = a_{11}a_{22} - a_{12}a_{21} = \frac{n_1}{n_3} \tag{2.49}$$

We observe that if the initial and final media of the light ray path are the same, the determinant of the system matrix is equal to one.

2.7.4 System matrix for a thin lens

In the case of a thin lens, we have $d \to 0$. If this lens is in the air, such that $n_1 = n_3 = 1$, its matrix assumes a particularly simple form:

$$\mathbf{L} = \begin{bmatrix} 1 & 0 \\ -1/f & 1 \end{bmatrix}, \tag{2.50}$$

where f is the lens focal length and

$$\frac{1}{f} = D_1 + D_2. \tag{2.51}$$

It can be verified that Eq. (2.51) is in accordance with Eq. (2.27).

Example 2.4

Using the matrix theory, obtain an expression for the focal length of a compound lens consisting of two thin lenses, with focal lengths f_1 and f_2, located in the air and spaced by a distance d.

Solution

The matrix corresponding to the compound lens, $\mathbf{L}_c$, is given by the product of three matrices: two of them similar to Eq. (2.50) and another similar to Eq. (2.42), corresponding to the ray transfer between the two thin lenses. Thus, one has

$$\mathbf{L}_c = \mathbf{LTL} = \begin{bmatrix} 1 & 0 \\ -1/f_2 & 1 \end{bmatrix} \begin{bmatrix} 1 & d \\ 0 & 1 \end{bmatrix} \begin{bmatrix} 1 & 0 \\ -1/f_1 & 1 \end{bmatrix}$$

$$= \begin{bmatrix} 1 - d/f_1 & d \\ d/f_1 f_2 - 1/f_1 - 1/f_2 & 1 - d/f_2 \end{bmatrix}$$

Comparing the previous matrix with that in Eq. (2.50), we find that the inverse of the focal length of the compound lens is given by

$$\frac{1}{f} = \frac{1}{f_1} + \frac{1}{f_2} - \frac{d}{f_1 f_2} \tag{2.52}$$

2.7.5 System matrix for image formation

The system matrix that represents the image formation by a single lens, as represented in Fig. 2.7, depends on the transfer matrix, $\mathbf{T_o}$, describing the light propagation from the object to the lens; the lens matrix itself, $\mathbf{L}$; and the transfer matrix from the lens to the image plane, $\mathbf{T_i}$, and is given by

$$\mathbf{S} = \mathbf{T_i L T_o} \tag{2.53}$$

The procedure that led to Eq. (2.53) can be generalized to any optical system. Thus, indicating the matrix of this system in the form

$$\mathbf{S} = \begin{bmatrix} A & B \\ C & D \end{bmatrix}, \tag{2.54}$$

we can write the following matrix equation:

$$\begin{bmatrix} x_i \\ \gamma_i \end{bmatrix} = \begin{bmatrix} A & B \\ C & D \end{bmatrix} \begin{bmatrix} x_0 \\ \gamma_0 \end{bmatrix}, \tag{2.55}$$

where x_0 (x_i) and γ_0 (γ_i) represent, respectively, the height and the angle relative to the optical axis of the object (image) ray. It becomes important to understand the different situations corresponding to the annulment of each element of the system matrix in Eq. (2.55).

i) When $D = 0$, one has $\gamma_i = Cx_0$ regardless of γ_o, meaning that all rays coming from the same point in the object plane, located at a distance x_0 from the optical axis, leave the system making the same angle with the optical axis. That is, the object is located in the *primary focal plane of* the system.

ii) When $A = 0$, one has $x_i = B\gamma_0$, and the situation is the reverse of the previous one. That is, all incident rays in the optical system, making the same angle γ_o with the optical axis, are focused at the same point in the image plane, located at a distance x_i from the optical axis. Such an image plane corresponds to the *secondary focal plane* of the system.

iii) When $B = 0$, one has $x_i = Ax_o$ regardless of γ_o, meaning that all rays coming from the same point in the object plane converge to the same point in the image plane. That is, the annulment of element B of the system matrix allows one to determine the position of the image point corresponding to a given object point. At the same time, it turns out that element A corresponds to the lateral magnification of this system.

iv) When $C = 0$, one has $\gamma_i = D\gamma_o$. That is, all the rays that enter the system in a given direction, parallel to each other, leave the system also parallel to each other, although in a different direction. This corresponds to the configuration of a telescope, its angular magnification being given by element D.

Example 2.5

Let us consider an object located in the air, at a distance of 40 cm from a block of glass ($n = 1.5$) 1 m long. The face of this block on the side of the object has a spherical convex shape, with a radius of curvature $R = 10$cm. Using the matrix method, characterize the image of this object formed in the glass block.

Solution

In this case, the system matrix consists of the product of three matrices, corresponding to (1) the ray transfer from the object to the block, (2) the refraction at the spherical surface of the block, and (3) the ray transfer in the glass to the image plane. Using Eqs. (2.37) and (2.42) for the refraction and transfer matrices, we have

$$S = \begin{bmatrix} 1 & s' \\ 0 & 1 \end{bmatrix} \begin{bmatrix} 1 & 0 \\ \dfrac{1-1.5}{10(1.5)} & \dfrac{1}{1.5} \end{bmatrix} \begin{bmatrix} 1 & 40 \\ 0 & 1 \end{bmatrix}$$

where s' is the distance between the refractive surface (vertex) and the image plane. By performing the product of the matrices, we obtain

$$S = \begin{bmatrix} 1 - \dfrac{s'}{30} & 40 - \dfrac{2s'}{3} \\ -\dfrac{1}{30} & -\dfrac{2}{3} \end{bmatrix}$$

According to point iii), the distance s' is determined by making $B = 0$, that is,

$$B = 40 - \frac{2s'}{3} = 0 \quad \text{or} \quad s' = 60 \text{ cm}$$

On the other hand, the lateral magnification of the image is given by element A of the system matrix:

$$A = 1 - \frac{s'}{30} = -1$$

We conclude that the image forms inside the glass block, 60 cm to the right of the vertex point (s' is positive), is inverted (A is negative) and has the same size as the object ($|A| = 1$).

Example 2.6

Consider an object with height h, situated at a distance s from a thin convergent lens, which has a focal length f. At a distance s' on the opposite side of the lens, there is a screen. Using the matrix formalism, obtain

a) the relationship between s, s', and f, such that the image of the object is observed on the screen;
b) the relationship between the lateral magnification of the system and the distances s and s'.

Solution

a) The system matrix consists of the product of three matrices, corresponding to (1) the ray transfer from the object to the lens, (2) the action of the lens itself, and (3) the ray transfer from the lens to the image plane. Using Eqs. (2.42) and (2.50) for the transfer and thin lens matrices, one has

$$S = \begin{bmatrix} 1 & s' \\ 0 & 1 \end{bmatrix} \begin{bmatrix} 1 & 0 \\ -\frac{1}{f} & 1 \end{bmatrix} \begin{bmatrix} 1 & s \\ 0 & 1 \end{bmatrix}$$

By performing the product of the matrices, we obtain

$$S = \begin{bmatrix} A & B \\ C & D \end{bmatrix} = \begin{bmatrix} 1 - \frac{s'}{f} & s' + s - \frac{ss'}{f} \\ -\frac{1}{f} & -\frac{s}{f} + 1 \end{bmatrix}$$

The distance at which the image forms, s', is determined by making $B = 0$, i.e.

$$B = s' + s - \frac{ss'}{f} = 0$$

or

$$\frac{1}{s} + \frac{1}{s'} = \frac{1}{f} \tag{2.56}$$

This result is in accordance with Eqs. (2.26) and (2.27).

b) The lateral magnification of the image is given, under the same circumstances, by element A of the system matrix:

$$A_l = A = 1 - \frac{s'}{f} = -\frac{s'}{s} \tag{2.57}$$

This result is in accordance with Eq. (2.28).

2.8 Location of Cardinal Points

A paraxial optical system is completely characterized by the four elements A, B, C, and D of its ray-transfer matrix **S**. Alternatively, such an optical system may be characterized by the locations of its *cardinal points.*

The locations of the four cardinal points are represented in Fig. 2.10. The limits of an optical system are defined by the input and output planes. The focal points F_1 and F_2 are located at distances p and q from the reference input and output planes and at distances f_1 and f_2 from the principal points H_1 and H_2, respectively. These points are at distances r and s from the input and output planes, respectively. Distances measured to the right of their reference planes are considered positive and to the left, negative.

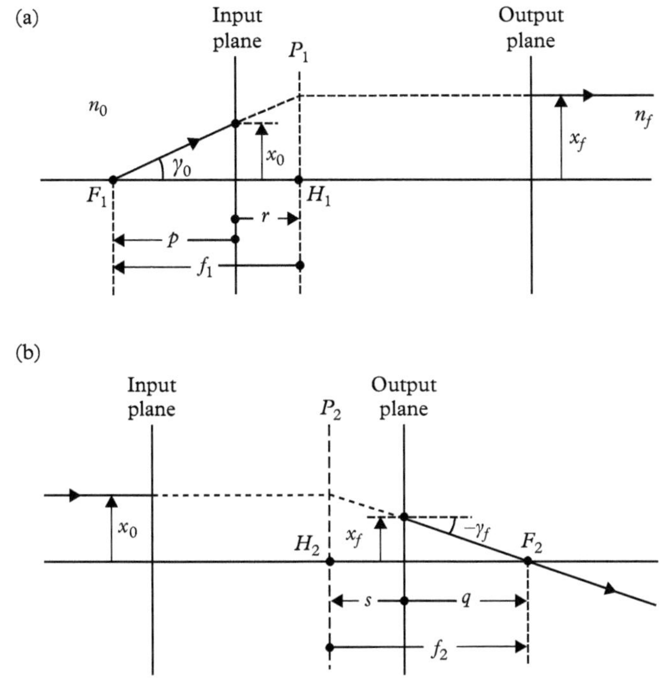

Figure 2.10 *Location of the four cardinal points of a paraxial optical system.*

The matrix equation relating an optical ray at the entrance and at the output of the optical system is

$$\begin{bmatrix} x_f \\ \gamma_f \end{bmatrix} = \begin{bmatrix} A & B \\ C & D \end{bmatrix} \begin{bmatrix} x_0 \\ \gamma_0 \end{bmatrix}, \tag{2.58}$$

where x_0 (x_f) and γ_0 (γ_f) represent, respectively, the height and the angle relative to the optical axis of the input (output) ray.

The determinant of matrix **S** is given by

$$|\mathbf{S}| = AD - BC \tag{2.59}$$

Regardless of the system concerned, this determinant is given by n_0/n_f, where n_0 and n_f are the refractive indices of the initial and final media, respectively.

Let us consider a ray that enters the system at a height x_0 and makes an angle γ_0 relatively to the optical axis, leaving the system with output coordinates $(x_f, 0)$. Thus, the equations for this ray become

$$x_f = Ax_0 + B\gamma_0 \tag{2.60}$$

$$0 = Cx_0 + D\gamma_0 \tag{2.61}$$

From Eq. (2.61), we obtain:

$$x_0 = -\frac{D}{C}y_0 \tag{2.62}$$

For small angles, one has from Fig. 2.9 that

$$y_0 = \frac{x_0}{(-p)} \tag{2.63}$$

where the negative sign indicates that F_1 is located at a distance p to the left of the input plane. From Eqs. (2.62) and (2.63), one obtains

$$p = -\frac{x_0}{y_0} = \frac{D}{C} \tag{2.64}$$

Similarly, one has

$$f_1 = -\frac{x_f}{y_0} = -\frac{Ax_0 + By_0}{y_0} = \frac{AD}{C} - B = \frac{AD - BC}{C} \tag{2.65}$$

Using Eq. (2.59), it is found that the focal distance f_1 can be written in the form:

$$f_1 = \frac{|S|}{C} = \frac{(n_0/n_f)}{C} \tag{2.66}$$

Finally, the positive distance r can be expressed in terms of p and f_1 as

$$r = p - f_1 = \frac{(D - (n_0/n_f))}{C} \tag{2.67}$$

where Eqs. (2.64) and (2.66) were used.

A similar procedure can be used to find the output distances q, f_2, and s, which are given by

$$q = -\frac{A}{C} \tag{2.68}$$

$$f_2 = -\frac{1}{C} \tag{2.69}$$

$$s = q - f_2 = \frac{1 - A}{C} \tag{2.70}$$

From Eqs. (2.66) and (2.69), one can verify that the first and second focal lengths of an optical system are equal in magnitude if the initial and final media have the same refractive indices.

2.9 Periodic Optical Systems

A periodic optical system consists of a sequence of identical unit systems. An example of a periodic optical system is given by a set of identical and equally spaced lenses. Another example may be a set of two mirrors, forming a resonant optical cavity. In these cases, rays cross the same unit system repeatedly.

Each unit stage that constitutes the periodic system is characterized by an *ABCD* matrix, as represented in Fig. 2.11. The determinant of such a matrix is given by the ratio between the refractive indices of the initial and final media, n_0/n_f. In many cases, these refractive indices are equal, so that $|\mathbf{S}| = 1$. We will assume such a case in the following discussion.

Let us consider a ray that enters the system at a height x_0 and makes an angle y_0 relatively to the optical axis. To determine the values (x_p, y_p) of these parameters at the output of the p-th stage, the *ABCD* matrix should be considered successively. The relationship between these parameters at the output of the p-th and (p+1)-th stages is given by

$$\begin{bmatrix} x_{p+1} \\ y_{p+1} \end{bmatrix} = \begin{bmatrix} A & B \\ C & D \end{bmatrix} \begin{bmatrix} x_p \\ y_p \end{bmatrix} \quad (2.71)$$

Alternatively, Eq. (2.71) can be replaced by the following relations:

$$x_{p+1} = Ax_p + By_p \quad (2.72)$$

$$y_{p+1} = Cx_p + Dy_p \quad (2.73)$$

We can derive an equation that governs the behaviour of the ray height, x_p, by eliminating the angle y_p between Eqs. (2.72) and (2.73). From Eq. (2.72) one has:

$$y_p = \frac{x_{p+1} - Ax_p}{B} \quad (2.74)$$

Replacing p with $p+1$ in Eq. (2.74), one obtains:

$$y_{p+1} = \frac{x_{p+2} - Ax_{p+1}}{B} \quad (2.75)$$

Substituting Eqs. (2.74) and (2.75) into Eq. (2.7), the following equation is derived:

$$x_{p+2} - (A + D)x_{p+1} + x_p = 0 \quad (2.76)$$

where $|\mathbf{S}| = AD - BC = 1$ has been assumed. Eq. (2.76) describes the evolution of the ray height, x_p, and can be solved iteratively on a computer.

With the aim of finding an explicit solution for Eq. (2.76), suppose that it has the form:

$$x_p = x_0 K^p \quad (2.77)$$

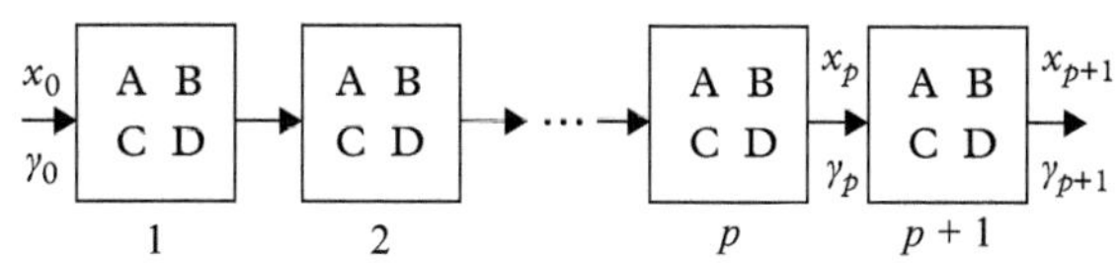

Figure 2.11 *A sequence of identical optical systems, each characterized by its ABCD matrix.*

where K is a constant. Substituting Eq. (2.77) into Eq. (2.76), the following algebraic equation is obtained for the parameter K:

$$K^2 - (A+D)K + 1 = 0 \tag{2.78}$$

whose solutions are

$$K = \frac{A+D}{2} \pm i\sqrt{1 - \left(\frac{A+D}{2}\right)^2} \tag{2.79}$$

We can define a new variable, ψ, in the form

$$\psi = \cos^{-1}\left(\frac{A+D}{2}\right) \tag{2.80}$$

Making the substitution

$$\frac{A+D}{2} = \cos\psi \tag{2.81}$$

in Eq. (2.79), we obtain

$$K = \exp(\pm i\psi) \tag{2.82}$$

In this way, the solutions given by Eq. (2.77) may be presented as

$$x_p = x_0 \exp(\pm ip\psi). \tag{2.83}$$

The general solution of Eq. (2.76) can be written as a linear combination of the previous particular solutions, corresponding to the plus and minus signs in Eq. (2.68). Such a general solution can be presented in the form:

$$x_p = x_{\max} \sin(p\psi + \psi_0) \tag{2.84}$$

where $x_{\max}$ and ψ_0 are constants to be determined from the initial conditions.

In order that x_p be a harmonic function, the variable $\psi = \cos^{-1}[(A+D)/2]$ must be real. This implies that the following condition is satisfied:

$$\frac{|A+D|}{2} \leq 1 \tag{2.85}$$

On the contrary, if $\frac{|A+D|}{2} > 1$, then ψ will be an imaginary number and the solution will be a hyperbolic function, whose value increases monotonously with p. A harmonic solution ensures that x_p remains limited for all values of p. Actually, Eq. (2.85) represents a *stability condition* for the ray trajectory.

Example 2.7

Consider a sequence of identical thin lenses, each with a focal length f, spaced by a constant distance d (Fig. 2.12). Find the maximum value of this spacing that ensures a stable trajectory for the paraxial rays.

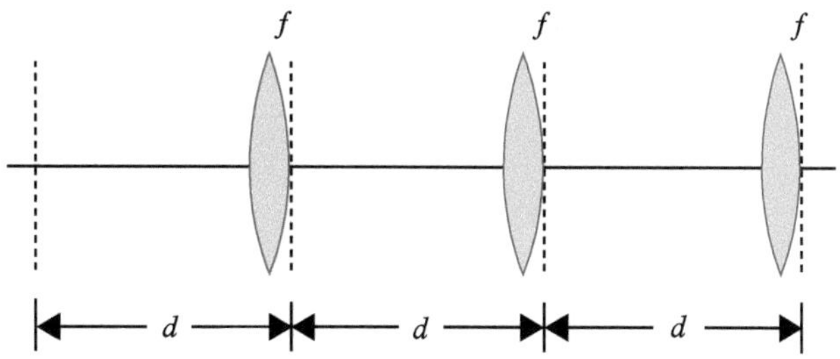

Figure 2.12 *A sequence of identical thin lenses with a focal length f, spaced by a constant distance d.*

Solution

The unit system of the lens sequence consists of a distance d, travelled in free space, followed by a lens. The matrix for this unit system is given by

$$\mathbf{S} = \begin{bmatrix} 1 & 0 \\ -\frac{1}{f} & 1 \end{bmatrix} \begin{bmatrix} 1 & d \\ 0 & 1 \end{bmatrix} = \begin{bmatrix} 1 & d \\ -\frac{1}{f} & 1-\frac{d}{f} \end{bmatrix} \equiv \begin{bmatrix} A & B \\ C & D \end{bmatrix}$$

One has $A = 1$ and $D = 1 - d/f$, so the condition of stability

$$\frac{|A+D|}{2} \leq 1$$

becomes

$$\left|1 - \frac{d}{2f}\right| \leq 1$$

This condition can be written in the form:

$$0 \leq d \leq 4f$$

Thus, the maximum value of the spacing between the lenses that ensures a stable trajectory for the paraxial rays is equal to four times the focal length of the lens.

2.10 Further Reading

Born, M., and E. Wolf, *Principles of Optics*, 7th Ed., Cambridge University Press, 2002.

Gerrard, A., and J. M. Burch, *Introduction to Matrix Methods in Optics*, Wiley, 1975; Dover, reprinted 2012.

Guenther, B., *Modern Optics*, 2nd Ed., Oxford University Press, 2015.

Hecht, E., *Optics*, 5th Ed. Pearson, 2016.

Kingslake, R., *Lens Design Fundamentals*, Academic Press, 1978.

Klein, M., and T. Furtak, *Optics*, 2nd Ed., John Wiley & Sons, 1986.
Lin, P. D., *Advanced Geometrical Optics*, Springer-Verlag, 2017.
Longhurst, R., *Geometrical and Physical Optics*, 2nd Ed. John Wiley & Sons, 1967.
Mouroulis, P., and J. Macdonald, *Geometrical Optics and Optical Design*, Oxford University Press, 1997.
Pedrotti, F., L. M. Pedrotti, and L. S. Pedrotti, *Introduction to Optics*, 3rd Ed., Cambridge University Press, 2018.
Rossi, B., *Optics*. Addison-Wesley Publishing Company, 1957.

2.11 Problems

2.1 An object is situated at a distance of 10 cm from the vertex of a convex mirror with a focal length of 8 cm. Determine the position and the lateral magnification of the image.

2.2 An object 1 cm high is at a distance of 3 cm from a concave mirror, with a radius of 12 cm. Determine the position and the lateral magnification of the resulting image, and make the corresponding ray diagram.

2.3 Derive an expression for the lateral shift of the transmitted ray when the incident ray strikes at an angle ϕ on a transparent plate with a refractive index n and a thickness d, located in the air. Obtain the results for the cases i) $\phi = 0$ and ii) $\phi = \pi/2$.

2.4 A glass sphere, with a refractive index n, has a concentric spherical cavity filled with air, with a radius r, equal to the thickness of the glass layer. On the inner surface of the sphere, there is a point object, which is observed from the outside in the opposite direction. Show that the point image is observed at a distance $r(n-1)/(3n-1)$ from the object position.

2.5 Consider a biconvex lens with a refractive index $n = 1.5$ and a thickness $d = 10$cm, limited by two spherical surfaces of radius $R = 5$ cm. A real object is positioned in the air, 20 cm from the lens. Find the position and nature of the image formed by such a lens.

2.6 A point source P illuminates a thin biconvex lens with refractive index n and surfaces with a curvature radius R. The second surface of the lens is covered by a reflective film. Find the axial distance between the point P and the lens in order that the reflected light, after exiting the lens, forms a collimated beam parallel to the system axis.

2.7 A glass sphere (n = 1.5) of radius R has half of its surface covered by a reflective film. A small object is placed on the axis of the sphere at a distance $2R$ from the unmirrored surface. Find the position of the final image formed by the refractive and mirrored surfaces.

2.8 Using Eqs. (2.34), (2.45), and (2.51), prove that the focal length of a thin lens in air is given by Eq. (2.27). Verify that the focal length is positive (negative) in the case of a biconvex (biconcave) lens, also discussing other possible geometries.

2.9 Consider a thin positive lens with focal length f. Show that the shortest distance between a real object and its (real) image is $4f$.

2.10 Consider a two-lens system formed by a converging lens and a diverging lens, both with a focal distance f = 15 cm, separated by a distance d = 75 cm. Let the object be 30 cm from the first (converging) lens. Find the position and nature of the image formed by the two-lens system (i) graphically and (ii) analytically.

2.11 Consider an optical system formed by a converging lens with a focal length f and a concave mirror with a focal length f'. Find the distance between the two elements in order that the focus of the global system coincides with the primary focus of the lens. Make a ray diagram illustrating this configuration.

2.12 Consider a sequence of identical thin lenses as represented in Fig. 2.12, each with a focal length f, spaced by a constant distance d, such that

i) $d = 2f$

ii) $d = f$.

Find the period and illustrate the paraxial ray propagation in each case.

2.13 Show that the reflection in a concave spherical mirror of radius R is described by the matrix:

$$\mathbf{S} = \begin{bmatrix} 1 & 0 \\ 2/R & 1 \end{bmatrix}$$

2.14 Find the system matrix for the set of N parallel planar transparent plates of refractive indices $n_1, n_2, \ldots, n_N$ and thicknesses $d_1, d_2, \ldots, d_N$, placed in the air ($n = 1$), normal to the z propagation axis.

2.15 A thin biconvex lens with a refractive index $n_g = 1.5$ has a radius of curvature of 20 cm and is placed at the entrance of a tank full of water ($n_w = 1.33$). Suppose that an object is placed on the outside of the tank at a distance of 100 cm from the lens. Determine the position and the magnification of the image using matrices.

3 Optical Instruments and Aberrations

Geometrical optics will be used in this chapter to describe some of the most common optical instruments, namely the magnifying glass, the compound microscope, and the telescope. The human eye is a particularly important natural optical system and will deserve particular attention, being considered first. The process of image formation in the eye is similar to that found in a conventional camera, whose basic characteristics will also be described.

A problem that occurs frequently when using optical systems concerns the difference between their actual performance and the predictions based on the paraxial approximation of geometrical optics. Such discrepancies are known as *aberrations* and will be briefly addressed in the final part of this chapter.

3.1 The Human Eye

The human eye constitutes a truly remarkable optical system, consisting of several elements admirably adapted to their specific purposes. Among its several distinctive characteristics, it is able to form images of objects at very different distances from the observer. It can scan a scene as vast as the overhead sky, as well as focus on very small details of an object. It can also adapt to an extraordinary range of intensities and distinguish subtle shades of colour.

Fig. 3.1 shows its essential elements. The constitution and functioning of the human eye may be considered similar to those of a camera, which will be described later.

The shape of the human eye is approximately spherical, with a diameter of about 22 mm. Light enters it through a variable aperture—the *pupil*—and is focused by the *corneal-crystalline system* on the *retina*, which covers the posterior surface of the eyeball. With the exception of the frontal area, the eye is contained by an opaque, fibrous, protective outer layer, called the *sclera*.

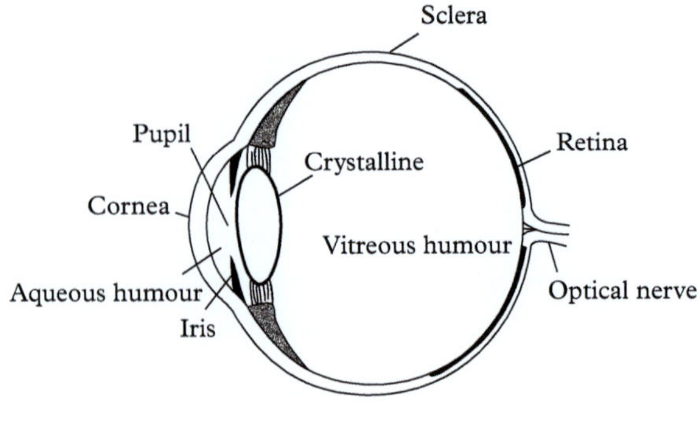

Figure 3.1 *Representation of the human eye.*

Lasers, Optics, and Photonics. Mário F. S. Ferreira, Oxford University Press. © Mário F. S. Ferreira (2026).
DOI: 10.1093/oso/9780198952770.003.0003

The deflection of light that reaches the eye occurs mainly on the anterior surface of the *cornea*, which has a refractive index of 1.38 and is roughly 12 mm in diameter and 0.6 mm thick in the centre. As for the *crystalline*, it is a convergent lens whose refractive index gradually varies from 1.41 in the centre to 1.38 in the periphery. Between the cornea and the lens, there is the *anterior chamber*, filled with a clear liquid, the *aqueous humour*. This liquid has a refractive index of 1.336, similar to that of water. Immersed in this liquid is the *iris*, a diaphragm that can increase or decrease the diameter of the pupil from 2 mm (in high light situations) to about 8 mm (in very low light situations). This mechanism enables the eye to respond to light irradiances that differ by about five orders of magnitude, an ability referred to as *adaptation*. It is the iris that gives colour to a person's eyes. Between the lens and the retina, corresponding to most of the eyeball, there is the *posterior chamber*. It is filled with a gelatinous substance, called *vitreous humour*, with a refractive index of 1.34, again similar to that of water. As for the *retina*, it is a layer with a thickness of about 0.1 mm that receives light energy and converts it into electrochemical energy. This function is performed by two types of photoreceptor cells: *rods* and *cones*. The rods respond to low-intensity light and contribute to the perception of low-resolution, black-and-white images; cones respond to high-intensity light and contribute to the perception of high-resolution, coloured images. There are about 7 million cones and 75 to 150 million rods that are linked to about 1 million fibres forming the *optic nerve*, which transmits all the information to the brain.

The optic nerves of the left and the right eye cross at the *optic chiasm*, which is located at the bottom of the brain, immediately below the hypothalamus. From the optic chiasm, nerve fibres coming from the right (left) half of each eye terminate in the right (left) half of the brain. Thus, the brain forms a single image, even though each half of the brain receives an image from *both* eyes. The slight differences between the two images from both eyes provide the possibility of judging the depth or position of objects in a three-dimensional field, an ability referred to as *stereoscopic vision*.

One of the most admirable properties of the human eye is related to the fact that the shape of the lens can be slightly adjusted by the action of the so-called *ciliary muscle*. For example, when an object approaches our eyes, the ciliary muscle contracts, which slightly increases the curvature of the lens and decreases its focal length so that the image of that object formed on the retina remains sharp. When observing a distant object, the ciliary muscle becomes relaxed, and the lens assumes its flattest shape, providing the least refraction of incident light rays. However, this process of adjusting the shape of the lens—commonly referred to as *accommodation*—is limited. For distances between the object and the eye below a certain value, the image begins to lose sharpness. The closest point at which the lens can still focus the image on the retina is called the *near point*. The distance from the eye to the near point is called the *minimum distance of distinct vision* (MDDV), which increases with advancing age. It is 7 to 10 cm for a teenager, increasing to 20 to 40 cm for a middle-aged person. The reference value of the MDDV is 25 cm for a middle-aged adult with normal vision.

The apparent size of an object is determined by the size of its image on the retina. This image is larger when the object is closer to the eye (Fig. 3.2), which is why the object seems bigger in this situation than when it is farther away. A convenient measure of the image size on the retina is given by the angle subtended by the object at the eye. In the case of an object with height h at the near point, we have:

$$\theta_0 \approx \tan^{-1}\theta_0 = \frac{h}{25\text{cm}} \tag{3.1}$$

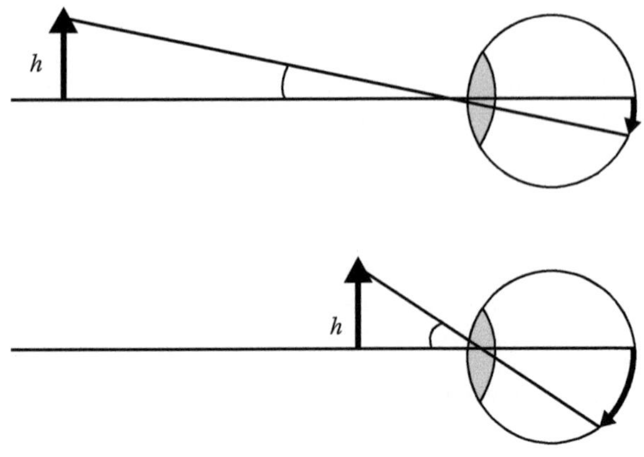

Figure 3.2 *Apparent size of an object.*

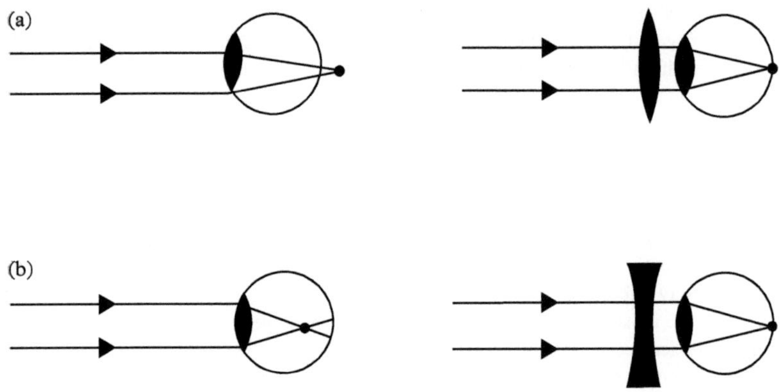

Figure 3.3 *Vision defects and their respective corrections: (a) hyperopia and (b) myopia.*

Vision defects are usually corrected by wearing glasses or contact lenses or by performing surgery. For example, if images of distant objects form behind the retina, it is said that the person suffers from *hyperopia*. Compared with the normal eye, the hyperopic eye is commonly found to be shorter in the axial distance. Correction of this defect is possible using a suitable converging lens (Fig. 3.3(a)). In general, the hyperopic eye can accommodate to see distant objects clearly. However, maximum accommodation occurs at a point that is farther away from the eye than the normal near point—the so-called hyperopic near point.

When the images of distant objects are located in front of the retina, it is said that the person suffers from *myopia*. Commonly, the myopic eye is longer than normal. This defect can be corrected using a diverging lens (Fig. 3.3(b)). Although the less remote far point is a clear disadvantage, it is found that the myopic eye is able to see clearly at points closer than the normal near point. Moreover, since angular magnification increases with proximity to the eye, the myopic eye provides improved vision of object details held close to the eye.

Presbyopia refers to the loss of accommodation that occurs due to age-related changes in the lens (decreased elasticity) and ciliary muscle (decreased strength). Presbyopia can be corrected using a converging lens, as for hyperopia. However, in order to restore clear vision of objects located at any distance from the presbyopic eye, it is necessary to use a lens with multiple powers. In the case of *bifocal* lenses, the upper and lower parts of the spectacle lens restore clear vision of distant and near objects, respectively. Using *progressive* lenses, in which the power gradually increases from the top to the bottom of the lens, clear vision can be restored over the full distance range.

When the cornea is not perfectly symmetrical, presenting different curvature radii for different meridional planes (those containing the optical axis), it is said that the person suffers from *astigmatism*. Such corneal asymmetry leads to different refractive powers, which cause light to be focused at different distances from the cornea, resulting in somewhat hazy vision. This vision defect is perhaps the most frequent and can be corrected in many cases using suitable cylindrical lenses.

Example 3.1

The minimum distance of distinct vision for a certain person is 1 m. Which lens should be used by this person to get clear vision of an object located 25 cm from the eye?

Solution

Let us consider the equation of thin lenses:

$$\frac{1}{s}+\frac{1}{s'}=\frac{1}{f}$$

where we substitute $s = 0.25$ m and $s' = -1$ m. The minus sign of s' is due to the fact that the image is on the same side of the lens where the object is located. In this way, we obtain the result

$$D=\frac{1}{f}=\frac{1}{0.25}+\frac{1}{(-1)}=+3$$

That is, a convergent lens with a power of +3 diopters should be used to get clear vision of the object.

3.2 The Simple Magnifier

The apparent size of an object can be increased by using a simple magnifying glass. The position of the object is adjusted to a distance s from the magnifying glass, so that its image is at a distance equal to or greater than the MDDV. Seen through the lens, the object appears larger, as the image on the retina is magnified by a factor θ/θ_0, where θ_0 is given by Eq. (3.1) and $\theta = h/s$ (Fig. 3.4). The ratio θ/θ_0 is the *angular magnification*, A_θ, given by

$$A_\theta=\frac{\theta}{\theta_0}=\frac{25}{s} \tag{3.2}$$

In the case where the image is located at the MDDV, the distance s can be found using the lens equation:

$$\frac{1}{s}+\frac{1}{(-25)}=\frac{1}{f} \tag{3.3}$$

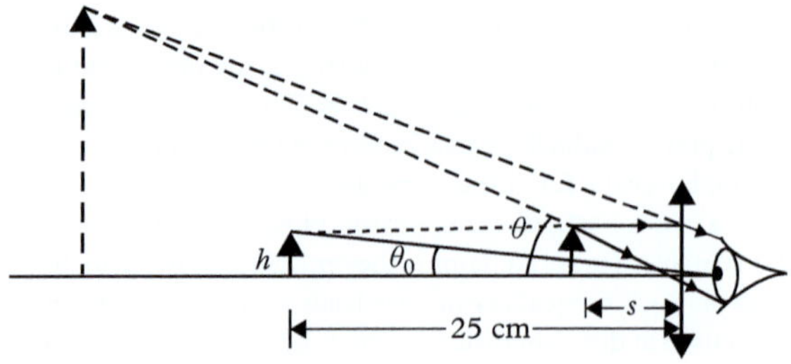

Figure 3.4 *Geometry for the analysis of a simple magnifying glass.*

where f is the focal length of the lens, which is assumed to be given in centimetres. That is,

$$s = \frac{25f}{25+f} \tag{3.4}$$

Thus, the angular magnification of the magnifying glass is given by

$$A_\theta = 1 + \frac{25}{f} \tag{3.5}$$

If the object is located in the focal plane of the magnifying glass ($s = f$), the corresponding image is formed at infinity and its magnification is given simply by

$$A_\theta = \frac{25}{f} \tag{3.6}$$

In the case of relatively small focal lengths, Eqs. (3.5) and (3.6) do not differ much, with Eq. (3.6) being generally used to indicate the magnification of the simple magnifying glass. For example, in case this magnifying glass has a focal length of 10 cm, its magnification is 2.5. This situation is represented in the form 2.5×, which means that the image dimension on the retina, with the object in the lens focus, is 2.5 times larger than with the object at the near point and observed without the help of that lens. The magnification of a simple magnifying glass is usually between 2× and 10×, although it is possible to obtain higher magnification values when the lens aberrations—which will be discussed in Section 3.6—are properly corrected.

3.3 The Compound Microscope

The earliest known examples of compound microscopes appeared in Europe around 1620, but their inventor is unknown. They are used to observe very small objects at short distances, increasing the magnification provided by the simple magnifying glass described in the previous section. Fig. 3.5 shows a schematic of a compound microscope in its simplest form. It consists of two positive lenses: an *objective lens*, with a small focal length f_{ob}, which is closer to the object, and an *eyepiece*, or *ocular* lens, with focal length f_{oc}, which is closer to the eye. The object is placed in a slightly earlier position relative to the focal plane of the objective lens, so that we have an enlarged image in the focal plane of the eyepiece. This plane is located at a distance $g + f_{ob}$ from the objective, where g is the so-called *length of the tube*. The eyepiece 'looks' at the real image formed by the objective.

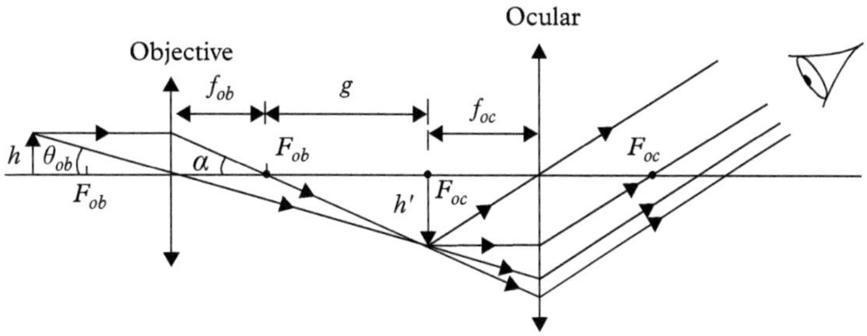

Figure 3.5 *Geometry for the analysis of a compound microscope.*

From Fig. 3.5, one has

$$\tan\alpha = \frac{h}{f_{ob}} = \frac{(-h')}{g} \tag{3.7}$$

where the minus sign is due to the fact that the image is inverted. Thus, the lateral magnification of the objective is

$$A_l = \frac{h'}{h} = -\frac{g}{f_{ob}} \tag{3.8}$$

The overall magnification of the compound microscope is equal to the product of the lateral magnification of the objective and the angular magnification of the eyepiece: $A_\theta = 25/f_{oc}$:

$$A_g = A_l A_\theta = -\frac{g}{f_{ob}}\frac{25}{f_{oc}} \tag{3.9}$$

In many microscopes, the value of g assumes a standard value of 16 cm. Focal lengths f_{ob} and f_{oc} generally correspond to actual values for lenses consisting of multiple elements whose aberrations are properly corrected.

Example 3.2

In a compound microscope, the objective has a focal length of 2 cm, whereas the eyepiece provides a magnification of 5×. Considering that the two lenses are spaced 25 cm apart, determine how far from the objective the object should be if the observer focuses the image at infinity. Calculate the overall magnification of the microscope.

Solution

The angular magnification of the eyepiece is

$$A_\theta = \frac{25}{f} = 5$$

continued

Example 3.2 continued

where its focal length is $f_{oc} = 5$ cm. The image formed by the objective is located in the focal plane of the eyepiece. Given that the separation between the lenses is $L = 25$ cm, the distance from that image to the objective is

$$s' = L - f_{oc} = 20\,\text{cm}$$

The object must be at a distance s from the objective, as given by the thin lens equation:

$$\frac{1}{s} + \frac{1}{20} = \frac{1}{2},$$

from which we obtain $s = 2.22$ cm.

On the other hand, given that the focal length of the objective is $f_{ob} = 2$ cm, the length of the tube is

$$g = L - f_{ob} - f_{oc} = 18\,\text{cm}$$

In this way, the lateral magnification of the objective is

$$A_l = -\frac{s'}{s} = -\frac{g}{f_{ob}} = -\frac{18}{2} = -9$$

Using Eq. (3.9), we find that the overall magnification of the microscope is given by

$$A_g = A_l A_\theta = -9 \times 5 = -45$$

3.4 The Optical Telescope

An optical telescope is an instrument using lenses, curved mirrors, or a combination of both, aiming to observe objects that are generally very distant from the observer. The first known practical telescopes were *refracting telescopes* with glass lenses and were invented in the Netherlands at the beginning of the 17th century. They were used for both terrestrial applications and astronomy. The *reflecting telescope*, which uses mirrors to collect and focus light, was invented within a few decades of the first refracting telescope. Fig. 3.6 illustrates some examples of both types of telescopes. The astronomical and the Galilean telescopes are examples of the first type, whereas the Newtonian and the Gregorian telescopes are examples of the second type. Telescopes that combine refracting and reflecting surfaces are called *catadioptric*.

Like the compound microscope described in the previous section, in its basic form the telescope is also made up of two lenses, as schematically represented in Fig. 3.7. The *objective* provides an intermediate image of a distant object in its secondary focal plane, which coincides with the primary focal plane of the *ocular*; the two lenses are separated by a distance equal to the sum of their focal lengths. The ocular forms a virtual and inverted final image at infinity of that intermediate image. That is, in this configuration, both the object and the final image are at infinity. This system can also be used to expand a laser beam.

Let θ_{ob} be the angle corresponding to the object when it is observed directly by view, which is equal to the angle of the same object seen by the objective. If θ_{oc} is the angle corresponding to the final image (Fig. 3.7), the magnification of the telescope is $A_\theta = \theta_{oc}/\theta_{ob}$. Considering the geometry of Fig. 3.7 and using the approximation of small angles, one has

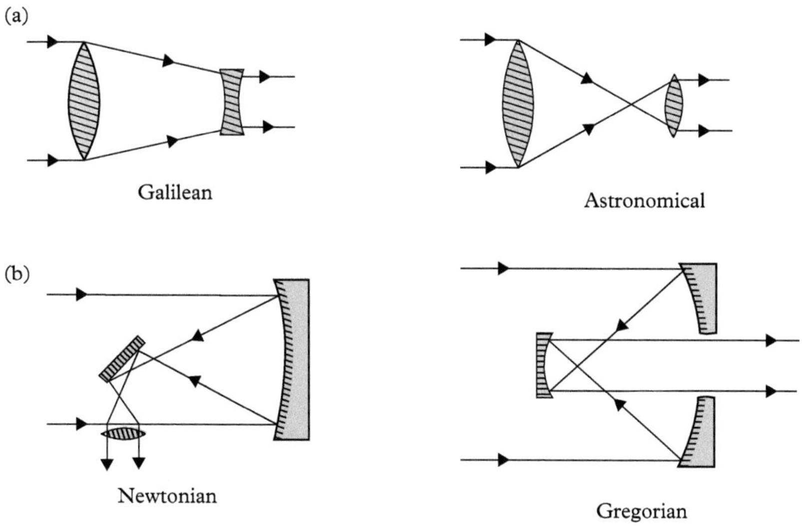

Figure 3.6 *Examples of (a) refracting and (b) reflecting telescopes.*

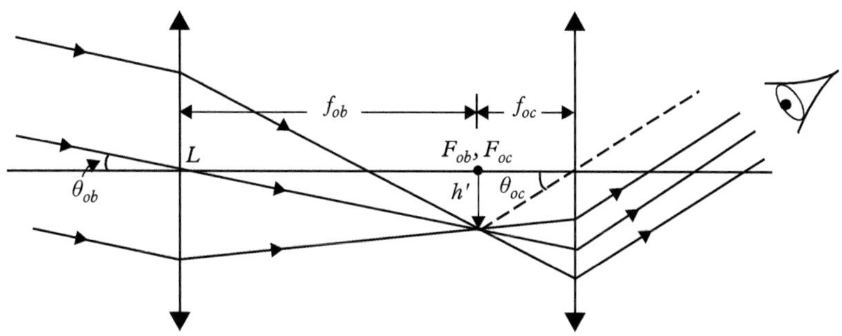

Figure 3.7 *Schematic representation of an astronomical telescope.*

$$\theta_{0b} \approx \tan\theta_{0b} = -\frac{h'}{f_{ob}} \tag{3.10}$$

and

$$\theta_{oc} \approx \tan\theta_{oc} = \frac{h'}{f_{oc}} \tag{3.11}$$

In Eq. (3.11), both h' and θ_{oc} are negative, which corresponds to an inverted image. The magnification of the telescope is given by

$$A_\theta = \frac{\theta_{oc}}{\theta_{ob}} = -\frac{f_{ob}}{f_{oc}} \tag{3.12}$$

Although the inversion of the image in the telescope shown in Fig. 3.7 is not a serious problem in the observation of astronomical objects, this characteristic is obviously inconvenient when it comes to observing terrestrial objects. For this reason, the ground telescope features an additional converging lens that allows one to obtain a final upright image. The Galilean telescope, illustrated in Fig. 3.6(a), produces an erect image by using a diverging lens as the ocular.

In order to increase the amount of light received from a distant object, the objective should be as wide as possible. However, the manufacture of large lenses presents obvious difficulties, in particular in terms of their homogeneity and chromatic characteristics. One possibility to overcome these difficulties is to build a telescope using mirrors, as Newton recognized as early as 1668. In the *Newtonian* telescope, a parabolic mirror is used to focus accurately all parallel rays to the same primary focal point. A plane mirror is used to divert the converging rays to a secondary focal point near the body of the telescope, where an ocular is located to view the image. In the Gregorian telescope, a concave ellipsoidal mirror is used as the secondary mirror. The primary and secondary focal points of this telescope are the foci of the ellipsoid.

The diameter of the mirrors used in the astronomical telescopes must be as large as possible in order to maximize the light received. For example, the mirrors of the Gemini telescopes installed in Hawaii and Chile have a diameter of 8 m. As for the famous Hubble Space Telescope, its diameter is only 2.3 m. However, given its location, this telescope has the great advantage of avoiding the effects of random absorption and refraction that occur when light passes through the atmosphere.

The construction of larger space telescopes involves the use of multiple elements, which should be transported separately and then assembled in space. The surface of mirrors constructed in this way must always have irregularities smaller than a small fraction of the wavelength, an objective which is difficult to achieve. On the other hand, it should be noted that it is not easy to correct any design errors in such circumstances. In the case of the Hubble Space Telescope, due to poor ground tests, it was launched with serious spherical aberration–related problems, which significantly affected its initial performance.

Example 3.3

Two refractive telescopes present the same objective lens with focal length f_{ob}. One of them is a Keplerian telescope, using a converging ocular with focal length f_c. The other one is a Galilean telescope, using a diverging ocular with focal length f_d. Obtain an expression for the ratio of the lengths of these telescopes in terms of their magnification A, assuming that it is the same for objects at infinity.

Solution

The magnification of a telescope is given by Eq. (3.12). Thus, we can write:

$$A_K = -\frac{f_{ob}}{f_c}$$

$$A_G = -\frac{f_{ob}}{f_d} = \frac{f_{ob}}{|f_d|}$$

where a negative magnification corresponds to an inverted image and a positive magnification corresponds to an erect one. Assuming that $-A_K = A_G = A$, we have $f_c = |f_d|$.

continued

Example 3.3 continued

For a telescope, the secondary focal point of the objective coincides with the primary focal point of the ocular. Thus, the lengths of the two telescopes are

$$L_K = f_{ob} + f_c$$

$$L_G = f_{ob} - |f_d|$$

The ratio of the two lengths is given by

$$\frac{L_K}{L_G} = \frac{f_{ob} + f_c}{f_{ob} - |f_d|} = \frac{A+1}{A-1}$$

Besides providing an erect image, a Galilean telescope is shorter than a Keplerian one for the same magnification.

3.5 The Camera

The simplest camera consists of a box with a small hole and without any lens (Fig. 3.8). This box corresponds to the *camera obscura* described by Leonardo da Vinci (1452–1519) and analysed in detail by Giovanni Battista Della Porta (1535–1615) in his work *Magia Naturalis* (1589). If the hole in the box is too large, no picture is formed; as its diameter decreases, the image becomes increasingly more defined. However, below a certain value, the image loses its definition. The diameter of the hole that provides an optimal setting is proportional to its distance from the image plane.

A pinhole camera becomes an actual camera when a light-sensitive film is placed on the image plane. The first photograph was taken using this device by Joseph Niépce (1765–1833) in 1826, after an exposure time of eight hours. In fact, the main problem of this first generation of cameras was the rather long exposure times that were needed. As a consequence, in order to obtain a good-quality photograph, it was necessary that both the camera and the photographed objects were at rest during all this time.

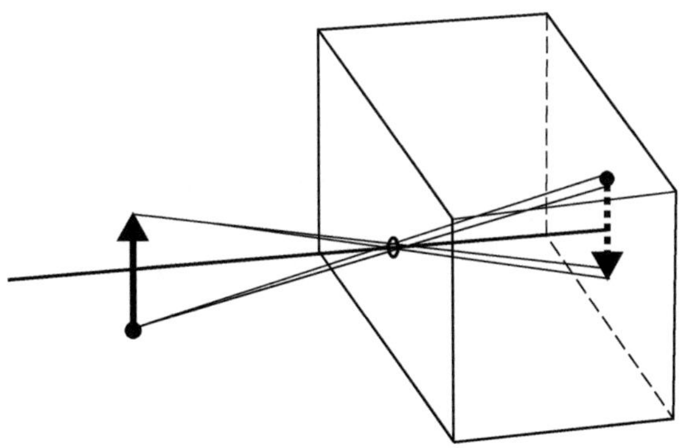

Figure 3.8 *The pinhole camera.*

Currently, a camera consists basically of a box with a set of lenses constituting the objective, a diaphragm, a moving mirror, an eyepiece, and a film sensitive to light (Fig. 3.9). Such photographic film is replaced by an electronic image sensor in modern digital cameras. The variable-aperture diaphragm allows one to control the exposure time, while the moving mirror deflects light in the direction of a focusing screen. The pentaprism allows one to conveniently orient the image that is observed through the eyepiece. At the time of the photograph, the diaphragm closes to a certain point, the mirror deviates, and the film becomes exposed to light. After exposure, the diaphragm opens completely, and the mirror returns to its initial position.

The exposure time is proportional to the square of the so-called *f*-number of the objective, *f*/#, given by

$$f/\# \equiv \frac{f}{d} \tag{3.13}$$

where *f* is the focal length and *d is* the diameter of the diaphragm. For example, an objective with a 50 mm focal length and a 25 mm aperture has an *f*/# = 2, which is represented in the form *f*/2. For lower values of the *f*-number, more light can reach the film. Objective diaphragms normally include a range of 1, 1.4, 2, 2.8, 4, 5.6, etc. That is, the *f*/# varies by $\sqrt{2}$, so the flux density that reaches the film varies in steps of 1/2.

The *f-number* is also referred to as objective speed. For example, it is said that an *f*/1.4 objective is twice as fast as an *f*/2 objective. In fact, the film receives the same amount of energy if the camera is set to *f*/1.4 and 1/500 of a second or to *f*/2 and 1/250 of a second.

Currently, there are several types of cameras, each suited to specific uses and offering unique capabilities. Large-format and medium-format cameras offer higher image resolution and are often used in professional and artistic photography. Digital cameras, which became prevalent in the late 20th and early 21st centuries, use electronic sensors to capture and store images. More recently, we have witnessed the widespread use of smartphone camera technology, profoundly influencing how society creates and shares visual content.

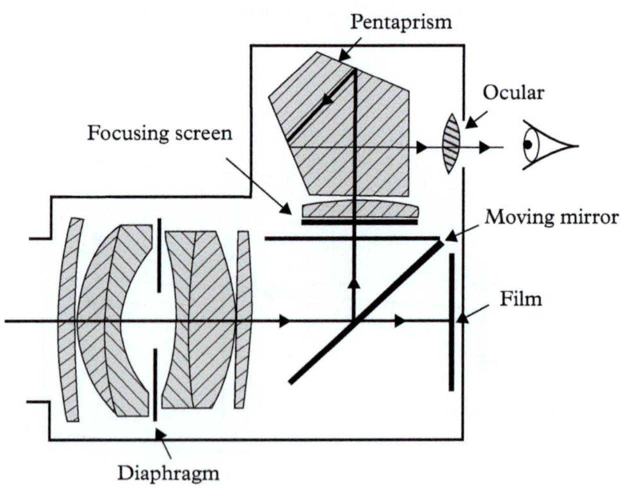

Figure 3.9 *Schematic representation of a camera.*

3.6 Aberrations

The expected performance of some instruments discussed in previous sections can be affected by different types of *aberrations*. Such aberrations appear as discrepancies between the rigorous tracing of rays through an optical system in relation to the predictions of paraxial theory, or first-order theory, presented in Chapter 2. Aberrations can be of two types: *chromatic* (due to the dependence of the refractive index on wavelength) or *monochromatic*, which occur even with monochromatic light.

3.6.1 Monochromatic aberrations

The sine function can be expanded in a Maclaurin series:

$$\sin\gamma = \gamma - \frac{\gamma^3}{3!} + \frac{\gamma^5}{5!} - \cdots \tag{3.14}$$

When considering only the first two terms in the right-hand member of Eq. (3.14), we have the so-called third-order theory. The deviations of this theory from the first-order theory give rise to the so-called *primary aberrations: spherical aberration*, *coma*, *astigmatism*, *field curvature*, and *distortion*. Including the terms of order higher than the third in the development of Eq. (3.14) allows one to consider higher-order aberrations. In the following, we will briefly address the primary aberrations in a qualitative way.

Spherical Aberration

Spherical aberration occurs when using both lenses and mirrors with spherical surfaces and basically consists of the dependence of the focal length on the distance of light rays from the optical axis. Light rays that strike a spherical surface off-centre are refracted or reflected more or less than those that strike close to the centre. This deviation reduces the quality of the images produced by optical systems.

Spherical aberration is illustrated in Fig. 3.10 for the case of a converging lens, in which the marginal rays intersect the optical axis before the paraxial rays. The opposite occurs in the case of a diverging lens. The distance between the focus of the marginal rays (FMR) and the focus of the paraxial rays (FPR) is called *longitudinal spherical aberration* (LSA). This aberration is positive in the case of a converging lens and negative in the case of a diverging lens.

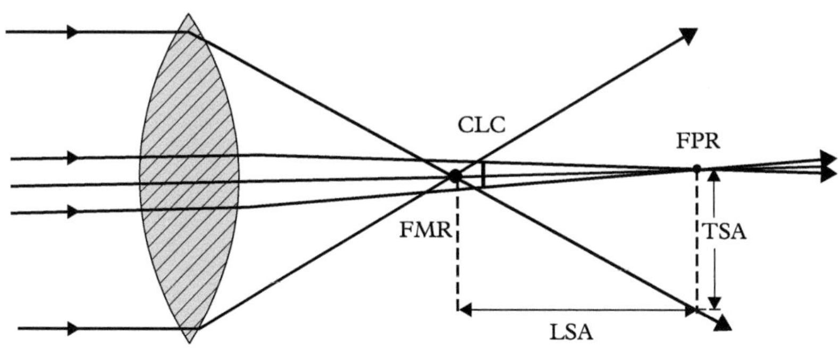

Figure 3.10 *Spherical aberration in the case of a converging lens. FPR: focus of the paraxial rays; FMR: focus of the marginal rays; LSA: longitudinal spherical aberration; TSA: transverse spherical aberration; CLC: circle of least confusion.*

As a result of spherical aberration, the image of a point object, observed on a screen, is not a point, as one could expect based on paraxial theory. Instead, there is a light spot consisting of a brighter central core, surrounded by a halo produced by marginal rays. The distance between the optical axis and the point at which the radius intersects the screen is called *transverse spherical aberration* (TSA). There is a position for which the diameter of the light spot on the screen is minimal, corresponding to the so-called *circle of least confusion* (CLC).

The rays that pass through consecutive areas of the lens intersect in a curve called a *caustic*. The rotation of this curve around the optical axis gives rise to a three-dimensional figure, the so-called *caustic surface*, whose vertex coincides with the paraxial focus.

Spherical aberration cannot be completely eliminated in the case of a simple lens. However, it can be reduced by limiting the entrance pupil to only nearly paraxial rays. This remedy is, of course, accompanied by a reduction in image brightness. Spherical aberration can also be reduced by making one or both lens surfaces have a non-spherical shape, or by performing an appropriate combination of converging and diverging lenses.

Spherical aberration makes the focus of telescopes and other instruments less than ideal. This is an important effect because spherical shapes are much easier to produce than aspherical ones. In many cases, it is cheaper to use multiple spherical elements to compensate for spherical aberration than to use a single aspheric lens.

Coma

Coma is an aberration that manifests in the case of point sources situated outside the optical axis, whose images have the shape of a comet. This effect is due to the variation of the lateral magnification with the distance from the optical axis to the point of incidence of the ray on the lens. In some cases, the marginal rays intersect the image plane closer to the axis than the paraxial rays, and then the coma is said to be negative. In other cases, the marginal rays intersect this plane farther from the axis than the paraxial rays (Fig. 3.11(a)), and then the coma is said to be positive. Each circular zone of the lens forms a circular image called *a comatic circle*. The set of all these comatic circles, whose radii increase with the radii of those zones, gives rise to a figure with the shape of a comet, as shown in Fig. 3.11(b).

Coma depends on the shape of the lens, as does spherical aberration. However, the two types of aberration differ in several aspects. In particular, comatic aberration depends on the angle of obliquity and does not occur for point objects located on the optical axis.

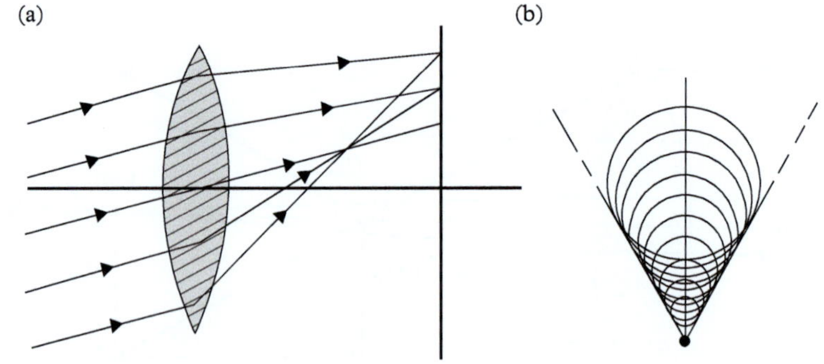

Figure 3.11 *(a) Illustration of a case of positive coma and (b) formation of a comatic image from a set of comatic circles.*

Coma can also affect the performance of telescopes, even those using parabolic mirrors. Actually, these mirrors are free from spherical aberration, since a bundle of rays parallel to the optical axis will be perfectly focused to a point, no matter where they strike the mirror. However, when the incoming rays strike the mirror at an angle to the optical axis, individual rays are not reflected to the same point, causing the image of a point source, such as a distant star, to be affected by comatic aberration. In refractive telescopes and other optical systems, especially those imaging a wide spectral range, coma can also be a function of wavelength, corresponding to a form of chromatic aberration.

Astigmatism

Astigmatism is an aberration that has to do with the asymmetry of the incident ray cone on the optical system when the point object is off the optical axis. For the description of this type of aberration, it is convenient to define two planes. One of them is the *tangential plane*, which contains both that object point and the system axis of symmetry. Rays that propagate in this plane are called *tangential rays*. The second plane is the *sagittal plane*, which is orthogonal to the tangential plane and contains the same object point. This plane contains the *principal ray*—the ray coming from the object point that passes through the axial point in the plane of the entrance pupil—but does not contain the optic axis. The rays propagating in this plane are called *sagittal rays*. In the case of a parallel beam of rays focusing obliquely on the optical system, the configuration of the rays is different for those two planes. The inclination of the rays is higher in the tangential plane, which determines a shorter focal length in this plane, compared to the focal length in the sagittal plane. The astigmatic difference between the two focal lengths increases rapidly with the obliquity of the light rays, that is, with the distance of the point object to the optical axis.

A point object outside the optical axis presents two linear images, as represented in Fig. 3.12: a *primary image*, situated in the plane containing the *tangential focus* F_t, and a *secondary image*, located in the plane containing the *sagittal focus* F_s. If a screen is placed between these two foci, perpendicular to the principal ray, an elliptic-shaped image is generally obtained. However, roughly halfway between F_t and F_s, this image will have a circular shape, corresponding to the so-called CLC.

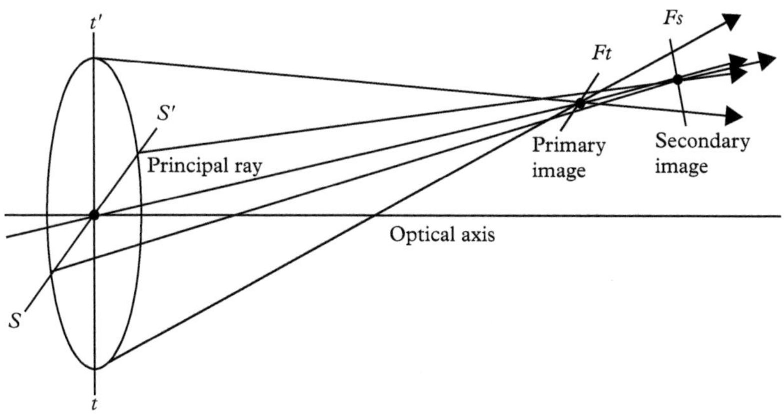

Figure 3.12 *Representation of the astigmatic images of a point object outside the optical axis.*

The astigmatic aberration described earlier occurs even if the optical system is perfectly symmetrical. However, there is a second form of astigmatism, which occurs when the optical system is not symmetric about the optical axis. Such asymmetry may be due to manufacturing errors in the surfaces of the components or misalignment of the components. In this case, astigmatism is observed even for rays from on-axis object points. This form of astigmatism is particularly important in vision science, since the shape of the cornea is often not symmetric, as seen in Section 3.1.

Field Curvature

Field curvature is an optical aberration in which a plane object normal to the optical axis cannot be brought properly into focus on a flat image plane. Actually, the image of a flat object perpendicular to the axis is only approximately flat in the paraxial zone. Outside this zone, the image forms on a curved surface. This primary aberration, also known as *Petzval curvature*, is illustrated in Fig. 3.13 for the case of a converging lens. The image of an object based on a spherical surface S_o forms on another spherical surface, S_i, both with the same centre O. If the surface S_o becomes flatter, approaching the surface S_o', the images of off-axis object points move towards the lens along their principal ray, forming a parabolic surface, S_p, called the *Petzval surface*. In the case of a diverging lens, such surface curves move away from the object plane. Field curvature can be avoided by properly combining converging and diverging lenses.

Considering a single-element lens system such as that represented in Fig. 3.13, all plane waves propagating parallel to the optical axis will be focused to a point at the focal distance from the lens. If a flat image sensor is placed at the focal plane of such a lens, image points near the optical axis will be in perfect focus, but off-axis rays will come into focus before the image sensor, dropping off as the cosine of the angle they make with the optical axis. We must observe that this aberration is less problematic when the imaging surface is spherical, as in the human eye.

Distortion

Distortion is an aberration occurring because magnification can be a function of the distance from the image to the optical axis. In fact, the different zones of the lens have distinct focal lengths, which determine different magnifications. This causes the image as a whole to be misshaped, even though each individual point is still sharply focused.

In the case of *positive distortion*, each image point is displaced radially outwards from its centre, with the most distant points moving the greatest amount because the magnification

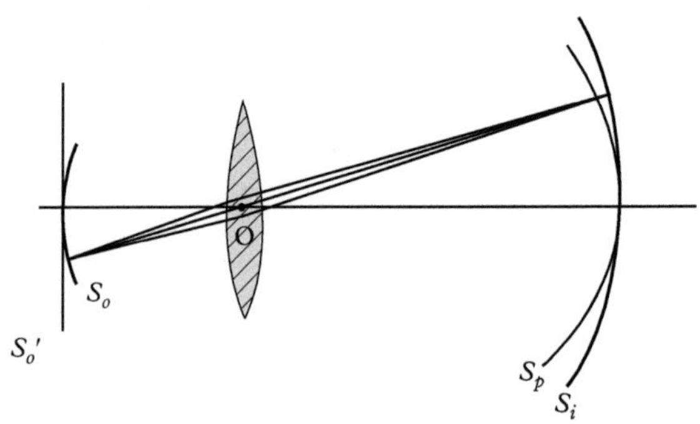

Figure 3.13 *Illustration of field curvature.*

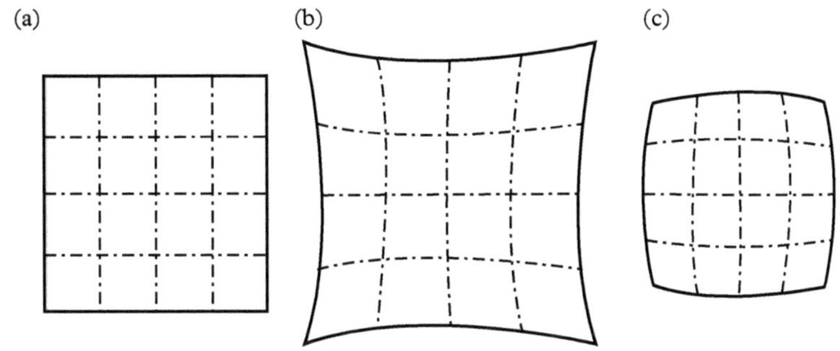

Figure 3.14 *Illustration of distortion in the case of the image of a square lattice (a). Distortion can be positive (b) or negative (c).*

increases with axial distance. As a consequence, lines that do not go through the centre of the image are bowed inwards, as illustrated in Fig. 3.14(b). On the other hand, the corners of squares form elongated points, as in a cushion. For this reason, positive distortion is also called *pincushion distortion*. With negative distortion, the magnification decreases with distance from the optical axis, causing each point on the image to move radially inwards. As a consequence, straight lines bulge outwards at the centre, as in a barrel (Fig. 3.14(c)). For this reason, negative distortion is also called *barrel distortion*. It is also possible to have a kind of distortion that is a mixture of positive and negative distortions, sometimes referred to as *moustache distortion* or *complex distortion*. Distortion is usually negligible in the case of thin lenses. However, thick lenses, converging or diverging, determine a positive or negative distortion, respectively.

3.6.2 Chromatic aberrations

Chromatic aberration results from dispersion: the refractive index of the lens elements varies with the wavelength of light. Actually, the refractive index of a transparent material decreases with increasing wavelength. Since the focal length of a lens is dependent on the index of refraction of the lens material, this variation in refractive index affects focusing. As a consequence, polychromatic light from a point object is imaged not as a point but as a series of points, one for each distinct wavelength.

The focal length of a thin lens located in air is given by

$$f = \left[(n-1)\left(\frac{1}{R_1} - \frac{1}{R_2}\right)\right]^{-1} = \frac{1}{(n-1)\alpha} \tag{3.15}$$

where

$$\alpha = \frac{1}{R_1} - \frac{1}{R_2} \tag{3.16}$$

R_1 and R_2 being the radii of curvature of the lens surfaces and n the refractive index of the material from which it is made. As n depends on the wavelength, so does the focal length (Fig. 3.15). The axial distance between two extreme focal points in a given frequency range is called *longitudinal chromatic aberration* (LCA). There is a position between these extreme focal points where the dimension of the light spot corresponding to the image of a distant point object is minimal. This position corresponds to the CLC.

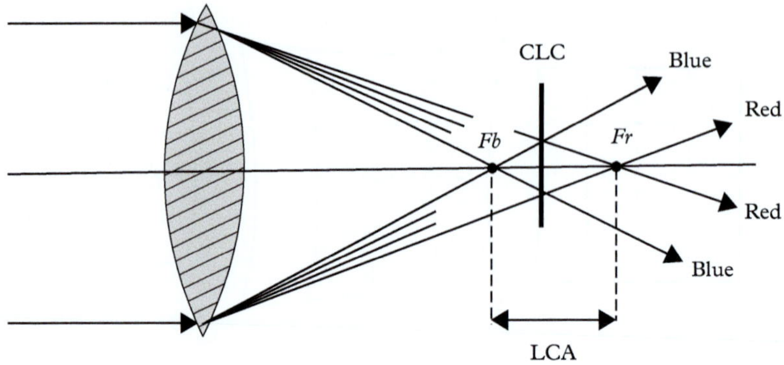

Figure 3.15 *Illustration of chromatic aberration for a converging lens. CLC, circle of least confusion; LCA, longitudinal chromatic aberration; Fb focus of the blue light; Fr, focus of the red light.*

In the case of a converging lens, as in Fig. 3.15, the focus corresponding to the blue light, *Fb*, is to the left of the focus corresponding to the red light, *Fr*, and the axial chromatic aberration is said to be positive. In the case of a diverging lens, the positions of these focal points are inverted, and the axial chromatic aberration is negative.

Example 3.4

A thin and biconvex *crown* glass lens has surfaces with curvature radii of 10.0 cm. When the lens is in the air, an image is formed at a distance of 20.0 cm from it using yellow light. Assuming that the refractive indices for blue and red light are n_B = 1.501 e n_R = 1.509, respectively, calculate the extent of the axial chromatic aberration corresponding to this range of wavelengths.

Solution

Using Eq. (3.15), the equation of a thin lens located in the air can be written in the form:

$$\frac{1}{s}+\frac{1}{s'}=(n-1)\left(\frac{1}{R_1}-\frac{1}{R_2}\right)$$

Differentiating this equation and considering that, for a given position of the object, s is constant, one has

$$\Delta s' = -s'^2\left(\frac{1}{R_1}-\frac{1}{R_2}\right)\Delta n$$

The minus sign in the previous result means that, as the refractive index n decreases when the wavelength increases, the distance s' should increase with the wavelength.

Using for s' the reference value corresponding to the yellow light ($s' = 20.0\,\text{cm}$) and considering $\Delta n = n_R - n_B = 8\times 10^{-3}$, $R_1 = +10\,\text{cm}$, and $R_2 = -10\,\text{cm}$, we find that the axial chromatic aberration is given by

$$|\Delta s'| = (0.2)^2\left(\frac{1}{(0.1)}-\frac{1}{(-0.1)}\right)\left(8\times 10^{-3}\right) = 6.4\times 10^{-3}\,\text{m}$$

Achromatic Doublet

Many types of glass have been developed to reduce chromatic aberration. These are low-dispersion glasses, most notably glasses containing fluorite. Chromatic aberration can be further minimized by using an achromatic lens, which corresponds to a compound lens formed by different lenses assembled together.

A lens system is said to be achromatic if its focal distance does not depend on wavelength. Actually, the coincidence of the extreme focal points F_B and F_R, in Fig. 3.15, is possible by conveniently associating a converging lens with a diverging lens, which results in a so-called achromatic doublet. If the two lenses are separated by a distance d, the focal length of the pair is given, as seen in Example 2.3, by

$$\frac{1}{f} = \frac{1}{f_1} + \frac{1}{f_2} - \frac{d}{f_1 f_2} \tag{3.17}$$

where f_1 and f_2 are the focal lengths of the two lenses. The condition of achromaticity of the pair for red and blue light implies:

$$\frac{1}{f_R} = \frac{1}{f_B} \tag{3.18}$$

Using the notation

$$f_j = \frac{1}{(n_j - 1)\,\alpha_j}, \quad j = 1, 2 \tag{3.19}$$

and assuming that the two lenses are in contact ($d = 0$), the condition (3.18) allows one to obtain the following result:

$$\frac{\alpha_1}{\alpha_2} = -\frac{n_{2B} - n_{2R}}{n_{1B} - n_{1R}} \tag{3.20}$$

The focal length of the doublet can be specified for yellow (Y) light, intermediate between the blue and the red light. For the yellow wavelength, we have:

$$\frac{\alpha_1}{\alpha_2} = \frac{(n_{2Y} - 1) f_{2Y}}{(n_{1Y} - 1) f_{1Y}} \tag{3.21}$$

By equating the second members of Eqs. (3.20) and (3.21), we get:

$$\frac{f_{2Y}}{f_{1Y}} = -\frac{(n_{2B} - n_{2R})(n_{1Y} - 1)}{(n_{1B} - n_{1R})(n_{2Y} - 1)} \tag{3.22}$$

Given that the refractive index decreases with the wavelength in the visible region of the spectrum, it is $n_B > n_R$. On the other hand, as $n_Y > 1$, we can conclude from Eq. (3.22) that one of the two lenses of the achromatic doublet must be divergent and the other convergent.

It is also possible to build an achromatic doublet with two lenses made of the same glass, separated from each other by a certain distance d. Making $n_{1R} = n_{2R} = n_R$ and $n_{1B} = n_{2B} = n_B$ and introducing again the yellow light reference, in particular through Eq. (3.21), the following result can be obtained for the distance between the two lenses:

$$d = \frac{(f_{1Y} + f_{2Y})(n_Y - 1)}{n_B + n_R - 2} \tag{3.23}$$

Assuming that $n_Y = (n_B + n_R)/2$, the previous result is reduced to

$$d = \frac{f_{1Y} + f_{2Y}}{2} \tag{3.24}$$

In order for an optical system to be considered free of chromatic aberrations, the red and blue rays must emerge from it parallel to each other—in order to cancel the lateral chromatic aberration—and must intersect the optical axis at the same point—in order to cancel the axial chromatic aberration. That is, the two rays must match, which can be achieved using thin lenses.

Example 3.5

Consider two equal converging thin lenses, separated by a distance d. Derive an expression for that distance

a) which ensures the same focal length of the pair for light with wavelengths λ_A and λ_B;
b) which satisfies the condition in (a) such that $\lambda_A = \lambda_0 - \Delta\lambda/2$ and $\lambda_B = \lambda_0 + \Delta\lambda/2$, with $\Delta\lambda << \lambda_0$.

Solution

a) The focal length of the pair of lenses is given by

$$\frac{1}{f(\lambda)} = \frac{2}{f_0(\lambda)} - \frac{d}{f_0^2(\lambda)} \tag{3.25}$$

where $f_0(\lambda)$ is the focal length of each lens for the wavelength λ. It is intended that $f(\lambda_A) = f(\lambda_B)$, so

$$\frac{2}{f_0(\lambda_A)} - \frac{d}{f_0^2(\lambda_A)} = \frac{2}{f_0(\lambda_B)} - \frac{d}{f_0^2(\lambda_B)} \tag{3.26}$$

From Eq. (3.26) we obtain the result

$$d = \frac{2f_0(\lambda_A)f_0(\lambda_B)}{f_0(\lambda_A) + f_0(\lambda_B)} \tag{3.27}$$

b) Assuming that $\lambda_A = \lambda_0 - \Delta\lambda/2$, $\lambda_B = \lambda_0 + \Delta\lambda/2$, and $\Delta\lambda << \lambda_0$, one has:

$$f_0(\lambda_A) \approx f_0(\lambda_0) - \frac{\Delta\lambda}{2}\left(\frac{df_0}{d\lambda}\right)_{\lambda=\lambda_0} \tag{3.28}$$

$$f_0(\lambda_B) \approx f_0(\lambda_0) + \frac{\Delta\lambda}{2}\left(\frac{df_0}{d\lambda}\right)_{\lambda=\lambda_0} \tag{3.29}$$

Substituting Eqs. (3.28) and (3.29) into Eq. (3.27) and neglecting the term in $(\Delta\lambda)^2$ in the numerator, one obtains the result

$$d = f_0(\lambda_0). \tag{3.30}$$

That is, the lenses must be separated by a distance equal to the focal length of a single lens, calculated at the central wavelength λ_0.

3.7 Further Reading

Hecht, E., *Optics*, 5th Ed., Pearson, 2016.
Kingslake, R., *Lens Design Fundamentals*, Academic Press, 1978.
Klein, M., and T. Furtak, *Optics*, 2nd Ed., John Wiley & Sons, 1986.
Malacara, D., *Geometrical and Instrumental Optics*, Academic Press, 1988.
Mourculis, M., and J. Macdonald, *Geometrical Optics and Optical Design*, Oxford University Press, 1977.
Pedrotti, F., L. M. Pedrotti, and L. S. Pedrotti, *Introduction to Optics*, 3rd Ed., Cambridge University Press, 2018.
Sasián, J., *Introduction to Aberrations in Optical Imaging Systems*, Cambridge University Press, 2013.

3.8 Problems

3.1 A magnifying glass with a focal length of 10 cm is used to observe an object by a person whose minimum distance of distinct vision is 25 cm. If that person places the magnifying glass near the eye, what is the best position of the object?

3.2 The minimum distance of distinct vision for a certain person is 50 cm. Find the power of the magnifying glass to be used by this person to get clear vision of an object located 20 cm from the eye.

3.3 A microscope has an objective with a focal length of 10.0 mm and an ocular with a focal length of 25.0 mm. Obtain the distance between the two lenses and the magnification of the microscope, considering that the distance between the object and the lens is 11 mm.

3.4 A microscope uses an objective and an ocular with focal lengths of 4 cm and 5 cm, respectively. Determine the magnification of the microscope, considering that the distance between the lenses is 16 cm.

3.5 A telescope consists of two lenses with focal lengths of 8 cm and 2 cm. Assuming that both the object and the final image are located at infinity, find: (a) the distance between these lenses and (b) the magnification of the telescope.

3.6 The distance between the objective and the ocular of a telescope is 76 cm when the telescope is used to observe a star. However, when the same telescope is used to observe a given building under similar conditions for the eye, the ocular must be moved a further 0.5 cm from the objective. Knowing that the focal length of the ocular is 2.5 cm, determine the distance at which such a building is located.

3.7 A tank full of water (refractive index 4/3) has a convex semi-spherical transparent window with a radius of 20 cm on one of its vertical walls. A collimated beam of light completely illuminates the semi-spherical window. Determine the distance between the focus of the marginal rays and the focus of the paraxial rays of such a beam (longitudinal spherical aberration).

3.8 Determine the diameter of the light spot observed in the focal plane of the paraxial rays in Problem 3.7 (transverse spherical aberration).

3.9 A *crown* glass lens and a *flint* glass lens are glued together to constitute an achromatic doublet. Show that the focal lengths of the *crown* and *flint* lenses satisfy the equation:

$$\kappa_c f_f + \kappa_f f_c = 0,$$

where

$$\kappa = \frac{1}{n_0 - 1}\frac{dn}{d\lambda}\Delta\lambda$$

n_0 being the refractive index of the glass at the centre of the visible spectrum and $\Delta\lambda$ the range of wavelengths of this spectrum.

3.10 For *crown* glass one has $\kappa_c = 0.0169$, whereas for *flint* glass it is $\kappa_f = 0.0384$. Using the result of Problem 3.9, show that one should use a *crown* glass converging lens and a *flint* glass diverging lens in order to obtain a converging achromatic pair.

3.11 Show that a set of two lenses with focal distances f_1 and f_2, made of the same glass, will constitute an achromatic doublet if they are separated by a distance $d = (f_1 + f_2)/2$.

3.12 The ocular of a microscope is constituted by two lenses having focal lengths of 6 cm and 3 cm, respectively. Determine (a) the equivalent focal length of the ocular and (b) the optimum separation between the two lenses in order to reduce the chromatic aberration.

Radiometry and Photometry

4

Radiometry addresses the problem of measuring the energy contained in optical radiation and how this energy is transported through optical systems. It is assumed, for this purpose, that the propagation of radiation can be described on the basis of the laws of geometrical optics. In contrast, photometry is restricted to that part of the electromagnetic spectrum that can induce a visual response. In fact, the purpose of photometry is to quantify the visual response produced by radiation.

In the past, there was some confusion regarding the nomenclature used in radiometry and photometry. Sometimes different names were used to indicate the same quantity. More recently, however, a progress has been observed towards using uniform designations and more logical units for these quantities, based on the International System (IS) of Units.

4.1 Radiometric Quantities

Radiometry is the science of the measurement of electromagnetic radiation. Radiometric quantities appear either without subscripts or with the subscript e (from *electromagnetic*). For example, the symbol Q_e represents the total energy emitted, transferred or received in a radiative process, which is called *radiant energy* (J). On the other hand, the symbol U_e represents the *density of radiant energy* $(\mathrm{J/m^3})$, corresponding to the radiant energy contained in a unit volume, given by

$$U_e = \frac{dQ_e}{dV} \tag{4.1}$$

The *radiant power* or *radiant flux*, Φ_e (W), is the rate of transfer of radiant energy, given by

$$\Phi_e = \frac{dQ_e}{dt} \tag{4.2}$$

One of the most important quantities in radiometry is the flux of radiant energy passing through a unit area of a given real or imaginary surface. Conventionally, a distinction is made between the surface of a source and other surfaces. The *radiant emittance* $(\mathrm{W/m^2})$ is defined as the flux emitted per unit area of a source:

$$M_e = \frac{d\Phi_e}{dA} \tag{4.3}$$

Similarly, the *irradiance*, E_e, is the incident flux per unit area of a real or imaginary surface:

$$E_e = \frac{d\Phi_e}{dA} \tag{4.4}$$

Lasers, Optics, and Photonics. Mário F. S. Ferreira, Oxford University Press. © Mário F. S. Ferreira (2026).
DOI: 10.1093/oso/9780198952770.003.0004

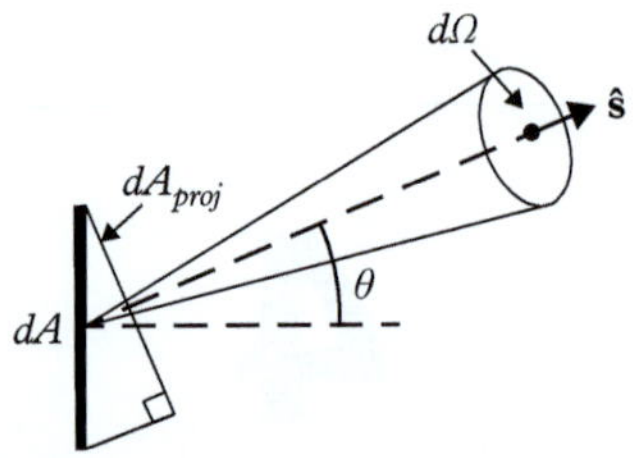

Figure 4.1 *Geometry used in the definition of radiance.*

The *radiance*, L_e, is the flux emitted per unit of projected area and per unit solid angle. If $d^2\Phi_e$ represents the flux emitted inside a solid angle $d\Omega$ by an element dA_{proj} of the projected area of the source, the radiance is then given by

$$L_e = \frac{d^2\Phi_e}{dA_{proj}d\Omega} \tag{4.5a}$$

where

$$dA_{\text{proj}} = dA\cos\theta = d\vec{A}\hat{\mathbf{s}} \tag{4.5b}$$

θ being the angle between the normal to the surface element dA and the direction of observation, as indicated by the unit vector $\hat{\mathbf{s}}$ (Fig. 4.1).

Another important quantity is the *radiant intensity*, I_e, which corresponds to the flux per unit solid angle emitted by a source in a given direction $\hat{\mathbf{s}}$. It is given by

$$I_e = \frac{d\Phi_e}{d\Omega} \tag{4.6}$$

where $d\Phi_e$ is the flux emitted within the solid angle $d\Omega$. The radiant intensity from a sphere radiating Φ_e W of power uniformly in all directions, for example, is $I_e = \Phi_e/4\pi$, since the total surrounding solid angle is 4π sr. Usually, the radiant intensity depends on the direction of $\hat{\mathbf{s}}$ and is of particular interest when the source is far away from the detector and all the rays that arrive at it are approximately parallel.

It should be noted that, in the field of physical optics, the word *intensity* is usually used to designate the energy flux per unit area, corresponding to the *irradiance* in the nomenclature of radiometry. Moreover, the *radiant power*, Φ_e, is usually denoted by P.

Example 4.1

A laser beam has a power of 3 mW and a divergence angle $\phi = 1$ mrad. The beam is emitted normally by a source area $\Delta A = 0.5\,\text{mm}^2$. Determine (i) the radiant intensity, I_e, and (ii) the *radiance*, L_e, of the light source.

Solution

i) The radiant flux is emitted within a solid angle $d\Omega$ given by

$$d\Omega = \pi\phi^2 = 3.14 \times 10^{-6}\ \text{sr}$$

The radiant intensity is given by Eq. (4.6):

$$I_e = \frac{d\Phi_e}{d\Omega} = \frac{3 \times 10^{-3}}{3.14 \times 10^{-6}} \approx 10^3\,\text{W/sr}$$

ii) The radiance is obtained from Eq. (4.5), considering that $\theta = 0$ rad and $dA_{proj} = \Delta A = 0.5\,\text{mm}^2$. So, using the result obtained in (i), one has:

$$L_e = \frac{d^2\Phi_e}{dA_{proj}d\Omega} = \frac{10^3}{0.5} = 2 \times 10^3\,\text{W/(sr.mm}^2)$$

4.2 Spectral Radiometric Quantities

The definitions presented in Section 4.1 refer to the total radiation corresponding to all wavelengths. However, in several situations, it is convenient to use radiometric quantities that take into account the spectral composition of light. For example, the *spectral radiant flux*, $\Phi_e(\lambda)d\lambda$, represents the flux in a band of wavelengths between λ and $\lambda + d\lambda$. By integration, one has:

$$\Phi_e = \int_0^\infty \Phi_e(\lambda)d\lambda \tag{4.7}$$

In some circumstances it is more convenient to use a frequency distribution. In this case, the radiant flux in the frequency range between v and $v + dv$ is given by $\Phi(v)dv$, and one has:

$$\Phi_e = \int_0^\infty \Phi_e(v)dv \tag{4.8}$$

If $d\lambda$ is the range of wavelengths corresponding to the frequency range dv, one has

$$\Phi(\lambda)d\lambda = \Phi(v)dv \tag{4.9}$$

That is,

$$\Phi(\lambda) = \Phi(v)\left|\frac{dv}{d\lambda}\right| = \Phi(v)\frac{v}{\lambda} \tag{4.10}$$

Spectral versions of other radiometric quantities can be defined similarly to those referred to for radiant flux.

Spectral quantities correspond to a given frequency v, which is associated with photons of energy hv, where h is Planck's constant. Thus, all spectral quantities involving energy can be converted into quantities involving photons, so it is enough to divide the expressions of those quantities by hv. For example, the flux of photons in the frequency band between v and $v + dv$ is given by

$$N(v)dv = \frac{\Phi_e(v)}{hv}dv \tag{4.11}$$

A beam of monochromatic light with frequency v and irradiance E_e has a photon flux density

$$\phi(\mathbf{r}) = \frac{E_e}{hv} \tag{4.12}$$

This relation converts a classical quantity into a quantum quantity. In the case of quasi-monochromatic light with a centre frequency $\bar{v}$, the aforementioned result can be used by substituting the frequency v with $\bar{v}$.

The average photon flux Φ (in units of photons/s) is obtained by integrating the average flux density over a given area, A:

$$\Phi = \int_A \phi(\mathbf{r})dA = \frac{\Phi_e}{h\bar{\nu}} \tag{4.13}$$

where

$$\Phi_e = \int_A E_e(\mathbf{r})dA \tag{4.14}$$

is the radiant power (W). For example, an optical power of 1 nW, at a wavelength $\lambda = 0.2\,\mu\text{m}$, corresponds to an average flux of $\Phi = 10^9$ photons per second. That is, an object illuminated by this beam will receive an average of one photon every nanosecond.

4.3 Radiation Sources: Special Cases

4.3.1 Isotropic sources

In the case of an isotropic spherical source with radius R_0, the total flux $\Phi_{e,tot}$ can be obtained from the emittance M_e by integration over the entire surface:

$$\Phi_{e,tot} = \int M_e dA = 4\pi R_0^2 M_e \tag{4.15}$$

A small detector with area dA_{det} located at a distance R from the source will receive a flux:

$$d\Phi_e = \Phi_{e,tot}\frac{dA_{\text{det}}}{4\pi R^2} = \frac{\Phi_{e,tot}}{4\pi}d\Omega \tag{4.16}$$

where $d\Omega$ is the solid angle element corresponding to the area dA_{det}. Using Eqs. (4.6), (4.12), and (4.13), we find that the radiant intensity of this source is given by

$$I_e = \frac{\Phi_{e,tot}}{4\pi} = R_0^2 M_e \tag{4.17}$$

4.3.2 Lambertian sources

A Lambertian source is, by definition, a source whose radiance does not depend on the direction of observation or the position on the source. From Eqs. (4.5) and (4.6), we obtain the following result for the radiant intensity of a flat Lambertian source in a direction that makes an angle θ with the normal:

$$I_e = \int L_e \cos\theta dA = I_{e0}\cos\theta \tag{4.18}$$

where $I_{e0} = L_e A_0$, L_e being the radiance of the source and A_0 its area.

The dependence of radiant intensity on $\cos\theta$ is typical of a flat and finite Lambertian source. The reduction of this intensity with θ is simply due to the decrease in the projected area of the source.

In the case of Lambertian sources, there is a particularly simple relationship between radiance and radiant emittance. Considering Eqs. (4.3) and (4.5) for these two quantities, one has:

$$M_e = \int L_e \cos\theta d\Omega = 2\pi L_e \int_0^{\pi/2} \sin\theta \cos\theta d\theta = \pi L_e \tag{4.19}$$

where L_e was assumed to be a constant and the solid angle element $d\Omega$ was used in the form ma $d\Omega = 2\pi \sin\theta d\theta$.

4.3.3 Irradiance from finite sources

In many cases, the calculation of radiometric quantities for finite sources involves relatively complicated integrals. However, this calculation becomes easier if the source is of Lambertian type and has a simple geometry. We will consider, in the following two examples, the calculation of the irradiance at a given point situated at a distance R from the centre of the source.

Lambertian Disc

Fig. 4.2 shows the geometry associated with a disc-shaped source. An elemental ring of this source with radius r has an area given by

$$dA_0 = 2\pi r dr = \pi d(r^2) = \pi d(\rho^2) \tag{4.20}$$

where $\rho^2 = R^2 + r^2$. The flux coming from this ring and reaching the area dA_1 is given by

$$d^2\Phi_e = L_e d\Omega dA_0 \cos\theta \tag{4.21}$$

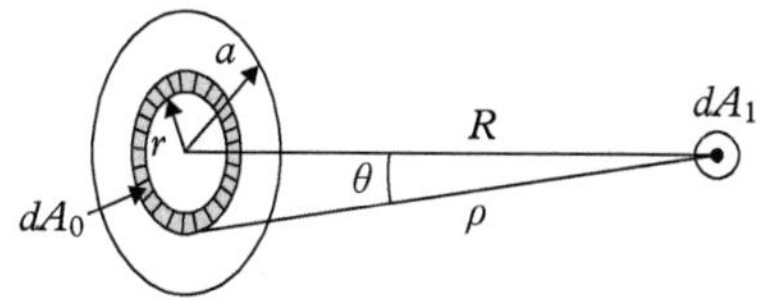

Figure 4.2 *Geometry for the case of a Lambertian disc.*

where $d\Omega = dA_1 \cos\theta/\rho^2$ is the solid angle corresponding to the area dA_1 at a given point of the elemental ring. Considering that $\cos^2\theta = R^2/\rho^2$, we can calculate the irradiance at dA_1 in the form:

$$E_e = \frac{d\Phi_e}{dA_1} = \pi L_e R^2 \int_{R^2}^{R^2+a^2} \frac{d\rho^2}{\rho^4} = \pi L_e \left(\frac{a^2}{a^2+R^2}\right) \tag{4.22}$$

It can be seen from Eq. (4.22) that, for a given value of R, E_e tends to πL_e as $a \to \infty$. On the other hand, for given finite values of a and $R >> a$, the disc can be considered a point source with intensity $I_e = L_e \pi a^2$ and irradiance given by

$$E_e \approx \frac{I_e}{R^2} \tag{4.23}$$

One can verify from Eq. (4.23) that the irradiance is inversely proportional to the square of the distance between the source and the detector.

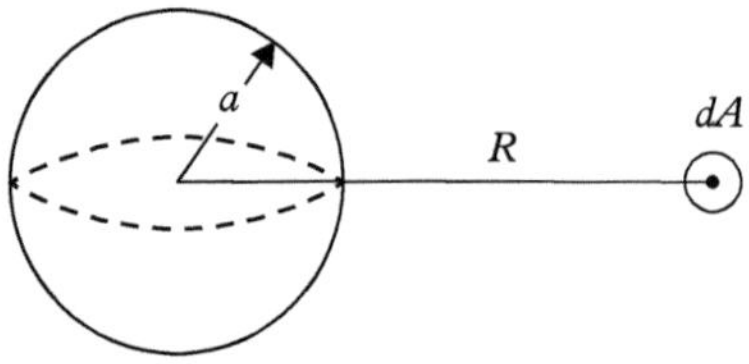

Figure 4.3 *Geometry for the case of a spherical Lambertian source.*

Spherical Lambertian Source

Fig. 4.3 illustrates a spherical Lambertian source with radius a and uniform irradiance L_e. The radiant emittance of the source is given by Eq. (4.19), and the total flux Φ_e is given by the product of this result and the spherical surface area:

$$\Phi_e = 4\pi a^2 M_e = 4\pi^2 a^2 L_e \tag{4.24}$$

The radiant intensity of this source can be found by considering that the emitted radiation is isotropic. In these circumstances, we have:

$$I_e = \frac{\Phi_e}{4\pi} = \pi a^2 L_e \tag{4.25}$$

The irradiance at a distance R from the source centre is given by

$$E_e = \frac{\Phi_e}{4\pi R^2} = \frac{\pi a^2 L_e}{R^2} = \frac{I_e}{R^2} \tag{4.26}$$

This result is similar to that given by Eq. (4.23) for the case of a disc of radius a. Actually, the spherical Lambertian source is similar to a uniform disc for a distant observer.

4.4 Radiance Theorem

The radiance theorem establishes that the radiance is conserved during propagation through a lossless medium, consisting of a sequence of homogeneous media.

Example 4.2

Show that the radiance is conserved during the propagation through a homogeneous lossless medium.

Solution

Let us consider two elements of areas dA_0 and dA_1, separated by a distance R, in a homogeneous lossless medium, as represented in Fig. 4.4. The solid angle corresponding to dA_1 at dA_0 is given by

$$d\Omega_0 = \frac{dA_1 \cos\theta_1}{R^2} \tag{4.27a}$$

whereas the solid angle corresponding to dA_0 at dA_1 is given by

$$d\Omega_1 = \frac{dA_0 \cos\theta_0}{R^2} \tag{4.27b}$$

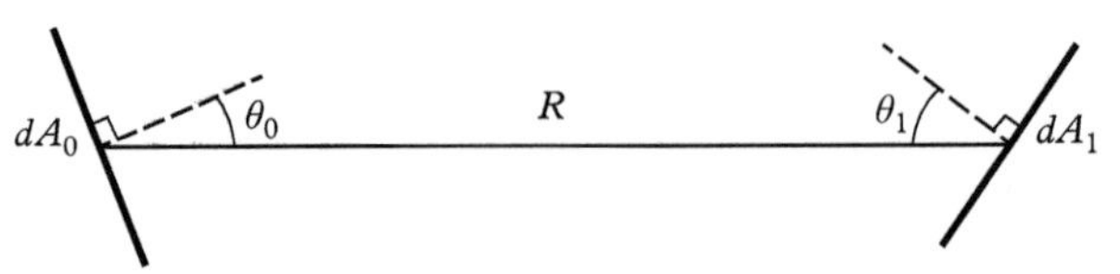

Figure 4.4 *Geometry for the conservation of radiance in a homogeneous medium.*

Example 4.2 continued

The radiance L_{e_1} measured at dA_1 in the direction of dA_0 is given by

$$L_{e1} = \frac{d^2\Phi_e}{dA_1 \cos\theta_1 d\Omega_1} \tag{4.28}$$

where $d^2\Phi_e$ is the flux transferred from dA_0 to dA_1, given by

$$d^2\Phi_e = L_{e0}(dA_0 \cos\theta_0)(d\Omega_0) \tag{4.29}$$

and L_{e0} is the radiance measured at dA_0 in the direction of dA_1.

Substituting Eq. (4.29) into Eq. (4.28) and using Eq. (4.27), it can be seen that

$$L_{e1} = L_{e0} \tag{4.30}$$

This result shows that the radiance is conserved during propagation in a homogeneous medium.

4.4.1 Transmission in the interface between two media

Let us now consider an area element dA located on the surface between two different optical media with refractive indices n and n', as represented in Fig. 4.5. Snell's law of refraction establishes the following relation between the angles of incidence, θ, and transmission,θ':

$$n \sin\theta = n' \sin\theta' \tag{4.31}$$

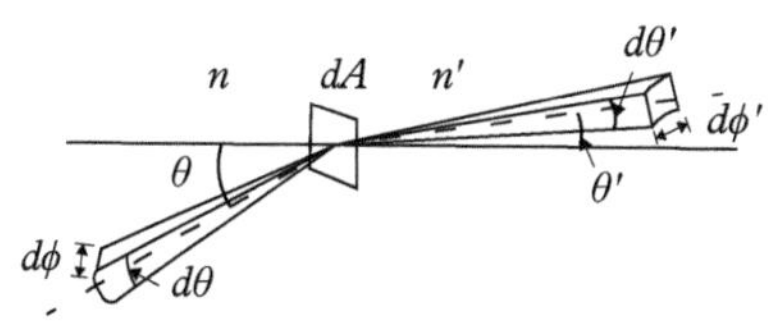

Figure 4.5 *Geometry for the conservation of radiance at the interface between two homogeneous media.*

Considering a polar coordinate system whose axis is normal to dA, the ratio between the elementary solid angles $d\Omega$ and $d\Omega'$ is given by

$$\frac{d\Omega}{d\Omega'} = \frac{\sin\theta d\theta d\phi}{\sin\theta' d\theta' d\phi'} = \frac{n' d\theta}{n d\theta'} \tag{4.32}$$

where Eq. (4.31) and the equality between the elementary azimuthal angles $d\phi$ and $d\phi'$, imposed by Snell's law, have been used. By differentiating Eq. (4.31), we obtain:

$$n \cos\theta d\theta = n' \cos\theta' d\theta' \tag{4.33}$$

Using Eq. (4.33), we can write Eq. (4.32) in the form:

$$\frac{d\Omega}{d\Omega'} = \frac{n'^2 \cos\theta'}{n^2 \cos\theta} \tag{4.34}$$

The flux of a beam with radiance L_e incident at dA within an elemental solid angle $d\Omega$ and making an angle θ with the normal is given by

$$d^2\Phi_e = L_e dA \cos\theta d\Omega \tag{4.35}$$

Using Eqs. (4.5), (4.34), and (4.35), we find that the radiance of the transmitted beam is given by

$$L'_e = \frac{d^2\Phi_e}{dA\cos\theta' d\Omega'}$$
$$= L_e \frac{dA\cos\theta d\Omega}{dA\cos\theta' d\Omega'} = L_e \frac{n'^2}{n^2} \tag{4.36}$$

From Eq. (4.33), we obtain the result:

$$\frac{L_e}{n^2} = \frac{L'_e}{n'^2} \tag{4.37}$$

Eq. (4.37) shows that the quantity L_e/n^2, which represents the so-called *basic radiance*, does not vary when the beam propagates across the interface between two homogeneous media. On the other hand, it was previously seen that the radiance L_e is conserved when the beam propagates in a homogeneous and lossless medium. Thus, we conclude that *the basic radiance of a beam propagating in a lossless optical system, consisting of a sequence of homogeneous media, is conserved.*

Example 4.3

Show that the radiance of the image of the sun produced in the focal plane of a lens is similar to that on the surface of the sun if the attenuation caused by the earth's atmosphere is neglected.

Solution

The sun can be considered as a Lambertian source in the form of a disc of area S and radiance L_e, located at a distance R from the lens. The power received by the lens is given by

$$\Phi_e = \frac{L_e SA}{R^2} \tag{4.38}$$

A being the lens area.

All the power received by the lens is concentrated in the image that it forms of the sun, with area S'. Neglecting the attenuation of such power in the atmosphere, it is found that the apparent radiance of the image, L_e', is given by

$$L'_e = \frac{\Phi_e}{S' d\Omega'} \tag{4.39}$$

where

$$d\Omega' = \frac{A}{f^2} \tag{4.40}$$

is the solid angle implied by the lens at its focal plane, and f is the focal length. Substituting Eq. (4.38) into Eq. (4.39) and using the equality

$$\frac{S}{R^2} = \frac{S'}{f^2}, \tag{4.41}$$

we conclude that

$$L_e = L_e' \tag{4.42}$$

4.5 Photometric Quantities

Photometry is distinguished from radiometry by the fact that the former is concerned only with the part of radiation that can induce a visual response. Photometric quantities are defined similarly to radiometric quantities, and this correspondence is shown in a clear manner in Table 4.1. In this table, the photometric quantities are represented by the same symbols used to indicate the radiometric quantities, distinguishing themselves only by the subscript *v*, instead of *e*.

The fundamental quantity of photometry is the *luminous flux*, Φ_v, which corresponds to the radiant flux Φ_e of radiometry and is given by the integral of the spectral radiant flux $\Phi_e(\lambda)$ over the visible spectrum. The luminous flux was chosen in 1977 by the International Committee of Weights and Measures as the standard photometric quantity. The unit of luminous flux Φ_v is the lumen, which represents the luminous flux of radiation with a wavelength of 555 nm corresponding to a radiant flux of 1/683 W.

Another important quantity of photometry is the *luminous intensity*, I_v, whose unit is the candela or lumen per steradian. A candela (1 cd) corresponds to a flux of one lumen (1 lm) through a solid angle of one steradian (1 sr).

Table 4.1 *Correspondence between radiometric quantities and photometric quantities*

Radiometry	Symbol	SI units
Radiant energy	Q_e	Joule (J)
Radiant density	U_e	J/m^3
Radiant flux	Φ_e	Watt (W)
Radiant emittance	M_e	W/m^2
Irradiance	E_e	W/m^2
Radiant intensity	I_e	W/sr
Radiance	L_e	W/sr·m^2
Photometry	**Symbol**	**SI units**
Luminous energy	Q_v	talbot = lm·s
Luminous density	U_v	talbot/m^3
Luminous flux	Φ_v	lumen (lm)
Luminous emittance	M_v	lux (lx) = lm/m^2
Illuminance	E_v	lx = lm/m^2
Luminous intensity	I_v	candela (cd) = lm/sr
Luminance	L_v	lm/sr·m^2

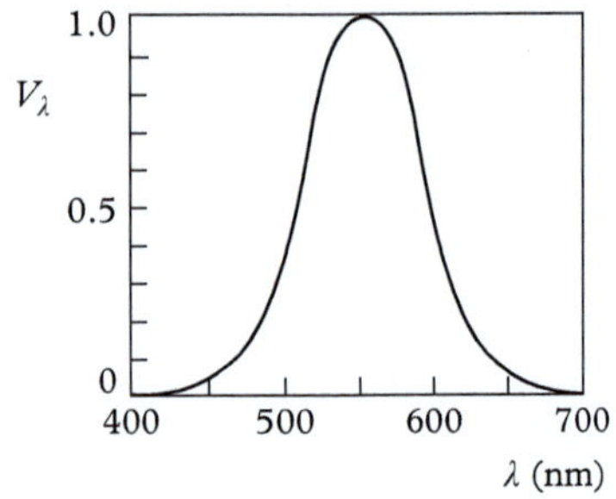

Figure 4.6 *Curve representing the spectral luminous efficiency.*

The *emittance* and *illuminance* (in lm/m^2 or lux) are quantities that represent the density of luminous flux emitted or received, respectively, by a given surface. On the other hand, the *luminance* L_v (lm/ (m^{2}sr)), as in the case of the radiance L_e in radiometry, is the flux per unit solid angle and per unit area of the projected source.

4.5.1 Conversion between luminous and radiant units

The flux of a light beam can be measured using radiometric units or photometric units. In order to establish a relationship between the two types of measurement, it is necessary to determine the visual response produced by radiation with different wavelengths. The so-called *spectral luminous efficiency* (V_λ) curve was obtained for such a purpose as a result of experiments carried out with many people. This curve is shown in Fig. 4.6 and has a maximum value at $\lambda = 555$ nm. The conversion at this wavelength is defined as follows:

$$K_m = 683\ \text{lm/W} \tag{4.43}$$

This means that a flux of 1 W of radiation, with wavelength $\lambda = 555$ nm, gives the same visual sensation as 683 lm. For other wavelengths, the conversion factor is

$$K = K_m V_\lambda = 683 V_\lambda\ \text{lm/W} \tag{4.44}$$

This means that, for $\lambda \neq 555$ nm, 1 W corresponds to a luminous flux of less than 683 lm.

One can use Eq. (4.44) to convert any radiometric spectral quantity into the corresponding photometric quantity. For example, if one has the spectral radiant flux $\Phi_e(\lambda)$, then the luminous flux is given by

$$\Phi_v = K_m \int_0^\infty V_\lambda \Phi_e(\lambda)\, d\lambda \tag{4.45}$$

Eq. (4.45) corresponds to the so-called *Abney's law*.

It is sometimes useful to define an average conversion factor $\overline{K}$, given by

$$\overline{K} = \frac{K_m \int_0^\infty V_\lambda \Phi_e(\lambda)\, d\lambda}{\int_0^\infty \Phi_e(\lambda)\, d\lambda} \tag{4.46}$$

Using this factor, one can write Eq. (4.45) in the form

$$\Phi_v = \overline{K} \Phi_e \tag{4.47}$$

where Φ_e is related to $\Phi_e(\lambda)$ through Eq. (4.8).

The human eye is sensitive to about 10^{-9} lx. Assuming that the pupil of the eye has a radius of 2 mm, the area receiving this light is $\sim 10^{-5}$ m^2, which means that the human eye can detect about 10^{-14} lm. This value corresponds to a visible power of about 1.5×10^{-17} W.

Example 4.4

A red lamp ($\lambda = 650\,\text{nm}$) emits 200 W of radiant power. Calculate the irradiance and the illuminance on a surface located 4 m far from the lamp.

Solution

The irradiance is given from Eq. (4.4) by

$$E_e = \frac{200}{4\pi(4)^2} \approx 1\,\text{W/m}^2$$

On the other hand, the curve V_λ in Fig. 4.6 shows that $V_\lambda\,(650\,\text{nm}) = 0.1$. Therefore, the illuminance is given by

$$E_v = 683\,V_\lambda E_e = 683 \times 0.1 \times 1 = 68.3\,\text{lx}$$

4.5.2 Some orders of magnitude

As an illustration, let us consider the values of the radiant intensity of some sources emitting in the band of optical frequencies:

Moon: 6 W/sr

Carbon filament lamp: 10^3 W/sr

Sun: 2×10^6 W/sr

Laser He–Ne (1 mW): 10^{10} W/sr

In a typical situation, a person may read without fatigue when the illuminance is between 30 and 150 lx. By way of comparison, the illuminances produced by some light sources are indicated as follows:

Laser: He - Ne (1 mW) : 10^6 lx

Sun (zenith): 1.2×10^5 lx

Clear sky: 1×10^4 lx

Moon (full): 0.2 lx

Star of first magnitude: 0.6×10^{-6} lx

It is interesting to note that the irradiance produced on the retina of the human eye by a He–Ne laser with a power of 1 mW is of the order of $10^7\,\text{W/m}^2$, whereas the irradiance produced by the sun is $5 \times 10^4\,\text{W/m}^2$, that is, 200 times less. Hence, one must take great care with any assembly using lasers!

4.6 Radiation Detectors

For a long time, the eye was the only detector used by man in the optical domain. Indeed, it was for this reason that photometry was developed, with its own definitions and units.

The human eye does have some notable features, as seen in Chapter 3. Among these characteristics is the ability to compare luminosities and to detect relatively weak luminosities. However, mainly due to its inability to store and process information, the human eye has lately given way to other types of detectors in several situations.

Most detectors may be classified as either *thermal detectors* or *photon detectors*. In a thermal detector, the receptor is typically a blackened surface that efficiently absorbs at all wavelengths, causing a rise in its temperature. In a photon detector, the light absorbed in the detector material directly creates charge carriers, giving rise to a photocurrent.

4.6.1 Thermal detectors

Thermal detectors transform incident radiation into heat, thereby increasing the temperature of some of their elements. This temperature increase is then converted into an electrical or mechanical signal. This kind of detector is capable of responding in a relatively wide band of wavelengths, without a significant variation in sensitivity. This property causes them to be sometimes used to calibrate other detectors.

Thermocouples and Thermopiles

A well-known thermal detector is based on the so-called *thermocouple*, also known as a '*thermoelectrical thermometer*'. It is an electrical device consisting of two dissimilar electrical conductors forming an electrical junction. The electric potential is found to be different on the two sides of the junction by an amount that varies with temperature as a result of the *Seebeck effect*. Typical thermocouple sensitivities around room temperature are $\sim 40\,\mu$V/°C. This sensitivity can be enhanced by using an array of such thermocouples in series, forming a *thermopile*. This detector has good linearity characteristics, although the response time is relatively long ($\sim$ 1s).

Bolometers and Thermistors

Thermal detectors also include bulk devices made of a material having a temperature-dependent electrical resistance. In general, two sensitive elements are used in adjacent arms of a bridge circuit, one of which is exposed to the incident radiation. Any change in resistance is indicated by a change of current in the bridge circuit. There are two types of such devices: *bolometers* and *thermistors*. A bolometer uses, as its sensitive element, a thin layer of metal connected to a body of constant temperature (thermal reservoir) through a thermal link. Any radiation impinging on the sensitive element raises its temperature above that of the reservoir. In the case of a thermistor, the sensitive element is a semiconductor whose resistance is strongly dependent on temperature. There are *negative-temperature-coefficient* (NTC) thermistors and *positive-temperature-coefficient* (PTC) thermistors. In NTC thermistors, the resistance decreases as temperature increases, usually due to an increase in conduction electrons bumped up by thermal agitation from the valence band. In PTC thermistors, the resistance increases with an increase in temperature, usually due to increased thermal lattice agitation.

Pyroelectric Detectors

Pyroelectric detectors are optical sensors which are based on the pyroelectric effect. Pyroelectricity is the ability of certain materials, such as lithium tantalate or triglycine sulphate (TGS), to generate a temporary voltage when they are heated or cooled. The change in temperature modifies the positions of the atoms slightly within the crystal structure, resulting in a permanent macroscopic polarization, which produces some pyroelectric voltage. For a constant optical power, this pyroelectric signal would eventually disappear due to leakage current. As a consequence, such detectors would not be suitable for measuring continuous-wave light. However, they are widely used for detecting light pulses, often in the infrared spectral region.

4.6.2 Photon detectors

This type of detector is based on the absorption of photons. The process is accompanied by the emission of an electron from the surface of a metal (external photoelectric effect) or by the production of a hole-electron pair inside a semiconductor (internal photoelectric effect). As examples of this type of photodetector, we have the *vacuum photodiode*, the *photomultiplier*, as well as the *photoconductive* and *photovoltaic* solid-state detectors.

When the measurable effect is the release of electrons from an illuminated surface, the device is called a *photoemissive detector*. The *vacuum photodiode* is an example of this type of detector, in which the incident radiation causes the emission of electrons by the photoelectric effect from the surface of the photocathode, which typically contains alkali metals. By applying an appropriate potential difference, these electrons are collected at the anode, giving rise to a current that is measured in an external circuit. The vacuum photodiode is used to detect relatively strong signals, for which it is not necessary to use the high sensitivity of photomultipliers.

A *photomultiplier* consists of the combination of a photodetector and a low-noise current amplifier, as illustrated in Fig. 4.7. The incidence of radiation on the photocathode determines the emission of electrons, which cause the emission of secondary electrons when they reach a second electrode, called the dynode. By focusing on a second dynode, each of these secondary electrons causes the emission of other electrons and so on. The set of all emitted electrons is collected in an anode. The measurement of the resulting current and the knowledge of the amplification of the system allow the quantification of the incident radiation. The photomultiplier is characterized by a very low detection threshold (a single photon can be detected), a wide spectral response, perfectly linear behaviour, and a very short response time (~ 1 ns).

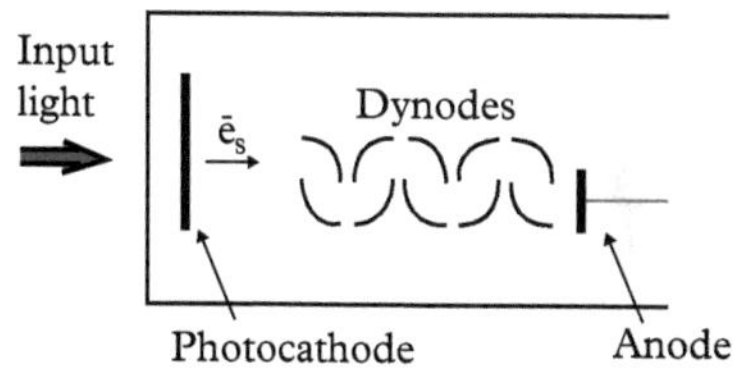

Figure 4.7 *Schematic representation of a photomultiplier.*

Photoconductive Detectors

Photoconductive detectors are often used as infrared detectors. They can simply consist of a piece of semiconductor material with two attached metallic electrodes for sensing the resistance. Without illumination, a bias voltage across such a material produces a small or 'dark' current. The absorption of incident light creates electron-hole pairs, and this reduces the electrical resistance across the two electrodes, resulting in a larger photocurrent. Semiconducting compounds, cadmium sulphide (CdS) and cadmium selenide (CdSe), are often used in the visible and near-infrared regions.

Photoconductors can use intrinsic or extrinsic semiconductor materials. In an *intrinsic* material, one exploits inter-band transitions, i.e. optical transitions across the band gap from the valence band to the conduction band. This is possible only when the photon energy is greater than the band gap energy. In an *extrinsic* material, one exploits additional energy levels introduced by doping with chemical elements. High doping concentrations are preferable to obtain strong absorption, so that rather thin layers can be used to realize a photoconductor. Since no *p–n* junction is required, the simpler device design and fabrication of photoconductors are an advantage compared with photodiodes.

Junction Photodiodes

A *semiconductor photodiode* is basically constituted by a *p–n* junction, which is obtained by doping adjacent regions of a semiconductor with donor (*n*) and acceptor impurities (*p*). The *p–n* junction will be discussed in Chapter 18. At room temperature, there are free holes in the valence band of a *p*-type material and free electrons in the conduction band of an *n*-type material. Charge neutrality is maintained at all points in each of these materials due to the presence of ionized impurities. When both types of materials are put into electrical contact, free electrons can diffuse from the *n*-type region to the *p*-type region, whereas holes can spread in the reverse direction. Both processes lead to the recombination of electrons with holes, giving rise to a region that is devoid of mobile charges, known as the depletion region. An electric field directed from the *n*-type to the *p*-type material is then created, which inhibits the diffusion of additional charges. When photons are absorbed in this region, the created electron-hole pairs are separated by the electric field, and a change in voltage across the *p–n* junction is observed. This corresponds to the so-called *photovoltaic mode* of operation of the photodiode, which is illustrated in Fig. 4.8(a). The photovoltaic effect is the basis for the operation of solar cells—a conventional solar cell is just a large-area photodiode. However, photodiodes usually operate in the *photoconductive mode*, as illustrated in Fig. 4.8(b). In this case, the *n*-type side of the junction is connected to the positive terminal of an external voltage source, whereas the *p*-type side is connected to the negative terminal. The reverse-biased external voltage increases the electric field in the junction region, which in turn increases the rate of separation of the electrons and holes generated by absorption of the incident radiation.

Some photodiodes use a region of *intrinsic* (undoped) material layered between the *p*-type and *n*-type materials. The intrinsic layer is a region of high electric field, which significantly increases the rate of separation of the electrons and holes produced by light absorption, thus decreasing the response time of the detector. Photodiodes using such *p-i-n* junctions are mainly used to detect light in the visible to near-infrared region of the spectrum.

Some photodiodes are available in the form of photodiode arrays of a one-dimensional or two-dimensional kind. Two-dimensional detector arrays, e.g., for use as image sensors, may be realized with photodiodes or with other kinds of photodetectors. In such a configuration, each photodiode responds to the incident radiation to provide one *pixel* (picture element) of output. For applications requiring a large number of pixels, a *charge-coupled device* (CCD) is used.

4.6.3 The photographic film

Images can be detected and recorded on a surface sensitive to incoming radiation. Until recently, *photographic film* was the most common medium used for image detection and recording. Photographic film was discovered in 1850 by H. Niepce, and at the basis of its operation is the conversion, by the action of light, of some Ag^{+} ions in the film into Ag

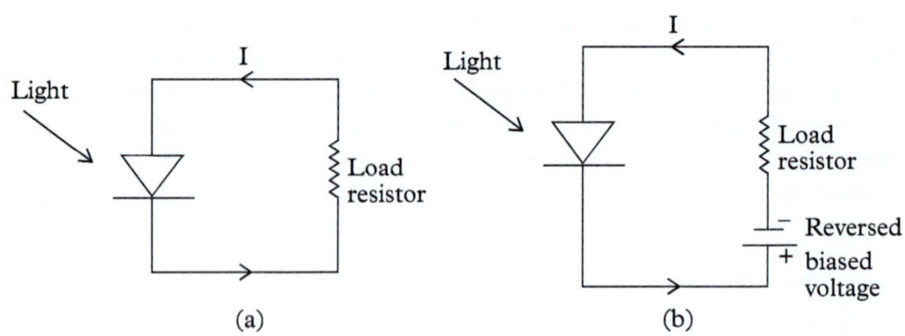

Figure 4.8 *Two modes of operation of a photodiode: (a) photovoltaic mode and (b) photoconductive (reverse-biased) mode.*

atoms. A later chemical treatment allows the image to be revealed by amplification of the aggregates of *Ag* atoms and fixes it by elimination of Ag^+ ions that have not undergone the photochemical effect. The density of the silver atoms, and thus the opacity of the film, is a measure of both the irradiance and the time of exposure, so that photographic film, unlike many other detectors, has the advantage of light-signal integration.

Photographic emulsions are available with spectral sensitivity that extends from the X-ray region into the near infrared. However, despite its capabilities, photographic film has some limitations. In particular, the range of irradiance values for which a linear photographic record is obtained is relatively restricted. On the other hand, it is found that the quantum yield, i.e. the number of silver grains darkened in relation to the number of incident photons, is relatively low (~ 0.8%). Actually, the photographic film is being replaced in most cases by other image-recording techniques. In particular, *CCDs*, in which the incoming radiation is detected by a two-dimensional array of photodiodes, are becoming the preferred choice for image detection.

4.7 Further Reading

Ahmed, S. N. *Physics and Engineering of Radiation Detection*, 2nd Ed., Elsevier, 2015.

Boyd, R. W., *Radiometry and the Detection of Optical Radiation*, John Wiley & Sons, 1983.

Klein, M., and T. Furtak, *Optics*, 2nd Ed., John Wiley & Sons, 1986.

Nabet, B., ed., *Photodetectors: Materials, Devices and Applications*, Elsevier-Woodhead, 2016.

Nicodemus, F. E., Radiometry, *in* R. Shanon and E. J. Wyant, Eds, *Applied Optics and Optical Engineering*, Vol. 19, p. 13, Academic Press, 1983.

Qimby, R S., *Photonics and Lasers—An Introduction*, John Wiley & Sons, 2006.

Pedrotti, F., L. M. Pedrotti, and L. S. Pedrotti, *Introduction to Optics*, 3rd Ed., Cambridge University Press, 2018.

4.8 Problems

4.1 A light source emits 100 W of radiant power at wavelength $\lambda = 500$ nm. Determine the irradiance on a surface located 5 m far from the lamp.

4.2 A laser beam is emitted normally by a source area $\Delta A = 0.25\,\text{mm}^2$. The beam has a power of 2 mW and a divergence angle of 1 mrad. Determine the radiance, L_e, of the laser beam.

4.3 A small, monochromatic light source of area 4 cm^2, radiating at 500 nm, has a power of 500 W. Assuming that the source radiates uniformly in all directions, determine (a) its radiant intensity and (b) its radiant emittance.

4.4 Considering the light source of Problem 4.3, determine (a) the irradiance on a screen situated normally 1 m from the source and (b) the radiant flux passing through a hole with a diameter of 2 cm.

4.5 When observed from the earth, the sun can be considered approximately a Lambertian source, with an apparent diameter $\theta = 31'$ and an emittance given by $M_e = \sigma T^4$, where $\sigma = 5.67 \times 10^{-8}\ \mathrm{W/m^2K^4}$ is Stefan's constant and $T(\approx 5800\ \mathrm{K})$ is the temperature of its surface. Assuming that the earth's atmosphere absorbs about a fifth of the radiation, calculate the radiance of sunlight on earth.

4.6 A light source is shaped like a square with 4 mm on a side and a power of 2 W. Determine the radiance of this source, assuming that it can be considered Lambertian. Also determine the radiant flux received by a lens 4 cm in diameter, located 12 cm from the source.

4.7 Consider a disc-shaped Lambertian source with a radius of 20 cm and a radiance of 600 W/sr.m^2. Determine the irradiance at (i) a point on the floor located 3 m directly below the source and (ii) at another point on the same floor, 2 m away from the previous one.

4.8 Two lamps, one of 25 W and the other of 75 W, are spaced 2 m apart. At what position should a screen be placed, normal to the axis passing through the lamps, so that the irradiance on both sides of it is the same?

4.9 A square table, with side b = 4 m and centred at point C, is illuminated by a point source, S. This source is located in the vertical passing through the midpoint, P, of one side of the table. The radiant intensity of the lamp is I_e.

   a) Express, in terms of, I_e, b and the angle $\theta = (C\hat{S}P)$, the irradiance at the centre point C of the table.

   b) Determine the value of θ for which the irradiance at C is maximum. Calculate the distance from the source to the table in this case.

4.10 The moon is at a distance of 4×10^5 km from the earth and has a diameter of 3500 km. It can be seen as a uniform disc when observed from the earth. Calculate the luminance of the moon, knowing that the light that comes to us is characterized by an illuminance of 0.2 lx.

Wave Theory

5

Wave theory is an important mathematical model in many areas of physics. The basis of this theory is the so-called wave equation, whose solutions can describe a wide variety of phenomena. Thus, the knowledge acquired in the context of the wave theory of optics can be easily applied to other domains of physics involving the propagation of waves.

This chapter will begin by deriving the wave equation in one dimension, after which its solutions, namely the simple harmonic wave, will be discussed. We will also show, using Fourier's theory, that an arbitrary wave can be regarded as the superposition of multiple single harmonic waves. Some special functions used in the wave optics model will also be described.

5.1 The Wave Equation

Let us consider the propagation of a wave along a string subject to a tension T and aligned in the direction of the x-axis. Fig. 5.1 represents a segment of such a string. If the local transverse displacement is sufficiently small, we have $\partial y/\partial x << 1$. In this case, the angle θ between the string and the direction of the x-axis is also small, and we can use the approximations $\cos\theta \approx 1$ e $\sin\theta \approx \tan\theta = \partial y/\partial x \equiv \gamma$. The component of the tension in the direction of the *x-axis* is

$$T_x = T\cos\theta \approx T \tag{5.1}$$

The components of the tension parallel to the x-axis, which act at the ends of the string segment considered, cancel each other, so the movement of that segment will take place only in the direction of the y-axis. The tension component in this direction is given by

$$T_y = T\sin\theta \approx T\gamma \tag{5.2}$$

The total tension in the direction of the y-axis is

$$\Delta T_y = T\sin\theta_2 - T\sin\theta_1 \approx T(\gamma_2 - \gamma_1) \tag{5.3}$$

Assuming that $\Delta x = x_2 - x_1$ is small enough, one can write

$$\gamma_2 = \gamma_1 + \frac{\partial y}{\partial x}\Delta x \tag{5.4}$$

In this way, one has from Eq. (5.3) that

$$\Delta T_y = T\frac{\partial \gamma}{\partial x}\Delta x = T\frac{\partial^2 y}{\partial x^2}\Delta x \tag{5.5}$$

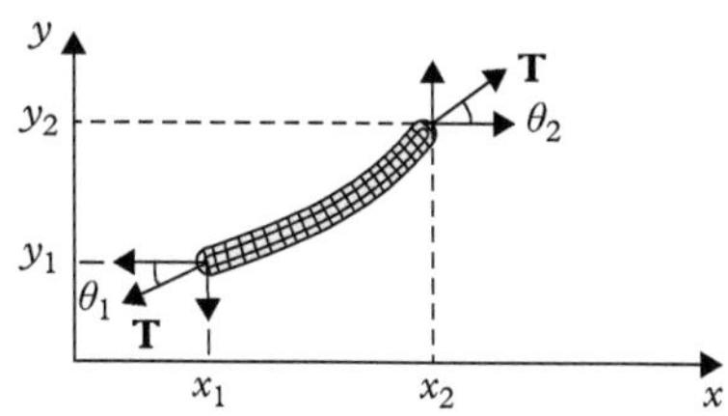

Figure 5.1 *Schematic representation of a string segment under a tension T and disturbed by the propagation of a wave.*

Lasers, Optics, and Photonics. Mário F. S. Ferreira, Oxford University Press. © Mário F. S. Ferreira (2026).
DOI: 10.1093/oso/9780198952770.003.0005

The string segment has a mass $m = \rho\Delta x$, where ρ is the mass per unit length. On the other hand, its acceleration in the direction of the y-axis is $a_y = \partial^2 y/\partial t^2$. Applying Newton's second law to this segment, one has:

$$T\frac{\partial^2 y}{\partial x^2}\Delta x = \rho\Delta x\frac{\partial^2 y}{\partial t^2} \tag{5.6}$$

Eliminating the factor Δx and dividing both sides of Eq. (5.6) by T, the following equation is obtained:

$$\frac{\partial^2 y}{\partial x^2} = \frac{1}{v^2}\frac{\partial^2 y}{\partial t^2} \tag{5.7}$$

where

$$v = \sqrt{\frac{T}{\rho}} \tag{5.8}$$

Eq. (5.7) is the so-called *wave equation* and describes not only the propagation of waves on a string, as has been considered, but also other types of waves: sound waves, seismic waves, electromagnetic waves, etc. The v parameter in Eq. (5.7) corresponds to the wave propagation speed, which is given by Eq. (5.8) in the case of waves on a string.

Eq. (5.7) can be written in the form:

$$\left(\frac{\partial}{\partial x} + \frac{1}{v}\frac{\partial}{\partial t}\right)\left(\frac{\partial}{\partial x} - \frac{1}{v}\frac{\partial}{\partial t}\right)y = 0 \tag{5.9}$$

This way of writing the wave equation immediately shows us that both

$$y(x,t) = f(x - vt) \tag{5.10a}$$

and

$$y(x,t) = g(x + vt), \tag{5.10b}$$

where f and g are arbitrary functions, are solutions of Eq. (5.7). The solution given by Eq. (5.10a) corresponds to a wave propagating in the positive direction of the x-axis, whereas the solution given by Eq. (5.10b) corresponds to a wave propagating in the opposite direction. Moreover, any linear combination of the previous solutions is given by

$$y(x,t) = af(x - vt) + bg(x + vt) \tag{5.11}$$

where a and b are constants, is also a solution of Eq. (5.7). This corresponds to the so-called *superposition principle*, which is the basis for the description, for example, of the phenomenon of interference to be dealt with in Chapter 8.

5.2 Harmonic Waves

Harmonic waves present special interest, since one can use them to describe an arbitrary wave. A harmonic wave propagating in the positive direction of the x-axis can be written in the form

$$y(x,t) = A\cos(kx - \omega t) \tag{5.12}$$

where A is the *amplitude* of the wave, ω is the *angular frequency*, and k is the so-called *propagation constant* or *wave number*. The argument $(kx - \omega t) \equiv \varphi$ represents the *phase* of the wave. In the most general case, this phase may have the form

$$\varphi = kx - \omega t + \varepsilon \tag{5.13}$$

where ε is a constant, called the *initial phase*.

The range of x values over which there is a phase variation of 2π defines the so-called *wavelength*, λ, which corresponds to the spatial period. In fact, if $x_2 = x_1 + \lambda$, one has:

$$kx_2 - \omega t = k(x_1 + \lambda) - \omega t = kx_1 - \omega t + 2\pi$$

So the following relation between the propagation constant and the wavelength is found:

$$k = \frac{2\pi}{\lambda} \tag{5.14}$$

The angular frequency ω is related to the *linear frequency* f or the *period* T (duration of an oscillation), in the form

$$\omega = 2\pi f = \frac{2\pi}{T} \tag{5.15}$$

A point of the wave with a certain phase moves with the velocity

$$v = \left.\frac{dx}{dt}\right|_{\varphi} = \frac{\omega}{k} = \frac{\lambda}{T} = f\lambda \tag{5.16}$$

which is therefore called the *phase velocity*.

Fig. 5.2 shows a representation of the harmonic wave given by Eq. (5.12) at a given instant $t = nT$, n being an integer.

It should be noted that a harmonic wave can be equally represented by the sine function rather than the cosine function used in Eq. (5.12). Furthermore, the sine and cosine functions form a complete set of functions; that is, any periodic function can be represented as a linear combination of sines and cosines, similar to Eq. (5.12). This linear combination is called a *Fourier series* and will be described in Section 5.6.

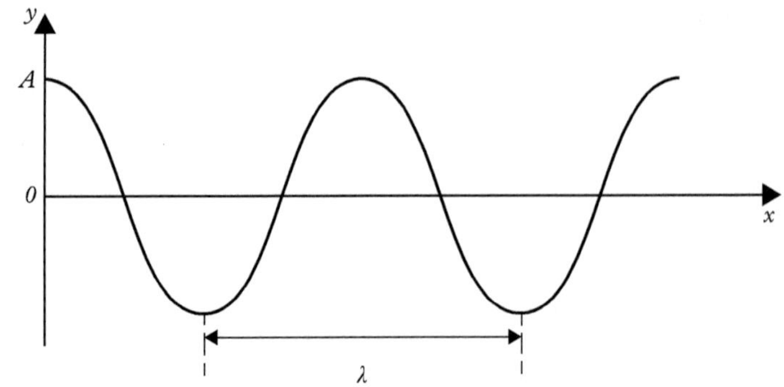

Figure 5.2 *Representation of the variation in the space of the harmonic wave given by Eq. (5.12).*

Example 5.1

A wave propagates according to the expression

$$y(x,t) = 0.6\cos\left(5\pi x - 8\pi t - \frac{\pi}{4}\right)$$

where x is in metres and t is in seconds. Determine the frequency, wavelength, initial phase, speed, and displacement for $x = 20$cm and $t = 0$.

Solution

Comparing the previous expression with Eq. (5.12), one has $\omega = 8\pi$ and $k = 5\pi$. Thus, the frequency and wavelength are, respectively,

$$f = \frac{\omega}{2\pi} = 4\,\text{Hz}$$

and

$$\lambda = \frac{2\pi}{k} = 0.4\,\text{m}$$

The initial phase (for $x = 0$ and $t = 0$) is $\varepsilon = -\pi/4$ rad. The phase velocity is

$$v = \lambda f = 1.6\,\text{m/s}$$

in the positive direction of the x-axis. Finally, the displacement for $x = 20$cm and $t = 0$ is

$$y(0.2,0) = 0.6\cos\left((5\pi)(0.2) - (8\pi)(0) - \frac{\pi}{4}\right) = -0.42\,\text{m}$$

5.3 Group Velocity

Since the refractive index of a medium in which the wave is propagating is frequency dependent, the phase velocity of such a wave is, in general, also a function of the frequency. This fact has important implications when the propagating wave is composed of several frequencies, as is the case in applications using modulated light. The velocity of the carrier and the velocity of the modulation will, in general, be different.

Let us consider the simple situation of a propagating plane wave containing only two frequencies. The total real electric field of such a wave can be written as the sum of the fields of the two waves, which we assume to propagate in the z direction and to have the same amplitude E_{01}:

$$E(z,t) = E_{01}\left[\cos(k_1 z - \omega_1 t) + \cos(k_2 z - \omega_2 t)\right] = 2E_{01}\cos(k_m z - \omega_m t)\cos(k_a z - \omega_a t) \tag{5.17}$$

where

$$\omega_a = \frac{1}{2}(\omega_1 + \omega_2), \qquad \omega_m = \frac{1}{2}(\omega_1 - \omega_2) \tag{5.18}$$

and

$$k_a = \frac{1}{2}(k_1 - k_2), \qquad k_m = \frac{1}{2}(k_1 - k_2) \tag{5.19}$$

The quantities ω_a and k_a are the average angular frequency and the average propagation constant, respectively, whereas the quantities ω_m and k_m are designated as the *modulation frequency* and the *modulation propagation constant*, respectively.

The total field can be regarded as a travelling (carrier) wave of frequency ω_a having a time-varying or modulated amplitude $E_0(z,t)$ such that

$$E(z,t) = E_0(z,t)\cos(k_a z - \omega_a t) \tag{5.20}$$

where

$$E_0(z,t) = 2E_{01}\cos(k_m z - \omega_m t) \tag{5.21}$$

The phase velocity of the carrier wave is given by

$$v = \frac{\omega_a}{k_a} \tag{5.22}$$

Fig. 5.3 illustrates the combination of the higher-frequency *carrier wave* with the lower-frequency *envelope wave*. The rate at which the envelope wave advances is known as the *group velocity*, v_g. It is obtained by considering the phase of the envelope $(k_m z - \omega_m t)$, being given by

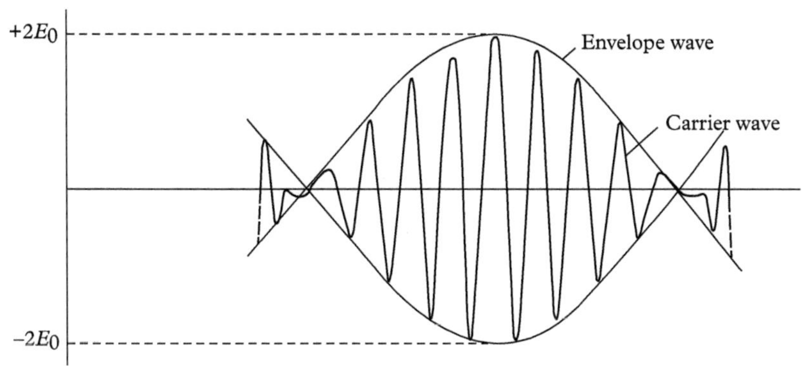

Figure 5.3 *Envelope and carrier waves resulting from the combination of two harmonic waves with different frequencies.*

$$v_g = \frac{\omega_m}{k_m} = \frac{\omega_1 - \omega_2}{k_1 - k_2} = \frac{\Delta\omega}{\Delta k}. \tag{5.23}$$

The function describing the dependence of ω on k, $\omega = \omega(k)$, is called the *dispersion relation*. When the frequency range $\Delta\omega$ is small, the ratio $\Delta\omega/\Delta k$ tends to the derivative of the dispersion relation, and the group velocity becomes

$$v_g = \frac{d\omega}{dk} \tag{5.24}$$

Since $\omega = kv$, Eq. (5.24) yields

$$v_g = v + k\frac{dv}{dk} \tag{5.25}$$

The modulation or signal propagates with a velocity v_g that may be greater than, equal to, or less than the phase velocity of the carrier, v. In non-dispersive media, v does not depend on k and $v_g = v$. In dispersive media, in which $n(k)$ is known, $\omega = kc/n$ and the group velocity can be written in the form:

$$v_g = \frac{c}{n} - \frac{kc}{n^2}\frac{dn}{dk} \tag{5.26}$$

The group velocity can be considered as the propagation velocity of a 'group' of waves with frequencies distributed over an infinitesimally small bandwidth centred at ω_a. In the presence of a broad frequency spectrum, the slope of the curve $\omega(k)$ may change over the range of the spectrum. As a consequence, different spectral components propagate at different group velocities, leading to signal distortion. This problem is called *group velocity dispersion* and will be discussed in Chapter 16 in the context of optical fibres.

5.4 Complex Notation

The use of complex notation greatly simplifies calculations related to the superposition of waves. For this reason, it is appropriate to carry out a summary review of the main aspects relating to the algebra of complex numbers.

A complex number is an ordered pair of two real numbers (a, b) or $a + ib$ where a and b are called, respectively, the *real part* and the *imaginary part* and $i = \sqrt{-1}$. Similarly, a complex variable is an ordered pair of two real variables:

$$z = (x, y) = x + iy \tag{5.27}$$

It is often useful to consider a graphical representation of the complex variable. The complex plane, or Argand plane, is illustrated in Fig. 5.3. The complex variable z corresponds to the coordinate point (x, y) in this plane.

Based on Fig. 5.4 one can write the following relations between Cartesian coordinates (x, y) and polar coordinates (r, θ)

$$x = r\cos\theta, \quad y = r\sin\theta \tag{5.28}$$

r and θ are called the *modulus* and the *argument* of z, respectively. We can reverse the previous relationships and write them in the form:

$$r = |z| = \sqrt{x^2 + y^2}, \quad \theta = \tan^{-1}\frac{y}{x} \tag{5.29}$$

Using Eqs. (5.27) and (5.28), one has:

$$z = r(\cos\theta + i\sin\theta) \tag{5.30}$$

Using Euler's formula, Eq. (5.30) can be written in the form

$$z = re^{i\theta} \tag{5.31}$$

The addition and subtraction of complex variables are more easily performed using the Cartesian representation given by Eq. (5.27), whereas multiplication, division, potentiation, or root calculation become easier using the polar representation given by Eq. (5.31). For example, using this latter representation one can easily verify that the modulus of the product of two complex numbers is equal to the product of the respective moduli:

$$|z_1 \cdot z_2| = |z_1| \cdot |z_2| \tag{5.32}$$

and that the product argument is equal to the sum of the arguments:

$$\arg(z_1 \cdot z_2) = \arg(z_1) + \arg(z_2) \tag{5.33}$$

It can also be seen that

$$z^n = \{r(\cos\theta + i\sin\theta)\}^n = r^n(\cos n\theta + i\sin n\theta) \tag{5.34}$$

This result is known as the *De Moivre formula* and is valid not only for integer and positive values but also for negative or fractional values of n. In the latter case, it is easily shown that

$$z^{1/n} = \sqrt[n]{z} = r^{1/n}\left(\cos\frac{\theta + 2k\pi}{n} + i\sin\frac{\theta + 2k\pi}{n}\right), k = 0, 1, 2, \ldots n - 1 \tag{5.35}$$

Switching i *by* $-i$ in Eq. (5.27) one obtains the complex conjugate of z, indicated by z^*:

$$z^* = x - iy \tag{5.36}$$

The product zz^* is given by

$$zz^* = (x + iy)(x - iy) = x^2 + y^2 = r^2 \tag{5.37}$$

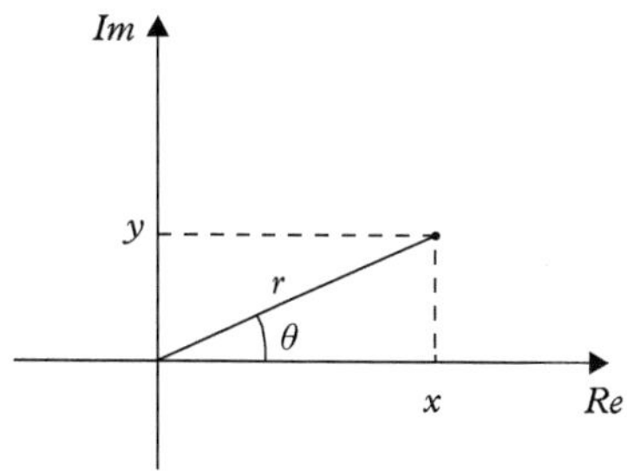

Figure 5.4 *Representation of a complex number in the Argand plane.*

Using Euler's formula, the harmonic wave given by Eq. (5.12) can be written in the form:

$$E(z,t) = E_0 \exp\{i(kz - \omega t)\} + c.c. \tag{5.38}$$

where $E_0 = A/2$ and *c.c.* indicates the complex conjugate of the previous term.

Often, the complex form is simply used

$$E(z,t) = E_0 \exp\{i(kz - \omega t)\} \tag{5.39}$$

In this case, it should be taken into account that only the real part of the final result of the calculations has physical significance.

Example 5.2

Let us consider two waves with the same amplitude, velocity, and frequency but with opposite propagation directions, overlapping in the same region of space, so that the resulting displacement is

$$y(z,t) = A\cos(kz - \omega t - \pi) + A\cos(kz + \omega t)y(x,t) = A\cos(\omega t - kx + \pi) + A\cos(\omega t + kx)$$

Using complex exponentials, show that

$$y(x,t) = -2A\sin(kz)\sin(\omega t)$$

This result is known as a *stationary wave*.

Solution

Using complex exponentials, the given wave function can be written in the form

$$\begin{aligned} y(z,t) &= A\left[\exp\{i(kz - \omega t - \pi)\} + \exp\{i(kz + \omega t)\}\right] \\ &= A\exp\{ikz\}\left[\exp\{-i\pi\}\exp\{-i\omega t\} + \exp\{i\omega t\}\right] \end{aligned}$$

By Euler's formula, it is $\exp\{-i\pi\} = \cos(-\pi) + i\sin(-\pi) = -1$. On the other hand, one has

$$\cos\theta = (\exp\{i\theta\} + \exp\{-i\theta\})/2$$

and

$$\sin\theta = (\exp\{i\theta\} - \exp\{-i\theta\})/(2i)$$

So the previous expression becomes

$$\begin{aligned} y(z,t) &= A\exp\{ikz\}\left[\exp\{i\omega t\} - \exp\{-i\omega t\}\right] \\ &= A\exp\{ikz\}(2i)\sin(\omega t) \\ &= A\left[2i\cos(kz)\sin(\omega t) - 2\sin(kz)\sin(\omega t)\right] \end{aligned}$$

The real part of this wave function is

$$y(z,t) = -2A\sin(kz)\sin(\omega t)$$

5.5 Three-Dimensional Waves

The one-dimensional wave equation given by Eq. (5.7) can be generalized to three-dimensional space, assuming the following form:

$$\frac{\partial^2 f(\mathbf{r},t)}{\partial x^2}+\frac{\partial^2 f(\mathbf{r},t)}{\partial y^2}+\frac{\partial^2 f(\mathbf{r},t)}{\partial z^2}=\frac{1}{v^2}\frac{\partial^2 f(\mathbf{r},t)}{\partial t^2} \tag{5.40}$$

where $\mathbf{r}=x\mathbf{i}+y\mathbf{j}+z\mathbf{k}$ is the position vector. One can write Eq. (5.40) in a more compact form as

$$\nabla^2 f(\mathbf{r},t)=\frac{1}{v^2}\frac{\partial^2 f(\mathbf{r},t)}{\partial t^2} \tag{5.41}$$

where

$$\nabla^2=\frac{\partial^2}{\partial x^2}+\frac{\partial^2}{\partial y^2}+\frac{\partial^2}{\partial z^2} \tag{5.42}$$

is the *Laplacian operator* written in Cartesian coordinates. A harmonic wave solution of Eq. (5.40) can be written in the form

$$f(\mathbf{r},t)=E(\mathbf{r})\cos\left[L(\mathbf{r})-\omega t\right] \tag{5.43}$$

At a given instant of time, the surfaces for which $L(\mathbf{r})=constant$ are called *wave-fronts*. A wave that presents a plane wave-front is called a *plane wave*, whereas, in the case of a spherical wave-front, we are in the presence of a *spherical wave*.

5.5.1 Plane waves

A plane wave is characterized by the fact that, at a given moment, its phase is constant in any plane perpendicular to the direction of propagation (Fig. 5.5). A plane wave can be represented in the form:

$$f(\mathbf{r},t)=A\cos(\mathbf{k}\cdot\mathbf{r}\pm\omega t) \tag{5.44}$$

where $\mathbf{k}$ is a vector normal to the wave-front, called the *wave vector*. In fact, the equation $\mathbf{k}\cdot\mathbf{r}=const$ represents a plane perpendicular to the vector $\mathbf{k}$. Eq. (5.44) can be written in complex form:

$$f(\mathbf{r},t)=E_0\exp\{i(\mathbf{k}\cdot\mathbf{r}\pm\omega t)\} \tag{5.45}$$

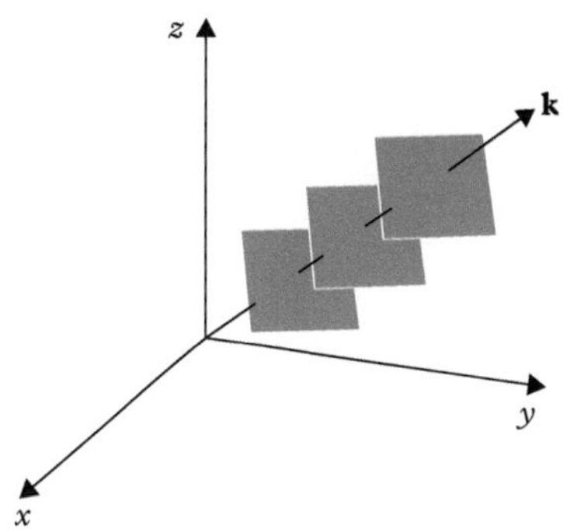

Figure 5.5 *Representation of plane-wave propagation.*

5.5.2 Spherical waves

The Laplacian of the field in a spherical coordinate system is given by

$$\nabla^2 f=\frac{1}{r^2}\frac{\partial}{\partial r}\left(r^2\frac{\partial}{\partial r}\right)f=\frac{1}{r}\frac{\partial^2}{\partial r^2}(rf) \tag{5.46}$$

where r is the radial coordinate and represents the distance to the origin of the coordinate system. Introducing this expression into Eq. (5.41) and multiplying the equation thus obtained by r, one has

$$\frac{\partial^2}{\partial r^2}(rf) - \frac{1}{v^2}\frac{\partial^2}{\partial t^2}(rf) = 0 \tag{5.47}$$

This is a one-dimensional wave equation for the quantity rf. The solution of this equation for rf can also be written in the form of Eq. (5.11):

$$rf(r,t) = g(r \pm vt) \tag{5.48}$$

That is, we have

$$f(r,t) = \frac{1}{r}g(r \pm vt) \tag{5.49}$$

The minus sign in the previous solution corresponds to a spherical wave expanding from the origin of the coordinate system, whereas the plus sign corresponds to a wave converging to that point.

Harmonic spherical waves are of special interest and can be represented in the following trigonometric form:

$$f(r,t) = \frac{A}{r}\cos(kr \pm \omega t) \tag{5.50}$$

In the complex form, this wave takes the following form:

$$f(r,t) = \frac{E_0}{r}\exp\{i(kr \pm \omega t)\} \tag{5.51}$$

It can be verified from Eqs. (5.49)–(5.51) that the amplitude of a spherical wave varies inversely with the distance to the origin. On the other hand, the curvature is reduced, and the spherical wave approaches a plane wave at points sufficiently far from the origin (Fig. 5.6).

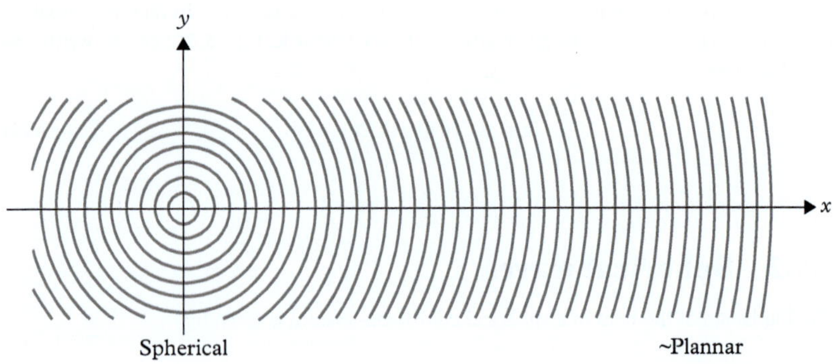

Figure 5.6 *A spherical wave approaches a plane wave at points sufficiently far from the origin.*

5.6 Fourier Series

A harmonic wave that extends infinitely, both in time and in space, is a theoretical abstraction. In practice, waves are distributed over a certain range of frequencies and travel in various directions. The frequency and angular spectra that should be considered in these circumstances can be conveniently included in the wave theory using the Fourier approach.

Let $f(x)$ be a function defined in the range $-L \leq x \leq L$ and determined outside this range by $f(x+2L)=f(x)$; that is, suppose that $f(x)$ is periodic with period $2L$. So the function $f(x)$ can be developed into a *Fourier series* in the form:

$$f(x) = \frac{a_0}{2} + \sum_{n=1}^{\infty} (a_n \cos(nkx) + b_n \sin(nkx)) \tag{5.52}$$

where $k = \pi/L$ and a_n, b_n are the so-called *Fourier coefficients*, given by

$$a_0 = \frac{1}{L} \int_{-L}^{L} f(x)\,dx \tag{5.53}$$

$$a_n = \frac{1}{L} \int_{-L}^{L} f(x) \cos(nkx)\,dx \tag{5.34}$$

$$b_n = \frac{1}{L} \int_{-L}^{L} f(x) \sin(nkx)\,dx \tag{5.55}$$

It should be noted that when $f(x)$ is an odd function ($f(-x) = -f(x)$), one has $a_n = 0$, whereas b_n can be written in the form

$$b_n = \frac{2}{L} \int_{0}^{L} f(x) \sin(nkx)\,dx$$

Similarly, when $f(x)$ is an even function ($f(-x) = f(x)$), one has $b_n = 0$, whereas a_n can be written in the form

$$a_n = \frac{2}{L} \int_{0}^{L} f(x) \cos(nkx)\,dx$$

The results in Eqs. (5.53)–(5.55) for the Fourier coefficients remain valid when the integration limits $-L$ and L are replaced by c and $c+2L$, respectively, c being an arbitrary real number.

The convergence conditions of the Fourier series were first formulated by the German mathematician P. G. L. Dirichlet (1805–1859) in 1829 and are known as the *Dirichlet conditions*. Under these conditions, in the case that $f(x)$ is discontinuous in x, the left-hand side of Eq. (5.52) must be replaced by the semi-sum of the lateral limits at this point: $\frac{f(x+)+f(x-)}{2}$.

Eq. (5.52) corresponds to the so-called *Fourier synthesis* of the function $f(x)$, performed by summing the components with the fundamental frequency and its harmonics, with different amplitudes. On the other hand, the reverse operation of obtaining the frequencies and amplitudes of the components of a periodic function is known as *Fourier analysis*.

In the case of a time function, $f(t)$, of period T, its Fourier series expansion is given by expressions similar to Eqs. (5.52)–(5.55), with $T/2$ in place of L and ω in place of k.

Example 5.3

The movement of a vibrating string fixed at the ends $x = 0$ and $x = L$ can be described by the following equation:

$$\frac{\partial^2 y(x,t)}{\partial t^2} = v^2 \frac{\partial^2 y(x,t)}{\partial x^2}$$

Consider that the solutions are given by

$$y(x,t) = \sum_{n=1}^{\infty} b_n(t) \sin(nkx)$$

Determine the coefficients $b_n(t)$, assuming the initial conditions:

$$y(x,0) = f(x), \quad y'(x,0) = g(x)$$

Solution

Substituting the given solution into the wave equation, one has

$$\sum_{n=1}^{\infty} b_n''(t) \sin(nkx) = -v^2 \sum_{n=1}^{\infty} b_n(t)(nk)^2 \sin(nkx)$$

That is, one has

$$b_n'' + v^2(nk)^2 b_n = 0$$

where $b_n'' \equiv \dfrac{d^2 b_n}{dt^2}$. The solution of the previous equation can be written in the form:

$$b_n(t) = A_n \cos(nkvt) + B_n \sin(nkvt)$$

Considering the initial conditions mentioned in the problem, one has

$$y(x,0) = f(x) = \sum_{n=1}^{\infty} b_n(0) \sin(nkx)$$

So one can write

$$b_n(0) = A_n = \frac{2}{L} \int_0^L f(x) \sin(nkx)\, dx$$

Considering the other initial condition, one has

$$y'(x,0) = g(x) = \sum_{n=1}^{\infty} b_n'(0) \sin(nkx)$$

So

$$B_n = \left(\frac{1}{nkv}\right) b_n'(0) = \frac{2}{nLkv} \int_0^L g(x) \sin(nkx)\, dx$$

Example 5.4

Obtain the Fourier series of the sawtooth wave, given by $f(x) = x,\ -\pi \le x \le \pi$

Solution

Using Eq. (5.53), one has

$$a_0 = \frac{1}{\pi}\int_{-\pi}^{\pi} x dx = \frac{1}{\pi}\frac{x^2}{2}\bigg]_{-\pi}^{\pi} = 0$$

If $n \ge 1$, one can obtain the coefficients a_n from Eq. (5.54):

$$a_n = \frac{1}{\pi}\int_{-\pi}^{\pi} x\cos(nx)dx = \frac{1}{\pi}\left[\frac{x\sin(nx)}{n} + \frac{\cos(nx)}{n^2}\right]_{-\pi}^{\pi} = 0$$

The aforementioned results were expected, given that the function $f(x)$ is odd. On the other hand, from Eq. (5.55), one has

$$b_n = \frac{1}{\pi}\int_{-\pi}^{\pi} x\sin(nx)dx = \frac{1}{\pi}\left[-\frac{x\cos(nx)}{n} + \frac{\sin(nx)}{n^2}\right]$$

$$= \frac{1}{\pi}\left[-\frac{\pi\cos(n\pi)}{n} - \frac{\pi\cos(-n\pi)}{n}\right] = -\frac{2}{n}\cos(n\pi) = \frac{2}{n}(-1)^{n+1}$$

Substituting these results in Eq. (5.52), one obtains the following Fourier series:

$$x = 2\left(\sin(x) - \frac{\sin(2x)}{2} + \frac{\sin(3x)}{3} - \cdots\right)$$

Fig. 5.7 shows $f(x)$ for $0 < x < \pi$ considering the sum of the first 4, 6, and 10 terms of the series. It can be seen that increasing the number of terms increases the accuracy of the sawtooth wave representation. As a curiosity, it can also be seen that setting $x = \pi/2$ in the previous series provides a derivation of Leibnitz's famous formula for π:

$$\frac{\pi}{4} = 1 - \frac{1}{3} + \frac{1}{5} - \frac{1}{7} + \frac{1}{9} - \cdots + \frac{(-1)^n}{2n+1}$$

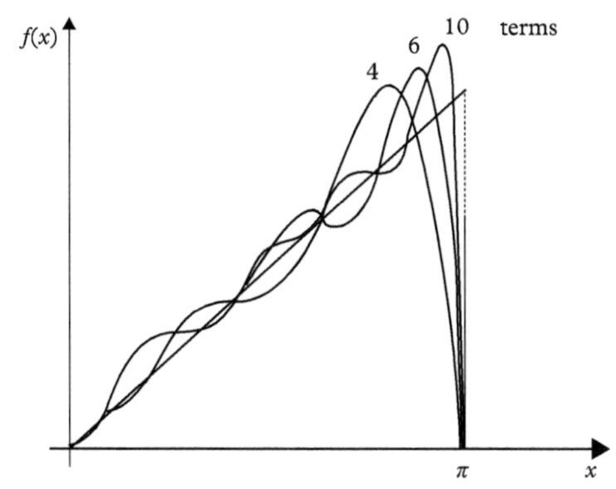

Figure 5.7 *Fourier approximation of the sawtooth wave with 4. 6, and 10 terms.*

5.6.1 Biographical note—Fourier

Jean Baptiste Fourier (1768–1830) was a French physicist and mathematician, professor at the Normal School (1795) and the Paris Polytechnic School between 1795 and 1798, the year he accompanied Napoleon I on his campaign in Egypt. When he returned from that campaign in 1802, Fourier published several works related to Egyptian antiquity and assumed the duties of mayor of the District of Isère, in south-eastern France. Napoleon granted him the title of baron in 1806. During these years, he also worked on heat conduction theory, having published, in 1822, his famous *Théorie Analytique de la Chaleur*, in which he extensively used the series that bear his name. These series eventually represented a significant contribution to the evolution of the concept of function. In fact, many of the most important mathematical discoveries that took place throughout the 19th century are directly related to Fourier's theory of series.

5.7 Complex Notation for Fourier Series

Using the identities

$$\cos\theta = \frac{\exp\{i\theta\} + \exp\{-i\theta\}}{2}$$

$$\sin\theta = \frac{\exp\{i\theta\} - \exp\{-i\theta\}}{2i} \tag{5.57}$$

where $i^2 = -1$, we can rewrite the Fourier series for $f(x)$ in the following way:

$$\begin{aligned} f(x) &= \frac{a_0}{2} + \sum_{n=1}^{\infty} (a_n \cos(nkx) + b_n \sin(nkx)) \\ &= \frac{a_0}{2} + \sum_{n=1}^{\infty} a_n \frac{\exp\{inkx\} + \exp\{-inkx\}}{2} + \sum_{n=1}^{\infty} b_n \frac{\exp\{inkx\} - \exp\{-inkx\}}{2i} \\ &= \frac{a_0}{2} + \sum_{n=1}^{\infty} \left(\frac{a_n - ib_n}{2} \exp\{inkx\} + \frac{a_n + ib_n}{2} \exp\{-inkx\} \right) \\ &= \sum_{n=-\infty}^{\infty} c_n \exp\{inkx\} \end{aligned} \tag{5.58}$$

where

$$c_0 = \frac{a_0}{2}, c_n = \frac{a_n - ib_n}{2}, c_{-n} = \frac{a_n + ib_n}{2} \tag{5.59}$$

The result

$$f(x) = \sum_{n=-\infty}^{\infty} c_n \exp\{inkx\} \tag{5.60}$$

represents the so-called *complex form of the Fourier series*. The coefficients c_n can be defined in the form:

$$c_n = \frac{1}{2L} \int_{-L}^{L} f(x) \exp\{-inkx\}\, dx \tag{5.61}$$

Example 5.5

Obtain the Fourier series in complex form of the function

$$f(x) = e^{\alpha x}, \quad -\pi < x < \pi.$$

Solution

Using Eq. (5.61) with $L = \pi$, one has

$$c_n = \frac{1}{2\pi}\int_{-\pi}^{\pi} e^{\alpha x}e^{-inx}dx = \frac{1}{2\pi}\int_{-\pi}^{\pi} e^{(\alpha-in)x}dx = \frac{1}{2\pi}\frac{e^{(\alpha-in)x}}{\alpha - in}\bigg|_{-\pi}^{\pi}$$
$$= \frac{1}{2\pi}\frac{e^{\alpha\pi}e^{-in\pi} - e^{-\alpha\pi}e^{in\pi}}{\alpha - in}$$

Considering that

$$e^{\pm in\pi} = \cos(\pm n\pi) = (-1)^n$$

one has

$$c_n = \frac{e^{\alpha\pi} - e^{-\alpha\pi}}{2\pi}\frac{(-1)^n}{\alpha - in} = \frac{\sinh(\alpha\pi)}{\pi}\frac{(-1)^n}{\alpha^2 + n^2}(a + in)$$

Using this result, the desired Fourier series is given by

$$e^{\alpha x} = \frac{\sinh(\pi\alpha)}{\pi}\sum_{n=-\infty}^{\infty}\frac{(-1)^n}{\alpha^2 + n^2}(\alpha + in)e^{inx}$$

5.8 Fourier Transforms

Fourier analysis is a technique that allows one to relate the shape of a function to its spectrum; for example, it provides a description of a given optical pulse in terms of its spectral components. The relationship established by Eq. (5.52) between a periodic wave and its discrete harmonic components should, in this case, be generalized to include a non-periodic function and a continuous spectrum.

The spectrum, $f(k)$, corresponding to a given function $f(x)$, is obtained by calculating the *Fourier transform* of this function, defined in the form:

$$F(k) = \frac{1}{\sqrt{2\pi}}\int_{-\infty}^{\infty} f(x)\exp\{ikx\}\,dx \tag{5.62}$$

On the other hand, the *inverse Fourier transform* is given by

$$f(x) = \frac{1}{\sqrt{2\pi}}\int_{-\infty}^{\infty} F(k)\exp\{-ikx\}\,dk \tag{5.63}$$

From a formal point of view, it can be seen that Eqs. (5.62) and (5.63) are almost symmetrical, differing only in the sign of i.

In the case of $f(x) = f_c(x)$ being an even function, Eq. (5.62) can be written in the form

$$\begin{aligned} f(k) &= \frac{1}{\sqrt{2\pi}} \int_{-\infty}^{\infty} f(x)\{\cos(kx) + i \sin(kx)\} dx \\ &= \sqrt{\frac{2}{\pi}} \int_{0}^{\infty} f_c(x) \cos(kx) dx \equiv F_c(k) \end{aligned} \tag{5.64}$$

Similarly, given that $\cos\omega t$ is an even function, we have from Eq. (5.63) that

$$f_c(x) = \sqrt{\frac{2}{\pi}} \int_{0}^{\infty} F(k) \cos(kx) \, dk \tag{5.65}$$

Eqs. (5.64) and (5.65) are known as *cosine Fourier transforms*.

In the case of $f(x) = f_s(x)$ being an odd function and using arguments similar to the previous ones, we have

$$F_s(k) = \sqrt{\frac{2}{\pi}} \int_{0}^{\infty} f_s(x) \sin(kx) dx \tag{5.66}$$

$$f_s(x) = \sqrt{\frac{2}{\pi}} \int_{0}^{\infty} F_s(k) \sin(kx) dk \tag{5.67}$$

Eqs. (5.66) and (5.67) are known as *sine Fourier transforms*.

Example 5.6

Obtain the Fourier transform of a pulse consisting of N cycles of an infinite harmonic wave $\sin(\omega_0 t)$, which can be represented in the form:

$$f(t) = \begin{cases} \sin(\omega_0 t), & |t| < \dfrac{N\pi}{\omega_0} \\ 0, & |t| > \dfrac{N\pi}{\omega_0} \end{cases} \tag{5.68}$$

Fig. 5.8 represents the case of a finite wave train with seven cycles.

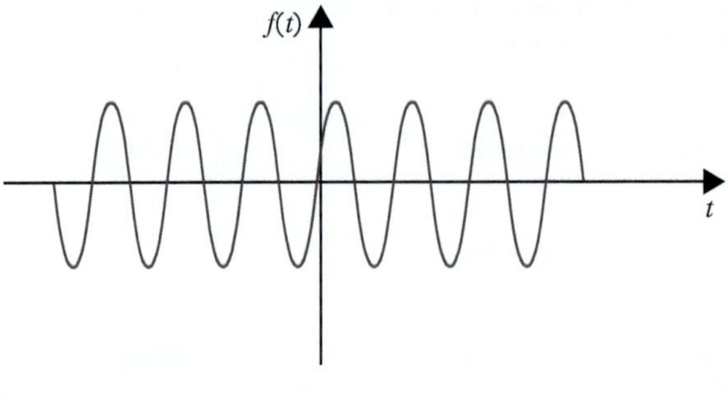

Figure 5.8 *Finite wave train with seven cycles.*

Example 5.6

Solution

Since $f(t)$ is an odd function, we can use Eq. (5.66) to calculate its Fourier transform, obtaining

$$f(\omega) = \sqrt{\frac{2}{\pi}} \int_0^{N\pi/\omega_o} \sin(\omega_0 t).\sin(\omega t)dt$$

$$= \sqrt{\frac{2}{\pi}} \left[\frac{\sin[(\omega - \omega_0)(N\pi/\omega_0)]}{2(\omega - \omega_0)} - \frac{\sin[(\omega + \omega_0)(N\pi/\omega_0)]}{2(\omega + \omega_0)} \right] \tag{5.69}$$

Considering $\omega \approx \omega_0$ and knowing that the frequency ω_0 has a very high value, only the first term in Eq. (5.69) is relevant. It is plotted in Fig. 5.9. The function $f(\omega)$ has a main maximum at $\omega = \omega_0$, given by

$$f(\omega_0) = \frac{1}{\sqrt{2\pi}} \frac{N\pi}{\omega_0} \tag{5.70}$$

and zeros for

$$\frac{\omega - \omega_0}{\omega_0} = \frac{\Delta\omega}{\omega_0} = \pm\frac{1}{N}, \pm\frac{2}{N}, \cdots \tag{5.71}$$

It can be seen from Eq. (5.71) that, if N is large (which corresponds to a long pulse), the frequency spectrum is narrow and vice versa. The ideal case of a monochromatic wave corresponds to $N \to \infty$.

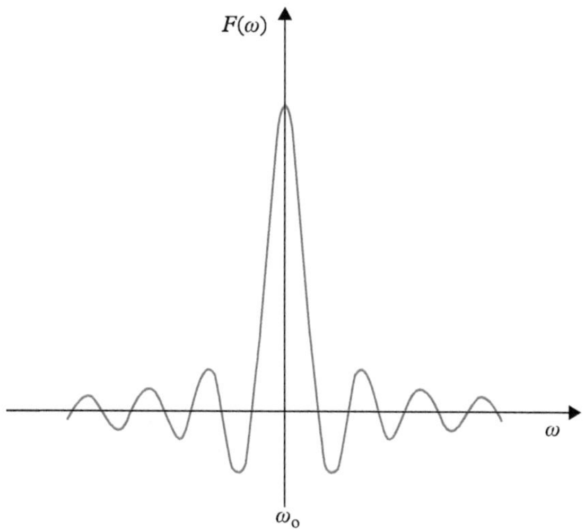

Figure 5.9 *Fourier transform of a finite wave train.*

In the case of a sinusoidal wave $\cos(\omega_0 t)$, modulated by an arbitrary function $g(t)$, its Fourier transform is given by

$$f(\omega) = \frac{1}{\sqrt{2\pi}} \int_{-\infty}^{+\infty} g(t)\cos(\omega_0 t)\exp(i\omega t)dt$$

$$= \frac{1}{\sqrt{2\pi}}\frac{1}{2} \int_{-\infty}^{+\infty} g(t)\left[\exp\left(i(\omega+\omega_0)t\right) + \exp\left(i(\omega-\omega_0)t\right)\right] dt$$

$$= \frac{1}{2}G(\omega+\omega_0) + \frac{1}{2}G(\omega-\omega_0) \tag{5.72}$$

where $G(\omega)$ is the Fourier transform of $g(t)$. That is, the spectrum of the modulated wave is given by the spectrum of the modulation function $g(t)$, centred on frequencies ω_0 and $-\omega_0$.

The pulse referred to in Example 5.6 can be considered as representing a sinusoidal wave $\sin(\omega_0 t)$, modulated by the function

$$g(t) = \begin{cases} 1, & |t| < N\pi/\omega_0 \\ 0, & |t| > N\pi/\omega_0 \end{cases}$$

5.9 The Dirac Delta Function

When a pulse becomes extremely short, its Fourier transform becomes extremely wide. To describe such a pulse, it is convenient to use the so-called *Dirac delta function*, $\delta(x)$. In the time domain, this function can describe, for example, an ultra-short pulse propagating in an optical fibre, assuming that its spectrum is infinitely wide.

The Dirac delta function is defined using the following properties:

$$\delta(x) = 0, \quad x \neq 0 \tag{5.73}$$

$$\int_{-\infty}^{\infty} \delta(x)\,dx = 1 \tag{5.74}$$

$$\int_{-\infty}^{\infty} \delta(x) f(x)\,dx = f(0) \tag{5.75}$$

assuming the continuity of $f(x)$ at $x = 0$.

In fact, there is no function with the aforementioned characteristics. However, it is possible to approximate the Dirac delta function by certain appropriate functions, for example:

$$\delta_n(x) = \frac{n}{\sqrt{\pi}}\exp\{-n^2x^2\} \tag{5.76a}$$

$$\delta_n(x) = \frac{\sin nx}{\pi x} = \frac{1}{2\pi}\int_{-n}^{n} \exp\{ixt\}dt \tag{5.76b}$$

It should be noted that the limits $\lim_{n\to\infty} \delta_n(x)$ do not exist. However, Eqs. (5.76a, b), with $n = 1, 2, 3, \ldots$, can be understood as sequences of standard functions such as the following:

$$\int_{-\infty}^{\infty} \delta_n(x)\, dx = 1 \tag{5.77}$$

The sequence of integrals has the limit

$$\lim_{n\to\infty} \int_{-\infty}^{\infty} \delta_n(x) f(x)\, dx = f(0) \tag{5.78}$$

In this way, one can define $\delta(x)$ consistently in the form

$$\int_{-\infty}^{\infty} \delta(x) f(x) dx = \lim_{n\to\infty} \int_{-\infty}^{\infty} \delta_n(x) f(x) dx \tag{5.79}$$

The Fourier transform of the Dirac delta function is given by

$$\Delta(k) = \frac{1}{\sqrt{2\pi}} \int_{-\infty}^{\infty} \delta(x) \exp\{ikx\} = \frac{1}{\sqrt{2\pi}} \tag{5.80}$$

where Eq. (5.75) was used. Starting from the previous result and using the inverse Fourier transform, a particularly useful representation can be obtained for $\delta(x)$:

$$\delta(x) = \frac{1}{\sqrt{2\pi}} \int_{-\infty}^{\infty} \left(\frac{1}{\sqrt{2\pi}}\right) \exp\{-ikx\}\, dk = \frac{1}{2\pi} \int_{-\infty}^{\infty} \exp\{-ikx\}\, dk \tag{5.81}$$

5.10 The Convolution Theorem

The convolution operation is useful in Fourier's analysis of any function, whatever variable the function depends on. This operation occurs naturally in the response of any optical instrument used to discriminate light in terms of its direction or wavelength. Such an optical instrument always has a limited resolving power, which ultimately degrades the image or the desired spectrum.

Consider two functions, $f(t)$ and $g(t)$, with Fourier transforms $f(\omega)$ and $G(\omega)$, respectively. The convolution of the functions f and g in the interval $]-\infty, \infty[$ is defined in the form:

$$h(t) = f(t) * g(t) \equiv \frac{1}{\sqrt{2\pi}} \int_{-\infty}^{\infty} g(s) f(t-s) ds \tag{5.82}$$

Using Fourier transforms, one has

$$\begin{aligned} h(t) &= \frac{1}{\sqrt{2\pi}} \int_{-\infty}^{\infty} g(s) f(t-s) ds = \frac{1}{2\pi} \int_{-\infty}^{\infty} g(s) \int_{-\infty}^{\infty} F(\omega) \exp\{-i\omega(t-s)\}\, d\omega ds \\ &= \frac{1}{2\pi} \int_{-\infty}^{\infty} F(\omega) \left[\int_{-\infty}^{\infty} g(s) \exp\{i\omega s\}\, ds \right] \exp\{-i\omega t\}\, d\omega \\ &= \frac{1}{\sqrt{2\pi}} \int_{-\infty}^{\infty} F(\omega) G(\omega) \exp\{-i\omega t\}\, d\omega \end{aligned} \tag{5.83}$$

Calculating the Fourier transform of both sides of Eq. (5.83), the following result is obtained:

$$H(\omega) = F(\omega)G(\omega) \tag{5.84}$$

where $H(\omega)$ is the Fourier transform of $h(t)$. Thus, it is verified that the Fourier transform of the convolution of two arbitrary functions is equal to the product of the Fourier transforms of these functions. This statement corresponds to the so-called convolution theorem, which will be used in Fraunhofer's diffraction theory developed in Chapter 11, namely to describe diffraction by an array of similar apertures.

The full description of a spectrum should include both the amplitude and phase of all components. However, in many cases it is necessary to consider only the intensity, which is proportional to the square of the amplitude and does not contain any information about the phase. The intensity distribution can also be expressed through a spectrum, commonly known as the *power spectrum*. For a given harmonic component $F(\omega)$, which is usually a complex quantity, the intensity is obtained by multiplying $F(\omega)$ by its complex conjugate, $F^*(\omega)$.

The convolution of a function $f(t)$ with its conjugate $f^*(t)$ is known as *autocorrelation*. Adapting the result given by Eq. (5.84), it can be verified that the Fourier transform of the autocorrelation of the $f(t)$ function gives the power spectrum $F(\omega)F^*(\omega)$.

5.11 Bessel Functions

Bessel functions find varied applications in physics and engineering, in particular in the domains of wave propagation, elasticity, and fluid motion. In this book, Bessel functions are used in Chapter 11 when describing Fraunhofer diffraction by a circular aperture and in Chapter 15 in the context of electromagnetic wave propagation in optical fibres.

5.11.1 Bessel functions of the first kind

The Bessel function of the first kind of order p is defined in the form:

$$J_p(x) = \frac{x^p}{2^p p!} \sum_{n=0}^{\infty} (-1)^n \frac{x^{2n}}{2^{2n} n!\,(p+1)\cdots(p+n)} = \sum_{n=0}^{\infty} (-1)^n \frac{(x/2)^{2n+p}}{n!\,(p+n)!} \tag{5.85}$$

The most commonly used Bessel functions of the first kind are those of order 0 and 1, given by

$$\begin{aligned} J_0(x) &= \sum_{n=0}^{\infty} (-1)^n \frac{1}{(n!)^2} \left(\frac{x}{2}\right)^{2n} \\ &= 1 - \frac{x^2}{2^2} + \frac{x^4}{2^2 \cdot 4^2} - \frac{x^6}{2^2 \cdot 4^2 \cdot 6^2} + \cdots \end{aligned} \tag{5.86}$$

and

$$J_1(x) = \sum_{n=0}^{\infty} (-1)^n \frac{1}{n!\,(n+1)!} \left(\frac{x}{2}\right)^{2n+1}$$

$$\frac{x}{2} - \frac{1}{1!2!}\left(\frac{x}{2}\right)^3 + \frac{1}{2!3!}\left(\frac{x}{2}\right)^5 - \cdots \tag{5.87}$$

The graphs of these functions are shown in Fig. 5.10. These graphs illustrate some interesting properties of the functions $J_0(x)$ and $J_1(x)$: both exhibit damped oscillatory behaviour, with an infinite number of zeros; these zeros occur alternately in a similar way to the $\cos(x)$ and $\sin(x)$ functions. Such similarity is reinforced by the following relation between the two functions:

$$J_0'(x) = -J_1(x) \tag{5.88}$$

We can define a function $J_{-p}(x)$, with $p > 0$, by substituting p by $-p$ in Eq. (5.85). If p is an integer, it can be shown that

$$J_{-p}(x) = (-1)^p J_p(x) \tag{5.89}$$

5.11.2 Bessel functions of the second kind

The second-kind Bessel function of order p is defined in the form

$$Y_p(x) = \frac{J_p(x)\cos(p\pi) - J_{-p}(x)}{\sin(p\pi)} \tag{5.90}$$

where p is assumed not to be an integer. In the case of p being an integer m, this function is defined in the form

$$Y_m(x) = \lim_{p \to m} Y_p(x) \tag{5.91}$$

The graph of $Y_0(x)$ is represented by the dashed curve in Fig. 5.4. This graph illustrates an important feature of the function $Y_p(x)$: for any value $p \geq 0$, it is not limited near the origin. This behaviour contrasts with that of the function $J_p(x)$, which remains limited in this region.

Figure 5.10 *Graphs of the Bessel functions $J_0(x)$, $J_1(x)$ and $Y_0(x)$.*

5.11.3 Recurrence formulas

It can be shown that the Bessel functions of the first kind satisfy the following relations:

$$\frac{d}{dx}\left[x^{p}J_{p}(x)\right]=x^{p}J_{p-1}(x) \tag{5.92}$$

$$\frac{d}{dx}\left[x^{-p}J_{p}(x)\right]=-x^{-p}J_{p+1}(x) \tag{5.93}$$

Calculating the derivatives in the previous relations and dividing by $e^{\pm p}$, we obtain:

$$\frac{d}{dx}J_p(x)+\frac{p}{x}J_p(x)=J_{p-1}(x) \tag{5.94}$$

$$\frac{d}{dx}J_p(x)-\frac{p}{x}J_p(x)=J_{p+1}(x) \tag{5.95}$$

Adding and subtracting Eqs. (5.94) and (5.95), the following relations are obtained:

$$2\frac{d}{dx}J_p(x)=J_{p-1}(x)-J_{p+1}(x) \tag{5.96}$$

$$\frac{2p}{x}J_p(x)=J_{p-1}(x)+J_{p+1}(x) \tag{5.97}$$

These results allow us to express the Bessel functions, as well as their derivatives, in terms of other Bessel functions. The Bessel functions of the second kind, $Y_p(x)$, also satisfy the relations mentioned earlier; it is sufficient to replace $J_p(x)$ by $Y_p(x)$.

5.12 Further Reading

Champeney, D. C., *Fourier Transforms and Their Applications*, Academic Press, 1973.
Goodman, J. W., *Introduction to Fourier Optics*, McGraw-Hill, 1968.
Hecht, E., *Optics*, 5th Ed., Pearson Education, 2016.
Pain, H. J., *The Physics of Vibrations and Waves*, 6th Ed., John Wiley & Sons, 2005.
Steward, E. G., *Fourier Optics: An Introduction*, John Wiley & Sons, 1987.
Webb, R. H., *Elementary Wave Optics*, Academic Press, 1969; Dover, reissued 2005.

5.13 Problems

5.1 (a) Show that the expression $y(x,t)=A\exp\left\{-(5t+2x)^2\right\}$ represents a progressive wave and determines its speed.

(b) Verify that this expression is a solution of the wave equation.

5.2 The index of refraction of a certain glass can be approximated by the relation $n(\lambda)=1.5255+\frac{4825}{\lambda^2}\text{nm}^2$. Find the index of refraction, the phase velocity, and the group velocity in this glass for a pulse with frequency components centred around 500 nm.

5.3 Write the function that describes the propagation of a plane wave in Cartesian coordinates in terms of co-sine directors $\cos\theta_x$, $\cos\theta_y$, $\cos\theta_z$, such that

$$k_x = k\cos\theta_x, \quad k_y = k\cos\theta_y, \quad k_z = k\cos\theta_z$$

where k is the modulus of the wave vector. Show that this function is a solution to the three-dimensional wave equation.

5.4 Consider the so-called *Ronchi grating*, consisting of strips of width a, alternately transparent and opaque. Assuming that this grating is illuminated normally by plane waves with amplitude A_0, obtain the Fourier series representing the transmitted field.

5.5 The wave equation describing a damped vibratory motion is given by:

$$\frac{\partial^2 y(x,t)}{\partial t^2} = v^2 \frac{\partial^2 y(x,t)}{\partial x^2} - k\frac{\partial y(x,t)}{\partial t}$$

Assuming the solutions are of the form:

$$y(x,t) = \sum_{n=1}^{\infty} b_n(t) \sin\frac{n\pi x}{L}$$

and considering the initial conditions:

$$y(x,0) = f(x), \quad y'(x,0) = g(x)$$

determine the coefficients $b_n(t)$ for (a) the weak damping regime and (b) the strong damping regime.

5.6 Show that the Fourier transform of the Gaussian function

$$f(x) = \exp\left(-ax^2\right)$$

is another Gaussian function. Use, for this purpose, the result $\int_{-\infty}^{\infty} \exp\left(-\beta^2\right) d\beta = \sqrt{\pi}$. We observe that the Gaussian function is often used in optics. It can represent, for example, pulses propagating along optical fibres or the irradiance profiles of a laser beam.

5.7 Determine the Fourier transform of the wave train given by

$$E(x) = \begin{cases} \cos^2 kx, & -L \leq x \leq L \\ 0, & |x| > L \end{cases}$$

Represent the transformed function when $L \to \infty$.

5.8 (a) Calculate the Fourier transform of the triangular pulse function

$$f(x) = \begin{cases} a(1 - a|x|), & |x| < 1/a \\ 0, & |x| > 1/a \end{cases}$$

(b) Show that this function gives rise to a sequence δ_a of Dirac deltas. Write δ_a as a Fourier integral and calculate the transform of $\delta(x)$ assuming $a \to \infty$.

5.9 In a resonant cavity, an electromagnetic oscillation of frequency ω_0 is attenuated according to the expression

$$A(t) = A_0 \exp\left(-\frac{\omega_0 t}{2Q}\right) \exp(-i\omega_0 t), \quad t > 0$$

with $A(t) = 0$ for $t < 0$. The parameter Q measures the ratio between the stored energy and the energy lost in a cycle. Calculate the frequency distribution of the oscillation, $a(\omega)a^*(\omega)$, where $a(\omega)$ is the Fourier transform of $A(t)$. Check that the higher the Q, the narrower the resonance line.

5.10 Perform the convolution of the function $f(x) = \delta(x-5) + \delta(x+2)$ with an arbitrary function $g(x)$.

Electromagnetic Optics

6

Light is an electromagnetic phenomenon, consisting of electric and magnetic fields that are solutions of Maxwell's equations. These equations provide the mathematical foundation used to model and evaluate the flow of electromagnetic energy in all situations. The purpose of this chapter is to review the fundamental concepts and properties related to light propagation in linear dielectric media. From Maxwell's equations we will derive the linear wave equation and discuss the main properties of electromagnetic waves. Moreover, the harmonic oscillator model will be used to describe the interaction between an optical wave and matter. Using such a model, the susceptibility, the refractive index, and the attenuation of an optical material are discussed. Additional information concerning the subject of this chapter can be found in many textbooks, namely those indicated in Section 6.10.

6.1 Maxwell's Equations

The mathematical foundation for the description of electromagnetic wave propagation in a dielectric medium is provided by *Maxwell's equations*. These equations are named after James Maxwell (1831–1879) and can be written as follows:

$$\nabla \cdot \mathbf{D} = \rho \tag{6.1}$$

$$\nabla \cdot \mathbf{B} = 0 \tag{6.2}$$

$$\nabla \times \mathbf{E} = -\frac{\partial}{\partial t}\mathbf{B} \tag{6.3}$$

$$\nabla \times \mathbf{H} = \mathbf{J} + \frac{\partial}{\partial t}\mathbf{D} \tag{6.4}$$

where $\mathbf{E}$ and $\mathbf{H}$ are the electric and magnetic field vectors, respectively, and $\mathbf{D}$ and $\mathbf{B}$ are the corresponding electric and magnetic flux densities, $\mathbf{J}$ is the current density vector, and ρ is the charge density. The electric flux density and the electric field are related in the form:

$$\mathbf{D} = \varepsilon\mathbf{E} = \varepsilon_0\mathbf{E} + \mathbf{P} \tag{6.5}$$

where ε is the permittivity of the medium, ε_0 is the vacuum permittivity, and $\mathbf{P}$ is the induced electric polarization. On the other hand, the relation between the magnetic flux density and the magnetic field is given by

$$\mathbf{B} = \mu\mathbf{H} = \mu_0\mathbf{H} + \mathbf{M} \tag{6.6}$$

where μ is the permeability of the medium, μ_0 is the vacuum permeability, and $\mathbf{M}$ is the induced magnetic polarization. In the following, we consider the case of a nonmagnetic

Lasers, Optics, and Photonics. Mário F. S. Ferreira, Oxford University Press. © Mário F. S. Ferreira (2026).
DOI: 10.1093/oso/9780198952770.003.0006

material, in which $\mathbf{M} = 0$. The constants μ_0 and ε_0 have the following values:

$$\mu_0 = 4\pi \times 10^{-7}\mathrm{Hm}^{-1} \tag{6.7}$$

$$\varepsilon_0 = \frac{10^{-9}}{36\pi}\mathrm{Fm}^{-1} \tag{6.8}$$

Eqs. (6.1) and (6.2) correspond to Gauss' law for the electric field and Gauss's law for the magnetic field, respectively, while Eq. (6.3) is Faraday's law of induction and Eq. (6.4) is Ampere's circuital law.

The electric and magnetic fields can be considered as two aspects of a sole physical phenomenon: the electromagnetic field. In the following, we analyse the main characteristics of such a field. We confine our analysis to isotropic, homogeneous, and sourceless materials, so that $\mathbf{J} = 0$ and $\rho = 0$.

6.2 Wave Equation for the Electromagnetic Field

The curl of Eq. (6.3) gives:

$$\nabla \times (\nabla \times \mathbf{E}) = \nabla \times \left(-\frac{\partial}{\partial t}\mathbf{B}\right) = -\frac{\partial}{\partial t}(\nabla \times \mathbf{B}) \tag{6.9}$$

The left-hand side of Eq. (6.9) can be simplified using the following vector identity:

$$\nabla \times (\nabla \times \mathbf{E}) = \nabla(\nabla \cdot \mathbf{E}) - \nabla \cdot \nabla\mathbf{E} \tag{6.10}$$

When $\rho = 0$ we have from Eqs. (6.1) and (6.5) that $\nabla \cdot \mathbf{E} = 0$. In such a case, Eq. (6.10) gives the result:

$$\nabla \times (\nabla \times \mathbf{E}) = -\nabla \cdot \nabla\mathbf{E} = -\nabla^2\mathbf{E} \tag{6.11}$$

and Eq. (6.9) becomes:

$$-\nabla^2\mathbf{E} = -\frac{\partial}{\partial t}(\nabla \times \mathbf{B}) \tag{6.12}$$

Using Eqs. (6.4)–(6.6), Eq. (6.12) can be written in the form:

$$\nabla^2\mathbf{E} - \mu\varepsilon\frac{\partial^2}{\partial t^2}\mathbf{E} = 0 \tag{6.13}$$

A similar procedure can be used to obtain an equation for **B**:

$$\nabla^2\mathbf{B} - \mu\varepsilon\frac{\partial^2}{\partial t^2}\mathbf{B} = 0 \tag{6.14}$$

Eqs. (6.13) and (6.14) are wave equations, with the wave velocity v given by

$$v = \frac{1}{\sqrt{\mu\varepsilon}} \tag{6.15}$$

The connection of the velocity of light with the electric and magnetic properties of a material was one of the most important results of Maxwell's theory. Considering Eqs. (6.7) and (6.8), the following value for the velocity of light in the vacuum is obtained:

$$v = c = 2.997924562 \times 10^8 \text{ m/s} \tag{6.16}$$

In a material, the velocity of light is less than c. The *index of refraction*, n, of the material is defined as the ratio of the speed of light in the vacuum, c, to its speed in the material, v:

$$n = \frac{c}{v} = \sqrt{\frac{\varepsilon\mu}{\varepsilon_0\mu_0}} \tag{6.17}$$

The index of refraction given by Eq. (6.17) corresponds to the real part of the complex refractive index, which will be discussed in Section 6.8. In the case of nonmagnetic materials, one has $\mu \approx \mu_0$ and the index of refraction is determined solely by the permittivity of the medium ε, which depends on the frequency of the incident electromagnetic wave.

Example 6.1

Consider a plane electromagnetic wave that propagates in a transparent medium and whose electric field is given, in IS units, by

$$E_x(z,t) = E_{0x} \cos\left[\pi\left(\frac{z}{3 \times 10^{-7}} - 6 \times 10^{14} t\right)\right]$$

and $E_y = E_z = 0$. Determine the wave propagation speed, the wavelength in the vacuum, and the refractive index of the medium at the frequency considered.

Solution

The phase of the wave is

$$\phi = kz - \omega t = \frac{\pi}{3 \times 10^{-7}} z - 6\pi \times 10^{14} t$$

In this way, one has

$$\omega = 6\pi \times 10^{14} \text{ Hz}, \qquad k = \frac{2\pi}{\lambda} = \frac{\pi}{3 \times 10^{-7}} \text{ m}^{-1}$$

The propagation speed and the wavelength in the medium are given, respectively, by

$$v = \frac{\omega}{k} = \left(6\pi \times 10^{14}\right) \frac{3 \times 10^{-7}}{\pi} = 1.8 \times 10^8 \text{ m/ms s}$$

and

$$\lambda = \frac{2\pi}{k} = (2\pi) \frac{3 \times 10^{-7}}{\pi} = 600 \text{ nm}$$

The refractive index of the medium is

$$n = \frac{c}{v} = \frac{3 \times 10^8}{1.8 \times 10^8} = 1.67$$

The wavelength in the vacuum is

$$\lambda_0 = n\lambda = 1000 \text{ nm}$$

6.2.1 Biographical note—James Maxwell

James Clerk Maxwell (1831–1879) was one of the most important scientists of the 19th century. He was a professor at the University of Aberdeen between 1856 and 1860 and later at the University of Cambridge, where he supervised the construction of the Cavendish laboratory. Maxwell has the merit of having demonstrated that light is made up of electromagnetic waves. In 1873 he published the work *Treatise on Electricity and Magnetism*, in which he presented, for the first time, the set of four differential equations that describe the nature of the electromagnetic field. Several aspects of Maxwell's theory were later experimentally verified by the German physicist Heinrich Rudolf Hertz. In addition to electromagnetism, Maxwell also developed the kinetic theory of gases and the principles of thermodynamics, having published the works *Matter and Motion* (1876) and *Theory of Heat* (1877).

6.3 Electromagnetic Waves

Eqs. (6.13) and (6.14) have the following solutions in the form of harmonic plane waves:

$$E = Re\left\{\mathbf{E}_0 e^{i(\mathbf{k}\cdot\mathbf{r}-\omega t+\phi)}\right\} \tag{6.18}$$

$$B = Re\left\{\mathbf{B}_0 e^{i(\mathbf{k}\cdot\mathbf{r}-\omega t+\phi)}\right\} \tag{6.19}$$

where $\mathbf{E}_0$ and $\mathbf{B}_0$ are constant vectors associated with the amplitude of oscillations, ω is the angular frequency, $\mathbf{k}$ is the wave vector, and *Re* indicates the real part. Hereafter we will drop the '*Re*', but it will be understood that the physical fields will be given by the real part of the complex field appearing in the equations. Considering Eqs. (6.1) (with $\rho = 0$), Eqs. (6.5) and (6.18), one has

$$\nabla\cdot\mathbf{E} = i\mathbf{k}\cdot\mathbf{E} = 0 \tag{6.20}$$

In a similar way, using Eqs. (6.2) and (6.19), one obtains:

$$\nabla\cdot\mathbf{B} = i\mathbf{k}\cdot\mathbf{B} = 0 \tag{6.21}$$

Eqs. (6.20) and (6.21) show that **E** and **B** must both be perpendicular to the direction of propagation, which is given by vector **k**.

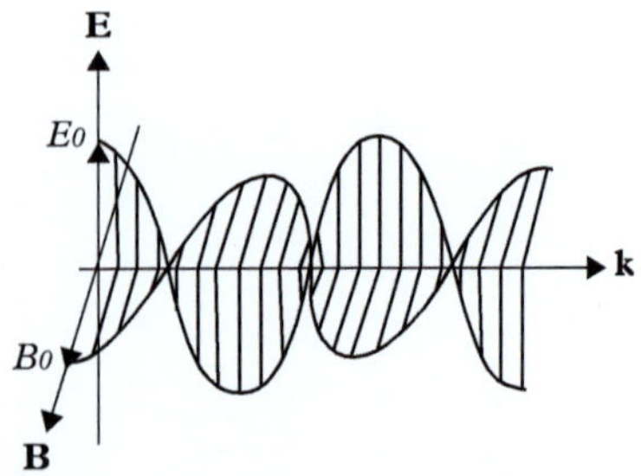

Figure 6.1 *Schematic representation of a propagating plane electromagnetic wave.*

Assuming that the electric and magnetic fields are given by Eqs. (6.18) and (6.19), Eq. (6.3) becomes:

$$i\mathbf{k}\times\mathbf{E} = i\omega\mathbf{B} \tag{6.22}$$

or

$$\mathbf{B} = \frac{1}{\omega}(\mathbf{k}\times\mathbf{E}) = \frac{1}{vk}(\mathbf{k}\times\mathbf{E}) \tag{6.23}$$

Thus, one can write the magnetic flux density in the form

$$\mathbf{B} = \frac{1}{v}(\hat{\mathbf{s}}\times\mathbf{E}) \tag{6.24}$$

where $\hat{\mathbf{s}} = \mathbf{k}/k$ is the unit vector in the propagation direction. Eq. (6.24) contains three important aspects concerning the electromagnetic waves:

i) **B** is perpendicular to **E**,
ii) **B** is in phase with **E**,
iii) the magnitudes of **B** and **E** are related as $B = E/v$.

Fig. 6.1 represents the propagation of a plane electromagnetic wave in a direction indicated by the wave vector **k**.

Example 6.2

Let us consider a plane and harmonic electromagnetic wave, which has an electric field given by

$$E_x(z,t) = E_{0x} \sin\left[\omega\left(\frac{z}{c} - t\right) + \varphi\right]$$

Obtain an expression for the magnetic flux density, **B**, of this wave.

Solution

Since $E_y = E_z = 0$, Eq. (6.3) gives

$$\frac{\partial E_x}{\partial z} = -\frac{\partial B_y}{\partial t}$$

That is, using the expression of the electric field,

$$\frac{\partial B_y}{\partial t} = -\frac{\omega}{c} E_{0x} \cos\left[\omega\left(\frac{z}{c} - t\right) + \varphi\right]$$

Integrating both members of the previous equation, one obtains

$$B_y(z,t) = \frac{1}{c} E_{0x} \sin\left[\omega\left(\frac{z}{c} - t\right) + \varphi\right] = \frac{1}{c} E_x(z,t)$$

It can be verified that the electric and magnetic fields are perpendicular to each other and that their magnitudes are related in the form $B = E/c$.

6.4 Energy Density and Flux

Electromagnetic theory allows one to obtain the following expression for the energy density associated with the electrical and magnetic fields:

$$U = \frac{1}{2}[\mathbf{D} \cdot \mathbf{E} + \mathbf{B} \cdot \mathbf{H}] \tag{6.25}$$

Using the constitutive relations given by Eqs. (6.5) and (6.6), one obtains

$$U = \frac{1}{2}\left[\varepsilon|\mathbf{E}^2| + \frac{|\mathbf{B}|^2}{\mu}\right] = \frac{1}{2}\left[\varepsilon + \frac{1}{\mu v^2}\right]|\mathbf{E}|^2 = \varepsilon|\mathbf{E}|^2 \tag{6.26}$$

In free space, one has

$$U = \varepsilon_0 |\mathbf{E}|^2 = \frac{|\mathbf{B}|^2}{\mu_0} \tag{6.27}$$

The presence of both an electric and magnetic field at the same point in space results in a flow of the field energy. The energy flux density is described by the *Poynting vector*, **S**, defined as

$$\mathbf{S} = \mathbf{E} \times \mathbf{H} = \frac{1}{\mu}\mathbf{E} \times \mathbf{B} \tag{6.28}$$

The energy flux density in a given direction, indicated by the unit vector $\hat{\mathbf{u}}$, is given by the scalar product $\hat{\mathbf{u}} \cdot \mathbf{S}$.

We will consider the case of a plane wave to determine some of the properties of the Poynting vector. Since **S** will involve terms quadratic in **E**, it will be necessary to use the real form of **E**:

$$\mathbf{E} = \mathbf{E}_0 \cos\phi, \quad \phi = \mathbf{k} \cdot \mathbf{r} - \omega t + \varphi \tag{6.29}$$

Also, from Eq. (6.23),

$$\mathbf{B} = \mathbf{B}_0 \cos\phi = \frac{1}{vk}(\mathbf{k} \times \mathbf{E}_0) \cos\phi \tag{6.30}$$

The Poynting vector becomes:

$$\mathbf{S} = \frac{1}{\mu}\mathbf{E}_0 \times \frac{1}{vk}(\mathbf{k} \times \mathbf{E}_0) \cos^2\phi = \frac{1}{\mu v}|\mathbf{E}_0|^2 \hat{\mathbf{s}} \cos^2\phi \tag{6.31}$$

Since the frequencies associated with light are very high (10^{14} to 10^{15} Hz), we normally do not detect the magnitude of **S** but rather its temporal average over a time T determined by the response time of the detector used. Considering that the time average of $\cos^2\phi$ over many periods is 1/2, the time-averaged value of the magnitude of the Poynting vector is given by

$$I \equiv \langle |\mathbf{S}| \rangle = \frac{1}{2\mu v}|\mathbf{E}_0|^2 \tag{6.32}$$

The quantity $I = \langle |\mathbf{S}| \rangle$ is the radiant *flux density* (units of $\mathrm{W/m^2}$) and corresponds to the *emittance* or *irradiance* defined in Chapter 4, depending on whether that flux moves away from or focuses respectively on a given surface. Unfortunately, as defined in Chapter 4, the standard symbol for irradiance, except for the subscript, is the same as that for the electric field. In the following, to avoid confusion, we will use the symbol I, rather than E_e, to denote the irradiance.

The energy density is given from Eq. (6.26) by

$$U = \varepsilon |\mathbf{E}_0|^2 \cos^2\phi \tag{6.33}$$

with a time average

$$\langle U \rangle = \frac{\varepsilon}{2} |\mathbf{E}_0|^2 \tag{6.34}$$

We may use the definition of the wave velocity, given by Eq. (6.15), to relate the irradiance I with the averaged energy density $\langle U \rangle$ in the form:

$$I = v \langle U \rangle \tag{6.35}$$

This is a particular case of a more general result:

Energy flux density = (energy density) × (propagation speed)

Example 6.3

A collimated light beam has a diameter of 2 mm and carries a power of 600 mW. Determine the irradiance of the beam and the amplitude of the electric and magnetic fields.

Solution

The area of the beam section is $\pi(10^{-3})^2$, which allows one to obtain the following result for the irradiance:

$$I = \frac{600 \times 10^{-3}}{\pi(10^{-3})^2} = 1.91 \times 10^5 \, \mathrm{W/m^2}$$

From Eq. (6.32) we find that the amplitude of the electric field is

$$E_0 = \sqrt{\frac{2I}{\varepsilon_0 c}} = 1.20 \times 10^4 \, \mathrm{V/m}$$

The amplitude of the magnetic flux density is given by

$$B_0 = \frac{E_0}{c} = 4.00 \times 10^{-5} \, \mathrm{T}$$

6.5 Laws of Reflection and Refraction

To study the phenomena associated with the reflection and transmission of a plane wave at the interface between two dielectric media is useful for exploiting and understanding the behaviour of light in dielectric waveguides. Of interest are not only the relations among the angles of incidence, reflection and refraction but also the fractions of optical power that are reflected and transmitted at the boundaries, as well as the phase shifts that occur on reflection.

Let us consider a monochromatic plane wave incident on a boundary between two media with refractive indices n_1 e n_2 (Fig. 6.2). The incident wave is given by

$$\mathbf{E}_i = \mathbf{E}_{0i} \exp\{i(\mathbf{k}_i \cdot \mathbf{r} - \omega t)\} \tag{6.36}$$

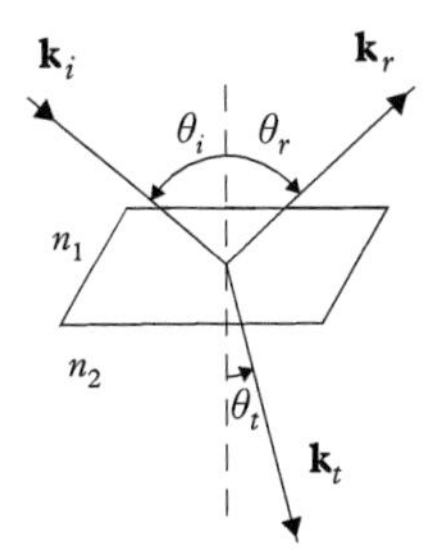

Figure 6.2 *Geometry illustrating Snell's laws.*

and can be decomposed into two waves with the same frequency: a reflected wave, $\mathbf{E}_r$, and a transmitted wave, $\mathbf{E}_t$, given by:

$$\mathbf{E}_r = \mathbf{E}_{0r} \exp\{i(\mathbf{k}_r \cdot \mathbf{r} - \omega t)\} \tag{6.37}$$

$$\mathbf{E}_t = \mathbf{E}_{0t} \exp\{i(\mathbf{k}_t \cdot \mathbf{r} - \omega t)\} \tag{6.38}$$

A relation among the three waves, valid for all points on the interface and for any instant of time, can be verified only if their phases are the same. This condition gives:

$$\mathbf{k}_i \cdot \mathbf{r} = \mathbf{k}_r \cdot \mathbf{r} = \mathbf{k}_t \cdot \mathbf{r} \tag{6.39}$$

From Eq. (6.39) we conclude that

$$\mathbf{k}_r - \mathbf{k}_i = b_1 \hat{\mathbf{N}} \tag{6.40}$$

$$\mathbf{k}_t - \mathbf{k}_i = b_2 \hat{\mathbf{N}} \tag{6.41}$$

where $\hat{\mathbf{N}}$ is a vector perpendicular to the interface, whereas b_1 and b_2 are scalars. Considering that $\mathbf{k}_i = k_0 n_1 \hat{\mathbf{s}}_i$, $\mathbf{k}_r = k_0 n_1 \hat{\mathbf{s}}_r$, and $\mathbf{k}_t = k_0 n_2 \hat{\mathbf{s}}_t$, the previous relations can be written in the form:

$$n_1 (\hat{\mathbf{s}}_r - \hat{\mathbf{s}}_i) = a_1 \hat{\mathbf{N}} \tag{6.42}$$

$$n_2 \hat{\mathbf{s}}_t - n_1 \hat{\mathbf{s}}_i = a_2 \hat{\mathbf{N}} \tag{6.43}$$

where a_1 and a_2 are scalars. The projection of these relations on the interface gives:

$$\theta_i = \theta_r \tag{6.44}$$

$$n_1 \sin\theta_i = n_2 \sin\theta_t \tag{6.45}$$

Eqs. (6.44) and (6.45) are known as Snell's laws of reflection and refraction, respectively, as seen in Chapter 2 using the Fermat minimum time principle.

6.6 Fresnel Equations

A wave encountering a boundary between dielectric media with different refractive indices will not only be refracted but will also be partly reflected. The ratios of the amplitudes of the reflected and transmitted waves to that of the incident wave are known as the *reflection* and *transmission coefficients*, r and t. For an electromagnetic wave, the *Fresnel equations* express the way in which these coefficients depend on the angles of incidence and refraction, and on the polarization of the wave, given by the direction of the electric field.

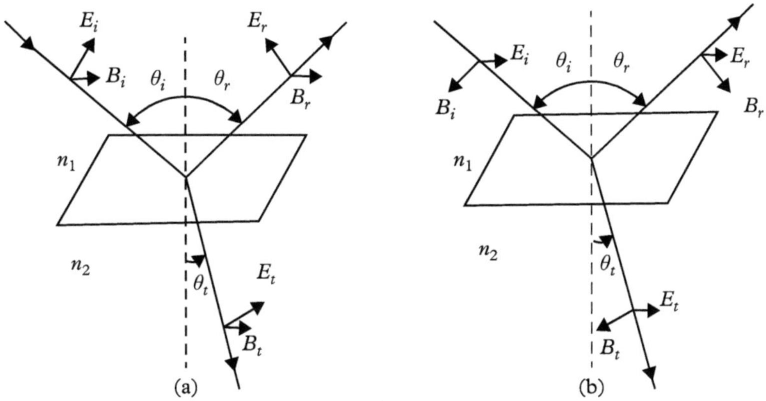

Figure 6.3 *Illustration of the incident, reflected, and refracted rays for two cases of the plane of polarization. The directions of the vector fields are shown for the E-vector: (a) in the plane of incidence and (b) perpendicular to it.*

In Fig. 6.3, the incident, reflected and refracted rays are shown for two cases of plane polarization, when the **E**–vector is (a) in the plane of incidence (TM polarization) and (b) perpendicular to the plane of incidence (TE polarization). The vectors **E** and **B** are related to the wave vector as given in Eq. (6.23).

The components of the **E**, **H**, and **B** fields parallel to the boundary between the two dielectric media must be the same on both sides of the boundary. Considering the case of Fig. 6.3(a), these boundary conditions give

$$B_i + B_r = B_t \tag{6.46}$$

$$E_i \cos\theta_i - E_r \cos\theta_r = E_t \cos\theta_t \tag{6.47}$$

Substituting in Eq. (6.46) $B = (n/c)E$ and combining with Eq. (6.47), one obtains the following results for the reflection and transmission coefficients:

$$r_{\|} = \left(\frac{E_r}{E_i}\right)_{\|} = \frac{n_2 \cos\theta_i - n_1 \cos\theta_t}{n_2 \cos\theta_i + n_1 \cos\theta_t} \tag{6.48}$$

$$t_{\|} = \left(\frac{E_t}{E_i}\right)_{\|} = \frac{2n_1 \cos\theta_i}{n_2 \cos\theta_i + n_1 \cos\theta_t} \tag{6.49}$$

where the subscript || refers to the case in which the **E** field is parallel to the plane of incidence. Using Snell's law of refraction, we can rewrite Eqs. (6.48) and (6.49) in terms of the angles only:

$$r_{\|} = \frac{\tan(\theta_i - \theta_t)}{\tan(\theta_i + \theta_t)} \tag{6.50}$$

$$t_{\|} = \frac{2\sin\theta_t \cos\theta_i}{\sin(\theta_i + \theta_t)\cos(\theta_i - \theta_t)} \tag{6.51}$$

A similar analysis for the case of Fig. 6.3(b), in which the electric field vector is perpendicular to the plane of incidence, gives:

$$r_\perp = \left(\frac{E_r}{E_i}\right)_\perp = \frac{n_1 \cos\theta_i - n_2 \cos\theta_t}{n_1 \cos\theta_i + n_2 \cos\theta_t} \tag{6.52}$$

$$t_\perp = \left(\frac{E_t}{E_i}\right)_\perp = \frac{2n_1 \cos\theta_i}{n_1 \cos\theta_i + n_2 \cos\theta_t} \tag{6.53}$$

Using Snell's law of refraction, the previous relations can be written in the form

$$r_\perp = -\frac{\sin(\theta_i - \theta_t)}{\sin(\theta_i + \theta_t)} \tag{6.54}$$

$$t_\perp = \frac{2\sin\theta_t \cos\theta_i}{\sin(\theta_i + \theta_t)} \tag{6.55}$$

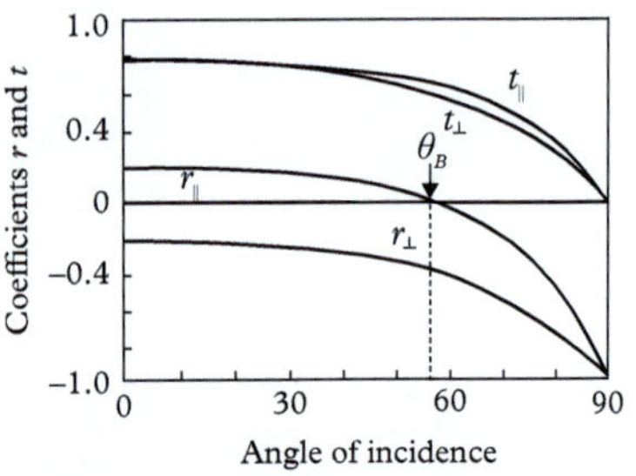

Figure 6.4 *Variation of the amplitude reflection and transmission coefficients with the angle of incidence for a boundary between air ($n_1 \approx 1$) and glass ($n_2 = 1.5$).*

Fig. 6.4 illustrates the variation of the reflection and transmission coefficients with the angle of incidence for a boundary between air ($n_1 \approx 1$) and glass ($n_2 = 1.5$).

It can be verified from Eq. (6.50) that $r_\|$ goes through zero when $\theta_i + \theta_t = \pi/2$. The angle of incidence $\theta_i = \theta_B$ corresponding to this situation is known as the *Brewster angle*. Light reflected at the Brewster angle becomes completely polarized, with the electric vector normal to the plane of incidence. On the other hand, a glass plate at the Brewster angle is completely transparent to light with the electric vector parallel to the plane of incidence.

The *reflectance R*, which is defined as the ratio of the reflected power to the incident power, is given by

$$R = \frac{P_r}{P_i} = r^2 \tag{6.56}$$

On the other hand, the *transmittance T* is defined as the ratio of the transmitted power to the incident power

$$T = \frac{P_t}{P_i} = \left(\frac{n_2 \cos\theta_t}{n_1 \cos\theta_i}\right) t^2 \tag{6.57}$$

Example 6.4

Derive expressions for the reflection and transmission coefficients in the case of normal incidence and calculate their values for an air–glass interface, in which one has $n_1 = 1$ and $n_2 = 1.5$

Solution

For normal incidence, one has $\theta_i \approx \theta_t \approx 0$ and $\cos\theta_i \approx \cos\theta_t \approx 1$. In this case, the reflection and transmission coefficients are independent of polarization and become

$$r \equiv r_\| = -r_\perp = \frac{n_2 - n_1}{n_1 + n_2} \tag{6.58}$$

continued

Example 6.4 continued

$$t = t_{\parallel} = t_{\perp} = \frac{2n_1}{n_1 + n_2} \tag{6.59}$$

It can be verified that the transmission coefficient is always positive, whereas the sign of the reflection coefficient depends on the relative magnitudes of the refractive indices n_1 and n_2.

The reflectance R and transmittance T are obtained from Eqs. (6.56)–(6.59):

$$R = r^2 = \left(\frac{n_2 - n_1}{n_1 + n_2}\right)^2 \tag{6.60}$$

$$T = \frac{n_2}{n_1} t^2 = \frac{4n_1 n_2}{(n_1 + n_2)^2} \tag{6.61}$$

We can verify from the aforementioned results that $R + T = 1$.

In the case of the interface between air ($n_1 = 1$) and glass ($n_2 = 1.5$), one has

$$r = \frac{1.5 - 1}{1.5 + 1} = 0.2 \tag{6.62}$$

$$t = \frac{2}{1.5 + 1} = 0.8 \tag{6.63}$$

The reflectance loss of 4% at normal incidence for a typical air/glass surface becomes a serious problem in the multicomponent lenses of several optical instruments. These losses can, however, be significantly reduced by coating the surface with a transparent layer of adequate refractive index and thickness.

6.6.1 Total internal reflection

Referring to Fig. 6.2 and noting that the geometry is the same if the ray direction is reversed, let us consider what happens if a ray inside the refractive medium meets the surface at a large angle of incidence θ_t, such that $\sin\theta_t > n_1/n_2$. In such a case, Eq. (6.45) would give $\sin\theta_i > 1$, which means that there can be no ray above the surface and that all the incident light is reflected. This phenomenon is known as *total internal reflection*. This phenomenon was already discussed in Chapter 2 and constitutes the basic mechanism that explains the propagation of light along planar waveguides and optical fibres.

Considering the geometry of Fig. 2.6, in which the refractive index of the incidence medium (n_1) is higher than that of the transmitting medium (n_2), the critical angle for total internal reflection is given by Eq. (2.17). For an angle of incidence greater than the critical angle, although there is no transmitted wave, a part of the electromagnetic field propagates in the lower refractive index medium as an *evanescent wave*. This field decays exponentially as a function of distance z into the lower-index medium, as represented in Fig. 6.5. It can be shown (Problem 6.15) that the variation of the field with z in this medium is given by

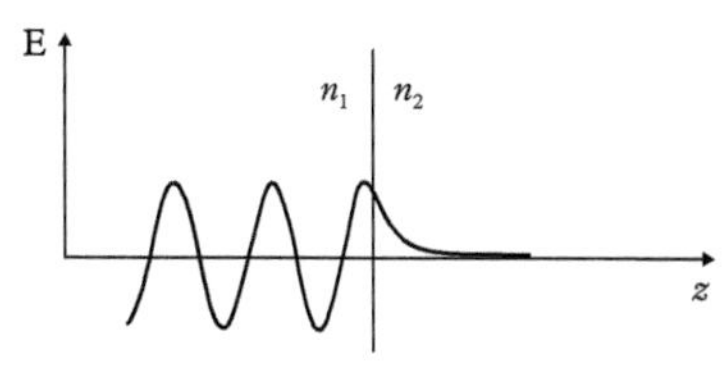

Figure 6.5 *The E field changes from oscillatory behaviour in the higher-index medium to exponentially decaying behaviour in the lower-index medium during total internal reflection.*

$$E(z) = E_0 e^{-\alpha z} \tag{6.64}$$

where

$$\alpha = k_0\sqrt{n_1^2 \sin^2\theta_i - n_2^2} \tag{6.65}$$

and $k_0 = 2\pi/\lambda$.

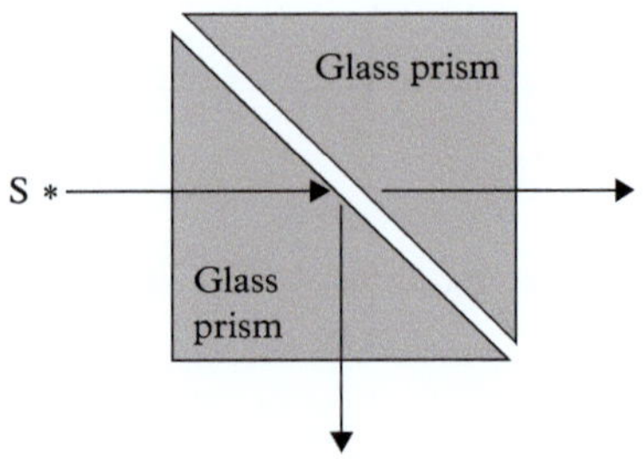

Figure 6.6 *A method for demonstrating the penetration of light into the lower refractive index medium during the phenomenon of total internal reflection.*

That the wave really penetrates into the lower refractive index medium can be demonstrated experimentally using two 45°–90°–45° prisms as represented in Fig. 6.6. These prisms are placed with their long faces close together, but not in contact. Light from a source S is found to be partially transmitted from the first prism to the second one, the amount being dependent on the separation between their long faces.

In the absence of any perturbation in the lower-index medium, the evanescent wave must return to the higher-index medium because there is total reflection of light energy. However, in the case of a narrow incident light beam, there is an apparent lateral displacement of the reflected beam, generally very small, which is known as the *Goos–Haenchen Shift*.

In the case of total internal reflection, it is found that the phase of the reflected wave is also shifted with respect to that of the incident wave. If the electric field vector is perpendicular to the plane of incidence (TE polarization), it can be shown (Problem 6.16) that this phase shift is given by

$$\phi_r = 2\tan^{-1}\left[\left(\frac{n_1}{n_2}\right)^2 \frac{\sqrt{\sin^2\theta_i - (n_2/n_1)^2}}{\cos\theta_i}\right] \tag{6.66}$$

The phase shift goes to zero at the critical angle and goes to π as $\theta_i \to 90°$. This expression will be useful in the analysis of modes in planar waveguides in Chapter 15.

6.7 The Harmonic Oscillator Model

When an optical wave propagates in a given medium, the relatively heavy nuclei of the atoms of the medium cannot follow its rather high frequency, and only the elastically bound valence electrons are excited. The electrons oscillate with the frequency of the optical field around their equilibrium state and, as a consequence, will emit waves of their own. Fig. 6.7 illustrates this oscillator; the light valence electron is represented by a mass m, whereas the forces between the atom and the valence electron are represented by a spring of constant k.

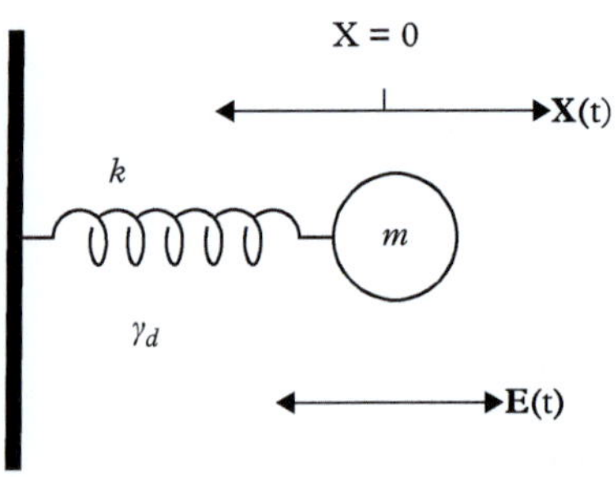

Figure 6.7 *Schematic illustration of a harmonic oscillator.*

The equation of motion of this oscillator can be written in the form

$$m\frac{d^2\mathbf{X}(t)}{dt^2} + \gamma_d\frac{d\mathbf{X}(t)}{dt} + k\mathbf{X}(t) = q\mathbf{E}(t) \tag{6.67}$$

where $\mathbf{X}$ is the displacement coordinate, q is the charge of the electron, γ_d is a damping coefficient, and $\mathbf{E}(t) = \mathbf{E}_0 e^{-i\omega t}$ is the external electric field. Since the displacement coordinate $\mathbf{X}$ will change periodically with the frequency ω of the electric field, it can be written as $\mathbf{X}(t) = \mathbf{X}_0 e^{-i\omega t}$. Substituting $\mathbf{X}(t)$ and $\mathbf{E}(t)$ in Eq. (6.67) and introducing the resonance frequency of the oscillator $\omega_0 = \sqrt{k/m}$, one obtains the result

$$\mathbf{X}_0 = \frac{q\mathbf{E}_0}{m}\frac{1}{(\omega_0^2 - \omega^2) - i\omega\gamma_m} \tag{6.68}$$

where $\gamma_m = \gamma_d/m$

The dipole moment of the oscillator is given by

$$\mathbf{p}(t) = q\mathbf{X}(t) = q\mathbf{X}_0 e^{-i\omega t} \tag{6.69}$$

Each atom or oscillator of the medium can oscillate in different modes and therefore has different resonance frequencies (ω_j) and damping coefficients (γ_{mj}). If N is the number of oscillators per unit volume and f_j is the number of possible oscillating states, each with resonance frequency ω_j, the polarization of the medium is given by

$$\mathbf{P}(t) = \left[\frac{Nq^2}{m}\sum_j \frac{f_j}{\left(\omega_j^2-\omega^2\right)-i\omega\gamma_m}\right]\mathbf{E}(t) = \varepsilon_0\chi(\omega)\mathbf{E}(t) \tag{6.70}$$

where f_i is also known as the oscillator strength, and χ is the linear electric susceptibility of the material. Solving for χ gives the result

$$\chi = \omega_p^2 \sum_j f_j \frac{\left(\omega_j^2-\omega^2\right)+i\omega\gamma_m}{\left(\omega_j^2-\omega^2\right)^2+\omega^2\gamma_m^2} \equiv \chi' + i\chi'' \tag{6.71}$$

where $\omega_p = \sqrt{Nq^2/m\varepsilon_0}$ is known as the plasma frequency. Fig. 6.8 illustrates the real and imaginary parts of the susceptibility for a material having two resonances of different strengths.

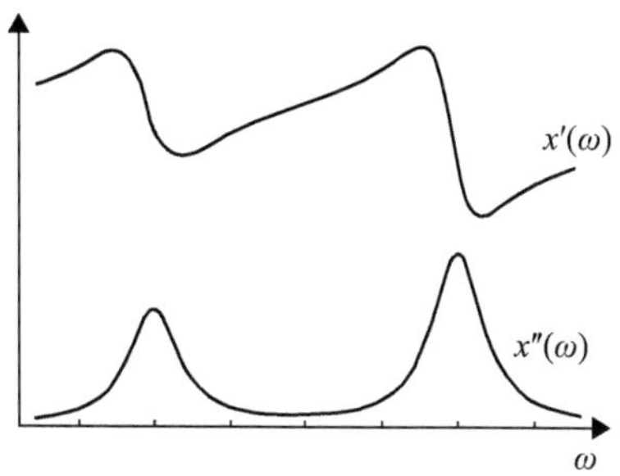

Figure 6.8 *Real and imaginary parts of the susceptibility of a material with two resonances.*

6.8 The Complex Refractive Index

The complex refractive index of the material is given by

$$\hat{n} = (1+\chi'+i\chi'')^{1/2} = n + in' \tag{6.72}$$

The imaginary part (n') of the complex refractive index is called the extinction coefficient. If $n' \ll n$, the real and imaginary parts of the complex refractive index can be approximated as

$$n \approx \sqrt{1+\chi'}, \qquad n' = \frac{\chi''}{2\sqrt{1+\chi'}} \tag{6.73}$$

The real part of the complex refractive index leads to a phase shift of the wave in the material, which depends on the frequency difference between the external field ω and the resonance frequency of the oscillator ω_j. The imaginary part is responsible for the absorption of the wave in the material, and it is particularly large near the resonance frequency, such that $\omega \approx \omega_j$. This explains why χ'' and χ' are sometimes called the absorptive and refractive index parts of the susceptibility, respectively.

Introducing the complex refractive index given in Eq. (6.72) into the solution of the wave equation corresponding to a plane wave that propagates in the positive z direction gives

$$\mathbf{E} = Re\left\{\mathbf{E}_0 e^{i(kz-\omega t)}\right\} = Re\left\{\mathbf{E}_0 e^{i((n+in')k_0 z-\omega t)}\right\} = Re\left\{\mathbf{E}_0 e^{-n'k_0 z} e^{i(nk_0 z-\omega t)}\right\} \tag{6.74}$$

The exponential function $\mathbf{E}_0 e^{-n' k_0 z}$ describes the decrease of the wave amplitude with distance. Since the irradiance is proportional to the square of the wave amplitude, it becomes

$$I = \frac{1}{2}\varepsilon_0 cn \left|\mathbf{E}_0 e^{-n' k_o z}\right|^2 = I_0 e^{-\alpha z} \tag{6.75}$$

where $I_0 = \varepsilon_0 cn|\mathbf{E}_0|^2/2$ and $\alpha = 2n' k_0$ is the so-called attenuation constant.

Eq. (6.74) shows that the real part of the refractive index (n) is indeed responsible for the alteration of the phase of the wave. The frequency dependence of the refractive index leads to different velocities for the spectral components of a propagating signal. Far from resonance, one has from Eq. (6.71) that $\chi'' \approx 0$ and

$$\chi' \approx \omega_p^2 \sum_j \frac{f_j}{\left(\omega_j^2 - \omega^2\right)} \tag{6.76}$$

where the attenuation of the oscillators was neglected ($\gamma_{mj} \approx 0$). Writing Eq. (6.76) in terms of wavelength, it becomes

$$\chi' \approx \frac{\omega_p^2}{(2\pi c)^2} \sum_j \frac{f_j \lambda_j^2 \lambda^2}{\left(\lambda^2 - \lambda_j^2\right)} = \sum_j \frac{A_j \lambda^2}{\lambda^2 - \lambda_j^2} \tag{6.77}$$

where A_j is the magnitude of the jth resonance, whereas λ and λ_j are the free-space wavelengths corresponding to ω and ω_j, respectively. From Eqs. (6.73) and (6.77), one can obtain the following result, known as the *Sellmeier formula* for the refractive index:

$$n^2 = 1 + \sum_j \frac{A_j \lambda^2}{\lambda^2 - \lambda_j^2} \tag{6.78}$$

Eq. (6.78), together with Fig. 6.8, shows that away from any resonance, or between resonances, the refractive index decreases with increasing wavelength. This behaviour is important for describing the material origins of group velocity dispersion.

6.9 The Limit of Geometrical Optics

Geometrical optics is an approximate description of the optical phenomena, which becomes acceptable in the limit as $\lambda \to 0$. In such a limit, Maxwell's equations provide an equation that describes the path taken by the normal to the wave surface of constant phase. This normal is the *optical ray* considered in the geometrical optics model, and the corresponding equation of motion is known as the *eikonal equation*.

Let us consider a linear, isotropic, sourceless but non-homogeneous medium, such that $\varepsilon = \varepsilon(r)$. Assuming that the electromagnetic fields vary with time with a frequency $f = \omega/2\pi$, Maxwell's equations can be written in the form:

$$\nabla \cdot (\varepsilon_r \mathbf{E}_e) = 0 \tag{6.79}$$

$$\nabla \cdot \mathbf{B}_e = 0 \tag{6.80}$$

$$\nabla \times \mathbf{E}_e = i k_0 c \mathbf{B}_e \tag{6.81}$$

$$\nabla \times \mathbf{B}_e = -i\frac{k_0}{c}\varepsilon_r \mathbf{E}_e \tag{6.82}$$

where $\mathbf{E}_e$ and $\mathbf{B}_e$ are the complex amplitudes of the $\mathbf{E}$ and $\mathbf{B}$ fields, respectively, such that

$$\mathbf{E}(r,t) = \mathbf{Re}\left\{\mathbf{E}_e(r)e^{-i\omega t}\right\} \tag{6.83}$$

$$\mathbf{B}(r,t) = \mathbf{Re}\left\{\mathbf{B}_e(r)e^{-i\omega t}\right\} \tag{6.84}$$

On the other hand,

$$k_0 = \omega\sqrt{\varepsilon_0\mu_0} = \frac{2\pi}{\lambda_0} \tag{6.85}$$

denotes the free-space propagation constant, whereas

$$\varepsilon_r = \frac{\varepsilon(r)}{\varepsilon_0} \tag{6.86}$$

is the relative permittivity.

Example 6.5

Assume that the complex amplitudes of the $\mathbf{E}$ and $\mathbf{B}$ fields are given, respectively, in the form

$$\mathbf{E}_e(r) = \mathbf{E}_0(r)e^{ik_0L(r)} \tag{6.87}$$

$$\mathbf{B}_e(r) = \mathbf{B}_0(r)e^{ik_0L(r)} \tag{6.88}$$

where $L(r)$ is a function describing the phase front of the wave, called the *eikonal*. Considering the limit of geometrical optics ($\lambda_0 \to 0$), show that the normal to such wave fronts, given by ∇L, satisfies the *eikonal equation:*

$$\nabla L = n\hat{\mathbf{s}} \tag{6.89}$$

where n is the refractive index and $\hat{\mathbf{s}}$ is a unit vector normal to the phase front.

Solution

Substituting Eqs. (6.87) and (6.88) into Eqs. (6.79)–(6.82), the following results are obtained:

$$\nabla L \cdot \mathbf{E}_0 = -\frac{1}{ik_0}(\varepsilon_r\nabla \cdot \mathbf{E}_0 + \nabla\varepsilon_r \cdot \mathbf{E}_0) \tag{6.90}$$

$$\nabla L \cdot \mathbf{B}_0 = -\frac{\nabla \cdot \mathbf{B}_0}{ik_0} \tag{6.91}$$

$$\nabla L \times \mathbf{E}_0 - c\mathbf{B}_0 = -\frac{\nabla \times \mathbf{E}_0}{ik_0} \tag{6.92}$$

$$\nabla L \times \mathbf{B}_0 + \frac{\varepsilon_r}{c}\mathbf{E}_0 = -\frac{\nabla \times \mathbf{B}_0}{ik_0} \tag{6.93}$$

continued

Example 6.5 continued

Considering the limit $\lambda_0 \to 0$, or $k_0 \to \infty$, one can obtain:

$$\mathbf{E}_0 \cdot \nabla L = \mathbf{B}_0 \cdot \nabla L = 0 \tag{6.94}$$

$$\mathbf{E}_0 = \frac{c}{\varepsilon_r}\mathbf{B}_0 \times \nabla L \tag{6.95}$$

$$\mathbf{B}_0 = \frac{\nabla L \times \mathbf{E}_0}{c} \tag{6.96}$$

Eliminating $\mathbf{E}_0$ from (6.96) with the help of Eq. (6.95) or eliminating $\mathbf{B}_0$ from (6.95) using Eq. (6.96), one obtains:

$$\varepsilon_r \mathbf{E}_0 = (\nabla L \times \mathbf{E}_0) \times \nabla L = |\nabla L|^2 \mathbf{E}_0 - \nabla L\,(\mathbf{E}_0 \cdot \nabla L) \tag{6.97}$$

$$\varepsilon_r \mathbf{B}_0 = \nabla L \times (\mathbf{B}_0 \times \nabla L) = |\nabla L|^2 \mathbf{B}_0 - \nabla L\,(\mathbf{B}_0 \cdot \nabla L) \tag{6.98}$$

Considering Eq. (6.94), the second terms on the right-hand side of Eqs. (6.97) and (6.98) vanish, and the following result is obtained:

$$|\nabla L|^2 = \varepsilon_r = n^2 \tag{6.99}$$

The *eikonal equation* given by Eq. (6.89) corresponds to Eq. (6.99).

The surfaces $L(r) = constant$ are surfaces of constant phase, i.e. phase fronts. The normal to these surfaces has the direction of ∇L and represents the rays of geometrical optics. The unit vector $\hat{\mathbf{s}}$ is normal to the phase front and has the same direction as the optical ray, being given by

$$\hat{\mathbf{s}} = \frac{\nabla L}{|\nabla L|} = \frac{\nabla L}{n} \tag{6.100}$$

Using Cartesian coordinates, Eq. (6.99) becomes

$$\left(\frac{\partial L}{\partial x}\right)^2 + \left(\frac{\partial L}{\partial y}\right)^2 + \left(\frac{\partial L}{\partial z}\right)^2 = n^2 \tag{6.101}$$

In the case of a homogeneous medium, the solution of Eq. (6.101) can be written as

$$L(r) = n(x\cos\theta_x + y\cos\theta_y + z\cos\theta_z) \tag{6.102}$$

where, $\cos\theta_x$, $\cos\theta_y$, and $\cos\theta_z$ are the direction cosines defining the direction of the light ray trajectory.

From Eqs. (6.28) and (6.95), the average Poynting vector can be written in the form:

$$<\mathbf{S}> = \frac{1}{2c\mu_0}(\mathbf{E}_0 \times (\nabla L \times \mathbf{E}_0)) \tag{6.103}$$

Using Eq. (6.99), one has

$$<\mathbf{S}> = \frac{n|\mathbf{E}_0|^2}{2c\mu_0}\hat{\mathbf{s}} \tag{6.104}$$

This result is similar to that shown in Eq. (6.32) and shows that the direction of the Poynting vector coincides with the optical ray direction, which is normal to the phase front.

6.10 Further Reading

Cook, D. M., *The Theory of the Electromagnetic Field*, Prentice-Hall, 1975.
Haus, H. A., *Waves and Fields in Optoelectronics*, Prentice-Hall, 1984.
Hecht, E., *Optics*, 5th Ed., Pearson Education, 2016.
Ida, N., *Engineering Electromagnetics*, 3rd Ed., Springer-Verlag, 2015.
Jackson, J. D., *Classical Electrodynamics*, 3rd Ed., John Wiley & Sons, 1999.
Klein, M. and T. Furtak, *Optics*, 2nd Ed., John Wiley & Sons, New York, 1986.
Ramo, S., J. R. Whinnery, and T. Van Duzer, *Fields and Waves in Communication Electronics*, 3rd Ed., John Wiley & Sons, 1994.
Sadiku, M. N., *Elements of Electromagnetics*, 6th Ed., Oxford University Press, 2014.
Staelin, D. H., A. W. Worgenthaler, and J. A. Kong, *Electromagnetic Waves*, Prentice-Hall, 1994.

6.11 Problems

6.1 The electric field of a plane electromagnetic wave has the following components:

$$E_x = 0, \qquad E_y = E_0 \exp\{i(\beta x - \alpha t)\}, \qquad E_z = 0$$

a) Obtain the components of the magnetic flux density **B**.

b) Derive the relation between α and β in order to satisfy Maxwell's equations.

6.2 A 100 W monochromatic point source is radiating equally in all directions in vacuum. Calculate the irradiance and the electric field amplitude 1 m away from the source.

6.3 Obtain an expression for a plane electromagnetic wave propagating in vacuum, with a wavelength of 600 nm and an irradiance of 20 mW/cm^2. Assume that the wave propagates in the z direction and that it is linearly polarized at an angle of 30° to the y-axis.

6.4 A 60 W lamp converts 3% of its electrical power into light. The radiation is evenly distributed over a solid angle of 1 sr. Calculate the irradiance and the electric field amplitude at a distance of 50 cm from the lamp. Assume that the light waves are harmonic.

6.5 A He–Ne laser ($\lambda = 632.8\,\text{nm}$) emits a cylindrical beam of light with a cross-section of 0.35 cm^2. The power received by a detector placed normal to the beam is 0.5 W.

a) Calculate the electric and magnetic field amplitudes.

b) What is the power emitted by an incandescent lamp of equal efficiency located at a distance of 2 m from the detector?

6.6 Derive the Fresnel Eqs. (6.48) and (6.49) for the reflectivity and transmissivity when the electric field is parallel to the plane of incidence (TM polarization).

6.7 Derive the result given by Eq. (6.65) for the attenuation constant of the evanescent field in the case of total reflection.

6.8 Derive the result given by Eq. (6.66) for the phase shift of the reflected field in the case of total reflection.

6.9 A laser beam is incident on the vertical glass wall of a rectangular tank filled with water, enters the tank, and strikes the water–air interface. For what range of angles of incidence on the vertical wall does the beam undergo total internal reflection at the water–air interface?

6.10 Determine the sum of the incident and refracted angles for the Brewster angle of incidence. Show that no Brewster angle exists when the electric field vector is perpendicular to the plane of incidence (TE polarization).

6.11 Light passes through a glass slab with parallel faces. Show that, if light is incident at Brewster's angle on the first (air–glass) interface, then there will also be no reflection at the second (glass–air) interface.

6.12 A laser beam is incident perpendicular to the surface of one of the short faces of a 45°–90°–45° prism, with a refractive index n = 1.5. Show that the light undergoes total internal reflection when it strikes the long face of the prism.

6.13 Suppose that the electric field vector of a linearly polarized wave makes an angle ϕ_i with the plane of incidence at the interface between two media. Show that, in such a case, the reflectance is given by

$$R = R_{||} \cos^2\phi_i + R_{\perp} \text{sen}^2\phi_i$$

where $R_{||}$ and $R_{\perp}$ are the reflectances of the electric field components parallel and perpendicular to the plane of incidence, respectively.

6.14 A light wave is incident from air on a thick glass slab of refractive index 1.6, with an angle of incidence of 60°. Assuming that the electric field is perpendicular to the plane of incidence, determine the fraction of light reflected from and transmitted through the air–glass interface. Verify that these two fractions add to unity.

6.15 Show that the evanescent electric field formed under the condition of total internal reflection in the lower refractive index medium varies with distance as given by Eqs. (6.54) and (6.65).

6.16 Show that the phase shift occurring under the condition of total internal reflection is given by Eq. (6.66).

6.17 Derive Eq. (6.104) for the average Poynting vector, starting from Eqs. (6.93) to (6.95) and using Eq. (6.100).

Light Polarization

7

As seen in Chapter 6, the fields associated with electromagnetic waves are vector quantities, such that the electric field, the magnetic field, and the direction of energy propagation are mutually perpendicular. In fact, both the magnetic field and the Poynting vector, and thus the entire electromagnetic wave, can be determined once the electric field is known. The direction of the electric field is known as the *polarization* of the wave.

This chapter deals with the superposition of two electromagnetic waves with the same frequency, but whose linear polarizations are orthogonal to each other. The resulting electric field does not necessarily have a linear polarization; in many cases, the direction of the respective electric field vector varies over time. The amplitudes and relative phase of the two waves determine the *polarization state* of the resulting field. Throughout this chapter, we describe these possible states of polarization, as well as various devices capable of altering and/or analysing them.

7.1 Polarization States of Light

Since the electric field is a vectorial quantity, its full description requires knowledge of both its magnitude and its direction. In the case of a plane wave with wave vector $\vec{k}$, the electric field $\vec{E}$ can have any direction situated in the plane perpendicular to **k**, whereas the direction of **B** is given by **k** × **E**. Wave polarization corresponds to the description of the **E** behaviour of the vector in the plane perpendicular to **k**.

The polarization plane is defined as the plane containing the wave vector and the electric field vector. If the direction of the vector **E** remains constant, the wave is said to be *linearly* polarized. On the other hand, if the direction of the vector **E** varies randomly over time, the wave is said to be *non-polarized.* If, for a given position in the direction of propagation, the tip of the vector **E** traces out a circle in a plane perpendicular to that direction, the wave is said to have *circular* polarization. In the more general case, the tip of the vector **E** traces out an ellipse in this plane, and the wave is then said to be *elliptically polarized.* If an electromagnetic field consists of the superposition of fields with many different polarizations, of which one is predominant, the field is said to be *partially polarized.*

Let us consider the case of a monochromatic wave that propagates along the z axis. Such a wave can be represented in the form:

$$\mathbf{E}(x, y, z, t) = E_x\mathbf{i} + E_y\mathbf{j} \tag{7.1}$$

where

$$E_x(x, y, z, t) = A_x \cos(kz - \omega t + \phi_x) \tag{7.2}$$

$$E_y(x, y, z, t) = A_y \cos(kz - \omega t + \phi_y) \tag{7.3}$$

Lasers, Optics, and Photonics. Mário F. S. Ferreira, Oxford University Press. © Mário F. S. Ferreira (2026).
DOI: 10.1093/oso/9780198952770.003.0007

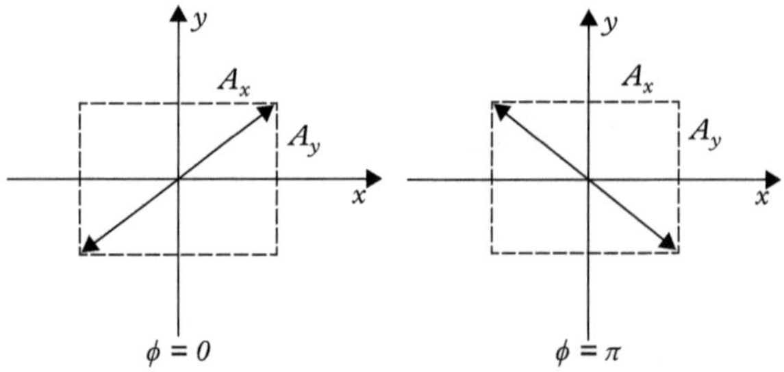

Figure 7.1 *Linearly polarized light.*

The classification of the different types of polarization depends on the relative phase

$$\phi = \phi_y - \phi_x \tag{7.4}$$

as well as on the relative values of the amplitudes A_x and A_y.

7.1.1 Linear polarization

Linear polarization of light occurs if

$$\phi = 2\pi p \tag{7.5a}$$

or

$$\phi = \pi(2p+1) \tag{7.5b}$$

p being an integer. In any of these cases, it can be verified from Eqs. (7.2) and (7.3) that E_y is proportional to E_x:

$$E_y = \frac{A_y}{A_x} E_x, \quad \text{if } \phi = 2\pi p \tag{7.6}$$

$$E_y = -\frac{A_y}{A_x} E_x, \quad \text{if } \phi = \pi(2p+1) \tag{7.7}$$

For a fixed z position, the electric field performs a simple harmonic motion along a line (Fig. 7.1). At a given t instant, the electric field shows a sinusoidal variation along the *z-axis*, located in the plane containing the z-axis and the line $y = \frac{A_y}{A_x}x$ (see Fig. 7.2) or $y = -\frac{A_y}{A_x}x$.

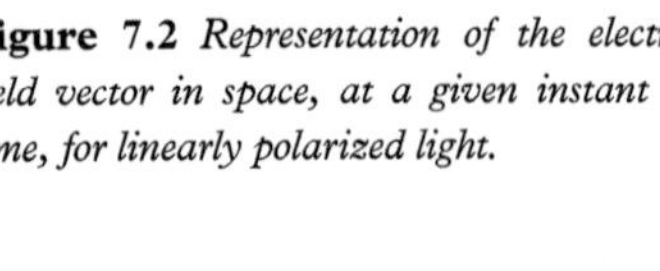

Figure 7.2 *Representation of the electric field vector in space, at a given instant in time, for linearly polarized light.*

7.1.2 Circular polarization

The circular polarization of light occurs when

$$A_x = A_y = A \text{ and } \phi = \pm\frac{\pi}{2} + 2\pi p \tag{7.8}$$

in which case Eqs. (7.2) and (7.3) take the form:

$$E_x(x,y,z,t) = A\cos(kz - \omega t + \phi_x) \tag{7.9}$$

$$E_y(x,y,z,t) = A\cos(kz-\omega t+\phi_x \pm \pi/2)$$
$$= \mp A\sin(kz-\omega t+\phi_x) \quad (7.10)$$

If $\phi = +\frac{\pi}{2}$, the electric field vector rotates clockwise in the x–y plane, and the light is then said to have *right circular polarization* (RCP). When $\phi = -\frac{\pi}{2}$, the electric field rotates counterclockwise and the light is said to have *left circular polarization* (LCP) (Fig. 7.3).

Circular polarization is also achieved when $\phi = \pm 3\frac{\pi}{2} + 2\pi p$, a situation in which Eq. (7.10) becomes

$$E_y(x,y,z,t) = \pm A\sin(kz-\omega t+\phi_x) \quad (7.11)$$

In this case, the minus sign corresponds to a clockwise rotation of the electric field (RCP), whereas the plus sign corresponds to a counterclockwise rotation (LCP).

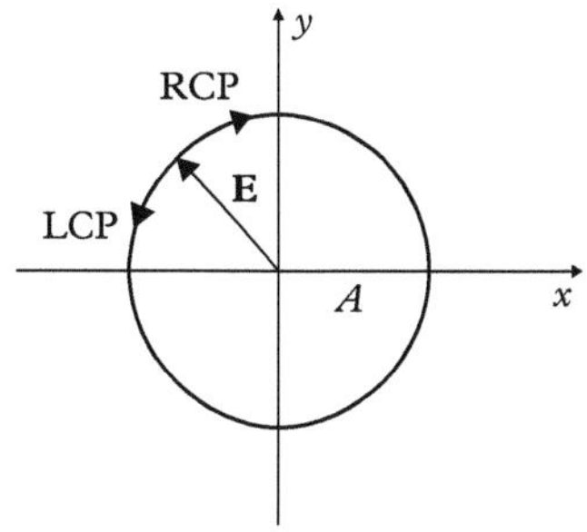

Figure 7.3 *Circularly polarized light. It is assumed that the light propagates towards the observer.*

Example 7.1

Describe the polarization state of the wave

$$\mathbf{E}(z,t) = \mathbf{i}A\cos(kz-\omega t+\pi/2) + \mathbf{j}A\cos(kz-\omega t)$$

Solution

The electric field can be written in the form

$$\mathbf{E}(z,t) = -\mathbf{i}A\sin(kz-\omega t+\pi/2) + \mathbf{j}A\cos(kz-\omega t)$$

Considering the position $z = 0$ and the instants of time $t = 0$, $T/4$, $T/2$, $3T/4$, and T, one has that $\mathbf{E}(0,t)$ is equal to $\mathbf{j}A$, $\mathbf{i}A$, $-\mathbf{j}A$, $-\mathbf{i}A$, and $\mathbf{j}A$, respectively. As the electric field vector has a constant magnitude A and rotates counterclockwise, it can be concluded that the wave has left circular polarization.

7.1.3 Elliptical polarization

When

$$\phi = \pm\frac{\pi}{2} + 2\pi p \text{ and } A_x \neq A_y \quad (7.12)$$

the two components of the electric field vector are given by

$$E_x(x,y,z,t) = A_x\cos(kz-\omega t+\phi_x) \quad (7.13)$$

$$E_y(x,y,z,t) = \mp A_y\sin(kz-\omega t+\phi_x) \quad (7.14)$$

From Eqs. (7.13) and (7.14), one can obtain the following equation:

$$\frac{E_x^2}{A_x^2} + \frac{E_y^2}{A_y^2} = 1 \quad (7.15)$$

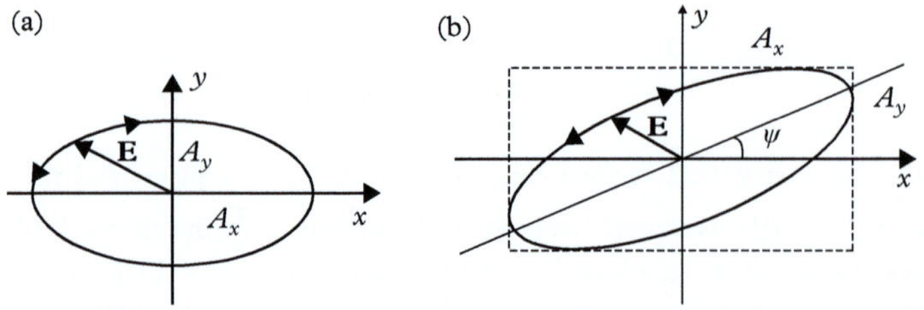

Figure 7.4 *Elliptically polarized light, with* $\phi = \pm\frac{\pi}{2}$ *(a) and* $\phi \neq \pm\frac{\pi}{2}$ *(b).*

which is recognized to be the equation of an ellipse located in the x–y plane, with semi-axes given by A_x and A_y. The electric field rotates clockwise if $\phi = +\frac{\pi}{2}$ and in the opposite direction if $\phi = -\frac{\pi}{2}$ (Fig. 7.4(a)).

The general case occurs when

$$\phi \neq \pm\frac{\pi}{2}, \pm\pi \tag{7.16a}$$

and

$$A_x \neq A_y \tag{7.16b}$$

In this case, using Eqs. (7.2) and (7.3) with $\phi = \phi_y - \phi_x$, one obtains the following equation:

$$\frac{E_y^2}{A_y^2} + \frac{E_x^2}{A_x^2} - \frac{2E_xE_y}{A_xA_y}\cos\phi - \sin^2\phi = 0 \tag{7.17}$$

Eq. (7.17) represents an ellipse inscribed in a rectangle with sides $2A_x$ and $2A_y$, as shown in Fig. 7.4(b). The ellipse's eccentricity depends on the phase difference $\phi = \phi_y - \phi_x$, being minimal if $\phi = \pm\frac{\pi}{2}$ and maximal if $\phi = \pm\pi$, 0. The angle ψ between the longer axis of the ellipse and the x-axis is given by

$$\tan 2\psi = \frac{2A_xA_y\cos\phi}{A_x^2 - A_y^2} \tag{7.18}$$

7.2 Jones Vectors

The components of the electric field indicated in Eq. (7.1) can be written using complex exponentials in the form

$$E_x(x, y, z, t) = A_x \exp\{i(kz - \omega t + \phi_x)\} \tag{7.19}$$

$$E_y(x, y, z, t) = A_y \exp\{i(kz - \omega t + \phi_y)\} \tag{7.20}$$

In this way, the electric field is given by

$$\begin{aligned}\mathbf{E} &= \left[\mathbf{i}A_x \exp\{i\phi_x\} + \mathbf{j}A_y \exp\{i\phi_y\}\right]\exp\{i(kz - \omega t)\}\\ &= \mathbf{A}\exp\{i(kz - \omega t)\}\end{aligned} \tag{7.21}$$

where

$$\mathbf{A} = \left[\mathbf{i}A_x \exp\{i\phi_x\} + \mathbf{j}A_y \exp\{i\phi_y\}\right] \tag{7.22}$$

represents the complex amplitude of the field.

The polarization state of light is completely determined by the amplitudes and relative phases of the two components of the electric field, contained in **A**, which can be written in the form of a column matrix or *jones vector*:

$$\mathbf{A} = \begin{bmatrix} A_x \exp\{i\phi_x\} \\ A_y \exp\{i\phi_y\} \end{bmatrix} \tag{7.23}$$

These matrices are normalized so that the sum of the squares of their elements is equal to one. For example, the column matrices

$$\begin{bmatrix} 1 \\ 0 \end{bmatrix}, \begin{bmatrix} 0 \\ 1 \end{bmatrix}, \frac{1}{\sqrt{2}}\begin{bmatrix} 1 \\ 1 \end{bmatrix} \tag{7.24}$$

represent, respectively, linearly polarized waves along the *x-axis,* the *y-axis,* and the axis that makes an angle of 45° with the *x-axis.*

In general, a linearly polarized wave along an axis making an angle α with the *x-axis* can be represented by the Jones vector

$$\mathbf{A} = \begin{bmatrix} \cos\alpha \\ \sin\alpha \end{bmatrix} \tag{7.25}$$

Alternatively, given a vector

$$\mathbf{A} = \begin{bmatrix} a \\ b \end{bmatrix} \tag{7.26}$$

where a and b are real numbers, not both zero, the inclination of the corresponding linearly polarized light is given by

$$\alpha = \tan^{-1}\left(\frac{b}{a}\right) \tag{7.27}$$

If $A_x = A_y = A$ and $\phi = \phi_y - \phi_x = \pm\frac{\pi}{2} + 2\pi p$ in Eq. (7.23), one has a circular polarization state of light. The case of *right-circularly polarized* (RCP) light is given by the normalized Jones vector

$$\mathbf{A} = \frac{1}{\sqrt{2}}\begin{bmatrix} 1 \\ -i \end{bmatrix} \quad \text{RCP} \tag{7.28a}$$

Similarly, the normalized Jones vector

$$\mathbf{A} = \frac{1}{\sqrt{2}}\begin{bmatrix} 1 \\ -i \end{bmatrix} \quad \text{LCP} \tag{7.28b}$$

represents *left-circularly polarized* (LCP) light, in which the light rotates *counterclockwise* when viewed head-on. Notice that one of the elements in the Jones vector for circularly polarized light is purely imaginary, and the magnitudes of the elements are the same.

When $A_x \neq A_y$ and $\phi = \pm\frac{\pi}{2} + 2\pi p$ the light wave has elliptical polarization, and the normalized Jones vectors have the form

$$\mathbf{A} = \frac{1}{\sqrt{A_x^2 + A_y^2}} \begin{bmatrix} A_x \\ iA_y \end{bmatrix} \quad \text{counter–clockwise rotation} \tag{7.29a}$$

$$\mathbf{A} = \frac{1}{\sqrt{A_x^2 + A_y^2}} \begin{bmatrix} A_x \\ -iA_y \end{bmatrix} \quad \text{clockwise rotation} \tag{7.29b}$$

In any case, the ellipse is oriented with its major axis along the x- or y-axis, as in Fig. 7.4(a). In the more general case of elliptical polarization, occurring for $\phi \neq \pm\frac{\pi}{2}, \pm\pi$, the normalized Jones vectors are given by

$$\frac{1}{\sqrt{A_x^2 + A_{yr}^2 + A_{yi}^2}} \begin{bmatrix} A_x \\ A_{yr} + iA_{yi} \end{bmatrix} \quad \text{counter – clockwise rotation} \tag{7.30a}$$

$$\frac{1}{\sqrt{A_x^2 + A_{yr}^2 + A_{yi}^2}} \begin{bmatrix} A_x \\ A_{yr} - iA_{yi} \end{bmatrix} \quad \text{clockwise rotation} \tag{3.30b}$$

with A_x and A_{yi} both positive. In this case, the ellipse is inclined with respect to the x-axis at an angle ψ as given by Eq. (7.18), where

$$A_y = \sqrt{A_{yr}^2 + A_{yi}^2} \quad \text{and} \quad \psi = \tan^{-1}\left(\frac{A_{yi}}{A_{yr}}\right) \tag{7.31}$$

Fig. 7.4(b) illustrates such an ellipse.

Jones's representation is particularly useful when it comes to adding two waves. For example, adding a wave with RCP to another with LCP results in a wave with linear polarization, as

$$\frac{1}{\sqrt{2}}\begin{bmatrix} 1 \\ i \end{bmatrix} + \frac{1}{\sqrt{2}}\begin{bmatrix} 1 \\ -i \end{bmatrix} = \sqrt{2}\begin{bmatrix} 1 \\ 0 \end{bmatrix} \tag{7.32}$$

Conversely, a wave with linear polarization can be considered as the sum of two waves with circular polarization, right and left.

It should be noted that Jones vectors apply only to completely polarized light. In the case of partially polarized light, the most appropriate method for its description makes use of the so-called *Stokes parameters*, which are functions not of the fields but of their irradiances.

Example 7.2

Indicate the polarization state represented by the Jones vector

$$\mathbf{A} = \begin{bmatrix} 2 \\ 1+i \end{bmatrix}$$

Solution

The amplitudes of the two components of the electric field are

$$A_x = 2, \quad A_y = \sqrt{1^2 + 1^2} = \sqrt{2}$$

On the other hand, the phase difference between these components is

$$\phi = \phi_y - \phi_x = \tan^{-1}\left[\frac{1}{1}\right] = 45^\circ$$

Thus, the Jones vector describes a state of elliptical polarization, the ellipse being described by the equation:

$$\frac{E_y^2}{A_y^2} + \frac{E_x^2}{A_x^2} - \frac{2E_xE_y}{A_xA_y}\cos\phi = \sin^2\phi$$

That is,

$$E_y^2 + \frac{E_x^2}{2} - E_xE_y = 1$$

The angle ψ between the longer axis of the ellipse and the *x-axis* is given by

$$\tan 2\psi = \frac{2A_xA_y\cos\phi}{A_x^2 - A_y^2}$$

Solving for the angle ψ, one obtains the result

$$\psi = \frac{1}{2}\tan^{-1}\left[\frac{(2)(2)\left(\sqrt{2}\right)\cos(45^\circ)}{4-2}\right] = 32^\circ$$

7.3 Polarizers

A polarizer is an optical device capable of modifying the polarization state of light. The different types of polarizers available are based on one of the following physical phenomena: *dichroism*, *reflection*, and *birefringence*.

7.3.1 Dichroic polarizers

Dichroism consists of the selective absorption of one of the components of the electric field. This phenomenon can occur naturally in various materials due to the anisotropy of their crystalline structures. One of these materials is tourmaline, whose crystal has a main axis determined by its atomic configuration. The component of the electric field perpendicular to the main axis is strongly absorbed by the crystal.

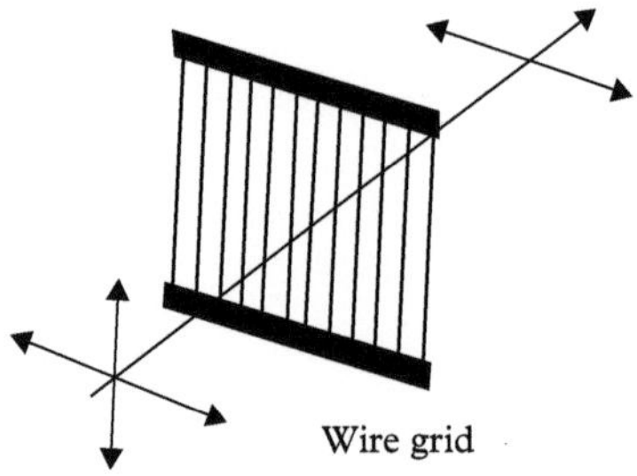

Figure 7.5 *The polarizing wire grid.*

The ability of dichroic materials to absorb light strongly with the electric field **E** along one direction and to transmit light easily with the field **E** along a perpendicular direction can be understood by reference to a device able to operate with microwaves, as represented in Fig. 7.5. This polarizer can be constructed by having a set of conductive wires parallel to each other, constituting the so-called wire grid. If the spacing of the vertical wire grid is much smaller than the wavelength of the microwaves (~1 mm to 1 m), it is found that the grid strongly attenuates the vertical component of the electric field, as this component induces currents in the wires, whose resistance dissipates the energy provided by the wave. The horizontal component of the field is practically unaffected when crossing the grid, so the light becomes polarized in that direction.

For optical wavelengths, the conduction paths analogous to the grid wires must be much closer together. The best-known dichroic polarizer in this wavelength range is the Polaroid, which was invented in 1938 by Edwin Herbert Land. It consists of a synthetic polymer impregnated with iodine, forming long, straight molecular chains parallel to each other. Iodine electrons can move freely along the polymer chains, which play a similar role to the conductor wires of the wire grid mentioned earlier. The electric field component of the incident wave orthogonal to the molecules is practically unaffected, whereas the parallel component displaces the electrons and is strongly absorbed. The energy of the incident wave is gradually dissipated as the wave advances through the absorber, so that the efficiency of the dichroic polarizer is a function of its thickness.

The devices described earlier can act as *polarizers*—when they are used to polarize or change the polarization state of light—or as *analysers*—when they are used to exploit this polarization state. The action of an analyser is shown in Fig. 7.6. Light coming from a polarizer P_1, with horizontal transmission axis, normally arrives at another polarizer P_2 (analyser), whose axis makes an angle θ with the direction of the polarizer P_1. The amplitude of the transmitted wave is reduced by a factor $\cos\theta$, so the transmitted irradiance is given by the so-called *Malus law:*

$$I(\theta) = I_0 \cos^2\theta \tag{7.33}$$

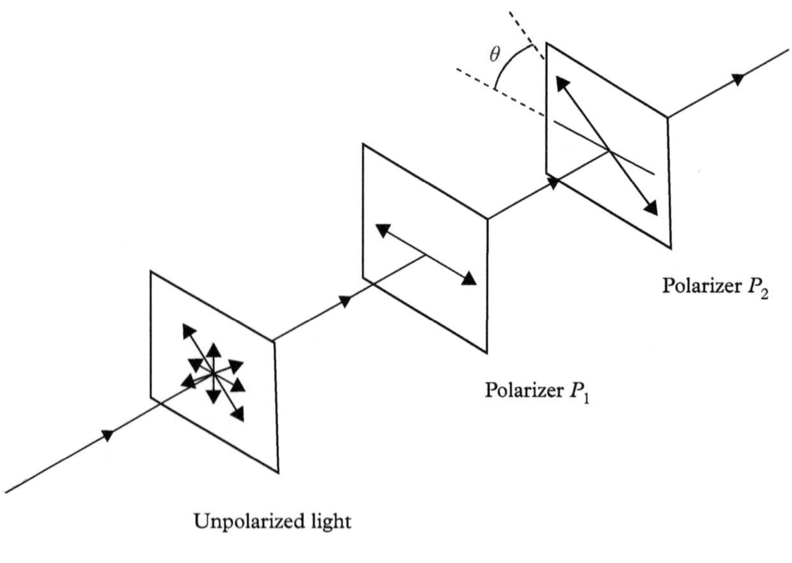

Figure 7.6 *Representation of the action of a polarization analyser.*

Example 7.3

A non-polarized beam of light with irradiance I_i passes successively through two linear polarizers. Explain what the relative orientation of these polarizers should be, so that the beam at the output of the second polarizer has an irradiance (a) $I_i/2$ and (b) $I_i/8$. Assume that the polarizers are perfect, that is, they do not cause any attenuation to the field components parallel to their transmission axes.

Solution

a) The incident beam can be decomposed into two components with orthogonal and incoherent linear polarizations, each of them presenting an irradiance $I_i/2$. In this way, the irradiance at the output of the first polarizer will be $I_i/2$. Malus' law applied to the incident and transmitted irradiances at the second polarizer is

$$I(\theta) = I_0 \cos^2\theta$$

where $I_0 = I_i/2$. In order that the irradiance at the output of the second polarizer is $I(\theta) = I_i/2 = I_0$, we must have $\theta = 0$, i.e. the transmission axes of the two polarizers must be parallel.

b) In this case,

$$I(\theta) = \frac{I_i}{8} = \frac{I_i}{2}\cos^2\theta$$

i.e. it should be $\cos\theta = \frac{1}{2}$ or $\theta = 60°$.

7.3.2 Reflection polarizers

Light can also be polarized by reflection, taking advantage of the fact, already mentioned in Chapter 6, that there is a particular angle of incidence, θ_B, for which the reflected light contains only the component of the electric field perpendicular to the plane of incidence. In fact, the reflectances for the components of the electric field parallel and normal to the plane of incidence are given by Eqs. (6.50) and (6.54) as

$$R_{\|} = r_{\|}^2 = \frac{\tan^2(\theta_i - \theta_t)}{\tan^2(\theta_i + \theta_t)} \tag{7.34a}$$

$$R_{\perp} = r_{\perp}^2 = \frac{\sin^2(\theta_i - \theta_t)}{\sin^2(\theta_i + \theta_t)} \tag{7.34b}$$

It is observed from Eq. (7.34a) that $R_{\|} = 0$ when its denominator becomes infinite, that is, when $\theta_i + \theta_t = 90°$. The angle $\theta_i = \theta_B$, for which this condition occurs, is called the *polarization* angle or *Brewster angle* and has already been discussed in Chapter 6.

When using this phenomenon to build an efficient polarizer, we verify that the reflected beam, although completely polarized, is not very intense. For example, in the reflection from air to glass, only 15% of the component is found in the reflected beam. In practice, to overcome this limitation, one can use a set of juxtaposed plates, constituting the so-called *pile-of-plates*, which is especially helpful in those regions of the infrared and ultraviolet spectrum where dichroic polarizers and calcite prisms are ineffective. Piles of plates are usually made of glass for the visible region, silver chloride for the infrared, and quartz for the ultraviolet. Reflections from the various plates are added successively to provide virtually total reflection.

On the other hand, the high value of the transmission coefficient found for the field component parallel to the plane of incidence can be used to produce perfectly non-reflective windows. These windows—called *Brewster windows*—are generally used in gaseous lasers, together with the mirrors that form the respective optical cavity, so that the emitted light is already completely polarized.

Example 7.4

Show that Brewster's angle is given by the expression

$$\theta_B = \tan^{-1}\frac{n_2}{n_1} \tag{7.35}$$

where n_1 and n_2 are the refractive indices of the incident and transmission media, respectively.

Solution

Starting from Snell's law

$$n_1 \sin\theta_i = n_2 \sin\theta_t$$

and using the condition $\theta_i + \theta_t = 90°$, one has

$$n_1 \sin\theta_i = n_2 \sin(90° - \theta_i) = n_2 \cos\theta_i$$

That is:

$$\tan\theta_i \equiv \tan\theta_B = \frac{n_2}{n_1}$$

Eq. (7.35) is known as *Brewster's law*.

7.3.3 Birefringent polarizers

A medium is *optically* isotropic if its refractive index is the same in all directions. In general, however, crystals are *anisotropic*. In this case, as discussed in Chapter 14, the phase speed of light depends on the direction of propagation in the crystal. Since the refractive index varies with direction, it is not characterized by a simple number but can be properly represented by a surface—the so-called *surface of refractive indices*. In addition, the refractive index in a given direction may depend on the polarization state of the wave, so an incident wave with arbitrary polarization will be divided into two components that experience different refractions. For this reason, the medium is said to be *birefringent.*

In the case of a *uniaxial birefringent crystal,* there is a single axis of symmetry, known as the *optical axis,* which also constitutes an axis of symmetry for the surface of refractive indices. Calcite ($CaCO_3$) is a good example of this type of crystal. A point source of non-polarized light located inside a calcite crystal will give rise to two wave fronts: one with spherical symmetry, known as the *ordinary wave,* and the other with ellipsoidal symmetry, known as the *extraordinary wave.* On the other hand, a non-polarized wave incident on the face of a calcite crystal will generally give rise to two refracted waves, with orthogonal linear polarizations and propagating in different directions within the crystal.

The refractive indices corresponding to the ordinary and extraordinary waves are usually indicated by n_o and n_e, respectively. The quantity $\Delta n = n_e - n_o$ is usually called *birefringence*. It is said that a material has *positive* or *negative birefringence* depending on the sign of Δn.

There is a class of polarizers that use double refraction (or birefringence) as a principle of operation. We can mention, by way of example, the Nicol, Glan–Thompson, and Wollaston polarizers.

Nicol's prism is made from a calcite crystal. This crystal is cut into two equal parts, which are subsequently glued together using Canadian balm. The refractive index of this balm is intermediate between the refractive index of the ordinary ray and that of the extraordinary ray. The angles of Nicol's prism are chosen in such a way that only the extraordinary ray is transmitted through its two parts. The ordinary ray undergoes total internal reflection at the first calcite–balm interface and is subsequently absorbed by the black coating on the prism faces.

Nicol's prism has some disadvantages, namely the lateral displacement suffered by the transmitted ray. Glan prisms can avoid these disadvantages. They consist of two calcite prisms, with parallel optical axes, agglutinated by a transparent glue or simply isolated by a layer of air (Fig. 7.7). The refractive index of the glue is such that, for the cut-off angle considered, the ordinary ray undergoes total internal reflection, while the extraordinary ray is transmitted with little attenuation.

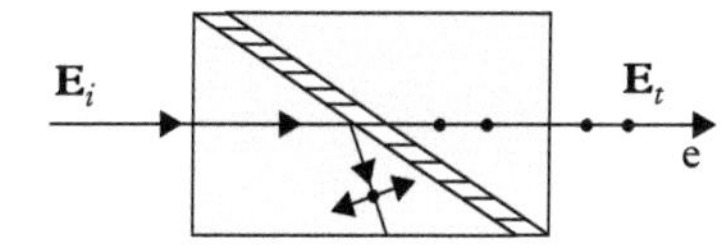

Figure 7.7 *Glan prism.*

The Wollaston prism (Fig. 7.8) consists of two orthogonal prisms of birefringent material—typically a uniaxial material such as calcite. These prisms are cemented together at their base to form two right triangular prisms with perpendicular optic axes. Wollaston's prism operates as a beam splitter that acts by polarization, as it lets both orthogonally polarized components pass through. The ordinary wave in the first part, vibrating perpendicular to the plane of the diagram, becomes the extraordinary wave in the second part, whereas the extraordinary wave in the first part, vibrating in the plane of the diagram, becomes the ordinary wave in the second part. If $n_e < n_o$ the rays are deflected, as shown in Fig. 7.8, whereas for $n_e > n_o$ the deviation will be in the opposite direction. The advantage of Wollaston's prism, in relation to Nicol and Glan prisms, lies in the symmetry of the configuration: both transmitted components have similar paths within the crystal, both experiencing the same absorption.

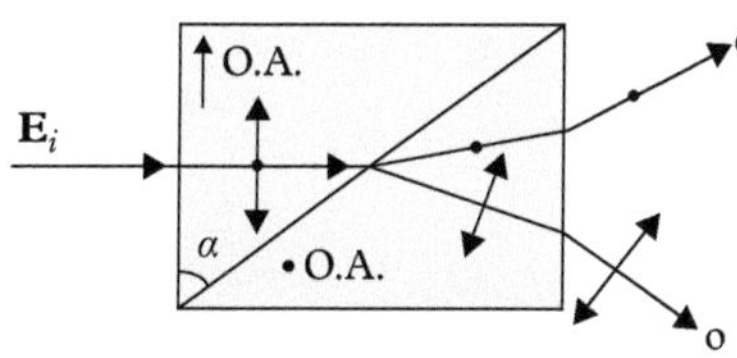

Figure 7.8 *Wollaston prism. O.A.: optical axis.*

7.4 Retarding Plates

The functioning of many optical components consisting of anisotropic materials depends on the phase variations that occur during propagation in these materials and not on differences in absorption or refraction. This occurs, in particular, in so-called *retarding plates*, which allow the polarization state of a wave to be modified.

Fig. 7.9 represents the case of a uniaxial crystal whose optical axis is parallel to the face of the plate. A linearly polarized wave is incident on this crystal, such that its electrical displacement vector **D** makes an angle θ with the optical axis. This wave is decomposed into two linearly polarized waves along orthogonal directions, corresponding to the ordinary and extraordinary waves within the birefringent crystal. These waves continue to propagate in the same direction but at different speeds.

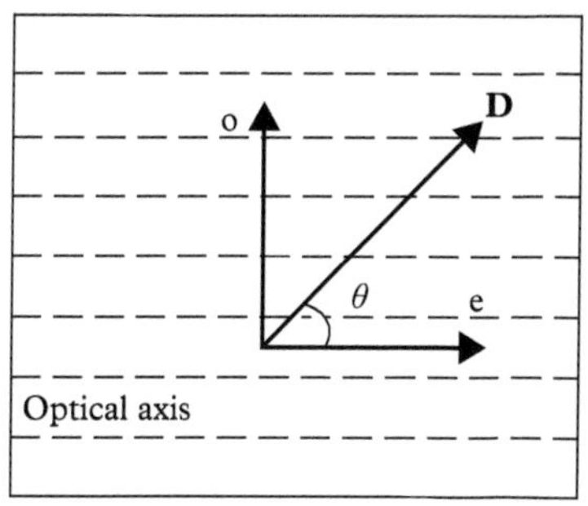

Figure 7.9 *Linearly polarized wave incident on a birefringent plate.*

The variation of the ordinary wave phase, after having propagated a distance d, is given by

$$\Delta\phi_o = -\frac{2\pi}{\lambda} n_o d \tag{7.36a}$$

whereas for the extraordinary wave one has

$$\Delta\phi_e = -\frac{2\pi}{\lambda} n_e d \tag{7.36b}$$

The phase difference introduced between the two waves is

$$\Delta\phi = \Delta\phi_e - \Delta\phi_o = -\frac{2\pi}{\lambda}(n_e - n_o)d \tag{7.37}$$

If $n_e < n_o$, then $\Delta\phi > 0$, and the extraordinary ray is advanced in phase relative to the ordinary ray. In this case, the polarization direction of the first wave corresponds to the so-called *fast axis*, whereas the polarization direction of the second wave coincides with the *slow axis*. When $n_e > n_o$, the opposite occurs.

7.4.1 Quarter-wave plate

When the relation

$$|n_e - n_o|\, d = \frac{\lambda}{4} + p\lambda \tag{7.38}$$

is verified, that is, when $|\Delta\varphi| = (2p+1)\pi/2$, p an integer, a relative shift of one quarter of a wavelength is achieved between the ordinary and extraordinary waves, and the plate is called a *quarter-wave plate* (QWP). Usually, mica or quartz is used to obtain QWPs, given the relative ease of achieving the desired thickness with these materials.

In the case of linearly polarized light focusing on a QWP, such that $\theta = 45°$ in Fig. 7.9, linear polarization is converted into circular polarization. In the case where the incident light has elliptical polarization and the larger and smaller axes of the ellipse coincide with the slow and fast axes of the QWP, the light at the exit will show linear polarization.

Example 7.5

Considering sodium light ($\lambda_0 = 589.3$ nm), determine the minimum thickness of a quartz plate that acts as a quarter-wave plate. Quartz is a uniaxial birefringent crystal whose ordinary and extraordinary refractive indices for that wavelength are $n_o = 1.5443$ and $n_e = 1.5534$, respectively.

Solution

The minimum thickness is obtained from Eq. (7.38), considering $p = 0$. Thus, one has

$$\begin{aligned} d &= \frac{\lambda_0}{4\,|n_e - n_o|} \\ &= \frac{589.3 \times 10^{-9}}{4\,|1.5534 - 1.5443|} = 1.62 \times 10^{-5}\,\text{m} = 0.0162\,\text{mm} \end{aligned}$$

7.4.2 Half-wave plate

When the thickness of the plate is such that the phase difference introduced between the two waves is an odd multiple of π, $|\Delta\phi| = (2p+1)\pi$, we have the so-called *half-wave plate* (HWP). This type of plate affects the orientation of the polarization, but not its shape.

In general terms, if the polarization of the incident wave makes an angle θ with the optical axis, then at the exit, the polarization plane will make an angle of $-\theta$ with this axis. In other words, the polarization plane will rotate by 2θ.

7.4.3 Babinet compensator

Because the expected phase delay introduced by a given retarder plate occurs only for a specific wavelength value, the thickness of such a plate will not be adequate when using light with other wavelengths. This fact suggests the great utility of a device that provides an adjustable phase variation and can be used with light of any wavelength. This device is called a *compensator.*

Fig. 7.10 shows the so-called *Babinet compensator,* consisting of two wedge-shaped prisms. The main effect is the production of a relative phase difference, which depends on the height at which the beam passes through the device. If the material used is calcite, the fast axis is in the plane of the figure. In the first part of the compensator, the phase variation of the extraordinary wave in relation to the ordinary wave is:

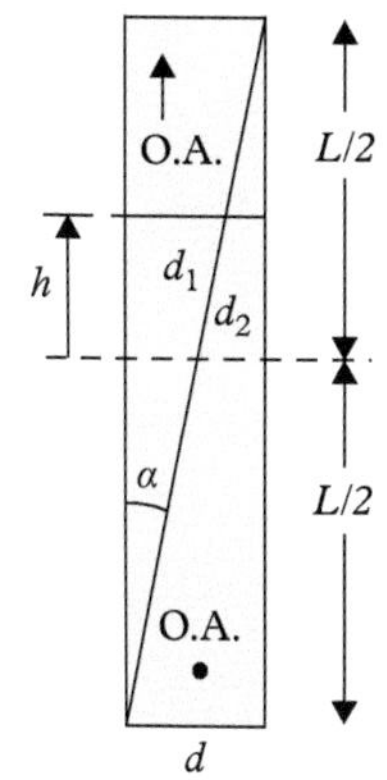

Figure 7.10 *Babinet compensator. O.A.: optical axis.*

$$\Delta\phi_1 = \frac{2\pi}{\lambda}(n_o - n_e)\, d_1 \tag{7.39}$$

In the second part of the compensator, the roles are reversed and one has

$$\Delta\phi_2 = -\frac{2\pi}{\lambda}(n_o - n_e)\, d_2 \tag{7.40}$$

The total phase variation is:

$$\begin{aligned}\Delta\phi &= \Delta\phi_1 + \Delta\phi_2 \\ &= \frac{2\pi}{\lambda}(n_o - n_e)(d_1 - d_2)\end{aligned} \tag{7.41}$$

In order to ensure proper operation of the compensator, the incident light must be in the form of a narrow beam, so that h has a well-defined value.

7.5 Jones Matrices

The representation of several devices, such as polarizers and retarding plates, can be done using 2×2 *jones matrices.* The effect on polarization of a series of such devices is represented by a 2×2 matrix, obtained by multiplying the matrices corresponding to each of them. For example, the matrices

$$\mathbf{H} = \begin{bmatrix} 1 & 0 \\ 0 & 0 \end{bmatrix}, \quad \mathbf{V} = \begin{bmatrix} 0 & 0 \\ 0 & 1 \end{bmatrix}, \quad \text{and} \quad \mathbf{O} = \frac{1}{2}\begin{bmatrix} 1 & 1 \\ 1 & 1 \end{bmatrix} \tag{7.42}$$

represent a linear polarizer in the horizontal direction (**H**), in the vertical direction (**V**), and in a direction of 45° (**O**), respectively.

The general form of a matrix representing a phase retarder is given by

$$\mathbf{M} = \begin{bmatrix} e^{i\varepsilon_x} & 0 \\ 0 & e^{i\varepsilon_y} \end{bmatrix} \tag{7.43}$$

where ε_x and ε_y represent the advance in phase of the E_x and E_y components of the incident light. As an example, let us consider QWP for which $|\varepsilon_y - \varepsilon_x| = \pi/2$. The case $\varepsilon_y - \varepsilon_x = \pi/2$ corresponds to the slow axis (SA) being vertical, whereas the case $\varepsilon_x - \varepsilon_y = \pi/2$ corresponds to the SA horizontal. In the first case, one has

$$\mathbf{M} = e^{-i\pi/4}\begin{bmatrix} 1 & 0 \\ 0 & i \end{bmatrix} \tag{7.44}$$

In arriving at the last matrix in Eq. (7.44), we used the relationship

$$e^{i\pi/4} = e^{-i\pi/4}e^{i\pi/2} = ie^{-i\pi/4}. \tag{7.45}$$

Similarly, the case of the SA being horizontal is represented by the matrix:

$$\mathbf{M} = e^{i\pi/4}\begin{bmatrix} 1 & 0 \\ 0 & -i \end{bmatrix} \tag{7.46}$$

The matrices corresponding to HWPs are given by

$$\mathbf{M} = e^{-i\pi/2}\begin{bmatrix} 1 & 0 \\ 0 & -1 \end{bmatrix} \quad \text{(SA vertical)} \tag{7.47}$$

$$\mathbf{M} = e^{i\pi/2}\begin{bmatrix} 1 & 0 \\ 0 & -1 \end{bmatrix} \quad \text{(SA horizontal)} \tag{7.48}$$

The elements of both matrices are identical in this case, since advancement of phase by π is physically equivalent to retardation by π.

7.6 Further Reading

Collett, E., *Polarized Light: Fundamentals and Applications*, Marcel Dekker, 1992.
Goldstein, D. H., *Polarized Light*, 3rd Ed., CRC Press/Taylor & Francis, 2010.
Hecht, E., *Optics*, 5th Ed., Pearson Education, 2016.
Huard, S., *Polarization of Light*, John Wiley & Sons, 1996.
Klein, M., and T. Furtak, *Optics*, John Wiley & Sons, 2nd Ed., 1986.
Kliger, D. S., J. W. Lewis, and C. E. Randall, *Polarized Light in Optics and Spectroscopy*, Academic Press, 1990.
Pedrotti, F., L. M. Pedrotti, and L. S. Pedrotti, *Introduction to Optics*, 3rd Ed., Cambridge University Press, 2018.

7.7 Problems

7.1 Write an expression for a wave that propagates in the direction of the x-axis with left circular polarization, such that its electric field points in the negative direction of the z *axis* at $x = 0$ and $t = 0$.

7.2 Describe the polarization state of the wave represented by

$$E_y = \mathbf{j}A \cos (kx - \omega t)$$

$$E_z = \mathbf{k}A \cos \left(kx - \omega t - \frac{\pi}{4}\right)$$

7.3 Write expressions for two waves that propagate along the x-axis, one with right circular polarization and the other with left circular polarization, the superposition of which gives an elliptically polarized wave rotating clockwise, with the longer axis in the direction of the z-axis.

7.4 Two perfect polarizers with vertical transmission axes are illuminated by natural light. Suppose that a third polarizer is placed between them, with a transmission axis making an angle of 30° with the vertical. Compare the irradiance at the output in this case with that which would exist in the absence of this third polarizer.

7.5 Show that the Brewster angles for external and internal reflection at the interface between two media are complementary to each other.

7.6 Find the sun's angular position above the surface of a calm lake, considering that the light reflected from its surface is completely plane-polarized. Assume that the refractive index of water is $n_w = 4/3$.

7.7 A certain fully polarized field is represented by the Jones vector $g\begin{bmatrix} \cos\beta \\ i\sin\beta \end{bmatrix}$, where g is a constant. Determine:

a) The irradiance of this field.

b) The possible polarization states of the field.

c) A Jones vector orthogonal to that given.

7.8 Write the Jones vectors and the expressions for the electric fields for the following cases:

a) A linearly polarized wave in the direction of the x-axis, and propagating in the yz plane, in a direction that makes an angle of 45° with the y-axis.

b) A wave with left elliptical polarization, propagating in the direction of the x-axis, such that the longer axis of the ellipse is parallel to the z-axis and its length is twice that of the smaller axis, the phase difference between the two components being $\pi/2$.

7.9 Consider a mica plate in which the optical axis is parallel to the surface. The refractive indices characteristic of the two component waves that propagate perpendicular to the plate are 1.5692 and 1.6049. Determine the thickness that a quarter-wave plate should have for a wavelength of 600 nm at normal incidence.

7.10 Calculate the angle between the ordinary and extraordinary rays emerging from a Wollaston prism (Fig. 7.8) consisting of calcite and whose prism angle is $\alpha = 20°$. Consider that the ordinary and extraordinary refractive indices in calcite are $n_o = 1.658$ and $n_e = 1.486$, respectively.

8 Interference

The solutions of the wave equation given by Eq. (6.13) satisfy the *superposition principle*, according to which the total electric field, **E**, at any point in space where two or more waves coexist, is equal to the vector sum of the electric fields of those waves. The phenomenon of interference consists of the interaction between these waves. This interaction results in a pattern in which the distribution of irradiance differs from the sum of the irradiances of the intervening waves, showing the existence of bright and dark bands called *fringes*. Bright fringes occur when the waves overlap such that there is a maximum irradiance of the resulting wave, a condition called *constructive interference*. The dark fringes correspond to *destructive interference* of the overlapping waves, situations in which there is a minimum irradiance of the resulting wave.

The first observation of the interference phenomenon was reported by Robert Boyle (1627–1691), when he observed the so-called Newton rings. However, the first experiments properly involving the effects of light interference were carried out by Thomas Young (1773–1829) in 1802. The results of these experiments contradicted Newton's corpuscular theory of light, so they were rejected by the generality of the scientists at that time. Ten years later, Fresnel conducted a new series of experiments that confirmed Young's results and led to the definitive consecration of the wave theory of light, to the detriment of the corpuscular theory.

8.1 Two-Beam Interference

According to the principle of superposition, if several fields, $\mathbf{E}_1$, $\mathbf{E}_2$, . . . , are present at a given point in space, the resulting optical field, **E**, is given by

$$\mathbf{E} = \mathbf{E}_1 + \mathbf{E}_2 + \cdots \tag{8.1}$$

In practice, the measured quantity is the irradiance, which is proportional to the temporal average of the field modulus squared (see Eq. (6.32)). If there are only two waves, it is given by

$$\left\langle |\mathbf{E}|^2 \right\rangle = \langle (\mathbf{E}_1 + \mathbf{E}_2) \cdot (\mathbf{E}_1 + \mathbf{E}_2) \rangle = \left\langle |\mathbf{E}_1|^2 \right\rangle + \left\langle |\mathbf{E}_2|^2 \right\rangle + 2\langle \mathbf{E}_1 \cdot \mathbf{E}_2 \rangle \tag{8.2}$$

Since we are only interested in the relative values of irradiance, it is considered in the following that the irradiance is given simply by $I = \left\langle |\mathbf{E}|^2 \right\rangle$. Eq. (8.2) can then be written in the following form:

$$I = I_1 + I_2 + 2\langle \mathbf{E}_1 \cdot \mathbf{E}_2 \rangle \tag{8.3}$$

Lasers, Optics, and Photonics. Mário F. S. Ferreira, Oxford University Press. © Mário F. S. Ferreira (2026).
DOI: 10.1093/oso/9780198952770.003.0008

where I_1 and I_2 are the irradiances of each of the interfering waves. All information about the interference is contained in the third term of Eq. (8.3). Interference may be constructive or destructive, depending on whether the sign of this term is positive or negative, respectively. If this term is zero at all points in space, it means that the waves do not interfere, so there is a uniform distribution of irradiance. A necessary condition for the interference term in Eq. (8.3) to be non-zero is that the electric fields of the two waves are not orthogonal. When the electric fields are parallel, on the other hand, the interference term makes its maximum contribution.

Let us consider two linearly polarized plane waves in the following forms:

$$\mathbf{E}_1 = \mathbf{E}_{01} \cos\left(\mathbf{k}_1 \cdot \mathbf{r} - \omega t + \varepsilon_1\right) \tag{8.4a}$$

$$\mathbf{E}_2 = \mathbf{E}_{02} \cos\left(\mathbf{k}_2 \cdot \mathbf{r} - \omega t + \varepsilon_2\right) \tag{8.4b}$$

where ε_1 and ε_2 are the initial phases. In this case, the interference term in Eq. (8.3) is given by

$$I_{12} = 2\langle \mathbf{E}_1 \cdot \mathbf{E}_2 \rangle = \mathbf{E}_{01} \cdot \mathbf{E}_{02} \cos\delta \tag{8.5a}$$

where

$$\delta = \mathbf{k}_1 \cdot \mathbf{r} - \mathbf{k}_2 \cdot \mathbf{r} + \varepsilon_1 - \varepsilon_2 \tag{8.5b}$$

It is the *phase difference* between the two interfering waves, resulting from the difference in optical paths (OPs), $(\mathbf{k}_1 \cdot \mathbf{r} - \mathbf{k}_2 \cdot \mathbf{r})$, and the initial phase difference, $(\varepsilon_1 - \varepsilon_2)$. If this last contribution is constant, $\varepsilon_1 - \varepsilon_2 = const.$, it is said that the sources of the two waves are *coherent*.

When the fields $\mathbf{E}_1$ and $\mathbf{E}_2$ are orthogonal, the interference term is null, so there is a uniform distribution of irradiance. When these fields are parallel, the irradiance distribution can be written from Eqs. (8.3) and (8.5) in the following form:

$$I = I_1 + I_2 + 2\sqrt{I_1 I_2}\cos\delta \tag{8.6}$$

If the two waves have the same amplitude, it is $I_1 = I_2$ and Eq. (8.6) becomes

$$I = 4I_1 \cos^2\frac{\delta}{2} \tag{8.7}$$

In this case, the resulting irradiance is $4I_1$ when there is a *totally constructive interference*, corresponding to a phase di[illegible]nce $\delta = p2\pi$ $(p = 0, \pm 1, \pm 2, \ldots)$ between the two waves. On the other hand [illegible]nce is zero in the case of a *totally destructive interference*, cor[illegible] $\delta = (2p+1)\pi$. The principle of energy conservation [illegible]f the beams is I_1, the resulting irradiance would [illegible]rence phenomenon, the spatial distribution of [illegible] a maximum value $(4I_1)$ and a minimum value

Example 8.1

The *visibility*, V, of a set of interference fringes is defined on the basis of the maximum and minimum values of the irradiance in the interference pattern in the following form:

$$V = \frac{I_{max} - I_{min}}{I_{max} + I_{min}} \tag{8.8}$$

Derive an expression for the visibility of the fringe pattern resulting from the interference of two waves whose electric fields are parallel. Find the visibility of this pattern when the amplitudes of the two waves are equal.

Solution

From Eq. (8.6), one has

$$I_{max} = I_1 + I_2 + 2\sqrt{I_1 I_2} \quad \text{and} \quad I_{min} = I_1 + I_2 - 2\sqrt{I_1 I_2}$$

So the visibility becomes

$$V = \frac{2\sqrt{I_1 I_2}}{I_1 + I_2}$$

If the amplitudes of the two waves are equal, it is $I_1 = I_2 = I_0$ and the visibility becomes

$$V = \frac{2\sqrt{I_0^2}}{2I_0} = 1$$

It is verified that the visibility is maximum in this case.

8.2 Complex Representation

In many cases, it is possible to ignore the vector nature of light waves. One of these cases occurs when all light waves propagate in the same direction and have the same vibration plane. In these circumstances, the analysis of the superposition of these waves can be facilitated by using complex representation for the electric field, as mentioned in Chapter 5.

Let us consider the situation in which the electric fields of the interfering waves are collinear, have the same frequency, and are given in the scalar form by

$$E_1 = E_{01} e^{-i(\omega t - \alpha_1)} \tag{8.9a}$$

$$E_2 = E_{02} e^{-i(\omega t - \alpha_2)} \tag{8.9b}$$

The resulting optical field is given by

$$E = E_1 + E_2 = \left(E_{01} e^{i\alpha_1} + E_{02} e^{i\alpha_2}\right) e^{-i\omega t} = E_0 e^{i\alpha} e^{-i\omega t} \tag{8.10}$$

where

$$E_0 e^{i\alpha} = E_{01} e^{i\alpha_1} + E_{02} e^{i\alpha_2}$$

is the complex amplitude of the resulting wave, given by the addition of the complex amplitudes of the initial waves.

The module, E_0, and phase, α, of the complex amplitude of the resulting field can be expressed according to the corresponding parameters of the individual waves. Given that

$$E_0^2 = \left(E_0 e^{i\alpha}\right)\left(E_0 e^{i\alpha}\right)^* \tag{8.12}$$

one has

$$\begin{aligned} E_0^2 &= \left(E_{01}e^{i\alpha_1} + E_{02}e^{i\alpha_2}\right)\left(E_{01}e^{i\alpha_1} + E_{02}e^{i\alpha_2}\right)^* \\ &= E_{01}^2 + E_{02}^2 + E_{01}E_{02}\left(e^{i(\alpha_1-\alpha_2)} + e^{-i(\alpha_1-\alpha_2)}\right) \end{aligned} \tag{8.13}$$

Using Euler's relation, one can write Eq. (8.13) in the form

$$E_0^2 = E_{01}^2 + E_{02}^2 + 2E_{01}E_{02}\cos\left(\alpha_1 - \alpha_2\right) \tag{8.14}$$

a result that is equivalent to that of Eq. (8.6).

On the other hand, by separating the real and imaginary parts of Eq. (8.11), one has

$$E_0 \cos\alpha = E_{01}\cos\alpha_1 + E_{02}\cos\alpha_2 \tag{8.15a}$$

$$E_0 \sin\alpha = E_{01}\sin\alpha_1 + E_{02}\sin\alpha_2 \tag{8.15b}$$

By dividing member by member these equations, one obtains the following result:

$$\tan\alpha = \frac{E_{01}\sin\alpha_1 + E_{02}\sin\alpha_2}{E_{01}\cos\alpha_1 + E_{02}\cos\alpha_2} \tag{8.16}$$

The fields given by Eqs. (8.9) and (8.10) are represented by vectors in the complex plane in Fig. 8.1.

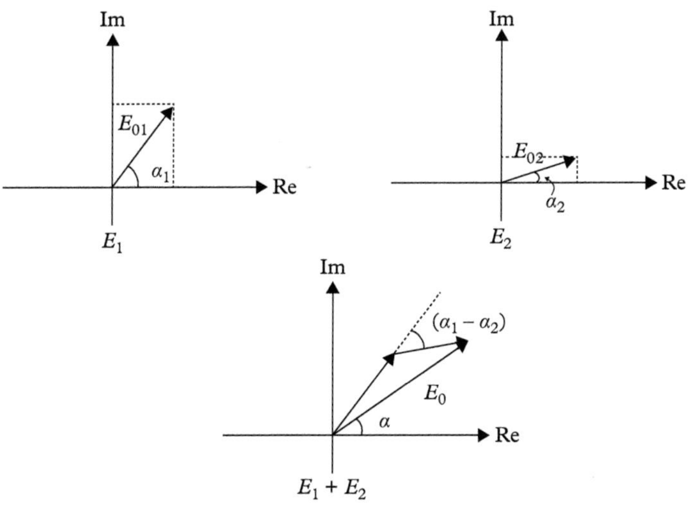

Figure 8.1 *Representation of the sum of fields E_1 and E_2 on the complex plane.*

Previous results may be generalized to the case of superposition of an arbitrary number, N, of harmonic waves. In this case, the complex amplitude of the resulting wave is given by

$$E_0 e^{i\alpha} = \sum_{j=1}^{N} E_{0j} e^{i\alpha_j} \tag{8.17}$$

where E_0 and α are given by

$$E_0^2 = \sum_{j=1}^{N} E_{0j}^2 + 2\sum_{i>j}^{N}\sum_{j=1}^{N} E_{0j}E_{0i}\cos(\alpha_j - \alpha_i) \tag{8.18}$$

and

$$\tan\alpha = \frac{\sum_{j=1}^{N} E_{0j}\sin\alpha_j}{\sum_{j=1}^{N} E_{0j}\cos\alpha_j} \tag{8.19}$$

Example 8.2

Obtain the irradiance resulting from the superposition of N waves with equal amplitudes in cases where (a) the waves are coherent and are in phase at the observation point and (b) the waves are incoherent.

Solution

a) The irradiance is proportional to the average value of E_0^2. In the case the waves are coherent, are in phase at the observation point, and present the same amplitude, Eq. (8.18) provides the result.

$$\begin{aligned} I \propto E_0^2 &= \sum_{j=1}^{N} E_{0j}^2 + 2\sum_{i>j}^{N}\sum_{j=1}^{N} E_{0j}E_{0i} \\ &= \left(\sum_{j=1}^{N} E_{0j}\right)^2 = N^2 E_{01}^2 \end{aligned} \tag{8.20}$$

We verify from this result that, in the case of superposition of in-phase coherent waves, *the resulting irradiance is proportional to the square of the sum of the amplitudes of these waves.*

b) When the superposing waves are incoherent, due to the rapid and random variation of the phase difference between such waves, the mean values of cosines in Eq. (8.18) are zero. In this case, assuming that the wave amplitudes are equal, the result is

$$I \propto E_0^2 = N E_{01}^2 \tag{8.21}$$

That is, when the overlapping waves are incoherent, *the resulting irradiance is proportional to the sum of the individual irradiances.*

8.3 Young's Experiment

One way to have two initially coherent waves is to obtain such waves from spatially distinct parts of the same original wavefront. This corresponds to the so-called *wavefront division interference*. To illustrate this type of interference, we will examine *Young's experiment*, which is illustrated in Fig. 8.2 and schematically represented in Fig. 8.3.

Let us consider a monochromatic plane wave incident on a narrow slit S. From this slit emerges a cylindrical wave, which illuminates two other narrow slits, S_1 and S_2, closely spaced from each other. These two slits obtain spatially distinct samples from the wavefront originating in S. In a symmetric situation, these samples are exactly in phase, and the two slits behave as two coherent secondary sources. We will analyse the interference produced by these two waves on a plane located at a distance D from the plane containing the slits S_1 and S_2.

The light coming from S_1 travels a distance r_1, whereas the light coming from S_2 travels a distance r_2, until they reach point P in the observation plane. Assuming the symmetry

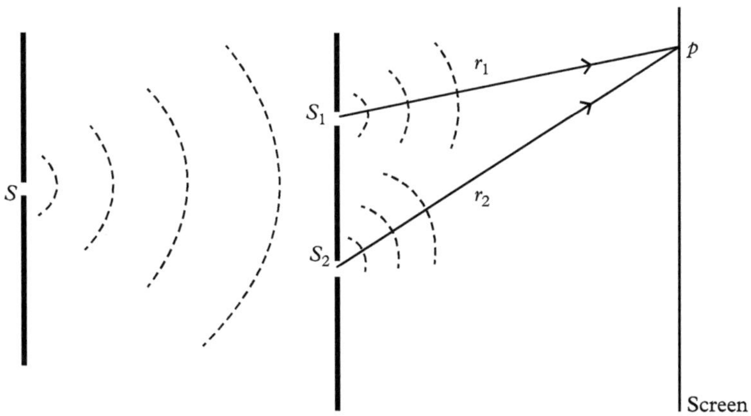

Figure 8.2 *Young's experiment.*

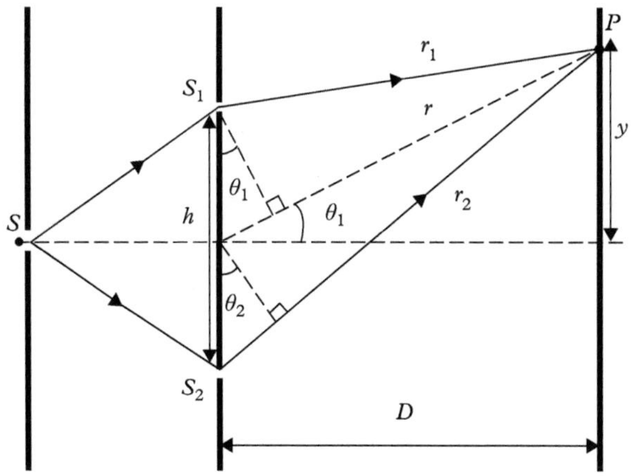

Figure 8.3 *Geometry to analyse Young's experiment.*

conditions mentioned earlier, the phase difference between the two waves at point P is due only to the different OPs, being given by

$$\delta = k(r_2 - r_1) \tag{8.22}$$

There is completely constructive interference if $\delta = p2\pi$ $(p = 0, \pm 1, \pm 2, \ldots)$, which corresponds to an OP difference:

$$r_2 - r_1 = p\lambda \tag{8.23}$$

Let us suppose that the distance D is much longer than the height y of the observation point P, measured relative to the x-axis. Under these conditions, the geometry of Fig. 8.3 allows us to write:

$$\sin\theta_1 = \frac{r - r_1}{\frac{h}{2}} \approx \tan\theta_1 = \frac{y}{D} \tag{8.24a}$$

and

$$\sin\theta_2 = \frac{r_2 - r}{\frac{h}{2}} \approx \tan\theta_2 = \frac{y + \frac{h}{2}}{D} \tag{8.24b}$$

Adding, member to member, the aforementioned equations and assuming that $h << y$, one has

$$r_2 - r_1 = \frac{yh}{D} \tag{8.25}$$

Assuming that the two slits are equal, the distribution of irradiance in the observation plane is given by Eqs. (8.7), (8.22), and (8.25) in the form

$$I = 4I_0 \cos^2\left(\frac{\pi h y}{\lambda D}\right) \tag{8.26}$$

where I_0 is the contribution from a single slit.

Considering Eqs. (8.24) and (8.24), a maximum of irradiance is found at a height.

$$y_p = \frac{D}{h} p\lambda \tag{8.27}$$

The spacing between two consecutive maxima is

$$\Delta y = y_{p+1} - y_p = \frac{D}{h}\lambda \tag{8.28}$$

One can verify from Eq. (8.28) that the spacing between the interference fringes is directly proportional to the distance at which the observation plane is located and inversely proportional to the separation between the slits S_1 and S_2 in Fig. 8.2.

The analysis of Young's experiment can easily be adapted to other interferometers based on the division of the wavefront, as will be seen in Chapter 9.

Example 8.3

A beam of light with wavelength λ illuminates the two slits in Young's experiment, which are spaced by 1.5 mm. The distance between the plane of these slits and the observation plane is 15 m, and the spacing between two consecutive bright fringes in this plane is 6.3 mm. What is the wavelength of light? Assuming that the two slits are equal, find an expression for the distribution of irradiance on the observation screen.

Solution

Using Eq. (8.28) one has

$$\lambda = \frac{h}{D}\Delta y = \frac{\left(1.5 \times 10^{-3}\right)\left(6.3 \times 10^{-3}\right)}{15} = 6.3 \times 10^{-7}\ \mathrm{m}$$

On the other hand, using Eq. (8.26), the distribution of irradiance on the observation screen is given by

$$I = 4I_0 \cos^2\left(\frac{\pi h y}{\lambda D}\right) = 4I_0 \cos^2\left(\frac{\pi\left(1.5 \times 10^{-3}\right)y}{\left(6.3 \times 10^{-7}\right)(15)}\right) = 4I_0 \cos^2(500y)$$

The result given earlier shows that there is a bright fringe at $y = 0$, corresponding to a point on the observation screen which is at the same distance from both slits.

8.4 Dielectric Films

Besides the wavefront division technique considered earlier, it is also possible to observe the phenomenon of interference between two waves coming from the same region of the initial wavefront. In this case, it is the wave amplitude that is divided, a process referred to as *interference by division of amplitude*.

A situation in which one can observe this type of interference occurs in the reflection of a wave incident on a dielectric film, as represented schematically in Fig. 8.4. The wave is incident on point P of the film, at which part of the amplitude is reflected towards point S, whereas the other part is transmitted to the inside of the film. Afterwards, this part is reflected at point Q and then proceeds to point R. The characteristics of the phenomenon of interference between the two reflected waves depend on the difference between the respective OPs, which are given by the products of the spatial distances and the refractive indices of the traversed media.

Figure 8.4 *Geometry for the study of interference by amplitude division in a dielectric film.*

The OP corresponding to the route from point P to point Q, and afterwards to point R, inside the dielectric film, which has a thickness d and a refractive index n_2, is given by

$$OP_{\overline{PQR}} = 2\frac{n_2 d}{\cos\theta_t} \tag{8.29}$$

On the other hand, the OP between points P and S in the medium with refractive index n_1 is given by

$$OP_{\overline{PS}} = n_1 \overline{PR} \sin\theta_i = n_2 \overline{PR} \sin\theta_t \tag{8.30}$$

where Snell's law was used for the refraction at point P. On the other hand, one has

$$\overline{PR} = 2d \tan\theta_t \tag{8.31}$$

So the difference in OPs, Λ, of the two rays becomes

$$\Lambda = OP_{\overline{PQR}} - OP_{\overline{PS}} = \frac{2n_2 d}{\cos\theta_t}(1 - \sin^2\theta_t) = 2n_2 d\cos\theta_t \tag{8.32}$$

The corresponding phase difference is given by

$$\delta = k_0\Lambda = \frac{4\pi d n_2 \cos\theta_t}{\lambda_0} \tag{8.33}$$

However, an additional phase difference must be taken into account due to the phenomenon of phase changes on reflection. Assuming that $n_3 < n_2 > n_1$ in Fig. 8.4, the reflection at point P occurs with light going from a lower index medium towards a higher index medium, a condition usually called *external reflection*. The reflection at point Q, on the other hand, occurs when light goes from a higher index medium towards a lower index medium, a condition called *internal reflection*. A relative phase shift of π rad occurs in the first case but not in the second one. As a result, the total phase difference between the first two waves reflected by the dielectric film is given by

$$\delta' = k_0\Lambda \pm \pi = \frac{4\pi n_2}{\lambda_0} d\cos\theta_t \pm \pi \tag{8.34}$$

A particularly important application of single-layer dielectric films is as *antireflecting coatings* on optical surfaces. In most cases, the light enters the film from air, so that $n_1 = 1$. Furthermore, if $n_2 < n_3$ no relative phase shift between the two reflected beams occurs and the optical-path difference alone determines the type of interference to be expected. Destructive interference occurs if the film thickness is an odd multiple of a half wavelength. However, the destructive interference will be complete only if the amplitudes of the two reflected waves are equal. To achieve this condition, the refractive index of the film must be chosen appropriately in order that the reflectances at the air/film and film/substrate interfaces are be equal.

Dielectric films can also be used to realize a multilayer stack of alternating high and low refractive indices (Fig. 8.5). If each film has an optical thickness of $\lambda/4$, it can be shown that all emerging beams are in phase. In such a case, the total reflected irradiance is increased and the quarter-wave stack performs as an efficient mirror.

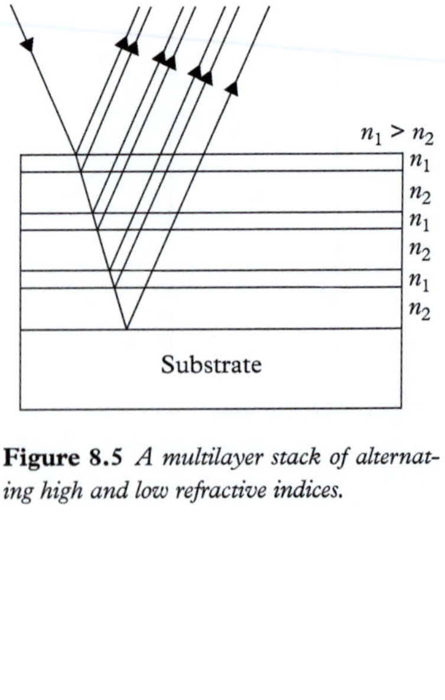

Figure 8.5 *A multilayer stack of alternating high and low refractive indices.*

8.4.1 Haidinger fringes

If the film thickness is uniform and it is illuminated by non-collimated light, provided by an extended light source, one can observe *fringes of equal inclination*, also known as *Haidinger fringes*. A bright or dark fringe will be associated with a particular angle of incidence which satisfies the condition for constructive or destructive interference, respectively. For example, a bright fringe occurs when $\delta' = p2\pi$, i.e. for an angle given by

$$\cos\theta_t = \frac{(2p+1)\lambda_0}{4n_2 d} \tag{8.35}$$

where the minus sign was considered in Eq. (8.34). This result can be written in terms of the angle of incidence using Snell's law:

$$\sin\theta_i = \sqrt{\left(\frac{n_2}{n_1}\right)^2 - \frac{(2p+1)^2\lambda_0^2}{16d^2n_1^2}} \tag{8.36}$$

Obviously, Haidinger fringes are not observed in the case of a point source, since every light ray coming from such a source arrives at the dielectric film with a different angle of incidence.

8.4.2 Fizeau fringes

Let us consider now a dielectric film with a non-uniform thickness which is illuminated by a collimated light beam. In this case, the angle of incidence is the same for the entire film and each interference fringe, bright or dark, corresponds to a given region of the film with a well-defined optical thickness. For example, the thickness corresponding to a given bright fringe can be obtained from Eq. (8.34) by making $\delta' = p2\pi$:

$$d = \frac{(2p+1)\lambda_0}{4n_2\cos\theta_t} \tag{8.37}$$

These fringes are called *fringes of equal thickness* or *Fizeau fringes*.

Example 8.4

A dielectric film with parallel faces and refractive index $n_2 = 1.5$ is illuminated by collimated light with wavelength $\lambda_0 = 565.7$ nm, having an incidence angle of 30°. What is the minimum thickness of the film that allows us to obtain a bright fringe at the observation point?

Solution

The condition for having a bright fringe at the point of observation is $\delta' = p2\pi$. In order to have a minimum film thickness, it must be $p = 0$. In this case, Eq. (8.34) allows us to write the condition.

$$\delta' = 0 = \frac{4\pi n_2}{\lambda_0}d\cos\theta_t - \pi$$

The corresponding film thickness is

$$d = \frac{\lambda_0}{4n_2\cos\theta_1} = \frac{\lambda_0}{4\sqrt{n_2^2 - \sin^2\theta_1}} = \frac{565.7\times10^{-9}}{4\sqrt{1.5^2 - \sin^2 30°}} = 0.1\,\mu\text{m}$$

8.4.3 Newton's rings

Being fringes of equal thickness, the contours of Fizeau fringes reveal directly any non-uniformities in the thickness of the film. For example, placing a convex lens in contact with a flat-faced plate, a dielectric film is obtained between the surfaces of the two elements. The fringes of equal thickness observed in this case have a circular shape, with centre at

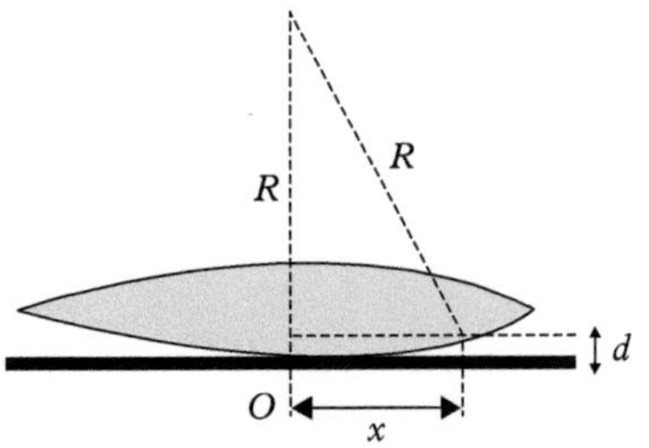

Figure 8.6 *Geometry for the description of Newton's rings.*

the contact point (point O in Fig. 8.6), being called *Newton's rings*. The regularity of these rings can be used to characterize the degree of perfection of the lens surfaces.

In Fig. 8.6, the radius of curvature of the lens is designated by R, whereas the radius of one of Newton's rings is indicated by x and the thickness of the film corresponding to that ring is per d. These three parameters can be related through the Pythagorean theorem:

$$x^2 = R^2 - (R - d)^2 \tag{8.38}$$

That is,

$$x^2 = 2Rd - d^2 \approx 2Rd \tag{8.39}$$

where we consider that $R >> d$. Assuming that the incidence is practically normal, one has $\cos\theta_t \approx 1$. The radius of the bright fringe of order p is obtained by combining Eq. (8.37) with Eq. (8.39) and given by

$$x_p = \sqrt{\frac{(2p+1)\lambda_0 R}{2n_2}} \tag{8.40}$$

Similarly, the radius of the dark fringe of order p is

$$x_p = \sqrt{\frac{p\lambda_0 R}{n_2}} \tag{8.41}$$

Generally, the film between the lens and the flat plate is constituted by air, so one has $n_2 = 1$. In such a case, the radiance of the central fringe (corresponding to $x_0 = 0$) is minimal.

Example 8.5

A biconvex lens with a radius of curvature R on both faces is placed on a flat surface and is illuminated normally with light of wavelength λ. The space between the lens and the flat surface is filled with a liquid with refractive index $n_l = 1.4$. Find the ratio of the radius of the 20th dark ring before and after the introduction of this liquid.

Solution

The radius of the dark ring of order p is given by Eq. (8.41). In case the space between the lens and the flat surface is empty ($n_2 = 1$), one has

$$x_{p,a} = \sqrt{p\lambda_0 R}$$

After the liquid has been introduced, the radius of the same ring becomes

$$x_{p,l} = \sqrt{\frac{p\lambda_0 R}{n_l}}$$

The ratio between the two values is then

$$\frac{x_{p,a}}{x_{p,l}} = \sqrt{\frac{p\lambda R}{p\lambda R/n_l}} = \sqrt{n_l}$$

That is, the ratio between the radius of the dark ring in both situations is given by $\sqrt{n_l}$, regardless of the ring order, the curvature radius of the lens faces, and the wavelength of light used. With the introduction of the liquid, an interference ring of a certain order decreases in radius.

8.5 Multiple-Reflection Interference

In the discussion of Section 8.4, multiple reflections on the faces of the dielectric film were neglected. However, this approximation is no longer valid when the reflectivity of the faces becomes significant. In this section we include the effect of such multiple reflections and show how the dielectric plate can constitute, in this case, a resonant cavity.

Fig. 8.7 shows the geometry for the study of the effect of multiple reflections in a dielectric plate with parallel faces. It is assumed that a wave of amplitude A is incident on the plate. The reflection coefficient on the front surface of the plate (external reflection) is indicated by r', whereas the reflection coefficient on the back surface of the plate (internal reflection) is represented by r. On the other hand, the parameters t and t' represent the transmission coefficients into and out of the plate, respectively.

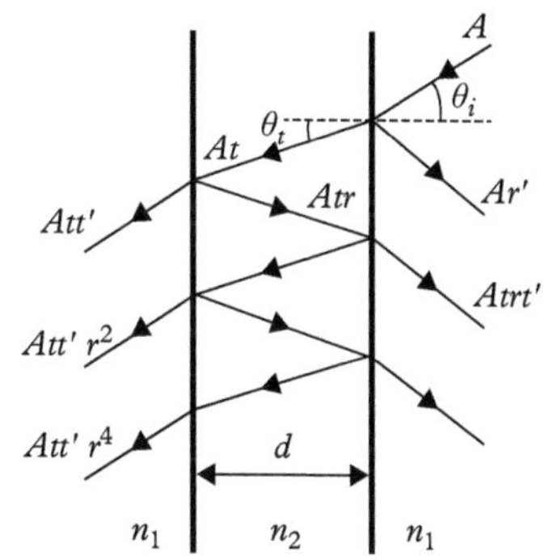

Figure 8.7 *Geometry to analyse the effect of multiple reflections in a parallel plate.*

The first wave transmitted through the plate can be written in the form:

$$E_0 = Att' e^{i(\mathbf{k}\cdot\mathbf{r}-\omega t)} \tag{8.42}$$

Compared with the previous wave, each wave subsequently transmitted has a factor r^2 in the amplitude and a constant phase difference given by Eq. (8.33). Thus, the transmitted wave of order n can be written in the form:

$$E_n = r^{2n} Att' e^{i(\mathbf{k}\cdot\mathbf{r}-\omega t)} e^{in\delta} \tag{8.43}$$

where $n = 0, 1, 2, \ldots$. By the superposition principle, the resulting field at the observation point is given by the sum of the fields corresponding to all transmitted waves. Such a sum can be written in the form:

$$E_t = E_0 \left(1 + r^2 e^{i\delta} + r^4 e^{i2\delta} + r^6 e^{i3\delta} + \cdots\right) \tag{8.44}$$

The geometric series in Eq. (8.44) converges since $r^2 e^{i\delta} < 1$. The result is given by

$$\begin{aligned} E_t &= E_0 \frac{1}{1 - r^2 e^{i\delta}} \\ &= E_0 \frac{1 - r^2 \cos\delta + ir^2 \sin\delta}{1 - 2r^2 \cos\delta + r^4} \end{aligned} \tag{8.45}$$

The irradiance transmitted through the plate is given by $[I_t = E_t E_t^*/2]$, that is,

$$I_t = \frac{I_i (tt')^2}{1 - 2r^2 \cos\delta + r^4} \tag{8.46}$$

where $I_i = A^2/2$ is the irradiance of the incident wave. If there is no energy absorption, it is $tt' = 1 - r^2$. Using this relation and the trigonometric identity $\cos\delta = 1 - 2\sin^2(\delta/2)$, Eq. (8.46) can be presented in the form:

$$I_t = I_i \frac{1}{1 + F \sin^2 \frac{\delta}{2}} \tag{8.47}$$

where

$$F = \frac{4r^2}{(1-r^2)^2} \tag{8.48}$$

is the so-called *finesse coefficient*. This parameter is a very sensitive function of the reflection coefficient r since, as r varies from 0 to 1, F varies from 0 to infinity.

The function $\left[1+F\sin^2(\delta/2)\right]^{-1} = I_t/I_i = T$ in Eq. (8.47) is the so-called *Airy function*, which is represented in Fig. 8.8 for several values of the reflectance $R = r^2$. It can be seen that the value of the Airy function is maximum (equal to one) for $\delta = 2\pi p$, whatever the value of r. When r approaches 1, the transmitted irradiance is reduced, except in the vicinity of the points where these maxima occur. Notice that, for real mirrors with absorption losses, the maximum transmittance is somewhat less than unity.

In the case of no energy absorption, the incident irradiance shall be equal to the sum of the reflected and transmitted irradiances:

$$I_i = I_r + I_t \tag{8.49}$$

The irradiance corresponding to the reflected beam can be obtained using Eqs. (8.47) and (8.49).

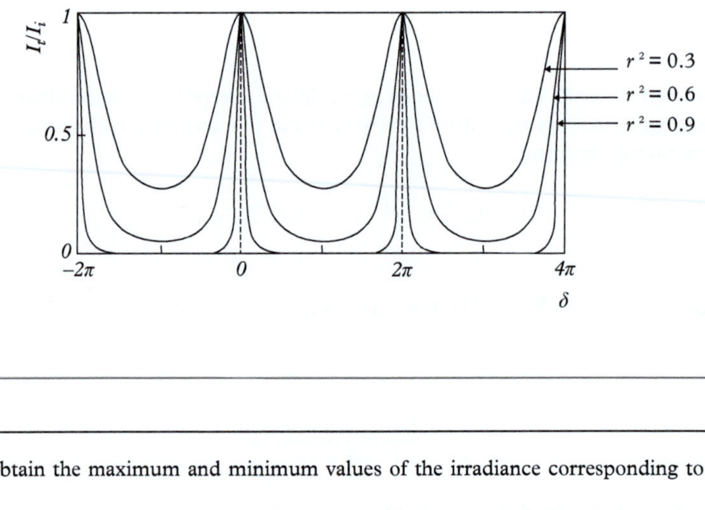

Figure 8.8 *Airy function for various values of reflectance $R = r^2$.*

Example 8.6

Taking into account Eqs. (8.47) and (8.49), obtain the maximum and minimum values of the irradiance corresponding to the transmitted and reflected beams.

Solution

From Eq. (8.47) it can be seen that the irradiance of the transmitted beam, I_t, is maximum when $\delta = 2\pi p$, in which case one has

$$(I_t)_{\max} = I_i \tag{8.50}$$

Under these conditions, I_r assumes a minimum value:

$$(I_r)_{\min} = 0 \tag{8.51}$$

continued

Example 8.6 continued

On the other hand, I_t is minimum and I_r is maximum for $\delta = (2p+1)\pi$, with the results:

$$(I_t)_{\min} = I_i \frac{(1-r^2)^2}{(1+r^2)^2} \tag{8.52}$$

and

$$(I_r)_{\max} = I_i \frac{4r^2}{(1+r^2)^2}. \tag{8.53}$$

In this case, the transmitted irradiance tends to zero, whereas the reflected irradiance tends to *Ii* when *r* approaches 1, as expected.

8.6 Further Reading

Born, M., and E. Wolf, *Principles of Optics*, 6th Ed., Pergamon Press, 1980.
Guenther, R., *Modern Optics*, John Wiley & Sons, 1990.
Hecht, E., *Optics*, 5th Ed., Pearson Education, 2016.
Klein, M., and T. Furtak, *Optics*, 2nd Ed., John Wiley & Sons, 1986.
Pedrotti, F., L. M. Pedrotti, and L. S. Pedrotti, *Introduction to Optics*, 3rd Ed., Cambridge University Press, 2018.

8.7 Problems

8.1 Laser light passes through two identical and parallel slits, 0.4 mm apart, in a Young's interference experiment. The light wavelength is $\lambda = 500$ nm and interference fringes are seen on a screen 1 m away. Find the distribution of irradiance in the observation plane and the separation between the second dark fringe and the central bright fringe. Consider that the irradiance due to one slit is I_0

8.2 Light of wavelength $\lambda = 600$ nm is used in a Young's interference experiment. When a thin film with refractive index $n = 1.5$ is placed behind one of the slits, the zero-order fringe moves to the position previously occupied by the fifth-order fringe. What is the film thickness?

8.3 Consider the interference pattern formed when the three slits represented in Fig. 8.9 are illuminated by a plane light wave.

a) Obtain an expression for the distribution of irradiance as a function of angle θ.

b) Determine the value of $\theta = \theta_1 \, (\neq 0)$ corresponding to the first secondary maximum.

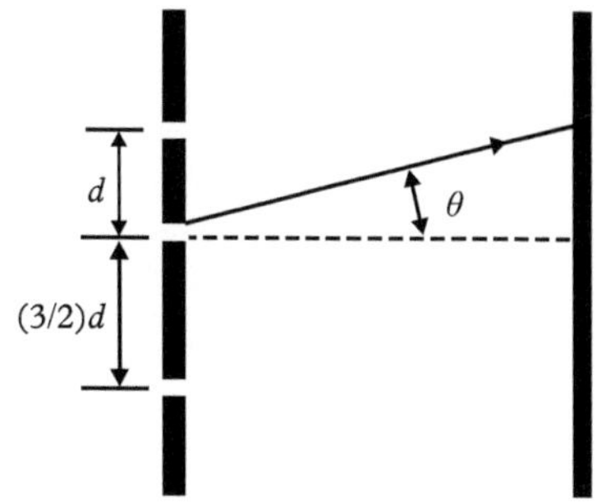

Figure 8.9 *Schematic of a three-slit interference experiment.*

8.4 It is intended to reduce the reflectivity of a glass surface (refractive index n_g) applying an appropriate dielectric film. Determine the refractive index $n_f \, (n_f < n_g)$ and the thickness d that this film should have. Consider, for the purpose of calculations, the wavelength $\lambda = 550$ nm, for which the sensitivity of the human eye is maximum.

8.5 An optical system consists of two layers of a dielectric (1) with refractive index $n_1 = 1.4$ and three layers of another dielectric (2) with refractive index $n_2 = 2.5$, deposited alternately on a glass substrate $(n_g = 1.5)$. The first layer to be deposited is of dielectric (2), as illustrated in Fig. 8.5. Show that the system constitutes a mirror of high reflectivity if the thickness of each layer is $\lambda/4$ and the incidence of light is normal.

8.6 Two rectangular glass plates are in contact along one of their edges, whereas on the opposite side they are separated by a 5 μm thick spacer (Fig. 8.10). The refractive indices of the upper and lower plates are, respectively, 1.5 and 2.0. Fringes of interference are formed by normally reflected light of wavelength $\lambda = 500$ nm. Calculate the number of fringes that are observed if

a) there is air between the plates;

b) there is oil with a refractive index $n = 1.8$ between the plates.

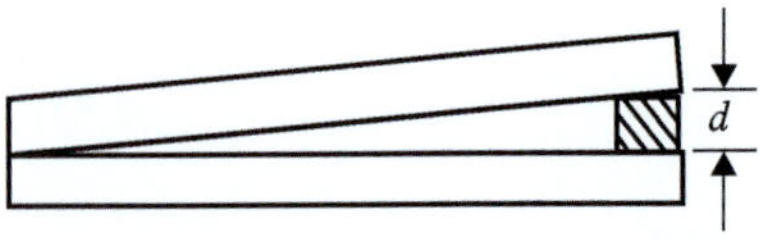

Figure 8.10 *Schematic of a non-uniform film formed between two rectangular glass plates.*

8.7 Two rectangular glass plates are in contact along one of their edges but make a small angle θ between them. There is water ($n = 4/3$) in the space between the plates, which are illuminated at normal incidence by light of wavelength $\lambda = 400$ nm. The observed fringes are spaced by 0.15 mm.

a) Determine the angle θ.

b) Determine the space between the plates at the fifth fringe position and the distance between this fringe and the contact edge.

8.8 Light with wavelength $\lambda = 500$ nm illuminates at normal incidence a soap film ($n = 4/3$) of variable thickness d. Find the intensities of the interference maxima and minima relative to the incident intensity.

8.9 A dielectric film ($n = 3/2$) with parallel faces and thickness $d = 200$ nm is illuminated normally with white light ($\lambda = 350 - 700$ nm). What is the wavelength of the reflected light?

8.10 A collimated beam of light with wavelength $\lambda = 400$ nm illuminates normally a soap film ($n = 4/3$). Constructive second-order interference of the reflected light is observed. What is the film thickness?

8.11 The diameter of the seventh bright ring in Newton's rings experiment decreases from 1.55 cm to 1.28 cm when a liquid is introduced between the lens and the flat glass plate on which it rests. Determine the refractive index of this liquid if sodium yellow light ($\lambda = 589.3$ nm) is used in this experiment.

8.12 A plano-convex lens is placed with its convex side down on a flat glass plate and illuminated from above with light of wavelength $\lambda = 650$ nm. The diameter of the third bright ring in the interference pattern is $D = 1.4$ mm. Find the radius of curvature of the lens convex surface.

8.13 A dielectric film with a thickness $d = 1$ mm and a refractive index $n = 1.5$ is illuminated with monochromatic light of wavelength $\lambda = 600$ nm. A pattern of circular Haidinger fringes is then observed. Determine the order p of the central fringe ($\theta_t = 0$). Is that fringe bright or dark?

8.14 Considering the superposition of all waves reflected in Fig. 8.7, obtain the irradiance of the resulting reflected field. Show that the result is the same as that obtained using Eqs. (8.47) and (8.49).

8.15 Obtain the full width at half maximum of multiple-reflection interference fringes when the reflectance of the dielectric plate is $R = r^2 = 0.90$.

Interferometry

9

Optical interferometers are systems or devices capable of measuring small variations in length, refractive index, or wavelength in any of a variety of ways by analysing changes in an interference pattern created by two or more beams. This general description reflects the wide variety of designs and applications of interferometers. They can be used, for example, to test the quality of a surface or a dielectric film; they can also be used to measure distances of the order of tens or hundreds of metres, as happens in the study of the stability of large structures.

Interferometers can use any of the approaches discussed in Chapter 8: *wavefront division* or *amplitude division*. The first approach includes interferometers using virtual sources, which can be analysed in a way similar to Young's experiment. The Michelson interferometer and other important interferometers are based on the amplitude-division technique. Some of these interferometers can be implemented using optical fibres.

9.1 Interferometers Using Virtual Sources

Young's experiment, described in Chapter 8, is an example of an interferometer that uses *wavefront division*. The same physical and mathematical considerations, referred to in connection with Young's experiment, can be applied to some other interferometers based on the division of the wavefront. Three of these interferometers are described in the following sections: Fresnel's mirrors, Fresnel's biprism, and Lloyd's mirror. Any of these arrangements uses only one real light source, which produces, through reflection or refraction, images that act as virtual light sources.

9.1.1 Fresnel's mirrors

Fresnel's mirrors consist of two flat mirrors that make a small angle θ between themselves (Fig. 9.1). The light coming from a real source S is reflected in the two mirrors, so that the reflected waves appear to come from two sources S_1 and S_2. These two virtual sources play the role of the slits S_1 and S_2 of Young's experiment represented in Fig. 8.2. The interference pattern can be observed in a plane located in the region where the two reflected beams superpose.

Sources S, S_1, and S_2 in Fig. 9.1 are situated on the circumference of centre O and radius $R = \overline{SO}$, such that

$$\overline{S_1S_2} = h \approx 2\theta R \tag{9.1}$$

The distance between the plane of the virtual sources S_1 and S_2 and the observation plane is $D = R + d$, where d is the distance from O to the plane of observation. Using Eq. (8.27), the spacing between the interference fringes is given by

$$\Delta y = \frac{D}{h}\lambda = \frac{(R+d)\lambda}{2\theta R} \tag{9.2}$$

Lasers, Optics, and Photonics. Mário F. S. Ferreira, Oxford University Press.
DOI: 10.1093/oso/9780198952770.003.0009

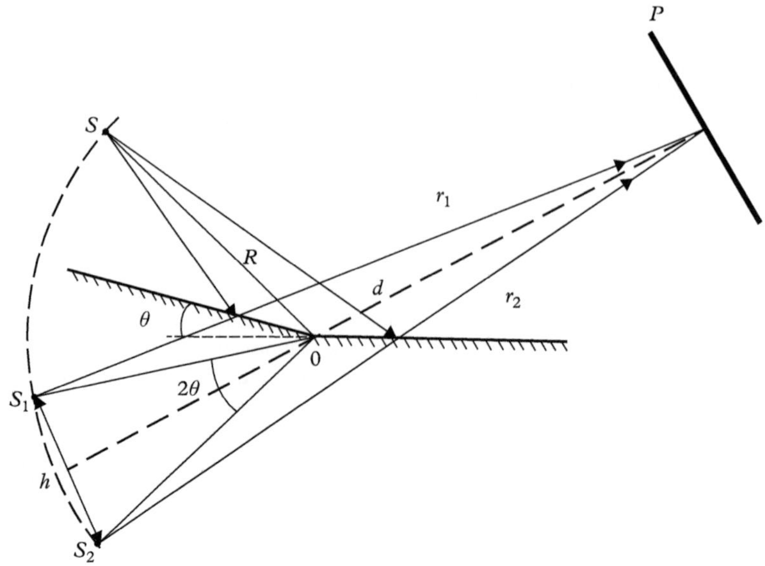

Figure 9.1 *Schematic representation of Fresnel's mirrors.*

Example 9.1

Find the angle θ between the two Fresnel's mirrors, assuming that the intersection of these mirrors is at a distance of 2 m from the source and 4 m from the plane of observation, whereas the wavelength of light is $\lambda = 600$ nm and the spacing between consecutive bright fringes is 2 mm.

Solution

In this case, the distance between the plane of the virtual sources and the observation plane is $D = R + d = 6\,\text{m}$. Using Eq. (9.2), the angle between the two Fresnel's mirrors is given by

$$\theta = \frac{(R+d)\lambda}{2R\Delta y}$$

$$= \frac{(2+4)\left(600 \times 10^{-9}\right)}{2(2)\left(2 \times 10^{-3}\right)} = 4.5 \times 10^{-4}\text{rad}$$

9.1.2 Fresnel's biprism

The Fresnel biprism consists of two prisms glued on their bases, as represented in Fig. 9.2. When a cylindrical wave falls on the two prisms, the upper part of the wavefront is refracted downward, making it appear that the refracted wave comes from a source located at S_1. As for the bottom of the original wavefront, it is refracted upwards, making it appear that the

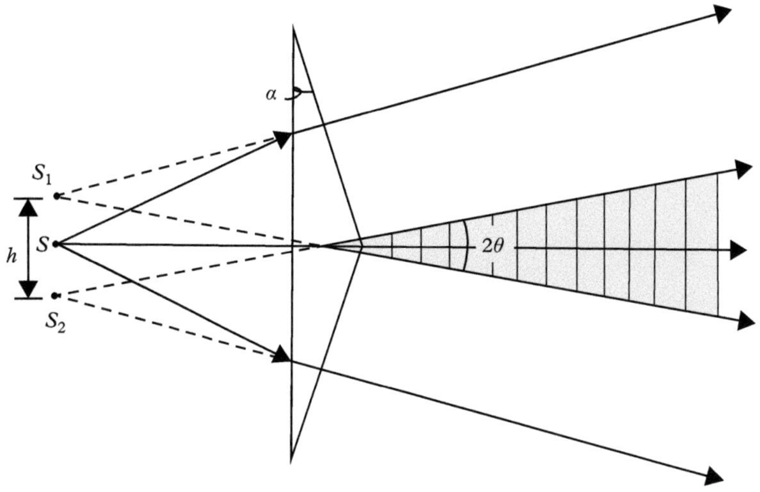

Figure 9.2 *Schematic representation of Fresnel's biprism.*

refracted wave comes from a source situated at S_2. The interference pattern is formed in the superposition region of the two refracted waves. The spacing between the interference fringes is also given by Eq. (9.2), where the angle θ is given by

$$\theta = (n - 1)\,\alpha \tag{9.3}$$

n being the refractive index and α the angle of the prism.

Example 9.2

Consider the Fresnel biprism represented in Fig. 9.2, with a refractive index n and a small angle α.

a) Assuming an incident ray normal to the vertical face of the prism, show that its angular deviation θ is given by Eq. (9.3).
b) Assuming that $n = 1,5$ and $\alpha = 0.006$ rad, obtain the fringe separation when the biprism is illuminated by a cylindrical wave with wavelength $\lambda = 600$ nm

Solution

a) The ray is not deviated when entering the biprism, but only when going out of it. The law of refraction applied to this case gives

$$n \sin\alpha = \sin(\alpha + \theta)$$

Considering the small-angle approximation, one has that $n\alpha = \alpha + \theta$, from which Eq. (9.3) follows.

b) The fringe separation is given by

$$\Delta y \sim \lambda/(2\theta) = 100\ \mu\text{m}$$

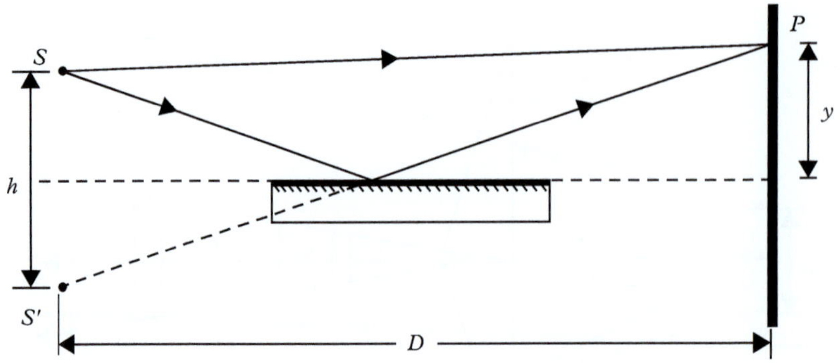

Figure 9.3 *Schematic representation of Lloyd's mirror.*

9.1.3 Lloyd's mirror

Lloyd's mirror consists of a single mirror, in which part of the wavefront emitted by a source S is reflected (Fig. 9.3). This source is usually quite close to the plane of the mirror. The interference pattern is obtained by the superposition of the wave reflected by the mirror and the wave coming directly from the source S. The spacing between the fringes in the observation plane is also given by Eq. (8.28), where h is the separation between the real source S and its image S' produced by the mirror.

Lloyd's mirror has a feature that distinguishes it from the other interferometers mentioned earlier. In fact, since the wave reflected by the mirror has a phase difference of π rad relative to the incident wave, Eq. (8.26) for the distribution of irradiance is changed and takes the form:

$$I = 4I_0 \sin^2\left(\frac{\pi h y}{\lambda D}\right) \tag{9.4}$$

If the observation plane is in contact with the mirror, a dark fringe is found at $y = 0$, instead of the bright fringe observed in Young's experiment.

9.2 Rayleigh's Interferometer

Rayleigh's interferometer is also based on the scheme of Young's experiment, as represented in Fig. 9.4. The light from the point source S is collimated before passing through the two slits S_1 and S_2. The beams emerging from these slits pass through the tubes T_1 and T_2, which are full of gas. The interference fringes are formed in the focal plane of lens L and move when the gas pressure in one of the tubes is changed. Counting the number of displaced fringes, N, allows one to measure the variation of the optical path through the tube and, therefore, the variation in the gas refractive index, Δn, using the relation

$$N = \frac{d\Delta n}{\lambda}, \tag{9.5}$$

where d is the length of the tube.

For a diluted gas, the refractive index differs from the unit by a value proportional to the density, so that n-1 is proportional to the pressure. The refractive index obtained

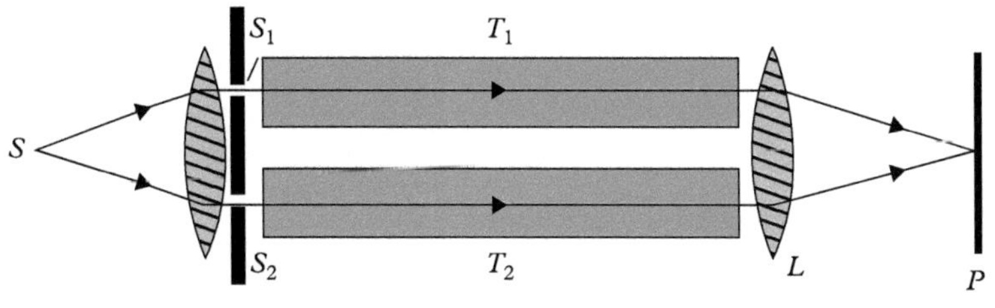

Figure 9.4 *Schematic representation of Rayleigh's interferometer.*

in a given measurement can then be used to calculate its value at a different pressure. This interferometer is particularly useful in the diagnosis of plasmas and has been used to measure variations in the refractive index on the order of 10^{-8}.

9.3 The Michelson Interferometer

Among the various interferometers based on amplitude division, the best-known and historically the most important is Michelson's interferometer, which was developed by the American physicist Albert Michelson (1852–1931). This interferometer can work with a more extended source than that used in Young's experiment, thereby increasing the irradiance of the interference pattern.

Fig. 9.5 presents a schematic representation of Michelson's interferometer. The beam of light coming from the source S is divided into two beams using a dielectric plate with a partially reflective layer on one of its faces, which acts as a beam-splitter, *BS*. The two beams then follow independent and orthogonal paths, reflect in mirrors $M1$ and $M2$, and meet again on the same beam-splitter, where they interfere.

It should be noted that one of the beams crosses the beam-splitter three times, whereas the other beam crosses it only once. For this reason, a compensation plate, CP, is introduced into the BS—$M2$ arm, parallel and with the same thickness as the *BS*, although without any reflective layer on its surfaces. The inclusion of this plate allows one to compensate for the dispersion effects of the beam in *BS* and ensures that any variation in the optical paths in the two arms of the interferometer depends only on changes in its geometry.

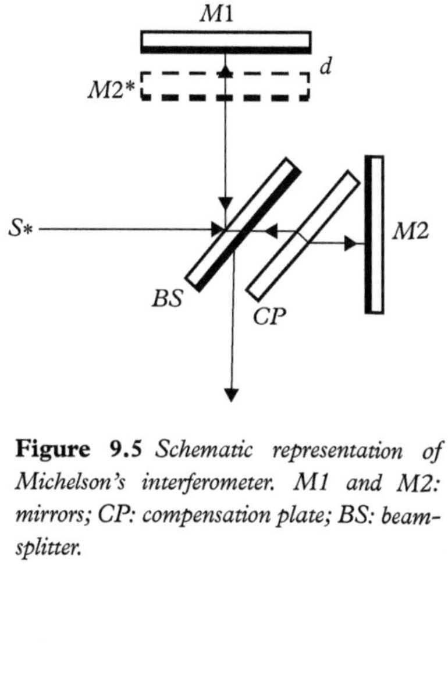

Figure 9.5 *Schematic representation of Michelson's interferometer. M1 and M2: mirrors; CP: compensation plate; BS: beam-splitter.*

Looking at the beam-splitter from the detector, an image of the mirror $M2$, indicated by $M2^*$, is observed near the mirror $M1$. The virtual mirror $M2^*$ and the mirror $M1$ form a dielectric film with a thickness d and a refractive index $n_2 = 1$. The difference in optical path between the rays reflected in $M1$ and $M2$ is $2d\cos\theta$, so the corresponding phase difference is $\frac{2\pi}{\lambda_0}2d\cos\theta$. However, there is an additional phase shift of π rad between the two rays, since one of them is reflected internally and the other externally in the beam-splitter plate. Therefore, the phase difference corresponding to a dark fringe in the observation plane is

$$\delta = p2\pi = \frac{2\pi}{\lambda_0}2d\cos\theta \tag{9.6}$$

where p is an integer. If the distance d is constant, equal-inclination circular fringes are observed. The maximum value of p occurs at the centre of this set of fringes, corresponding to $\theta = 0$:

$$p_{\max} = \frac{2d}{\lambda_0} \tag{9.7}$$

When d increases, the fringes move away from the pattern centre, where successive fringes of higher order appear. The irradiance at this point varies according to the expression:

$$I = I_0\left[1 - \cos\left(\frac{2\omega d}{c}\right)\right] \tag{9.8}$$

where a 50:50 beam-splitter is assumed and the relation is used.

$$\frac{\omega}{c} = \frac{2\pi}{\lambda_0} \tag{9.9}$$

has been used. The irradiance is thus given by the sum of a constant term and an oscillatory one.

The Michelson interferometer can be used to perform displacement measurements with great precision. In fact, when one of the mirrors is translated by $\lambda_0/2$, each fringe moves so as to take the place of the adjacent one. By counting the number of fringes, N, that pass through a given reference point, it is possible to determine the displacement, Δd, of the mirror, which is given by

$$\Delta d = N\frac{\lambda_0}{2} \tag{9.10}$$

Conversely, Eq. (9.10) allows us to calculate the light wavelength, λ_0, once the mirror displacement and the corresponding number N of passing fringes are known.

Example 9.3

A transparent plate with a refractive index n and thickness L is introduced into one of the arms of the Michelson interferometer, perpendicularly to the beam. Determine the displacement that should be given to the mirror so that the fringe system in the observation plane remains the same as before the plate is introduced.

Solution

When the transparent plate is introduced, the optical path corresponding to its thickness varies from $1L$ to nL. Considering that this element is crossed twice, the change in the optical path resulting from the introduction of the plate is $2(n-1)L$. In order to compensate for this variation in the optical path, the mirror must move by Δd, such that,

$$2\Delta d = 2(n-1)L.$$

That is,

$$\Delta d = (n-1)L$$

9.3.1 Biographical note—Michelson

Albert Abraham Michelson (1852–1931) was born in Strzelno, now belonging to Poland but then included in Germany, and went as a child to the United States. He studied at the United States Naval Academy and the Universities of Berlin, Heidelberg, and Paris. He was professor of physics at Clark University from 1889 to 1892 and at the University of Chicago between 1892 and 1929. Michelson measured the speed of light with great precision, using instruments designed by himself. In 1887, he used the interferometer he invented, along with the American chemist Edward Williams Morley, to perform an experiment that became famous. This experiment showed that ether—whose existence many scientists accepted to explain the propagation of light waves—did not exist after all. The results of this experiment were also important for the development of the theory of relativity. Michelson was the first American to receive the Nobel Prize in physics in 1907, in recognition of the instruments and experiments he developed. Among his most relevant publications are the works *Velocity of Light* (1902) and *Studies in Optics* (1927).

9.4 The Twyman–Green Interferometer

The Twyman–Green interferometer resulted from a development of the Michelson interferometer, distinguishing itself particularly by the fact of using a point source together with a collimating lens $L1$, so that all rays entering the interferometer are parallel to the optical axis. The parallel rays emerging from the interferometer produce an interference pattern which is observed in the focal plane of the lens $L2$ (see Fig. 9.6(a)). With this configuration, the light ray path can be compared throughout the entire field, making it possible to measure very small inclinations of the mirror $M2$.

As discussed in Section 9.3, when Michelson's interferometer is aligned in such a way that the air film between $M2$ and $M1$ has a uniform thickness, a pattern of circular fringes is observed. Under the same conditions, the Twyman–Green interferometer provides a uniform field. A slight inclination of one of the mirrors produces a system of parallel straight fringes through the field of view, and any small imperfection of the film formed between the mirror $M1$ and the virtual image of the mirror $M2$ can be detected through deviations in the geometry of these fringes.

The Twyman–Green interferometer allows us to measure the parallelism and quality of the surfaces of the optical elements located on the arm with mirror $M2$. Incidentally, this mirror can be replaced by a test lens, $L3$, and by a convex spherical mirror, $M3$, the centre of curvature of which coincides with the focus of the lens (see Fig. 9.6(b)). This setting allows us to analyse the quality of the lens $L3$ since any imperfection of this lens becomes evident through the irregularity of the interference pattern formed in the observation plane.

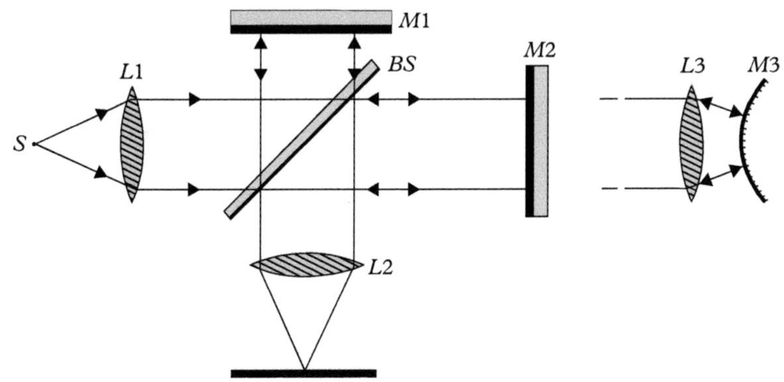

Figure 9.6 *(a) Schematic representation of Twyman–Green's interferometer and (b) a test lens, with a convex spherical mirror. M1, M2, and M3: mirrors; L1, L2, and L3: lenses; BS: beam-splitter.*

9.5 The Michelson–Morley Experiment

The Michelson–Morley experiment was first carried out in 1887 with the aim of measuring the speed of translation of the earth. At that time, it was accepted that light needed a medium, called ether, for its propagation through space. The properties of this medium had been the subject of various investigations, some of which were carried out by Maxwell. In the case that this ether exists, the speed of light, measured on earth, should depend on the direction in which the earth moved. A Michelson interferometer, presenting one of its arms in the direction of this movement and the other arm in the perpendicular direction, would allow one to verify this effect through the shifting of the interference fringes. Fig. 9.7 shows a schematic representation of the Michelson–Morley experiment.

For an observer moving at the speed of the earth in its orbit, the ether will appear to move in the opposite direction at that same speed (v). Whereas the earth moves in the direction *BSM*2, as indicated in Fig. 9.7, the time it takes for the light to go from *BS* to *M*2 and return to *BS* will be

$$t_{\parallel} = \frac{d}{c+v} + \frac{d}{c-v} = \frac{2cd}{c^2 - v^2} \tag{9.11a}$$

On the other hand, the time spent by the light to go from *BS* to M1 and return to *BS* will be

$$t_{\perp} = 2\frac{d}{\sqrt{c^2 - v^2}} \tag{9.11b}$$

Each arm of the interferometer built by Michelson and Morley had an effective length $d = 11$ m. The interferometer was mounted on a stone table floating on mercury, and the fringes were observed as the system rotated from 90°. This rotation allowed the two arms of the interferometer to be alternately parallel and perpendicular to the earth's motion.

The number of fringes, N, passing through a certain point of reference is given by the quotient of the difference of optical paths, $c\Delta t$, and the wavelength λ. Due to the 90° rotation of the system, the relative delay of the two beams is doubled, so it is

$$N = \frac{2c\Delta t}{\lambda} \approx \frac{2dv^2}{c^2\lambda} \tag{9.12}$$

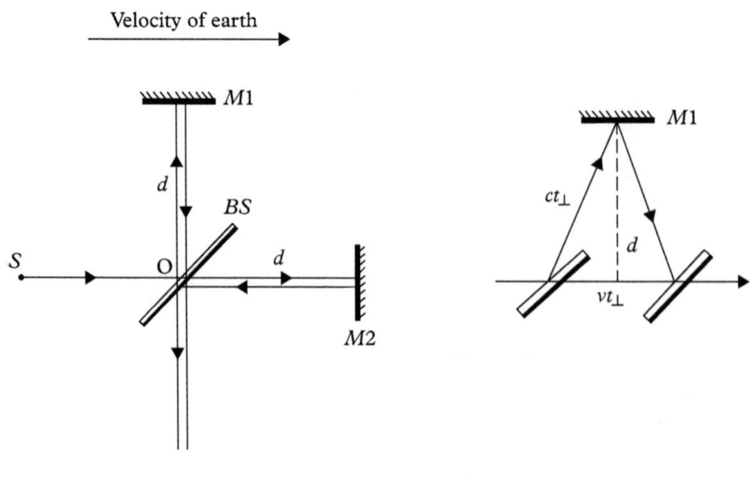

Figure 9.7 *Schematic representation of Michelson–Morley's experiment. M1 and M2: mirrors, BS: beam splitter.*

Considering $d = 11$ m, $\lambda = 590$ nm (corresponding to sodium light) and an orbital speed of the earth $v = 30$ km/s, it is verified from Eq. (9.12) that the fringe deviation would be $N = 0.37$. However, this deviation was not actually observed.

Michelson–Morley's experience was repeated with the utmost precision in 1930 by Joos and in 1958 by Cedarholm et al., using the newly invented ammonia maser for this purpose. In either case, the result was the same as that obtained by Michelson and Morley.

9.6 The Mach–Zehnder Interferometer

The Mach–Zehnder interferometer is another apparatus based on amplitude division. This interferometer is named after the physicists Ludwig Mach and Ludwig Zehnder; it was first proposed by Zehnder in 1891, refined by Mach in 1892. As illustrated in Fig. 9.8, the incident beam is divided into two beams at beam-splitter $BS1$, each of them being afterwards totally reflected by mirrors $M1$ and $M2$. The two beams are finally recombined at beam-splitter $BS2$. The Mach–Zehnder interferometer is a highly configurable instrument. Compared to the Michelson interferometer, it is more flexible, since the paths of the beams incident and reflected at mirrors $M1$ and $M2$ are different. This makes the alignment of this device relatively difficult. However, for the same reason, it has found some very useful applications. If the source has a low coherence length, then great care must be taken to equalize the two optical paths. When an object is placed in one of the arms of the interferometer, the difference in optical paths between the two arms is changed and the interference pattern is affected.

Mach–Zehnder interferometer is commonly used to study the flux of gas in wind tunnels in the presence of obstacles. One of the beams passes through the wind tunnel, while the other passes through suitable CPs. This study is based on the fact that the refractive index is, in many situations, proportional to the density of the gas. The beam inside the tunnel passes through regions with different densities and, therefore, different refractive indices, which determines the formation of contour fringes in the interference pattern.

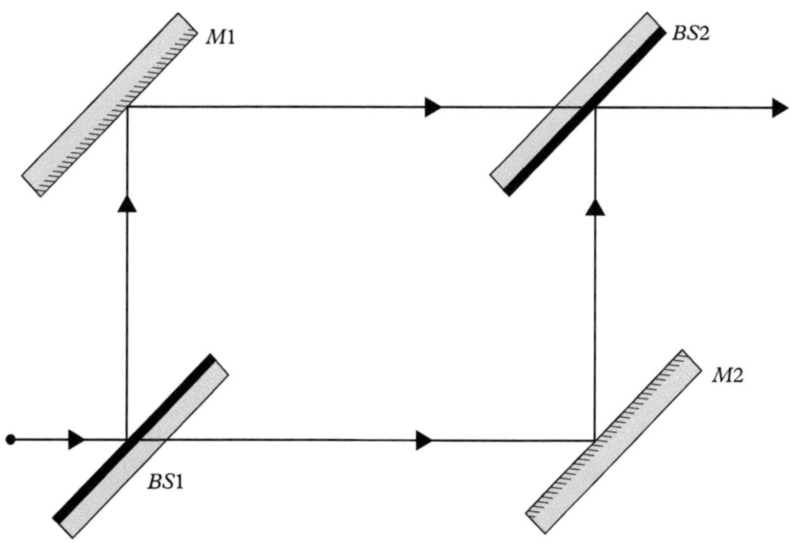

Figure 9.8 *Schematic representation of the Mach–Zehnder interferometer. M1 and M2: mirrors; BS1 and BS2: beam splitters.*

9.7 The Sagnac Interferometer

The Sagnac interferometer, named after the French physicist Georges Sagnac (1869–1928), was first implemented in 1913 and is characterized by the two beams following the same path but in opposite directions. Actually, it is an example of the so-called *common-path interferometers*: a class of interferometers in which the reference and sample beams travel along the same path. A common-path interferometer is generally more robust to environmental vibrations than a 'double-path interferometer' such as the Michelson interferometer or the Mach–Zehnder interferometer. The Sagnac interferometer can display three or more mirrors (see Fig. 9.9), and it is relatively stable and easy to align. A small variation in the orientation of one of the mirrors determines the appearance of a parallel fringe pattern in the observation plane.

The use of an interferometer typically requires the possibility of imposing variations only on one of its beams. However, in the case of the Sagnac interferometer, since its two beams overlap, it cannot be used in the conventional mode. Moreover, this interferometer is insensitive to any effect other than rotation around an axis perpendicular to the plane of the device. Rotation causes a shortening of the optical path for one of the beams and an elongation of the path for the other beam.

Assuming that the ring formed by the Sagnac interferometer has a medium radius R, the time spent by the light to take a walk around the device in the absence of rotation is $2\pi R/c$. However, if this ring is rotating at an angular speed ω, the optical path of one of the beams is increased by

$$ds = \frac{2\pi R}{c}\omega R \qquad (9.13)$$

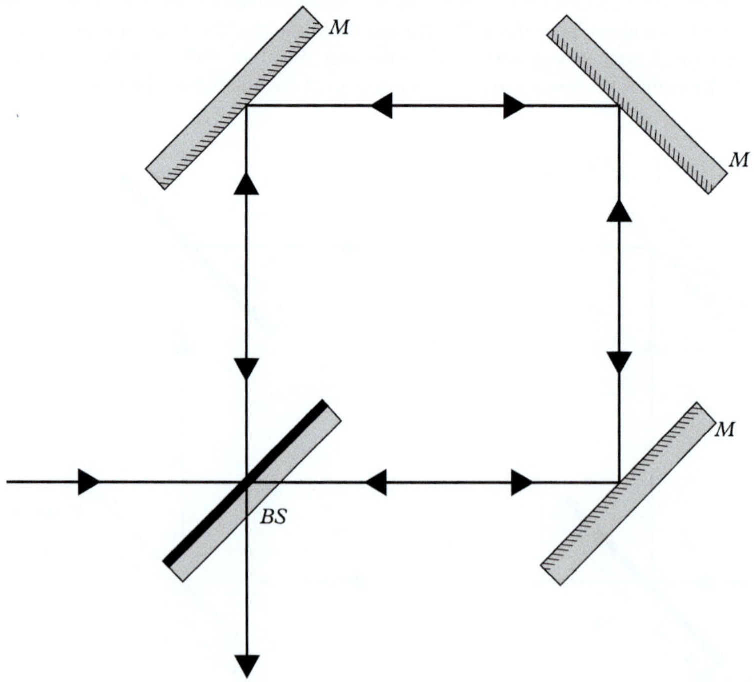

Figure 9.9 *Schematic representation of the Sagnac interferometer. M: mirror, BS: beam splitter.*

whereas the optical path of the other beam is shortened by the same value. That is, considering a given reference point in the ring, it is found that the first beam reaches that point with a delay in time of ds/c, whereas the second beam reaches the same point with an advance of ds/c. The difference in travel time is then given by

$$\Delta t = 2\frac{ds}{c} = 4\frac{\pi R^2 \omega}{c^2} = 4\frac{A\omega}{c^2} \tag{9.14}$$

where A is the ring area.

Assuming that the wavelength of light used in the interferometer is λ, the corresponding period is $T = \lambda / c$, and the displacement of the interference fringes caused by the rotation will be given by

$$\Delta N = \frac{\Delta t}{T} = 4\frac{A\omega}{c\lambda} \tag{9.15}$$

It is a result that has been confirmed experimentally. This method was used by Michelson and Gale in 1925 to determine the angular velocity of the earth.

9.8 Interferometers with Optical Fibres

As will be seen in Chapter 15, light can propagate over enormous distances in glass fibres that have a very low attenuation. Thus, the idea of using these optical fibres to guide the light beams corresponding to the two arms of some of the interferometers described earlier came naturally. The interferometers thus realized are characterized by their very simple configuration, by presenting a high sensitivity, and by being quite stable.

For example, the Mach–Zehnder interferometer, described in Section 9.6, can be performed using optical fibres in its two arms, as shown in Fig. 9.10. In this case, the variation

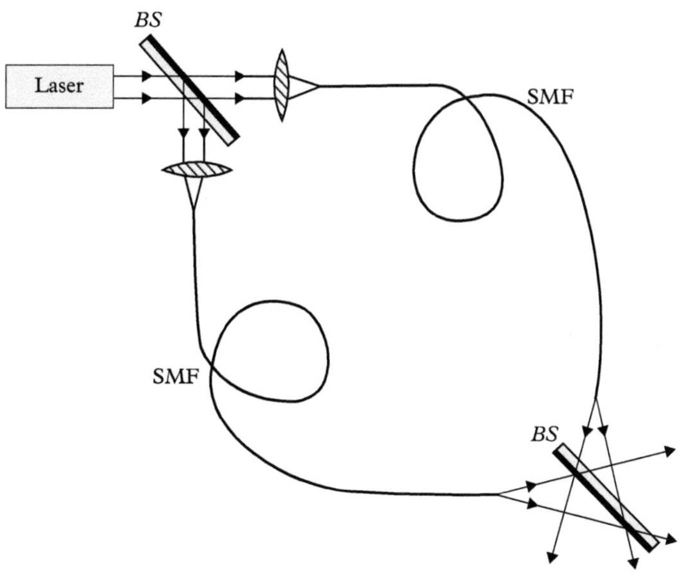

Figure 9.10 *Schematic representation of the Mach–Zehnder interferometer using optical fibres. BS: Beam splitter; SMF: single-mode fibre.*

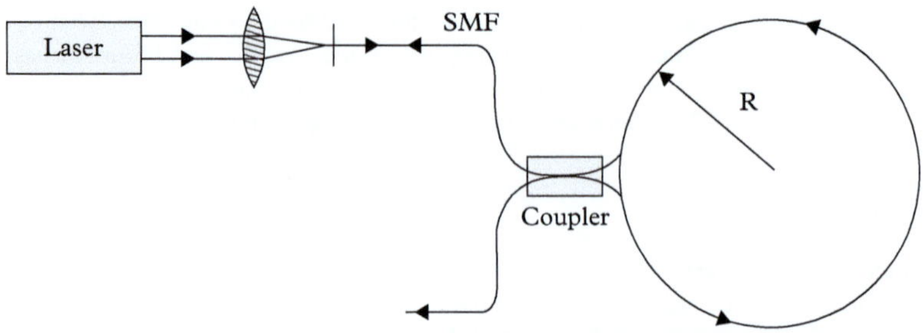

Figure 9.11 *Schematic representation of the Sagnac interferometer in optical fibre. SMF: single-mode fibre.*

in the phase of light that propagates in the optical fibre is caused by some external parameter (temperature, pressure, voltage, magnetic field, etc.) and can be measured quite rigorously through its effects on the interference pattern. In fact, the Mach–Zehnder interferometer implemented using optical fibres is a sensor with characteristics far superior to those of many conventional sensors.

The beams emerging from the two fibres in Fig. 9.10 can be considered as divergent spherical waves, which overlap using a beam-splitter. However, in contrast to what happens with a point source that emits uniformly in all directions, in this case, the interfering beams present a radially Gaussian distribution of amplitude. Consequently, the observed circular fringe pattern presents a global distribution of irradiance which is also Gaussian. When the two fibre ends approach each other, the circular fringes expand. Conversely, the fringes contract when these fibre ends move away from each other.

The Sagnac interferometer, described in Section 9.7, can also be performed with optical fibres and used as a gyroscope, as shown in Fig. 9.11. The system makes use of a fibre-optic coil, along which the two beams of light propagate in opposite directions, and is sensitive to rotation around the axis of the coil. The difference between the travel times of the two beams is given by Eq. (9.14), corresponding to a phase difference given by

$$\Delta\phi = k_0 c \Delta t = \frac{8\pi\omega NA}{\lambda c} \tag{9.16}$$

where A is the area and N number of fibre turns.

Example 9.4

Calculate the phase difference due to the rotation of the earth using light of wavelength $\lambda = 0.6328$ μm and a coil with radius $R = 0.4$ m and 500 fibre-optic turns.

Solution

The angular frequency corresponding to the rotation of the earth is $\omega = 2\pi/(24 \times 3600) = 7.27 \times 10^{-5}$ rad/s. Using the area $A = \pi(0.4)^2$ m^2 and the other parameter values in Eq. (9.16), one obtains a phase difference $\Delta\phi = 2.4 \times 10^{-3}$ rad.

9.9 The Fabry–Perot Interferometer

The Fabry–Perot (FP) interferometer consists of two surfaces with high reflectance, generally separated by air, as represented in Fig. 9.12. Such surfaces are the inner ones of

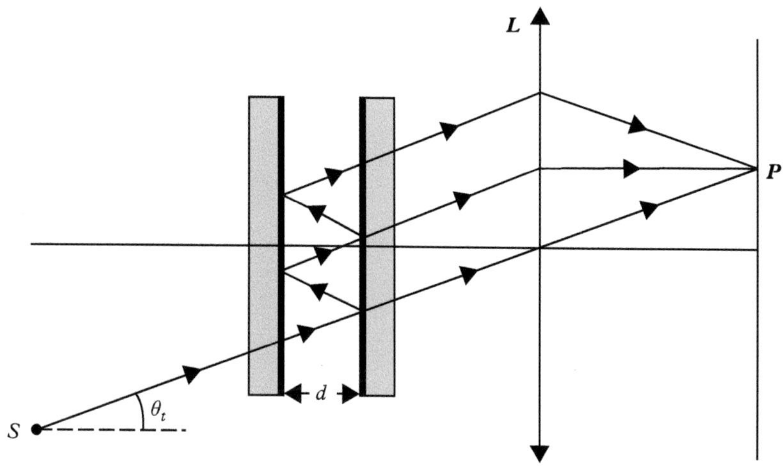

Figure 9.12 *Schematic representation of the Fabry–Perot interferometer. L: lens.*

thick glass or quartz plates, whose reflectance is usually increased by the application of partially reflective metal coatings. This interferometer has proven to be a powerful tool in a wide variety of applications. It has been used, for example, in precision wavelength measurements and in the analysis of the fine structure of spectral lines. The description of the Fabry–Perot interferometer is based on the theory presented in Section 8.5 describing the plane parallel plate.

The definition of the fringes in the interference pattern formed by the light transmitted through the Fabry–Perot cavity can be characterized by the width at half maximum of these fringes, corresponding to $I_t = (I_t)_{\max}/2$, where I_t is the irradiance of the transmitted field. It can be seen from Eq. (8.47) that such a value of irradiance occurs when $\delta = \delta_{\max} \pm \delta^*$, where $\delta_{\max} = 2\pi p$ and

$$\delta^* = 2\sin^{-1}\left(\frac{1}{\sqrt{F}}\right) \tag{9.17}$$

where F is the *finesse coefficient*, given by Eq. (8.48). Since F is generally very high, one has that $\sin^{-1}(1/\sqrt{F}) \approx 1/\sqrt{F}$, so the full width at half maximum (FWHM), $\gamma = 2\delta^*$, is given by

$$\gamma = \frac{4}{\sqrt{F}} \tag{9.18}$$

Fig. 9.13 illustrates this result. From Eqs. (9.17), (9.18), and (8.48) it can be seen that for higher values of the reflectance $R = r^2$, the bright fringes in the transmission pattern become narrower, as observed in Fig. 8.6.

Another figure of merit of the FP interferometer is the *finesse* of the cavity, Φ, which is not to be confused with the coefficient of finesse F. The finesse is given by the ratio of the distance between adjacent maxima to the full width at half maximum, given by Eq. (9.18):

$$\Phi = \frac{2\pi}{\gamma} = \frac{\pi}{2}\sqrt{F} \tag{9.19}$$

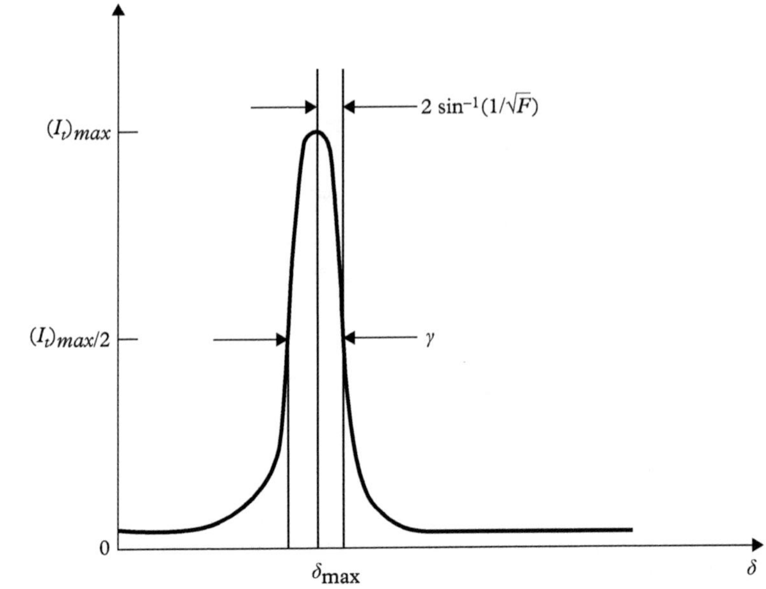

Figure 9.13 *A Fabry–Perot interferometer fringe and its width.*

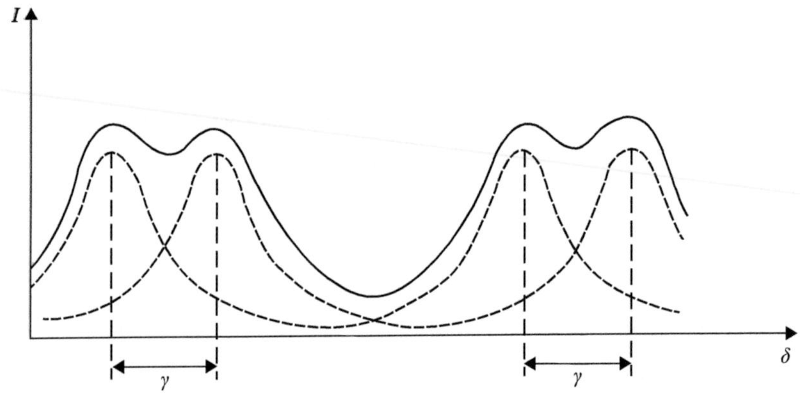

Figure 9.14 *Graph of irradiance for two resolvable wavelengths at the minimum separation.*

The resolving power, $\Re$, of the interferometer is given by $\lambda_0/\Delta\lambda_0$, where λ_0 is the average wavelength of light and $\Delta\lambda_0$ is the minimum value of the difference between two resolvable wavelengths. As a criterion of resolution, it is established that two wavelengths, λ_1 and λ_2, are resolvable provided that the Airy functions corresponding to these wavelengths intersect below the half maximum point of the irradiance peak. In the situation of minimal separation, the total irradiance at the saddle point is equal to the maximum irradiance at either wavelength alone (Fig. 9.14). In this case, the phase variation when going from the maximum irradiance at λ_1 up to the maximum irradiance at λ_2 is $\Delta\delta = \gamma$, given by Eq. (9.18). On the other hand, differentiating Eq. (8.33) yields:

$$\Delta\delta = \frac{-4\pi n d \sin\theta_t}{\lambda_0}\Delta\theta_t \tag{9.20}$$

A bright fringe occurs when the following condition is verified:

$$2nd\cos\theta_t = p\lambda_0 \tag{9.21}$$

Differentiating this equation, one has

$$-2nd\sin\theta_t\Delta\theta_t = p\Delta\lambda_0 \tag{9.22}$$

From Eqs. (9.18), (9.20), and (9.22), one has

$$\Delta\delta = \gamma \Leftrightarrow 2\pi p\frac{\Delta\lambda_0}{\lambda_0} = \frac{4}{\sqrt{F}} \tag{9.23}$$

The resolving power is then given by

$$\Re = \frac{\lambda_0}{\Delta\lambda_0} = \frac{p\pi}{2}\sqrt{F} \tag{9.24}$$

From Eq. (9.21) it can be seen that the maximum value of p occurs when $\cos\theta_t = 1$, being given by

$$p_{\max} = \frac{2nd}{\lambda_0} \tag{9.25}$$

In this case, Eq. (9.24) gives:

$$\Re = \frac{nd\pi}{\lambda_0}\sqrt{F} = p_{\max}\Phi \tag{9.26}$$

Large resolving powers are, of course, desirable. For the scanning Fabry–Perot interferometer, we see that large values of the resolving power occur when the mode number is large and for large values of the finesse. To maximize the mode number p, Eq. (9.25) requires that the plate separation d be as large as possible. On the other hand, the largest values of the finesse occur for reflection coefficients close to unity. Good Fabry–Perot interferometers may be expected to have resolving powers of a million or more.

Besides a minimum wavelength separation that can be resolved with a Fabry–Perot interferometer, there is also a maximum wavelength separation that can be resolved in an unambiguous manner. Actually, if the wavelength separation is too large, the transmittance peak associated with the mode number $p+1$ of λ_1 will overlap the transmittance peak with mode number p associated with λ_2.

The *free spectral range* of an interferometer is the difference in wavelengths $(\Delta\lambda)_{fsr}$ corresponding to a variation of the distance d of $\lambda_0/2$ or a variation in p of one unit, that is, a variation of the phase difference $\Delta\delta = 2\pi$. The free spectral range is the maximum wavelength difference that can be measured unambiguously by the interferometer.

Taking into account Eq. (9.23), we can write:

$$(\Delta\lambda)_{fsr} = \frac{\lambda_0}{p} \tag{9.27}$$

Substituting in this equation the maximum value of p, given by Eq. (9.25), the minimum free spectral range is obtained:

$$(\Delta\lambda)_{fsr} = \frac{\lambda_0^2}{2nd} \tag{9.28}$$

It should be noted that the ratio of the free spectral range to the minimum value of the difference of resolvable wavelengths is equal to the finesse Φ, defined in Eq. (9.19). In fact, using Eqs. (9.24), (9.25), and (9.28) one obtains:

$$\frac{(\Delta\lambda)_{fsr}}{\Delta\lambda_0} = \frac{\pi}{2}\sqrt{F} = \Phi \tag{9.29}$$

In the visible spectrum, the finesse of the Fabry–Perot cavities is usually on the order of 30.

Example 9.5

A Fabry–Perot interferometer consists of an air cavity ($n = 1$), bounded by two surfaces with a reflection coefficient $r = 0.95$, separated by 2 cm. Considering a wavelength $\lambda_0 = 500$ nm, determine the maximum interference order, finesse coefficient, minimum wavelength resolution range, and chromatic resolving power.

Solution
Using Eqs. (9.25), one has that the maximum interference order is

$$p_{\max} = \frac{2d}{\lambda_0} = \frac{2\left(2 \times 10^{-2}\right)}{500 \times 10^{-9}} = 80000$$

On the other hand, from Eq. (8.48) we obtain the finesse coefficient.

$$F = \frac{4r^2}{\left(1 - r^2\right)^2} = \frac{4(0.95)^2}{\left(1 - 0.95^2\right)^2} = 380$$

Eq. (9.23) allows us to obtain the minimum value of the difference between two resolvable wavelengths:

$$\Delta\lambda_0 = \frac{2\lambda_0}{p_{\max}\pi\sqrt{F}} = \frac{2(500 \times 10^{-9})}{(80000)\pi\sqrt{380}} = 2 \times 10^{-13}\ \text{nm}$$

Finally, using Eq. (9.24), the resolving power of the FP interferometer is

$$\Re = \frac{\lambda_0}{\Delta\lambda_0} = \frac{500 \times 10^{-9}}{2 \times 10^{-13}} = 2.5 \times 10^6$$

9.10 Further Reading

Francon, M., *Optical Interferometry*, Academic Press, 1966.
Guenther, R., *Modern Optics*. John Wiley & Sons, 1990.
Hariharan, P., *Optical Interferometry*, Academic Press, 1985.
Hariharan, P., *Basics of Interferometry*, Academic Press, 1992.
Tolansky, S., *An Introduction to Interferometry*, John Wiley and Sons, 1973.
Tyson, R., *Principles of Adaptive Optics*, Academic Press, 1991.

Pedrotti, F., L. M. Pedrotti, and L. S. Pedrotti, *Introduction to Optics*, 3rd Ed., Cambridge University Press, 2018.
Vaughan, J. M., *The Fabry–Perot Interferometer*, CRC Press, 1989.

9.11 Problems

9.1 A Young's experiment is set up using light of wavelength $\lambda = 600$ nm. When a film of transparent material is placed in front of one of the slits, the zero-order fringe moves to the position previously occupied by the bright fringe of order 5. Determine the film thickness, considering that its refractive index is $n = 1.5$.

9.2 A Fresnel biprism with an angle $\alpha = 3°$ and a refractive index $n = 1.5$ is illuminated using light of wavelength $\lambda = 500$ nm. Considering that the light source is at a distance of 20 cm and the observation screen is located 80 cm from the biprism, determine the spacing between fringes in the interference pattern.

9.3 Show that the spacing, h, between the virtual sources S_1 and S_2 in Fresnel's biprism (Fig. 9.2) is given by $h = 2s(n-1)\alpha$, where s is the distance between the plane of the sources and the biprism.

9.4 Fig. 9.15 illustrates an experiment with Lloyd's mirror. The point source S is located at a distance h from the mirror plane AB ($l = 30$ cm) and emits light with wavelength $\lambda = 500$ nm. Assume that $h \ll d$ and $d = 1$cm.

a) Obtain an expression for the spacing between fringes and the extent of the interference pattern in the observation plane P, which is perpendicular to mirror AB.

b) An image of plane P is formed using a converging lens with a focal length of 8 cm, located at a distance of 10 cm from P. The spacing between fringes in the image plane is 1 mm. Obtain the value of h and the number of bright fringes observed.

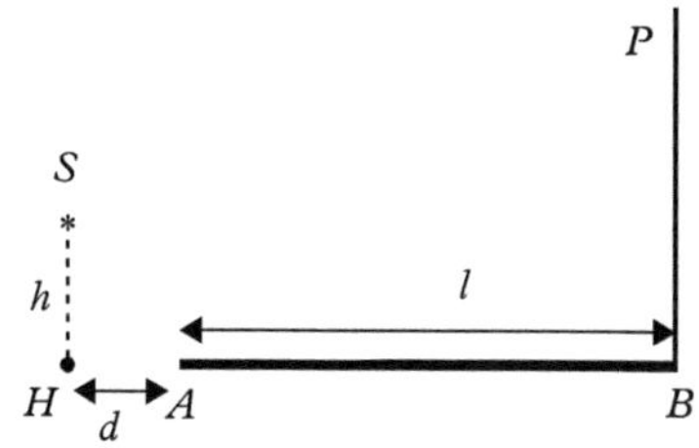

Figure 9.15 *Schematic representation of Lloyd's mirror set-up.*

9.5 A Michelson interferometer using a light source of wavelength $\lambda = 400$ nm is adjusted to obtain a pattern of concentric circular fringes.

a) Determine the displacement that must be given to the moving mirror so that 500 fringes appear at the centre of the interference pattern.

b) If the central fringe is bright, calculate the angle corresponding to the first dark ring.

9.6 The moving mirror in a Michelson interferometer is translated by 0.09 mm and a shift of 300 fringes is observed. Determine (a) the light wavelength used in the Michelson interferometer and (b) the displacement of the fringe system when a plate of glass ($n = 1.5$) with a thickness of 0.005 mm is placed in one arm of the interferometer, normal to the light beam.

9.7 The emerging beam at the exit of a Michelson interferometer is collected by a thin lens of focal length f. Interference rings are observed in the focal plane of this lens. Show that the area between two consecutive bright rings is constant, being independent of the order of these rings.

9.8 In each arm of a Michelson interferometer is placed a tube 10 cm long, in which a vacuum was made. The interferometer is illuminated with yellow sodium light ($\lambda = 589.3$ nm) and aligned so as to observe fringes of equal inclination. When air is introduced into one of the tubes, 12 fringes pass through a given point in the observation plane. Determine the variation of air pressure in this tube, assuming that the refractive index of air, n, relates to the pressure, P, in the form $n = kP$, where $k = 3 \times 10^{-4}$ atm^{-1}.

9.9 A Fabry–Perot interferometer has a finesse $\Phi = 2000$ and is tuned for maximum transmission at the wavelength $\lambda = 400$ nm. Find the displacement to be applied to one of the mirrors of the interferometer in order to reduce the output irradiance to half of this maximum value.

9.10 Demonstrate the result given by Eq. (9.17) for the half width at half maximum of a bright fringe of the transmitted field by a Fabry–Perot cavity.

9.11 A Fabry–Perot interferometer has a finesse $\Phi = 40000$ and a separation between the mirrors $d = 4$ cm. The space between the mirrors is filled with air. Calculate the reflection coefficient (r) of the mirrors, the free spectral range $(\Delta\lambda)_{fsr}$, the width of a spectral line, γ, and the chromatic resolution power of the interferometer, $\mathfrak{R}$.

9.12 A Fabry–Perot interferometer has a separation between the mirrors $d = 2$ cm and a reflection coefficient $r = 0.98$. Determine its finesse, its minimum resolvable wavelength interval, and its resolving power, considering a wavelength $\lambda = 600$ nm.

9.13 Determine the coefficient of finesse F, the finesse, and the mirror reflectivity r for a Fabry–Perot cavity which has a minimum transmittance $T_{\min} = 0.05$.

Coherence

10

In previous chapters, devoted to the phenomenon of interference, it was generally assumed that light was monochromatic and, in most cases, produced by a point source. However, no source is truly punctual, nor is the emitted light in any circumstances strictly monochromatic, so that it can be described by a sinusoidal wave extending infinitely in space and time. This means that, in practice, light is never truly coherent. In a broad sense, coherence describes the statistical similarity of a field, such as an electromagnetic field, at different points in space or time. This implies the stability of the phase of a wave, both in space and time. By stability in space, we mean a fixed relationship between the phases of two distinct waves, whereas by stability in time we mean the permanence of the phase of a single wave.

The degree of coherence between two waves can readily be measured by the interference visibility, which looks at the size of the interference fringes relative to the input waves as the phase offset is varied; a precise mathematical definition of the degree of coherence is given by means of correlation functions.

The coherence concept was originally introduced in connection with Young's double-slit experiment in optics, as described in Chapter 8, but it is now used in any field that involves waves such as acoustics, electrical engineering, neuroscience, and quantum mechanics. The property of coherence is the basis for commercial applications such as holography, the Sagnac gyroscope, radio antenna arrays, optical coherence tomography, and telescope interferometers.

10.1 Length of Coherence and Coherence Time

Quasi-monochromatic light can be considered as consisting of a large number of wave trains of finite extension and arbitrary relative phases. Light coming from two points separated from each other by a distance shorter than the length of an elemental wave train is coherent and can interfere, as with a monochromatic wave. However, if the distance between these points is greater than the length of a wave train, then the light coming from them is incoherent and will not interfere. The length of an elemental wave train is called the *coherence length*, l_c. On the other hand, we call the *coherence time*, τ_c, to the time that this wave train takes to pass through a given point. The two quantities are related as

$$l_c = c\tau_c \tag{10.1}$$

The width of a spectral line in terms of wavelength, $\Delta\lambda$, and in terms of frequency, $\Delta\nu$, relates to the coherence length, l_c, in the form

$$\frac{\lambda^2}{\Delta\lambda} = \frac{c}{\Delta\nu} \approx l_c \tag{10.2}$$

where c is the speed of light. Eq. (10.2) shows that the coherence length becomes infinite only if the spectral line width tends to zero, which never happens in reality.

Lasers, Optics, and Photonics. Mário F. S. Ferreira, Oxford University Press. © Mário F. S. Ferreira (2026).
DOI: 10.1093/oso/9780198952770.003.0010

Using Eqs. (10.1) and (10.2), it can be concluded that

$$\Delta \nu \cdot \tau_c \approx 1 \tag{10.3}$$

This result corresponds to the so-called *uncertainty principle*.

The typical values of the coherence length are in a range that goes from 1 μm, as with white light, to a few tens of kilometres, as with the light emitted by some lasers. The light emitted by a sodium or mercury lamp has a coherence length on the order of 1 cm.

Example 10.1

Determine the coherence length and coherence time of white light.

Solution

The spectrum of white light is between 780 nm (corresponding to the red colour) and 390 nm (corresponding to the violet colour). In terms of frequency, these limits correspond to 384 THz and 770 THz, respectively. That is, white light has a spectral width given by

$$\Delta \nu = 770 \text{ THz} - 384 \text{ THz} = 386 \text{ THz}$$

The coherence time is given from Eq. (10.3) by

$$\tau_c \approx \frac{1}{\Delta \nu} = \frac{1}{386 \times 10^{12}} = 2.6 \times 10^{-15} \text{ s}$$

As for the length of coherence, it is given from Eq. (10.1) by

$$l_c = c\tau_c = 780 \text{ nm}$$

It should be noted that this value, which corresponds to the length of an elemental wave train of white light, is of the same order as the wavelength.

10.2 Mutual Coherence Function and Degree of Coherence

Consider a spatially extended but reduced-bandwidth source that produces an optical field $E(\mathbf{r}, t)$. The effects of polarization will not be included in the discussion that follows, so a scalar representation of the field is sufficient. Let $E(\mathbf{r}_1, t) \equiv E_1(t)$ and $E(\mathbf{r}_2, t) \equiv E_2(t)$ be the optical fields at points P_1 and P_2. A detector placed at the exit of an interferometer, which combines the light coming from these two points, allows one to measure the corresponding irradiance, I, which is proportional to the average squared value of the modulus of the resulting field. Ignoring the proportionality factor, one has

$$I(\tau) = \langle \{E_1(t+\tau) + E_2(t)\}\{E_1(t+\tau) + E_2(t)\}^* \rangle \tag{10.4}$$

where the asterisk indicates the conjugate complex, $\langle\ \rangle$ denotes an average in time, and τ represents a relative delay that will normally exist between the two combined fields. Developing the product in the previous expression, one has

$$I(\tau) = \langle E_1(t+\tau)E_1^*(t+\tau)\rangle + \langle E_2(t)E_2^*(t)\rangle + \langle E_1(t+\tau)E_2^*(t)\rangle + \langle E_1^*(t+\tau)E_2(t)\rangle \quad (10.5)$$

The first two terms in Eq. (10.5) simply represent the irradiances at points P_1 and P_2, whereas the third and fourth terms represent the interference between the two fields. These last two terms are the complex conjugates of each other, so they both contain the same information. Conventionally, the third term is used to define the so-called *mutual coherence function*, $\Gamma_{12}(\tau)$, as

$$\Gamma_{12}(\tau) = \langle E_1(t+\tau)E_2^*(t)\rangle \quad (10.6)$$

It should be noted that if the points P_1 and P_2 coincide and $\tau = 0$, the mutual coherence function is reduced to the respective irradiances

$$\Gamma_{11}(0) = I_1, \qquad \Gamma_{22}(0) = I_2 \quad (10.7)$$

Using Eqs. (10.6) and (10.7), one can write the irradiance at the output of the interferometer in the form

$$I(\tau) = I_1 + I_2 + 2\mathrm{Re}\{\Gamma_{12}(\tau)\} \quad (10.8)$$

The mutual coherence function can be normalized in the form

$$\gamma_{12}(\tau) = \frac{\Gamma_{12}(\tau)}{\sqrt{\Gamma_{11}(0)\Gamma_{22}(0)}} \quad (10.9)$$

this quantity being called the *degree of complex coherence*. Using this definition and Eq. (10.7), we can write Eq. (10.8) in the form

$$I(\tau) = I_1 + I_2 + 2\sqrt{I_1 I_2}\mathrm{Re}\{\gamma_{12}(\tau)\} \quad (10.10)$$

This result represents the *general principle of interference for partially coherent light*.

The module of the degree of complex coherence, $|\gamma_{12}|$, is known as the *degree of coherence*. Eq. (10.9) and Schwarz's inequality show that $0 \le |\gamma_{12}(\tau)| \le 1$. The case $|\gamma_{12}| = 1$ corresponds to *total coherence* of the interfering fields, whereas the case $|\gamma_{12}| = 0$ corresponds to *total incoherence*. Finally, the case $0 < |\gamma_{12}| < 1$ corresponds to a situation of *partial coherence*.

As discussed in Chapter 8, the *visibility*, V, of a set of interference fringes is defined in terms of the maximum and minimum values of the irradiance in the interference pattern and is given by

$$V = \frac{I_{\max} - I_{\min}}{I_{\max} + I_{\min}} \quad (10.11)$$

Using Eq. (10.10), the visibility can be expressed in terms of the degree of coherence $|\gamma_{12}(\tau)|$ in the form

$$V = \frac{2\sqrt{I_1 I_2}}{I_1 + I_2}|\gamma_{12}(\tau)| \quad (10.12)$$

When the two fields have the same irradiance, one has

$$V = |\gamma_{12}(\tau)| \tag{10.13}$$

i.e. the *visibility of the fringes provides a direct measure of the degree of coherence of the interfering fields.*

Example 10.2

Derive the result given by Eq. (10.12) for the visibility of a pattern of interference fringes.

Solution

Equation (10.10) can be written in the form

$$\begin{aligned} I(\tau) &= I_1 + I_2 + 2\sqrt{I_1 I_2}\mathrm{Re}\{\gamma_{12}(\tau)\} \\ &= I_1 + I_2 + 2\sqrt{I_1 I_2}\,|\gamma_{12}(\tau)|\,\cos\delta(\tau) \end{aligned}$$

where $\delta(\tau)$ is the phase of $\gamma_{12}(\tau)$. Based on this expression, we have

$$I_{\text{máx}}(\tau) = I_1 + I_2 + 2\sqrt{I_1 I_2}\,|\gamma_{12}(\tau)|$$

and

$$I_{\min}(\tau) = I_1 + I_2 - 2\sqrt{I_1 I_2}\,|\gamma_{12}(\tau)|$$

Eq. (11.12) is obtained simply by substituting the aforementioned results into Eq. (10.11).

10.3 Temporal Coherence

When the points P_1 and P_2, referred to in the previous section are coincident, the mutual coherence function, given by Eq. (10.6), is reduced to the so-called *self-coherence function* at this point, such that $\Gamma_{12}(\tau) = \Gamma_{11}(\tau) = \Gamma_{22}(\tau) = \Gamma(\tau)$. The function $\Gamma(\tau)$ depends only on the difference τ between two instants of time. This is the case, for example, in Michelson's interferometer, where τ is equal to the ratio of the difference in optical paths in the two arms of the interferometer to the speed of light, c.

The self-coherence function $\Gamma(\tau)$ is given by

$$\Gamma(\tau) = \langle E_1(t+\tau)E_1^*(t)\rangle = \lim_{T\to\infty}\frac{1}{T}\int_{-T/2}^{T/2} E_1(t+\tau)E_1^*(t)dt \tag{10.14}$$

This function can be normalized, resulting in the so-called *complex degree of temporal coherence*:

$$\gamma(\tau) = \frac{\Gamma(\tau)}{\Gamma(0)} \tag{10.15}$$

Given that $\Gamma(0) = I_1$ is the maximum value (real) of the modules of $\Gamma(\tau)$, one has that $|\gamma(\tau)| \leq 1$. The *complex degree of temporal coherence*, given by Eq. (10.15), is also known as the *first-order correlation function*, $g^{(1)}(\tau)$, because it is based on the properties of the first power of the electric field.

Using the previous definitions, one can write the irradiance in the plane of observation as

$$I(\tau) = 2I_1 + 2\text{Re}\{\Gamma(\tau)\} = 2I_1\left[1 + \text{Re}\{\gamma(\tau)\}\right] \tag{10.16}$$

Example 10.3

Derive the self-coherence function, the degree of temporal coherence, and the visibility of a pattern of interference fringes in the case of a harmonic wave, given by

$$E_1(t) = E_0 e^{-i\omega t} \tag{10.17}$$

Solution

The function of self-coherence corresponding to the harmonic wave (10.17) is given by

$$\Gamma(\tau) = \lim_{T\to\infty}\frac{1}{T}\int_{-T/2}^{T/2}|E_0|^2 e^{i\omega t}e^{-i\omega(t+\tau)}dt = I_1 e^{-i\omega\tau} \tag{10.18}$$

Based on Eq. (10.15), the degree of temporal coherence is given by

$$\gamma(\tau) = \frac{\Gamma(\tau)}{\Gamma(0)} = e^{-i\omega\tau} \tag{10.19}$$

The visibility of the pattern of interference fringes is given by

$$V = |\gamma(\tau)| = 1 \tag{10.20}$$

Fig. 10.1 illustrates the harmonic wave and the visibility of a pattern of interference fringes produced by such a wave.

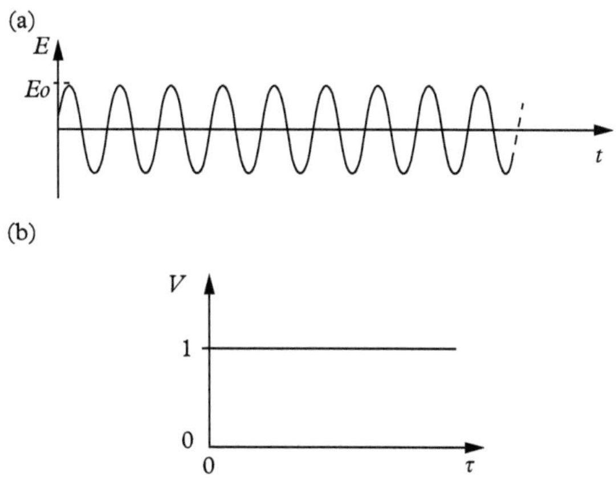

Figure 10.1 *(a) Electric field amplitude and (b) visibility function in the case of a harmonic wave.*

The result given by Eq. (10.20) corresponds to a limit case, in which the light is completely coherent. This situation occurs approximately in the case of light emitted by a stabilized single-mode laser. At the opposite limit, we have the case of completely incoherent light, characterized by $|\gamma(\tau)| = 0$ for $\tau \neq 0$. Sunlight or light from a filament lamp are examples in which this situation occurs approximately.

Example 10.4

Obtain the self-coherence function, the degree of temporal coherence, and the visibility of the interference fringes in the case of the superposition of two harmonic waves with equal amplitudes but different frequencies, represented as follows:

$$E(t) = E_0 e^{-i\omega_1 t} + E_0 e^{-i\omega_2 t} \tag{10.21}$$

Solution

Substituting Eq. (10.21) into Eq. (10.14), the self-coherence function becomes

$$\Gamma(\tau) = \lim_{T\to\infty} \frac{1}{T} \int_{-T/2}^{T/2} \left(E_0^* e^{i\omega_1 t} + E_0^* e^{i\omega_2 t}\right)\left(E_0 e^{-i\omega_1 (t+\tau)} + E_0 e^{-i\omega (t+\tau)}\right) dt$$

$$= |E_0|^2 \left(e^{-i\omega_1 \tau} + e^{-i\omega_2 \tau}\right) \tag{10.22}$$

Thus, the degree of temporal coherence, defined by Eq. (10.15), is given by

$$\gamma(\tau) = \frac{1}{2}\left(e^{-i\omega_1 \tau} + e^{-i\omega_2 \tau}\right) \tag{10.23}$$

The visibility of the interference fringes is given by

$$V = |\gamma(\tau)| = \frac{1}{2}\left|e^{-i\omega_1 \tau} + e^{-i\omega_2 \tau}\right| = \left|\cos\left(\frac{\omega_2 - \omega_1}{2}\right)\tau\right| \tag{10.24}$$

In contrast to the result given by Eq. (10.20), it turns out that in the case of the superposition of two harmonic waves with different frequencies, the visibility is a periodic function of the time delay τ. Fig. 10.2 illustrates this result.

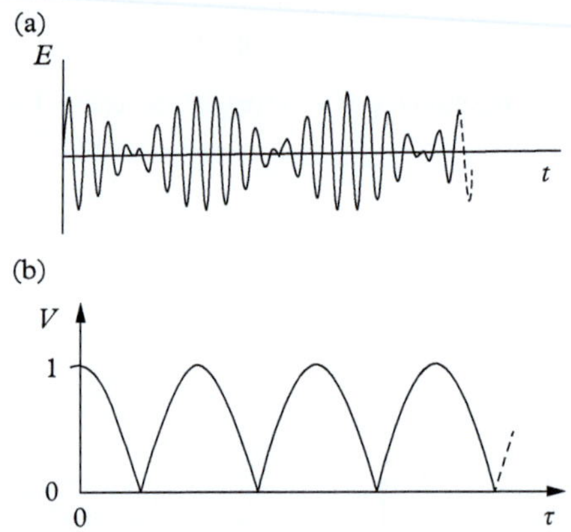

Figure 10.2 *(a) Total electric field and (b) visibility in the case of superposition of two harmonic waves with different frequencies.*

In the case of many light sources, whether natural or artificial, the visibility of the interference fringes decreases monotonously with time delay, τ. When this decay is equal to $1/e$, the corresponding time delay is the *coherence time*, τ_c, which relates to the *coherence length*, l_c, in the form given by Eq. (10.1). This coherence length can be measured using a Michelson interferometer.

It is given by the difference in optical paths. When using white light, the difference in optical paths between the two arms in this interferometer must be less than one wavelength in order to have good visibility of the fringe pattern. However, if light emitted by a laser source is used, given the high value of its coherence length, the visibility of such fringes remains high, even if the lengths of the interferometer arms are very different.

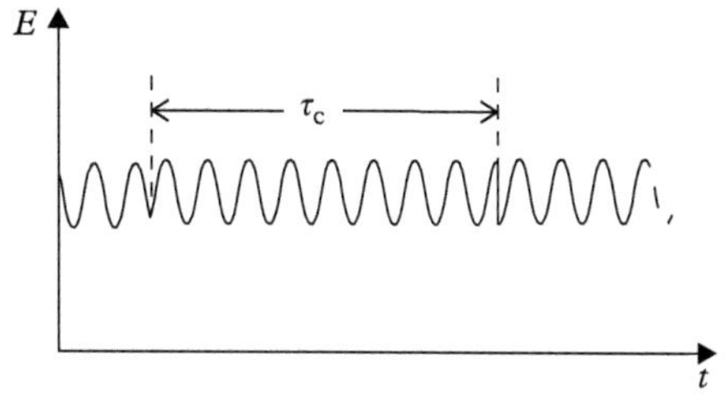

Figure 10.3 *The wave train can be characterized by an average time between phase interruptions, the coherence time τ_c.*

In the case of quasi-monochromatic light, the electric field varies sinusoidally for a certain time τ_c (coherence time) and then changes abruptly, as represented in Fig. 10.3. Such an electric field can be represented in the form:

$$E(t) = E_0 e^{-i\omega t} e^{i\varphi(t)} \tag{10.25}$$

where the phase $\phi(t)$ is a random step function, as illustrated in Fig. 10.4. The phase change that occurs each coherence time is randomly distributed between 0 and 2π and this behaviour repeats indefinitely.

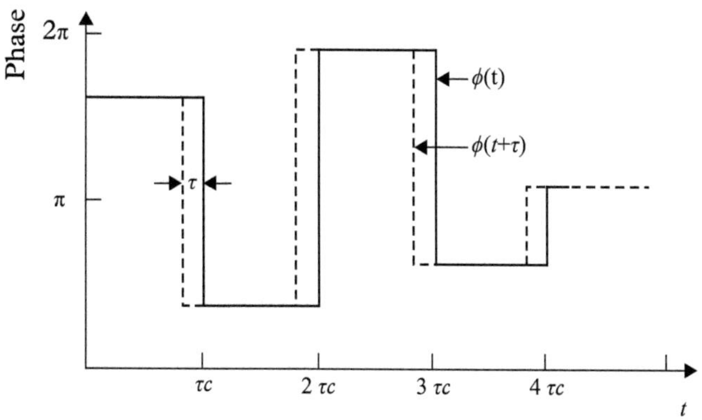

Figure 10.4 *Representation of the phase of a quasi-monochromatic source.*

Consider a beam of light whose field is represented by Eq. (10.25), which is divided into two beams of equal amplitude and subsequently superimposed. The complex degree of temporal coherence can be calculated using Eqs. (10.14) and (10.15), providing the following result (Problem 10.7):

$$\gamma(\tau) = \begin{cases} (1-\tau/\tau_c)e^{iw\tau} & \tau < \tau_c \\ 0 & \tau \geq \tau_c \end{cases} \tag{10.26}$$

The graph of $|\gamma(\tau)|$, which corresponds to the visibility V, is shown in Fig. 10.5. Obviously, the fringe visibility drops to zero if τ exceeds the coherence time, τ_c.

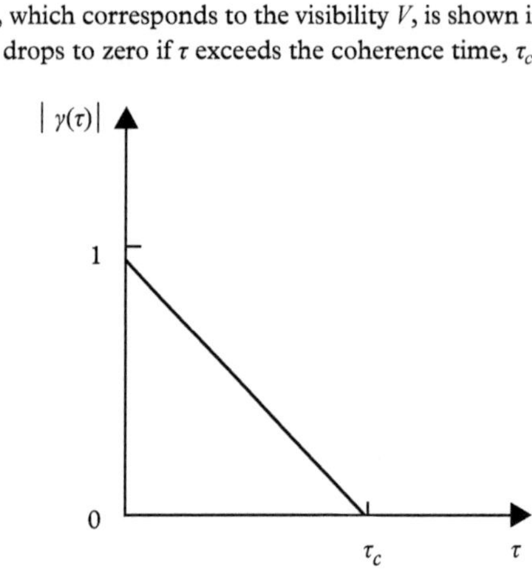

Figure 10.5 *Graph of the coherence of a quasi-monochromatic source.*

10.4 Spatial Coherence

The interference characteristics are adversely affected when the dimensions of the light source increase. In this case, it is important to consider the interference of light coming from different points of the extended source. This problem will be analysed in this section, based on Young's experiment discussed in Chapter 8.

Let us consider Young's slits illuminated by light from an extended source with diameter $2d$, instead of the point source S considered in the analysis of Section 8.2. The extended source can be seen as a set of independent point sources. Waves emitted from point S of the source, situated on the axis, reach the slits S_1 and S_2 with the same phase, giving rise to a system of fringes in the observation plane with a maximum on the optical axis. However, an off-axis point S' of the same source (Fig. 10.6) will give rise to a system of fringes deviated laterally from the previous one. Actually, distances $l_1 = \overline{S'S_1}$ and $l_2 = \overline{S'S_2}$ are not equal in this case, so there is a phase difference between the two waves originating in S_1 e S_2:

$$\varepsilon_1 - \varepsilon_2 = \frac{2\pi}{\lambda}(l_1 - l_2) \tag{10.27}$$

In this case, the interference maximum of order zero in the plane of observation is deviated from the optical axis by

$$y_m = \frac{\Delta y}{\lambda}(l_1 - l_2) \tag{10.28}$$

where Δy is the spacing between the fringes, given by Eq. (8.28).

Let us now suppose that the points S and S' emit simultaneously. If the extended source is incoherent, the phase difference between the waves generated at S and S' varies randomly.

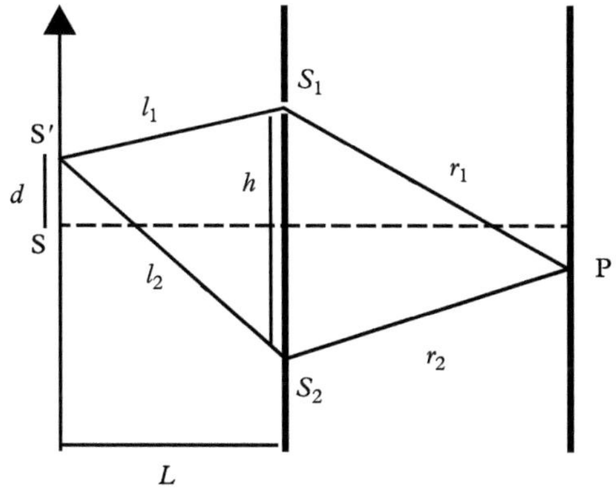

Figure 10.6 *Geometry of Young's experiment in the case of an extended source.*

In this case, the observation screen presents a distribution of irradiance given simply by summing the irradiances of the fringe systems corresponding to each of the two emitting points. Thus, in order for the interference fringes to be visible, the interference pattern generated by S and S' must not be too deviated from each other. More precisely, it should not happen that the maximum of one pattern coincides with the minimum of the other pattern. This condition requires

$$|y_m| < \frac{\Delta y}{2} \tag{10.29}$$

Taking into account Eq. (10.28), this means that

$$|l_1 - l_2| < \frac{\lambda}{2} \tag{10.30}$$

Considering the geometry of Fig. 10.6, one has

$$l_1 = \sqrt{(d - h/2)^2 + L^2} \approx L + \frac{4d^2 + h^2 - 4hd}{8L} \tag{10.31a}$$

$$l_2 = \sqrt{(d + h/2)^2 + L^2} \approx L + \frac{4d^2 + h^2 + 4hd}{8L} \tag{10.31b}$$

so

$$|l_1 - l_2| = \frac{hd}{L} \tag{10.32}$$

Using this result, Eq. (10.30) becomes

$$\frac{hd}{L} < \frac{\lambda}{2} \tag{10.33}$$

Eq. (10.33) gives the condition of spatial coherence in the case of an incoherent extended source.

When one considers the interference pattern in the observation plane near the optical axis, the distances between this region and the two slits S_1 and S_2 are almost the same. In these circumstances, the interference pattern reflects the relation between the fields $E(\mathbf{r}_1, t)$ and $E(\mathbf{r}_2, t)$ in both slits S_1 and S_2, respectively, which can be represented by the correlation function:

$$\Gamma(\mathbf{r}_1, \mathbf{r}_2, 0) = \Gamma_{12}(0) = \langle E(\mathbf{r}_1, t) E^*(\mathbf{r}_2, t)\rangle \tag{10.34}$$

The function given by Eq. (10.34) is called the *spatial coherence function*.

10.5 The Michelson Stellar Interferometer

The spatial coherence properties of light can be used to determine the diameter of stars or the separation of binary stars. The latter case is represented in Fig. 10.7, considering that the two stars constitute two incoherent point sources that illuminate Young's two slits. Each star gives rise to a fringe system, which is displaced relative to the fringe system produced by the other star. This displacement depends on the separation h between Young's slits.

The difference in paths between the stars and Young's two slits is given by Eq. (10.32). This difference in paths leads to a relative displacement of the two fringe systems, proportional to the separation h between the slits. If $h \to 0$ the fringe systems corresponding to each of the stars coincide, then, by increasing the value of h, one arrives at a value, h_c, for which the maximum of one of the fringe systems coincides with the minimum of the other fringe system, so the visibility vanishes. This situation occurs when:

$$\frac{h_c d}{L} = \frac{\bar{\lambda}}{2} \tag{10.35}$$

where $\bar{\lambda}$ is the average light wavelength.

From Eq. (10.35) it is found that

$$h_c = \bar{\lambda}\frac{L}{2d} = \frac{\bar{\lambda}}{2\theta} \tag{10.36}$$

where

$$\theta = \frac{d}{L} \tag{10.37}$$

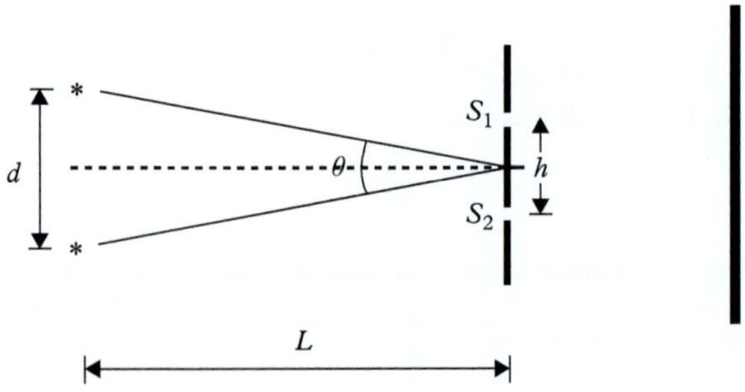

Figure 10.7 *Young's experiment in the context of stellar interferometry.*

is the angular separation of the two stars observed from the earth. Whereas θ is usually quite small, the spacing h_c becomes quite large. In order to make the measurement of θ more practical, the so-called Michelson stellar interferometer is generally used.

The Michelson stellar interferometer is schematically represented in Fig. 10.8 and was originally implemented on the 2.5 m telescope at Mount Wilson, near Los Angeles, in the 1920s.

The light from a bright star is collected by two mirrors $M1$ and $M2$ that are separated by a distance h. The light reflected by these mirrors goes to two other mirrors, $M3$ and $M4$, then passes through the apertures S_1 and S_2 of an opaque screen, and is finally collected by the lens of a telescope. The two apertures S_1 and S_2 give rise to the interference patterns typical of Young's experiment in the focal plane of the lens.

As mentioned earlier, the separation between the fringes of any of the patterns depends only on the distance between the apertures, but visibility depends on the distance h between the mirrors $M1$ and $M2$. When this distance is given by Eq. (10.36), the maxima of one of the fringe systems coincide with the minimums of the other system and the visibility of the resulting pattern vanishes. In this case, it is enough to measure the distance h to determine the angular separation θ between the stars of the pair.

The angular size of a star can also be calculated using Michelson's stellar interferometer, in a manner similar to that described previously. The visibility of the fringe pattern is related to the degree of coherence of the optical field in the mirrors $M1$ and $M2$ of the interferometer. The star is considered for this purpose as a circular disc with uniform brightness. Using the *van Cittert–Zernike theorem*, the visibility of the fringe pattern is given by the Fourier transform of the source distribution function and the following result is obtained:

$$V = 2\left|\frac{J_1\left(\frac{\pi h\theta}{\bar{\lambda}}\right)}{\frac{\pi h\theta}{\bar{\lambda}}}\right| \tag{10.38}$$

where $J_1(u)$ is the first-order Bessel function. The lowest value of h for which V vanishes is given by

$$h = 1.22\frac{\bar{\lambda}}{\theta} \tag{10.39}$$

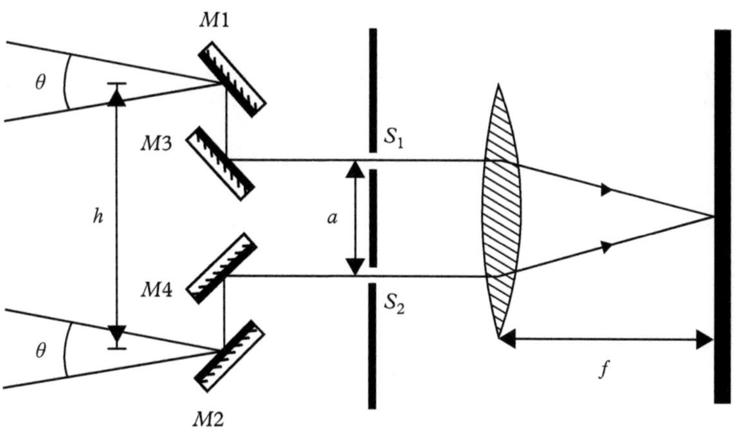

Figure 10.8 *Representation of Michelson's stellar interferometer. M1, M2, M3 and M4: mirrors.*

The measurement of h thus allows one to obtain the angular diameter, θ, of the star. Actually, only the biggest stars—namely the so-called red giants—can be measured using this technique.

Example 10.5

Using the technique described earlier on the star Betelgeuse, situated in the constellation Orion, Michelson found the first minimum of visibility when $h = 308$ cm. Considering an average wavelength $\bar{\lambda} = 570$ nm, what is the angular diameter of the star?

Solution

Using Eq. (10.39), one obtains

$$\theta = \frac{1.22\bar{\lambda}}{h} = \frac{1.22\left(570 \times 10^{-9}\right)}{3.08} = 2.26 \times 10^{-7} \text{ rad.}$$

Since the constellation Orion is situated at a distance of approximately 1.8×10^{15} km from the earth, the diameter of the star is $s = r\theta = 4.1 \times 10^{8}$ km, which is greater than the diameter of the earth's orbit around the sun!

10.6 Correlation Interferometry

In Michelson's stellar interferometer, the observation of a fringe pattern indicates the existence of a correlation between the amplitudes of the electric field in mirrors $M1$ and $M2$. However, since it is not possible to measure the amplitude of the electric field directly, it will be interesting to know to what extent this correlation also exists between the fluctuations of irradiance received in these mirrors. This question was at the origin of a set of remarkable experiments carried out by Hanbury Brown and Twiss between 1952 and 1956, while working at the Jodrell Bank telescope in England. Their interferometer was designed to be an improvement on the Michelson stellar interferometer.

One can consider the correlation of irradiances at two points P_1 and P_2, whose location is indicated by $\mathbf{r}_1$ and $\mathbf{r}_2$, in the form

$$K(\mathbf{r}_1, \mathbf{r}_2, \tau) = K_{12}(\tau) = \langle I_1(t+\tau)I_2(t)\rangle \tag{10.40}$$

Assuming that the light sources emit randomly, according to Gaussian statistics, it can be shown that

$$K_{12}(\tau) = \langle I_1\rangle\langle I_2\rangle + |\Gamma_{12}(\tau)|^2 = \langle I_1\rangle\langle I_2\rangle\left(1 + |\gamma_{12}(\tau)|^2\right) \tag{10.41}$$

In many cases, it is the fluctuation of the irradiance, given by

$$\Delta I(t) = I(t) - \langle I\rangle \tag{10.42}$$

which is measured. Thus, one has

$$\langle\Delta I_1(t)\rangle = \langle\Delta I_2(t)\rangle = 0 \tag{10.43}$$

On the other hand, the correlation function of these fluctuations is given by

$$R_{12}(\tau) = \langle \Delta I_1(t+\tau)\Delta I_2(t)\rangle = \langle I_1(t+\tau)I_2(t)\rangle - \langle I_1\rangle\langle I_2\rangle$$
$$= |\Gamma_{12}(\tau)|^2 = \langle I_1\rangle\langle I_2\rangle|\gamma_{12}(\tau)|^2 \qquad (10.44)$$

It can be verified from the previous results that the measurement of $K_{12}(\tau)$ or $R_{12}(\tau)$ allows us to determine $|\Gamma_{12}(\tau)|$.

This technique is called *correlation interferometry* and allows us to measure the diameter of stars or the distances between binary stars in a similar but simpler way to that used in Michelson's stellar interferometer. The first star to be studied with this method was Sirius, and a value of 0.0069 arcseconds was obtained for its angular diameter.

10.6.1 The second-order correlation function

Considering that the irradiances at points P_1 and P_2 are equal, one can write a normalized version of Eq. (10.40) and introduce the *second-order correlation function* of the light, which is defined as

$$g^{(2)}(\tau) = \frac{\langle E^*(t)E^*(t+\tau)E(t)E(t+\tau)\rangle}{\langle E^*(t)E(t)\rangle\langle E^*(t+\tau)E(t+\tau)\rangle} = \frac{\langle I(t)I(t+\tau)\rangle}{\langle I(t)\rangle\langle I(t+\tau)\rangle} \qquad (10.45)$$

The function $g^{(2)}(\tau)$ quantifies the intensity fluctuations of light. It is also known as the *degree of second-order coherence* and becomes particularly useful for characterizing the *temporal* coherence of the source.

It can be shown that, if $\tau >> \tau_c$, the intensity fluctuations at times t and $t+\tau$ will be completely uncorrelated with each other and $g^{(2)}(\tau) = 1$. On the other hand, if $\tau << \tau_c$, there will be correlations between the fluctuations at the two times. In particular, if $\tau = 0$, one has

$$g^{(2)}(0) = \frac{\langle I(t)^2\rangle}{\langle I(t)\rangle^2} \qquad (10.46)$$

It can be shown that, for any time dependence of $I(t)$, it will always be the case that

$$g^{(2)}(0) \geq 1 \qquad (10.47)$$

and

$$g^{(2)}(0) \geq g^{(2)}(\tau) \qquad (10.48)$$

In the case of a perfectly coherent monochromatic source with a time-independent irradiance I_0, one has

$$g^{(2)}(\tau) = 1 \qquad (10.49)$$

In the case of a light source with a time-varying irradiance, it can be seen from Eq. (10.46) that $g^{(2)}(0) > 1$ because there are equal intensity fluctuations above and below the average,

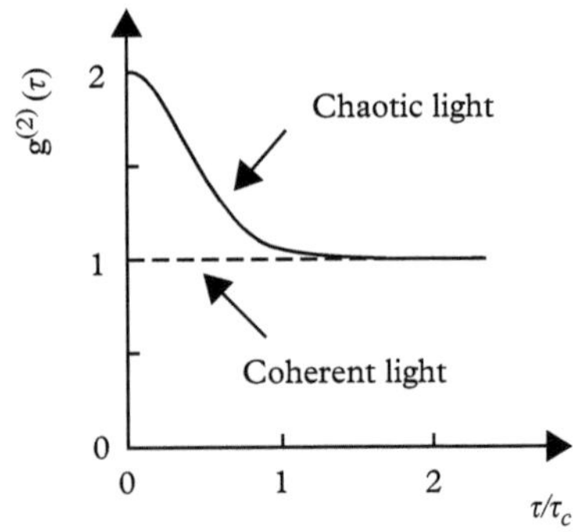

Figure 10.9 *Second-order correlation function for chaotic and perfectly coherent light.*

and the squaring process exaggerates the fluctuations above the mean value. Thus, we can conclude that, for any source with a time-varying irradiance, we expect $g^{(2)}(\tau)$ to decrease with τ, reaching the value $g^{(2)}(\tau) = 1$ for large values of τ.

Let us consider the case of chaotic light emitted by an atomic discharge lamp. Assuming that the source has a Doppler-broadened spectral line with a Gaussian shape, it can be shown that the second-order correlation function is given in this case by

$$g^{(2)}(\tau) = 1 + \exp\left[-\pi(\tau/\tau_c)^2\right] \quad (10.50)$$

This function is compared to that of perfectly coherent light in Fig. 10.9. It can be verified that $g^{(2)}(0) = 2$, the function $g^{(2)}(\tau)$ decreasing with τ and reaching the value $g^{(2)}(\tau) = 1$ for large values of τ.

10.7 Further Reading

Born, M., and E. Wolf, *Principles of Optics*, 6th Ed., Pergamon Press, 1980.
Guenther, R., *Modern Optics*. John Wiley & Sons, 1990.
Hecht, E., *Optics*, 5th Ed., Pearson Education, 2016.
Loudon, R., *The Quantum Theory of Light*, Oxford University Press, 1983.
Mandel, L., and E. Wolf, *Optical Coherence and Quantum Optics*, Cambridge University Press, 1995.
Marathay, A. S., *Elements of Optical Coherence Theory*, John Wiley & Sons, 1982.
Pedrotti, F., L. M. Pedrotti, and L. S. Pedrotti, *Introduction to Optics*, 3rd Ed., Cambridge University Press, 2018.
Lauterborn, W., and T. Kurz, *Coherent Optics: Fundamentals and Applications*, 2nd Ed., Springer-Verlag, 2003.

10.8 Problems

10.1 Two filters can be used in an interference experiment with yellow light $(\bar{\lambda} = 590\,\text{nm})$, the transmission bandwidth of them being 100 nm and 10 nm, respectively. Compare the coherence lengths of the light transmitted by each filter and indicate the best option for this experiment.

10.2 A Michelson interferometer uses light wavelength $\lambda = 600$ nm and spectral width $\Delta\lambda = 0.001$ nm. Assuming that, initially, the optical paths of the two arms of the interferometer are equal, what displacement must be given to one of its mirrors for the interference fringes to disappear?

10.3 In an interference experiment, a beam of light with wavelength 600 nm and spectral width 0.05 nm is divided into two beams of equal amplitude. These two beams travel different paths and are then superimposed with a path difference of 2 mm. Determine the visibility of the interference fringes. How does the visibility vary when the difference in optical paths doubles?

10.4 Suppose a lamp emitting quasi-monochromatic light with an average wavelength $\bar{\lambda} = 500$ nm is located behind a screen that has a circular hole of diameter 0.2 mm. Light from that hole illuminates the two apertures of Young's experiment, situated in a plane 1.5 m from the screen. What is the smallest separation, h, between these apertures for which the visibility of the fringes vanishes?

10.5 Consider the overlapping of two harmonic waves with different frequencies, ω_1 and ω_2, and different amplitudes, E_{01} and E_{02}. Obtain the amplitude of the resulting field, the self-coherence function $\Gamma(\tau)$, the degree of complex temporal coherence $\gamma(\tau)$, and the V visibility of the interference fringes.

10.6 Obtain an expression for the self-coherence function $\Gamma(\tau)$ in the case of the overlap of a high number, N, of harmonic waves with different frequencies and amplitudes.

10.7 Derive Eq. (10.26) for the complex degree of temporal coherence of quasi-monochromatic light.

10.8 When observed from the surface of the earth, the sun has an angular diameter of 0.5°. Considering a screen with two holes illuminated by sunlight, determine the maximum separation between these holes, so that it is possible to observe fringes of interference on another screen located behind the first. Assume an average value $\bar{\lambda} = 550$ nm for the wavelength of sunlight.

10.9 Show that Eq. (10.43) can be obtained from Eq. (10.41).

10.10 Evaluate the second-order correlation function $g^{(2)}(\tau)$ for a monochromatic light wave with a sinusoidal intensity modulation such that $I(t) = I_0(1 + A\sin\,\omega t)$, with $|A| \leq 1$.

11 Fraunhofer Diffraction

The optical diffraction phenomenon consists of any deviation from geometrical optics resulting from the interaction between a wavefront of light and some obstacle, transparent or opaque. In reality, there is no substantial difference between *interference* and *diffraction*, and the distinction between the two phenomena is somewhat arbitrary. Interference is referred to when considering the superposition of a small number of waves, while the term 'diffraction' is applied in cases where the number of interfering waves is high.

An accurate description of the diffraction phenomenon can be done based on Maxwell's equations and the consideration of the boundary conditions associated with the diffractive obstacle. However, the application of this theory is usually a complicated task, so in many cases it is convenient to use an approximate scalar theory. This theory is based on the so-called *Huygens' principle*, according to which *each point of a given wavefront can be considered as a source of elementary spherical wavelets, with the same frequency as the primary wave. The amplitude of the optical field at any point in space at a later time is given by the superposition of all these elemental wavelets.*

Fig. 11.1 illustrates the use of the Huygens' principle to describe the propagation of a spherical wave.

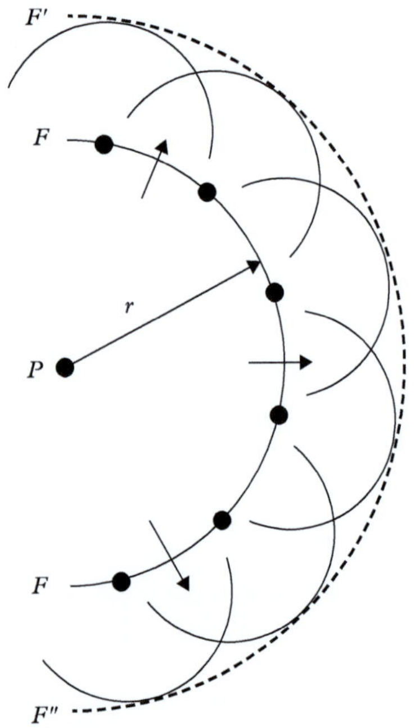

Figure 11.1 *Illustration of the Huygens principle for a spherical wave.*

The use of this scalar theory results in the so-called *diffraction integral.* One can obtain analytical solutions for this integral using two different approximations. In one of these approximations, it is assumed that both the observation plane and the light source are effectively far enough from the obstruction plane so that wavefronts arriving at the aperture and at the observation screen may be considered plane. In such case, we speak of *Fraunhofer* or *far-field diffraction.* When some of these conditions are not satisfied and the curvature of the wavefront must be taken into account, we have the so-called *Fresnel*, or *near-field, diffraction.*

11.1 The Diffraction Integral

Let us consider a pinhole in a screen illuminated by a plane wave, given by

$$\mathbf{E}_i(\mathbf{r}, t) = \mathbf{E}_i(\mathbf{r})\, e^{-i\omega t} \tag{11.1}$$

The wave emerging from the pinhole is a spherical wave, the complex amplitude of which can be written in the form:

$$E(\vec{r}) = \frac{C}{r} e^{i\mathbf{k}\cdot\mathbf{r}} \tag{11.2}$$

where C is a constant, proportional to the amplitude of the incident wave, E_i, and to the pinhole area, Δs.

Let us consider Fig. 11.2, which illustrates the case of two pinholes: one situated in P_1, at a distance $r_{01} = |\mathbf{r}_0 - \mathbf{r}_1|$ from the observation point P_0, and another located

Lasers, Optics, and Photonics. Mário F. S. Ferreira, Oxford University Press. © Mário F. S. Ferreira (2026).
DOI: 10.1093/oso/9780198952770.003.0011

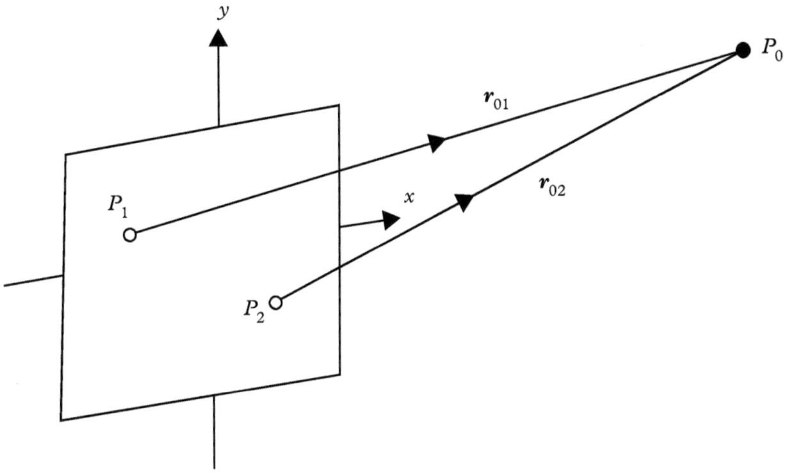

Figure 11.2 *Geometry for the application of Huygens' principle to the case of two pinholes.*

in P_2, at a distance $r_{02} = |\mathbf{r}_0 - \mathbf{r}_2|$ from P_0. The optical field at P_0 due to the pinhole $P_j\,(j = 1, 2)$ is

$$E_j(\mathbf{r}_0) = Q_j \frac{E_i(\mathbf{r}_j)}{r_{0j}} e^{i\mathbf{k}\cdot\mathbf{r}_{0j}} \Delta s_j \tag{11.3}$$

where Q_j is the so-called *obliquity factor* that depends on the angle, φ, between $\mathbf{r}_{0j}$ and the direction of the incident radiation. This dependence on the angle φ can be described in the form

$$Q(\varphi) = Q(0)\,\frac{1 + \cos(\varphi)}{2} \tag{11.4}$$

According to Eq. (11.4), the obliquity factor has a maximum value in the frontal direction ($\varphi = 0$) and vanishes in the opposite direction ($\varphi = \pi$), thus making it impossible for the wave to propagate in the direction of the source.

The effect of the obliquity factor on the secondary wavelets originating at points of the original wavefront was introduced by Fresnel to correct the original theory of Huygens. Actually, according to such a theory, a point source of secondary wavelets could radiate with equal effectiveness in all directions. This would produce new wavefronts in both forward and backward directions of a propagating wavefront, although the backward wave is not observed.

The total field at P_0 is given by the superposition of waves coming from P_1 and P_2:

$$E(\mathbf{r}_0) = E_1(\mathbf{r}_0) + E_2(\mathbf{r}_0) \tag{11.5}$$

In case there are N pinholes, Eq. (11.5) is generalized and assumes the form:

$$E(\mathbf{r}_0) = \sum_{j=1}^{N} E_j(\mathbf{r}_0) = \sum_{j=1}^{N} Q_j \frac{E_i(\mathbf{r}_j)}{r_{0j}} e^{i\mathbf{k}\cdot\mathbf{r}_{0j}} \Delta s_j \tag{11.6}$$

In the case that Δs_j tends to zero, the pinhole becomes an elementary Huygens source, with an elementary area ds. Actually, one can consider any aperture as a set of these elementary sources and convert the sum in Eq. (11.6) into an integral. In this process, $\mathbf{r}_{0j}$ is replaced by $\mathbf{R}$, representing the position of the Huygens elementary source relative to the observation point P_0, and $\mathbf{r}_j$ is replaced by $\mathbf{r}$, representing the position of the same elementary source relative to the origin of the coordinate system. Moreover, assuming that the aperture is relatively small, the variation of the obliquity factor Q_j is negligible and one can treat this factor as a constant, Q. Under these conditions, Eq. (11.6) becomes

$$E(\mathbf{r}_0) = Q \iint_{\Sigma} \frac{E_i(\mathbf{r})}{R} e^{i\mathbf{k}\cdot\mathbf{R}} ds \tag{11.7}$$

This result constitutes the so-called *diffraction integral* and shows that the field at P_0 can be seen as the result of overlapping multiple elemental waves emitted by the imaginary pinholes that make up the aperture $\sum$. This is the essence of *Huygens principle.*

11.1.1 Biographical note—Christian Huygens

Christian Huygens (1629–1695) was born in The Hague, Netherlands, and was educated at Leyden University. He became a close friend of Descartes, who was often invited to his parents' house. Huygens' reputation spread throughout Europe, and King Louis XIV succeeded in attracting him to France in 1666. During his stay in France, from where he returned only in 1681, Huygens contributed to the foundation of the French Academy of Sciences. In 1673, Huygens published the *Horologium Oscillatorium*, in which he dealt with the problem of simple and compound pendulums. In this context, Huygens also invented the pendulum clock. In 1690, after returning from a visit he made to England, during which he had the opportunity to meet Isaac Newton, Huygens published the work *Traité de la Lumiére*, in which he described his wave theory of light. Based on this theory, he was able to explain the laws of reflection and refraction, as well as the phenomenon of double refraction in calcite. Unlike Newton, Huygens correctly concluded that light should slow down when travelling in optically denser media. Huygens also devoted himself to astronomy, being the first to recognize Saturn's rings and discover Titan, one of the satellites of the same planet.

11.2 The Obliquity Factor

As mentioned in the previous section, the variation in the obliquity factor is negligible in many situations. The magnitude of this factor can be found by comparing the result obtained from Eq. (11.7) with the result predicted by geometrical optics.

Let us consider the aperture $\sum$ in Fig. 11.3 illuminated by a plane wave of amplitude E_i, which propagates parallel to the *z-axis*. The area of Huygens' elemental source in P_1 is given, in polar coordinates, by $ds = rdrd\phi$. Given that $E(r) = E_i = const.$, for $z = 0$, one has that the diffraction integral, which provides the optical field at the point P_0 in Fig. 11.2, can be written in the form:

$$E(z_0) = QE_i \int_0^{2\pi} \int_0^{r_m} \frac{e^{i\mathbf{k}\cdot\mathbf{R}}}{R} rdrd\phi \tag{11.8}$$

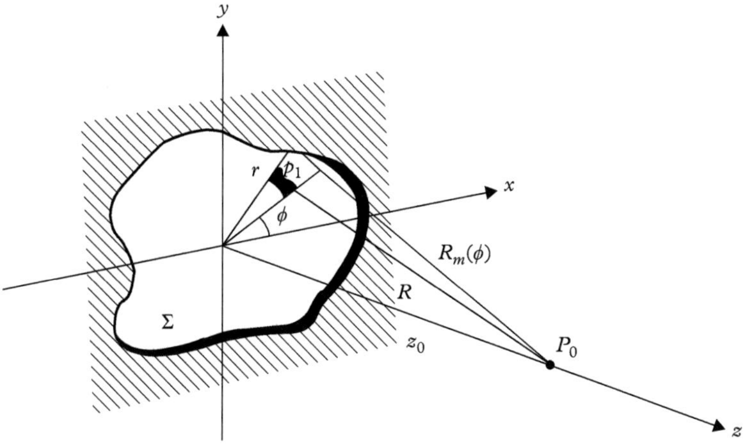

Figure 11.3 *Geometry for calculating the magnitude of the obliquity factor.*

where $r_m(\phi)$ is the maximum value of r, corresponding to the boundary of the aperture. From Fig. 11.2 one has that $z_0^2 + r^2 = R^2$, so the integration variable r in Eq. (11.8) can be written in terms of R using the relation $rdr = RdR$. The lower and upper limits of integration are given, respectively, by

$$R = z_0 \tag{11.9a}$$

and

$$R = R_m(\phi) = \sqrt{z_0^2 + r_m^2(\phi)}. \tag{11.9b}$$

So Eq. (11.8) becomes

$$E(z_0) = QE_i \int_0^{2\pi} \int_{z_0}^{R_m(\phi)} e^{i\mathbf{k}\cdot\mathbf{R}} dR d\phi \tag{11.10}$$

Performing the integration over R, one obtains:

$$E(z_0) = \frac{QE_i}{ik} \left[\int_0^{2\pi} e^{ikR_m(\phi)} d\phi - e^{ikz_0} \int_0^{2\pi} d\phi \right] \tag{11.11}$$

The second contribution in Eq. (11.11) corresponds to the sum of the waves diffracted by the boundary of the aperture. This contribution can generally be neglected, since the product $kR_m(\phi)$ varies by several multiples of 2π when integrating around the aperture. Thus, the final result of Eq. (11.11) can be written in the form:

$$E(z_0) = -\frac{2\pi QE_i}{ik} e^{ikz_0} = i\lambda QE_i e^{ikz_0} = i\lambda QE_{go} \tag{11.12}$$

where

$$E_{go} = E_i e^{ikz_0} \tag{11.13}$$

is the result predicted by geometrical optics for the optical field at the point P_0, located on the $z-axis$, at a distance z_0 from the aperture. In order to satisfy the equality $E(z_0) = E_{go}$, one must have:

$$i\lambda Q = 1 \Leftrightarrow Q = -\frac{i}{\lambda} \tag{11.14}$$

The fact that Q is an imaginary number means that there is a phase difference of $\pi/2$ between the incident wave and the elementary Huygens waves generated at the aperture. This phase shift corresponds to a correction first introduced by Fresnel to Huygens' original theory.

11.3 Babinet's Principle

Considering that the diffraction formula, given by Eq. (11.7), expresses the field $E(\mathbf{r}_0)$ in the form of an integral over the aperture, allows one to obtain a very important result, known as *Babinet's principle*. Actually, if the aperture is considered as the sum of two complementary parts, $\sum_1$ and $\sum_2$, such that $\sum = \sum_1 + \sum_2$, then the field at the point P_0 will be given, using Eq. (11.7), by the sum of two integrals: one calculated over $\sum_1$ and the other calculated over $\sum_2$. The integral over $\sum_1$ gives the field $E_1(\mathbf{r}_0)$ that would exist at the point P_0 if the aperture were reduced to just that part, while the integral over $\sum_2$ gives the field $E_2(\mathbf{r}_0)$ that there would be at that same observation point if there were only this part of the aperture. In this way, one has

$$E(\mathbf{r}_0) = E_1(\mathbf{r}_0) + E_2(\mathbf{r}_0) \tag{11.15}$$

That is, the field corresponding to the total aperture is obtained by adding the fields corresponding to the two complementary apertures.

Eq. (11.15) is particularly useful when one of the apertures, for example $\sum_1$, has a relatively complicated shape but can be expressed as the difference between two simpler apertures, $\sum_1 = \sum - \sum_2$. In this case, the field corresponding to the aperture $\sum_1$ can be calculated by subtracting the field corresponding to the total aperture $\sum$ and the field due to the complementary aperture $\sum_2$.

A situation of special interest occurs when the field at the point P_0 corresponding to the aperture $\sum$ is zero or negligible. In this case, Eq. (11.15) gives

$$E_1(\mathbf{r}_0) = -E_2(\mathbf{r}_0) \tag{11.16}$$

Since the irradiance is proportional to the square of the field modulus, the irradiances corresponding to the two complementary apertures become equal in such a case.

11.4 Conditions for Fraunhofer Diffraction

Fraunhofer diffraction requires that both the light source and the observation point P_0, are sufficiently distant from the aperture so that it may be considered that both the incident wave and the diffracted wave are made up of plane waves. Considering the geometry of

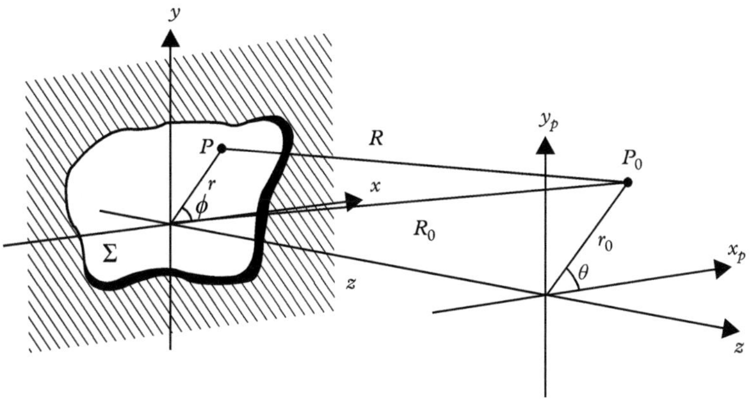

Figure 11.4 *Geometry for the study of Fraunhofer diffraction.*

Fig. 11.4, we have that the distance between the point P at the aperture and the observation point P_0 is given by

$$R^2 = (x - x_p)^2 + (y - y_p)^2 + z^2 \tag{11.17}$$

On the other hand, the distance R_0 between the centre of the aperture and the observation point P_0 is

$$R_0^2 = x_p^2 + y_p^2 + z^2 \tag{11.18}$$

From Eqs. (11.17) and (11.18) one can write the following result for the difference $R_0 - R$:

$$\begin{aligned} R_0 - R = \frac{R_0^2 - R^2}{R_0 + R} &= \left(\frac{xx_p + yy_p}{R_0} - \frac{x^2 + y^2}{2R_0}\right)\left(\frac{1}{1 - (R_0 - R)/2R_0}\right) \\ &\approx \left(\frac{xx_p + yy_p}{R_0} - \frac{x^2 + y^2}{2R_0}\right) \end{aligned} \tag{11.19}$$

In obtaining the final result in Eq. (11.19) it was assumed that

$$|R_0 - R| \ll R_0 \tag{11.20}$$

Using the result of Eq. (11.19), the diffraction integral given by Eq. (11.7) becomes

$$E_P = A \iint_\Sigma \exp\left[-ik\left(\frac{xx_p + yy_p}{R_0} - \frac{x^2 + y^2}{2R_0}\right)\right] dxdy \tag{11.21}$$

where

$$A = \frac{QE_i}{R_0} e^{i\mathbf{k}\cdot\mathbf{R}_0} \tag{11.22}$$

In obtaining Eq. (11.21), R has been replaced by R_0 in the denominator of Eq. (11.7).

When the observation point P_0 is sufficiently far from the aperture, we have

$$\frac{x^2 + y^2}{2\lambda R_0} << 1 \tag{11.23}$$

In this case, one can neglect the second term in the exponential of Eq. (11.21) and get the result

$$E_P = A \iint_{\Sigma} \exp\left[-ik\left(\frac{xx_p + yy_p}{R_0}\right)\right] dxdy \tag{11.24}$$

Eq. (11.24) satisfies the so-called *Fraunhofer condition*, according to which the phase of the diffracted field must be a linear function of the coordinates of the aperture, corresponding to a set of plane waves.

In the case where the aperture is covered by a film with a transmission function $f(x, y)$, the field E_i in Eq. (11.7) should be replaced by $f(x, y)\, E_i$, which represents the field transmitted by this aperture. In the general case, the function $f(x, y)$ is complex, describing both the amplitude and the phase variations experienced by the field transmitted through the film applied to the aperture. Under these circumstances, Eq. (11.24) can be generalized and written in the form

$$\begin{aligned} E_P &= A \iint_{\Sigma} f(x, y) \exp\left[-ik\left(\frac{xx_p + yy_p}{R_0}\right)\right] dxdy \\ &= A \iint_{\Sigma} f(x, y) \exp\left[-i\left(k_x x + k_y y\right)\right] dxdy \end{aligned} \tag{11.25}$$

where

$$k_x = \frac{kx_p}{R_0},\ k_y = \frac{ky_p}{R_0} \tag{11.26}$$

The integrations in Eq. (11.25) can be performed between $-\infty$ and $+\infty$, assuming that the transmission function $f(x, y)$ is extended in such a way that it vanishes outside the aperture region. Considering the definition of the Fourier transform presented in Chapter 5, it can be concluded from Eq. (11.25) that the *Fraunhofer diffraction pattern for the amplitude of the electric field is proportional to the two-dimensional Fourier transform of the aperture transmission function*. In the following sections, Eq. (11.25) will be used to obtain the Fraunhofer diffraction pattern determined by different types of apertures.

11.5 Diffraction by a Rectangular Aperture

Let us consider a rectangular aperture with a transmission function given by

$$f(x, y) = \begin{cases} 1, & |x| \leq a/2 \wedge |y| \leq b/2 \\ 0, & |x| > a/2 \vee |y| > b/2 \end{cases} \tag{11.27}$$

Substituting Eq. (11.27) into Eq. (11.25), one has

$$E_P = A \int_{-a/2}^{a/2} e^{-ik_x x} dx \int_{-b/2}^{b/2} e^{-ik_y y} dy$$
$$= SA \frac{\sin \alpha_x}{\alpha_x} \frac{\sin \alpha_y}{\alpha_y} \tag{11.28}$$

where

$$\alpha_x = \frac{k_x a}{2} = \frac{kax_p}{2R_0} \tag{11.29}$$

$$\alpha_y = \frac{k_y b}{2} = \frac{kby_p}{2R_0} \tag{11.30}$$

and $S = ab$ is the area of the aperture.

The distribution of irradiance in the Fraunhofer diffraction pattern produced by the rectangular aperture can be written in the form

$$I_p = I_0 \sin c^2 \alpha_x \sin c^2 \alpha_y \tag{11.31}$$

where

$$I_0 = (S|A|)^2 \tag{11.32}$$

represents the maximum irradiance, and sinc (w) is the cardinal sine function, defined as

$$\text{sinc}\,(\text{w}) \equiv \frac{\sin\,(w)}{w} \tag{11.33}$$

Fig. 11.5 shows the graph of sinc (w), as well as $\text{sinc}^2(w)$ functions.

The zeros of the sinc(w) function occur for

$$w = m\pi, \qquad (m \neq 0) \tag{11.34}$$

where m can be any integer, except zero. On the other hand, using the rule of L'Hospital, one has

$$\lim_{w\to 0}\left(\frac{\sin\,(w)}{w}\right) = \lim_{w\to 0}\left(\frac{\cos\,(w)}{1}\right) = 1 \tag{11.35}$$

which corresponds to the central maximum of the sinc (w) function. The locations of other extrema are obtained by applying the condition

$$\frac{d\,(\text{sinc}\,(w))}{dw} = 0$$

that gives,

$$\tan\,(w) = w \tag{11.36}$$

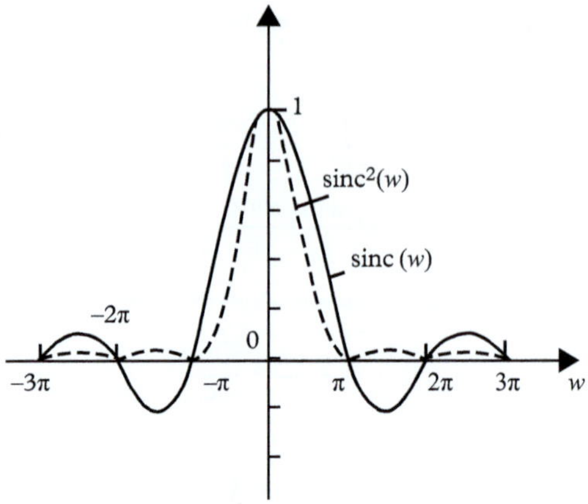

Figure 11.5 *Graphical representation of the functions sinc(w) and* $\text{sinc}^2(w)$.

The solutions of this transcendental equation can be determined graphically or numerically and are given by $w = \pm 1.4303\pi, \pm 2.4590\pi, \pm 3.4707\pi \ldots$ Actually, the secondary maxima of the function $\text{sinc}^2(w)$ do not quite fall at the midpoints between zeros, even though this condition is more nearly approached as w increases. It can be verified that the maximum value of the nearest secondary peaks, located at $w = \pm 1.4303\pi$, is only 4.7% of that of the central peak.

The maximum irradiance given by Eq. (11.31) occurs for $\alpha_x = \alpha_y = 0$. Irradiance minima occur for $\alpha_x = n\pi$ and $\alpha_y = m\pi$, where n and m are non-zero integers. The dimensions of the diffraction pattern are given by the coordinates x_p and y_p of the observation plane corresponding to $m = n = 1$, and it can be verified that they are inversely proportional to the dimensions of the aperture. Of course, the dimensions of the diffraction pattern increase uniformly with distance R_0 and also depend on the wavelength, as $k = \frac{2\pi}{\lambda}$ in Eqs. (11.19) and (11.20). Fig. 11.6 shows an example of the Fraunhofer diffraction pattern produced by a rectangular aperture.

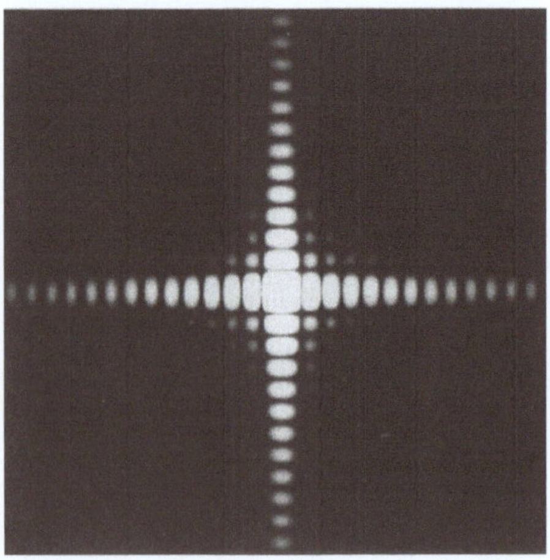

Figure 11.6 *Fraunhofer diffraction pattern produced by a rectangular aperture.*

11.5.1 Diffraction by a single slit

The problem of diffraction by a single slit can be considered as a particular case of diffraction by a rectangular aperture, as considered earlier. Let us suppose that the slit has a width a and is parallel to the y-axis, such that $b>>\lambda$. In this case, the factor $\mathrm{sinc}^2\alpha_y$ in Eq. (11.31) decreases rapidly when $y_p \neq 0$, so the light diffracted by the slit is predominantly distributed in the plane xz. The distribution of the irradiance given by Eq. (11.31) is then reduced to the form

$$I = I_0 \frac{\sin^2\alpha}{\alpha^2} \tag{11.37}$$

where

$$\alpha = \frac{ka}{2}\sin\theta, \qquad \sin\theta = \frac{x_p}{R_0} \tag{11.38}$$

θ being the angle measured relative to the plane yz. Fig. 11.7 shows an example of the Fraunhofer diffraction pattern produced by a single slit.

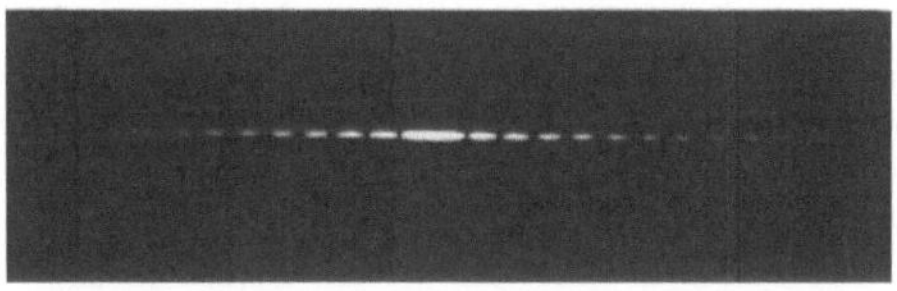

Figure 11.7 *Fraunhofer diffraction pattern produced by a single slit.*

Example 11.1

A single slit is illuminated by light with wavelength $\lambda = 650\,\mathrm{nm}$ and produces a Fraunhofer diffraction pattern in which first-order minima lie in a direction that makes an angle $\theta = 5°$ relative to the slit normal. Determine the width of the slit.

Solution

From Eqs. (11.37) and (11.38) we find that the minimum irradiance in the diffraction pattern occurs for

$$\alpha = \frac{\pi a}{\lambda}\sin\theta = m\pi, \qquad m\,\text{integer} \tag{11.39}$$

where a is the width of the slit. Considering $m = 1$, one has

$$a = \frac{\lambda}{\sin\theta} = \frac{6.5\times10^{-7}}{\sin(5°)} = 7.45\times10^{-6}\,\mathrm{m} \tag{11.39}$$

Example 11.2

A slit of width $2L$ is in the plane (x, y), centred at point (0,0) and parallel to the y-axis. It is covered by a mask whose amplitude transmission function $f(x)$ reduces linearly with $||x||$, as shown in Fig. 11.8. Find the diffracted far-field when the slit is illuminated normally by a monochromatic plane wave and compare with the result corresponding to the absence of the mask.

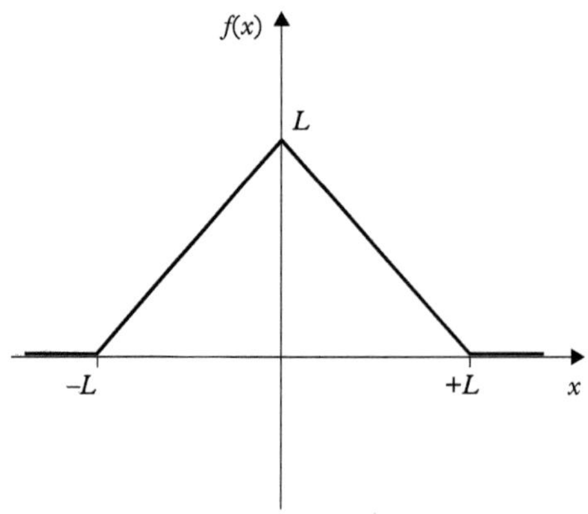

Figure 11.8 *Transmittance of the mask covering the slit.*

Solution

The mask transmission function $f(x)$ represented in Fig. 11.8 is given by

$$f(x) = \begin{cases} L + x & -L < x < 0 \\ L - x & 0 < x < L \\ 0 & |x| > L \end{cases} \tag{11.40}$$

The diffracted field is proportional to

$$F(k) = \sqrt{\frac{2}{\pi}} \int_0^L (L - x)\cos kx dx = \sqrt{\frac{2}{\pi}} \frac{2}{k^2}(1 - \cos kL) \tag{11.41}$$

In obtaining Eq. (11.41), the fact that $f(x)$ is an even function was used. Using trigonometric transformations, we can write the previous result in the form:

$$F(k) = \sqrt{\frac{2}{\pi}} \frac{4}{k^2} \sin^2\left(\frac{kL}{2}\right) = \sqrt{\frac{2}{\pi}} L^2 \operatorname{sinc}^2\left(\frac{kL}{2}\right) \tag{11.42}$$

In the absence of a mask, the aperture transfer function is given by

$$f(x) = \begin{cases} 1, & |x| < L \\ 0, & |x| > L \end{cases} \tag{11.43}$$

And the field diffracted in Fraunhofer's approximation is proportional to

$$F(k) = \sqrt{\frac{2}{\pi}} L \operatorname{sinc}(kL) \tag{11.44}$$

Example 11.2

The process of masking an aperture, usually for the purpose of reducing the secondary peaks of the diffraction pattern, is known as *apodization*. By comparing the results given by Eqs. (11.42) and (11.44), we can verify that masking the slit determines a reduction of the secondary maxima in the diffracted far-field.

11.6 Array Theorem

Let us consider a set of N identical apertures located along the x-axis, $f(x)$ being the transmission function of one of these apertures centred at the origin of the plane. The transmission function of an aperture centred at point x_n can be expressed in terms of the transmission function $f(x)$ using the relation:

$$f(x - x_n) = \int_{-\infty}^{+\infty} f(x')\,\delta(x - x_n - x')\,dx' \tag{11.45}$$

where δ represents the Dirac delta function. The integral in Eq. (11.45) can be understood as the convolution of the function $f(x)$ with the function $\delta(x - x_n)$, as defined in Chapter 5. The transmission function corresponding to the set of all apertures is given by

$$g(x) = \sum_{n=1}^{N} f(x - x_n) = \sum_{n=1}^{N} \int_{-\infty}^{+\infty} f(x')\,\delta(x - x_n - x')\,dx' \tag{11.46}$$

The Fraunhofer diffraction pattern corresponding to this set of apertures is given by the Fourier transform of $g(x)$:

$$G(k_x) = F\{g(x)\} \tag{11.47}$$

Considering the function $g(x)$ given by Eq. (11.46) and using the convolution theorem referred to in Chapter 5, one obtains:

$$G(k_x) = \sum_{n=1}^{N} F\{f(x)\}F\{\delta(x - x_n)\} = F\{f(x)\}F\left\{\sum_{n=1}^{N} \delta(x - x_n)\right\} \tag{11.48}$$

In two dimensions, this result takes the form:

$$G(k_x, k_y) = F\{g(x,y)\} = F\{f(x,y)\}F\left\{\sum_{n=1}^{N} \delta(x - x_n)\,\delta(y - y_n)\right\} \tag{11.49}$$

where

$$g(x,y) = \sum_{n=1}^{N} f(x - x_n, y - y_n) \tag{11.50}$$

Eq. (11.49) corresponds to the so-called *array theorem*. According to this theorem, the diffraction pattern determined by a given spatial distribution of identical apertures is equal to the product of the diffraction pattern due to only one of these apertures and the diffraction pattern corresponding to an array of point sources with an identical distribution. The first factor is known as the *form factor*, whereas the second one is called the *network factor*.

11.6.1 Diffraction by N slits

As an example of application, the array theorem can be used to calculate the diffraction pattern produced by N identical slits, each of width a and spaced from each other by a distance d. The distribution of irradiance corresponding to one of these slits is given by Eq. (11.37). On the other hand, the network function is

$$D(x) = \sum_{n=1}^{N} \delta(x - x_n) \tag{11.51}$$

where $x_n = (n-1)\,d$. The Fourier transform of Eq. (11.51) is given by

$$F\{D(x)\} = \frac{\sin N\beta}{\sin\beta} \tag{11.52}$$

where

$$\beta = \frac{kd}{2}\sin\theta \tag{11.53}$$

Based on the matrix theorem, the distribution of irradiance in the diffraction pattern corresponding to the N slits can be written in the following form:

$$I = I_0 \frac{\sin^2\alpha}{\alpha^2} \frac{\sin^2 N\beta}{\sin^2\beta} \tag{11.54}$$

In Eq. 11.54, the first factor represents the diffraction envelope of the resultant irradiance, whereas the last factor describes interference between light coming from different slits. Actually, this interference factor describes a series of sharp irradiance peaks (*principal maxima*), which occur when

$$\frac{\sin N\beta}{\sin\beta} = N \tag{11.55}$$

These peaks occur for $\beta = 0, \pm\pi, \pm 2\pi, \ldots$ that is, when

$$d\sin\theta_m = m\lambda \tag{11.56}$$

The irradiance at a principal maximum is proportional to N^2. In between successive principal maxima there are *secondary peaks*. For an arbitrary value N, there will be N-2 secondary peaks between two principal maxima. It is found that as the number N of slits increases, the increase in irradiance at the peaks of the principal maxima is accompanied by an overall decrease in irradiance at the secondary peaks.

A device with multiple slits, such as the one considered in this section, is an example of a *diffraction grating*. In general, this designation applies to any periodic structure consisting of diffractive elements, be they apertures or obstacles, which may alter the amplitude and/or the phase of a wave incident on it. Eq. (11.55) is sometimes called the *diffraction grating equation*, and m is identified as the *order* of diffraction.

Example 11.3

a) Obtain the angular dispersion of the first-order spectrum for white light normally incident on a transmission grating that has 1000 lines per centimetre.

b) Calculate the angular dispersion for a small band of wavelengths, given by $D = \frac{\Delta\theta}{\Delta\lambda}$, for the first order of dispersion across the same grating, operating in the visible region of the spectrum.

Solution

a) The spacing between consecutive lines is

$$d = \frac{10^{-2}}{1000} = 1 \times 10^{-5}\,\mathrm{m}$$

On the other hand, first-order dispersion corresponds to $m = 1$ in Eq. (1156). White light extends in a band of wavelengths between $\lambda_1 = 390\,\mathrm{nm}$ and $\lambda_2 = 780\,\mathrm{nm}$, so we have:

$$\sin\theta_1 = \frac{390 \times 10^{-9}}{1 \times 10^{-5}} = 0.039$$

$$\sin\theta_2 = \frac{780 \times 10^{-9}}{1 \times 10^{-5}} = 0.078$$

That is, one has $\theta_1 = 2°14'$, $\theta_2 = 4°28'$ and the angular dispersion becomes

$$\Delta\theta = \theta_2 - \theta_1 = 2°14'$$

b) By differentiating Eq. (11.50), one obtains

$$D = \frac{\Delta\theta}{\Delta\lambda} = \frac{m}{d\cos\theta_m}$$

Considering $\lambda \approx 500\,\mathrm{nm}$, one has $d >> \lambda$, so Eq. (11.56) allows us to conclude that $\cos\theta_m \approx 1$ for $m = 1$. Thus, the angular dispersion is $D \approx 1 \times 10^5\,\mathrm{rad/m}$.

11.6.2 Diffraction by two slits

In the case of only two slits, Eq. (11.54) can be written in the form

$$I = I_0 \frac{\sin^2\alpha}{\alpha^2} \cos^2\beta \tag{11.57}$$

If the width (a) of the slits tends to zero, then $\frac{\sin\alpha}{\alpha} \approx 1$ and Eq. (11.57) represents the irradiance due to a pair of linear sources, as considered in Young's experiment. On the other hand, if $d = 0$ (the two slits coincide), then $\beta = 0$ and Eq. (11.57) is reduced to

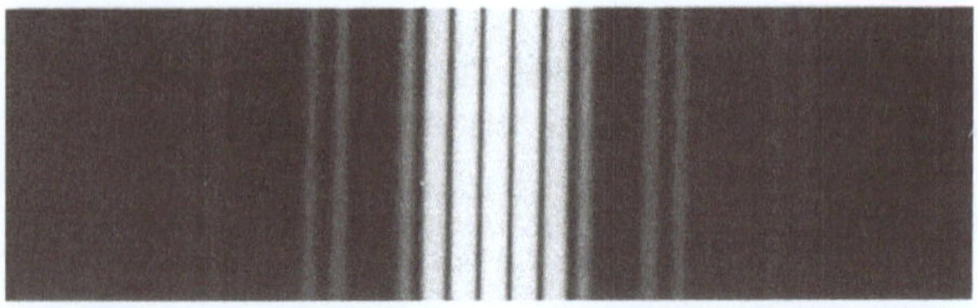

Figure 11.9 *Fraunhofer diffraction pattern produced by a double slit.*

Eq. (11.37), which describes diffraction by a single slit. It can therefore be considered that Eq. (11.57) is composed of an interference term, $\cos^2\beta$, modulated by a diffraction term, $\left(\sin^2\alpha\right)/\alpha^2$.

If the width of the slits is reduced, the diffraction pattern of each of the slits is uniform over an extensive central region, within which Young interference fringes form (Fig. 11.9). For angular positions corresponding to

$$\alpha = \pm\pi, \pm 2\pi, \dots \tag{11.58}$$

the diffraction effects are such that no light reaches the observation screen, and one does not observe the interference fringes that would be expected at these positions.

Using Eq. (11.53), we find that interference maxima occur when

$$d\sin\theta = m\lambda, \quad m\,\text{integer} \tag{11.59}$$

On the other hand, using Eq. (11.38), the minima in the diffraction pattern produced by a simple slit occur when

$$a\sin\theta = n\lambda, \quad n\,\text{integer} \tag{11.60}$$

When both conditions are simultaneously verified, one has

$$\frac{d}{a} = \frac{m}{n} = M \tag{11.61}$$

Thus, if the slit separation is an integral multiple of the slit width, the condition for missing order is met exactly.

Fig. 11.10 illustrates the distribution of the irradiance corresponding to diffraction by two identical slits for $d = 3a$. In this case, it turns out that the interference fringes of order $m = 3, 6, 9, \dots$ are absent.

Example 11.4

The Fraunhofer diffraction pattern of a double slit illuminated by light of wavelength $\lambda = 500\,\text{nm}$ appears in the secondary focal plane of a lens with focal length $f = 120\,\text{cm}$. The spacing between consecutive bright fringes is 1 mm, and the fourth maximum is absent. Determine the width of each slit and the distance between them.

Example 11.4 continued

Solution

The spacing between fringes in Young's experiment is given by $\Delta y = D\lambda/d$. Using this relation and considering that $D = f = 120\,\text{cm}$, we obtain the following result for the spacing between the slits:

$$d = \frac{D\lambda}{\Delta y} = \frac{(120 \times 10^{-2})(500 \times 10^{-9})}{1 \times 10^{-3}} = 0.6\,\text{mm}$$

In this case, one has from Eqs. (11.59) and (11.60) that $m = 4$ and $n = 1$, so $M = m/n = 4$. Thus, the width of the slit is given from Eq. (11.61) by

$$a = \frac{d}{4} = 0.15\,\text{mm}$$

11.7 Diffraction by a Circular Aperture

In the case of diffraction by a circular aperture, it is convenient for calculating the integrals in Eq. (11.25) to use polar coordinates, both in the aperture plane and in the observation plane. In the aperture plane, one considers

$$x = r\cos\phi, \quad y = r\sin\phi \tag{11.62}$$

and the elementary area is given by $rdrd\phi$. In the observation plane one has

$$x_p = r_0\cos\theta, \quad y_p = r_0\sin\theta \tag{11.63}$$

Considering that the aperture has a radius a and its transmission function is $f(r, \phi) = 1\,(r \leq a,\ 0 \leq \phi \leq 2\pi)$, the substitution of Eqs. (11.62) and (11.63) in Eq. (11.25) gives:

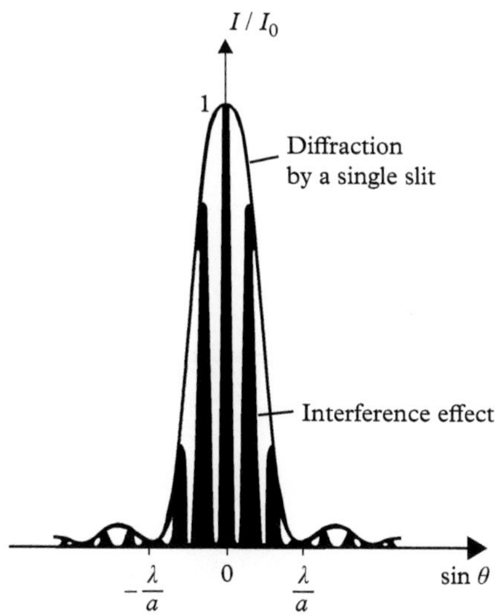

Figure 11.10 *Distribution of irradiance in the diffraction pattern produced by two equal slits, in the case d = 3a.*

$$E_P = A \int_0^a \int_0^{2\pi} \exp\left[ik\frac{rr_0}{R_0}\cos(\theta-\phi)rdrd\phi\right] \tag{11.64}$$

The integral in Eq. (11.64) can be calculated analytically in terms of the Bessel functions, providing the result:

$$E_P = 2\pi a^2 A\left[\frac{R_0}{kar_0}J_1\left(\frac{kar_0}{R_0}\right)\right] \tag{11.65}$$

where $J_1(u)$ is the Bessel function of the first kind and order one. The irradiance in the observation plane is given by

$$I = 4|A|^2 S^2\left[\frac{J_1(kar_0/R_0)}{(kar_0/R_0)}\right]^2 \tag{11.66}$$

where S is the area of the circular aperture. Since

$$\lim_{u\to 0}\frac{J_1(u)}{u} = \frac{1}{2} \tag{11.67}$$

one has that the irradiance at the centre of the diffraction pattern is

$$I_0 = |A|^2 S^2 \tag{11.68}$$

This result is similar to that given by Eq. (11.36), obtained in the case of a rectangular aperture. Fig. 11.11 represents the functions $\frac{2J_1(u)}{u}$ and $\left[\frac{2J_1(u)}{u}\right]^2$.

Eq. (11.66) should be compared with Eq. (11.37) to appreciate the analogous role played by the Bessel function. Like $(\sin x)/x$, the Bessel function approaches a maximum as x approaches zero so that the irradiance is greatest at the centre of the pattern.

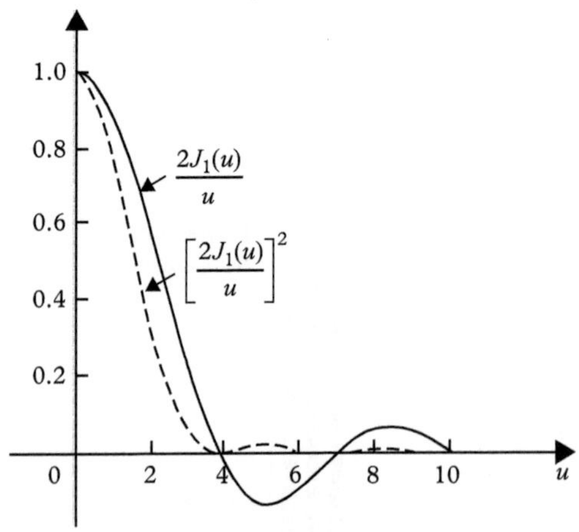

Figure 11.11 *Representation of functions $2J_1(u)/u$ (full curve) and $[2J_1(u)/u]^2$ (dashed curve).*

The pattern is symmetrical about the optical axis through the centre of the circular aperture and has its first zero when $u = 3.832$ as indicated in Fig. 11.11.

The diffraction pattern described by Eq. (11.66) is known as the *Airy pattern*, consisting of a bright central disc and a set of concentric rings, alternately dark and bright (Fig. 11.12). The radius of the first dark ring is commonly used to characterize the size of this diffraction pattern and is given by

$$r_{01} = 1.22\frac{R_0\lambda}{2a} \tag{11.69}$$

These results are of great practical importance. It should be noted, for example, that the image of a point source, formed by an ideal optical system consisting of mirrors or circular lenses, is not a point, as one would like to see, but has the appearance of an Airy pattern. If two distinct point objects are not sufficiently separated, their image patterns overlap, and it becomes difficult to resolve such images.

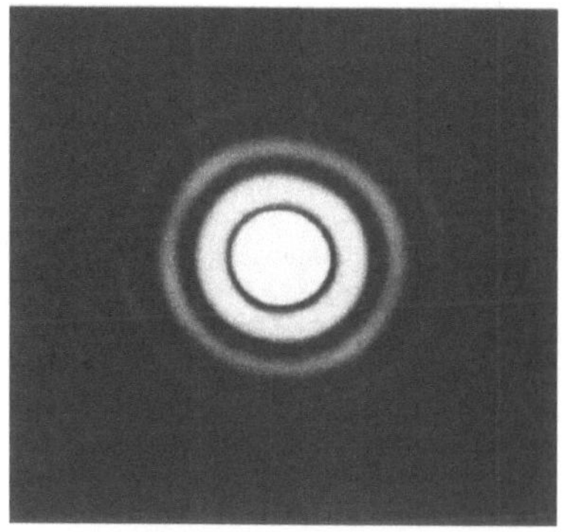

Figure 11.12 *Fraunhofer diffraction pattern produced by a circular aperture.*

In order to have resolvable images, it is usually required that the angular separation of the centres of the image patterns be no less than the angular radius of the Airy disc. This condition corresponds to the so-called *Rayleigh criterion* and implies that the minimum angular separation between the centres of such image patterns be, using Eq. (11.69):

$$\Delta\theta_{\min} = 1.22\frac{\lambda}{D} \tag{11.70}$$

where D is the diameter of the lens. According to Eq. (11.70), the minimum resolvable angular separation of two object points may be reduced by increasing the lens diameter and decreasing the wavelength.

As discussed in Section 3.1, the human eye may be approximated by a circular aperture (pupil), a lens (crystalline), and a screen (retina). As such, it also experiences the limits of resolution due to diffraction. On a bright day, the pupil diameter may be 2 mm and Eq. (11.70) gives $\Delta\theta_{\min} = 33.6 \times 10^{-5}$ rad for an average wavelength of 550 nm. At night, the pupil diameter increases and can be around 8 mm, which provides a higher resolution than in daylight.

Example 11.5

Determine the radius of the central disc, formed in the retina of the human eye, in the image of a distant point object on a bright day. Assume that the pupil has a radius $a = 1$ mm, that its distance to the retina is $d = 2$ cm, and the light wavelength is $\lambda = 600$ nm.

Solution

The pupil of the eye constitutes a circular aperture, forming an Airy diffraction pattern in the retina. From Eq. (11.69) one has that the radius of the central disc of this pattern is given by

$$\begin{aligned} r_{01} &= 1.22\frac{R_0\lambda}{2a} \\ &\approx 1.22\frac{\left(2\times10^{-2}\right)\left(6\times10^{-7}\right)}{2\left(1\times10^{-3}\right)} = 7.32\times10^{-6}\,\text{m} \end{aligned}$$

11.8 Further Reading

Born, M., and E. Wolf, *Principles of Optics*, 6th Ed., Pergamon Press, 1980.
Cowley, J. M., *Diffraction Physics*, 3rd Ed., Elsevier, 1995.
Guenther, R., *Modern Optics*. John Wiley & Sons, 1990.
Heavens, O., and R. Ditchburn, *Insight into Optics*, John Wiley & Sons, 1991.
Hecht, E., *Optics*, 5th Ed., Pearson, Education 2016.
Klein, M., and T. Furtak, *Optics*, 2nd Ed., John Wiley & Sons, 1986.
Pedrotti, F., L. M. Pedrotti, and L. S. Pedrotti, *Introduction to Optics*, 3rd Eds, Cambridge University Press, 2018.
Ufimtsev, P., *Fundamentals of the Physical Theory of Diffraction*, 2nd Ed., John Wiley & Sons, 2014.

11.9 Problems

11.1 A monochromatic beam of light with wavelength $\lambda = 500\,\text{nm}$ illuminates a single slit, and its diffraction pattern is observed on a screen that is at a distance of 1 m from the slit. Assuming that the first and third minima of the diffraction pattern are spaced by 4 mm, determine the width of the slit.

11.2 A monochromatic beam of light with wavelength $\lambda = 500\,\text{nm}$ illuminates a single slit located in front of a convergent thin lens with focal length $f = 80$ cm. Assuming that the first minima (adjacent to the central spot) of the Fraunhofer diffraction pattern are spaced 6 mm, calculate the width of the slit.

11.3 Show that the irradiance patterns produced by two complementary screens, one of which is opaque and contains a narrow slit of width a, are identical, except near the centre. This problem illustrates a particular case of Babinet's theorem.

11.4 Let us consider a square aperture of side $2L$, centred on the point with coordinates $(a, b, 0)$ and whose sides are parallel to the x and y axes. This aperture is illuminated by monochromatic plane waves, whose direction of propagation, situated in the plane xz, makes an angle θ with the z-*axis*. Determine, in the Fraunhofer approximation, the distribution of irradiance in the plane of observation. Compare this result with that obtained in the cases $a = b = 0$ and $\theta = 0$.

11.5 a) Determine the irradiance corresponding to the Fraunhofer diffraction produced by two parallel slits, one of width a and the other of width b, separated by h.

b) Use the result obtained in (a) to describe the diffraction pattern in the case $a = b$.

11.6 a) Show that the irradiance in the Fraunhofer diffraction pattern produced by two parallel slits of width d and spaced apart by h is given by Eq. (11.57).

b) Explain the result given by Eq. (11.57) in the cases: (i) $h \to 0$ and $d > 0$; (ii) $d \to 0$ and $h > 0$.

c) Assuming that the two slits have width $d = 0.1$ mm and are spaced by $h = 0.40$ mm, find the interference maxima that will be absent from the diffraction pattern.

11.7 The Fraunhofer diffraction pattern of a double slit illuminated with light of wavelength $\lambda = 600$ nm is observed in the secondary focal plane of a converging lens with focal length $f = 60$ cm. The spacing between the slits is $d = 1$ mm and the fifth bright fringe is absent. Determine (i) the spacing between consecutive dark fringes and (ii) the width of each slit.

11.8 Consider a monochromatic beam of light with wavelength λ that illuminates a single slit of width $a = 7\lambda$.

a) Obtain an expression for the distribution of irradiance in a region far from the slit and make the respective graphical representation. Calculate the positions of the first minimum and the first secondary maximum.

b) Consider that the slit is divided into two equal parts, each of width $a/2$. We want to change the phase of light passing through one of these parts by π rad. Explain how that can be accomplished.

c) Obtain an expression to describe the distribution of irradiance for the situation considered earlier in a region far from the slit. Calculate the positions of the first two minima and the first maximum in this case.

11.9 A transmission diffraction grating with 1000 elements per mm is illuminated at normal incidence by sodium D-lines. The angular separation between such lines in the second-order spectrum is 5 minutes of arc. Determine the wavelength separation between the D-lines, assuming that one of the sodium D-lines has a wavelength $\lambda = 589.592$ nm.

11.10 Calculate the angular separation between the D-lines of sodium ($\lambda = 589.592$ nm and $\lambda = 588.995$ nm) in the first-order spectrum determined by a transmission grating with 5000 lines per centimetre, assuming normal incidence.

11.11 Find the diameter of the Airy disc at the centre of the diffraction pattern formed on a screen at a distance of 2 m from a uniformly illuminated circular aperture of diameter 0.4 mm. Assume that the illuminating light has wavelength $\lambda = 600$ nm. Compare the beam spread to that from a slit of width equal to the aperture diameter.

11.12 A plane wave of wavelength λ illuminates normally an opaque screen with five parallel and equal slits of width a and spaced by d.

a) Find the angle with the normal at which the interference maxima form.

b) Find the angular width of the interference maxima.

c) Find the ratio of the irradiance at an interference maximum to the irradiance for a single slit.

d) Sketch the interference pattern as a function of angle.

11.13 A collimated beam of monochromatic light, with wavelength $\lambda = 500$ nm, illuminates a convergent lens with a diameter of 1.5 cm and a focal length $f = 60$ cm. Calculate the angular diameter of the central disc in the diffraction pattern formed in the focal plane of the lens.

11.14 Determine the diameter of the central Airy disc formed on the retina of the human eye corresponding to a distant point object both (a) in daylight and (b) at night. Consider that the pupil diameter in the two cases is 2 mm and 8 mm, respectively. Assume a reference wavelength $\lambda = 500\,\text{nm}$ and a distance between the pupil and the retina of 2 cm.

11.15 Two lamps are separated by 1 m. Calculate the maximum distance at which they can be resolved by a person with normal vision at night. Assume a reference wavelength of 550 nm and a pupil diameter of 8 mm.

Fresnel Diffraction

12

As seen in the previous chapter, Fraunhofer diffraction occurs when the aperture is illuminated by plane waves and the observation point is quite far from the aperture. In these circumstances, the phase of the diffracted field varies linearly with the aperture coordinates. In this chapter, we will consider the case where either the source or the plane of observation, or both, are relatively close to the aperture, so account must be taken of the curvature of the wavefront.

Fresnel diffraction patterns can be understood as lying in the continuity between two extreme cases: the pattern corresponding to geometrical optics, on the one hand, and the Fraunhofer diffraction pattern, on the other hand. In geometrical optics, it is assumed that light propagates along rectilinear trajectories, so it is expected that one observes a clear image of the aperture in the observation plane. In practice, an image of this type is observed only if the distance between the plane of observation and the plane of the aperture is quite short. In Fraunhofer diffraction, where this distance must indeed be very large, the image corresponds to a fringe pattern that bears no apparent resemblance to the shape of the aperture. Fresnel diffraction corresponds to an intermediate situation. In this case, the diffraction pattern appears as an image of the aperture, but the edges are fringed.

12.1 Fresnel Zones

In this section we use the Huygens–Fresnel principle to analyse the free propagation of a spherical wave. This description proves to be quite useful, not only because it provides a deeper understanding of such a principle but also because it provides a valuable technique to describe Fresnel diffraction by circular apertures or obstacles.

Let us consider a spherical wavefront emitted from a point source located at P_1 that at instant t has a radius R_s, as represented in Fig. 12.1. In order to determine the field at the observation point P_0 due to this wavefront, a set of spherical surfaces of radius $R = R_0$, $R_0 + \lambda/2$, $R_0 + 2\lambda/2, \ldots, R_0 + j\lambda/2, \ldots$ is considered, where R_0 is the distance between the wavefront and the point P_0, measured along the line connecting P_1 and P_0. These spherical surfaces divide the wavefront into a number of ring-shaped zones, called *Fresnel zones*.

Figure 12.1 *Geometry for the construction of the Fresnel zones corresponding to a spherical wave.*

Each Fresnel zone can be considered as an aperture illuminated from the left-hand side by a spherical wave of the form:

$$E_i(\mathbf{R}_s) = \frac{E_0 e^{i\mathbf{k}\cdot\mathbf{R}_s}}{R_s} \tag{12.1}$$

Let us consider an elemental ring with radius $R_s \sin\theta$ and width $R_s d\theta$ within a given Fresnel zone. The field at the point P_0 due to this ring is given by

$$dE(P_0) = E_i(\mathbf{R}_s) Q \frac{e^{i\mathbf{k}\cdot\mathbf{R}}}{R} dS \tag{12.2}$$

Lasers, Optics, and Photonics. Mário F. S. Ferreira, Oxford University Press. © Mário F. S. Ferreira (2026).
DOI: 10.1093/oso/9780198952770.003.0012

where Q is the obliquity factor and dS is the area of the elemental ring, given by

$$dS = 2\pi R_s^2 \sin\theta d\theta \tag{12.3}$$

Considering the geometric construction of Fig. 12.1, one has:

$$R^2 = R_s^2 + (R_s + R_0)^2 - 2R_s(R_s + R_0)\cos\theta \tag{12.4}$$

Differentiating this expression yields:

$$2RdR = 2R_s(R_s + R_0)\sin\theta d\theta \tag{12.5}$$

Eq. (12.5) can be substituted into Eq. (12.3), giving:

$$dS = 2\pi \frac{R_s}{R_s + R_0} RdR \tag{12.6}$$

Substituting Eq. (12.6) into Eq. (12.2), one has that the field at point P_0 due to the Fresnel zone of order j can be written in the form:

$$E_j(P_0) = 2\pi E_i(\mathbf{R}_s)Q_j \frac{R_s}{R_s + R_0} \int_{R_{j-1}}^{R_j} e^{i\mathbf{k}\cdot\mathbf{R}} dR \tag{12.7}$$

where $R_j = R_0 + j\lambda/2$ and $R_{j-1} = R_0 + (j-1)\lambda/2$. When writing Eq. (12.7), the obliquity factor was considered to be approximately constant over the Fresnel zone, i.e. $Q(\varphi) = Q_j$. The calculation of the integral in Eq. (12.7) provides the following result:

$$E_j(P_0) = 2i(-1)^j \frac{Q_j E_0 \lambda}{R_f + R_0} e^{ik(R_s + R_0)} \tag{12.8}$$

The signal alternation of the field corresponding to consecutive Fresnel zones is due to the phase difference between the waves coming from these zones, since the propagation distance corresponding to similar points of adjacent zones differs precisely by $\lambda/2$. Thus, the contributions to the field at point P_0 due to adjacent Fresnel zones tend to cancel each other. However, this cancellation is not total, as the obliquity factor varies, albeit slightly, from one zone to another.

The total field at the point P_0 due to N Fresnel zones is given by the superposition principle:

$$E = E_1 + E_2 + E_3 + \cdots + E_N \tag{12.9}$$

The field amplitude corresponding to successive Fresnel zones shows a gradual decrease with (n) due both to the inverse-square law dependence on the distance from P_0 and to the obliquity factor reduction.

Taking into account the alternation of the signal in consecutive terms of the aforementioned sum, as given by Eq. (12.9), one has

$$|E| = |E_1| - |E_2| + |E_3| - \cdots + |E_N| \tag{12.10}$$

The elements of the sum in Eq. (12.10) may be grouped as follows:

$$|E| = \frac{|E_1|}{2} + \left(\frac{|E_1|}{2} - |E_2| + \frac{|E_3|}{2}\right) + \left(\frac{|E_3|}{2} - |E_4| + \frac{|E_5|}{2}\right) + \cdots \tag{12.11}$$

Since the variation of the distance to P_0 and of the obliquity factor when considering adjacent zones is negligible, each of the parentheses in Eq. (12.11) is approximately null. In this approximation, the result for the total field modules depends on the number of terms in the sum being even or odd, as given by

$$|E| \approx \begin{cases} (|E_1| + |E_N|)/2 & N\,odd \\ (|E_1| - |E_N|)/2 & N\,even \end{cases} \tag{12.12}$$

We can conclude from Eq. (12.12) that, for N small, the resultant amplitude is essentially that of the first zone alone if N is odd, whereas it is near zero if N is even. For N large, as in the case of an unlimited aperture, E_N approaches zero for either N even or odd, and Eq. (12.12) gives

$$|E| \approx \frac{|E_1|}{2} \tag{12.13}$$

This surprisingly simple result shows that *the total field at the observation point P_0, produced by a freely propagating spherical wave, is equal to half the contribution due to the first Fresnel zone.*

12.2 Diffraction by Circular Apertures or Obstacles

The analysis carried out in the previous section can be easily applied to the case of Fresnel diffraction by circular apertures or circular obstacles. These two cases will be discussed in the following, considering that such apertures are illuminated by spherical waves or by plane waves.

12.2.1 Spherical waves

The analysis carried out in the previous section can be easily applied to the case of Fresnel diffraction by circular apertures. Considering a point of observation P_0 on the axis of the aperture and assuming that it is entirely filled by an integer number, N, of Fresnel zones, two situations are possible. If N is even, one has

$$E = (|E_1| - |E_2|) + (|E_3| - |E_4|) + \cdots + (|E_{N-1}| - |E_N|) \approx 0 \tag{12.14}$$

since the adjacent contributions are practically equal in modules. On the other hand, if N is odd, one has

$$E = |E_1| - (|E_2| - |E_3|) - (|E_4| - |E_5|) - \cdots - (|E_{N-1}| - |E_N|) \approx |E_1| \tag{12.15}$$

Figure 12.2 *Fresnel diffraction pattern produced by a circular aperture.*

This result may seem surprising when compared to that of Eq. (12.13). In fact, it turns out that the interposition of an opaque screen with only a small hole, despite almost completely blocking the wavefront, increases by four times the irradiance at P_0 compared to the situation of free propagation of the wave. Obviously, the principle of conservation of energy requires that there be other points at which the irradiance has decreased. Given the symmetry of the configuration, it is to be expected that a pattern of circular fringes will be observed in a plane perpendicular to the axis of the aperture, as shown in Fig. 12.2.

12.2.2 Plane waves

In the previous discussion, it was considered that the wave incident at the aperture was a spherical wave. In the case where the incident wave is a plane wave, the analysis of the problem becomes simpler. In this case, the Fresnel zones correspond to planar rings defined in the plane of the wavefront.

Using the definition given in Section 12.1 for the Fresnel zones and the Pythagorean theorem, one has that the radius of the $n-$ th zone, r_n, relates to the distance R_0 between the aperture and the observation point in the form

$$r_n^2 = \left(R_0 + n\frac{\lambda}{2}\right)^2 - R_0^2$$

That is,

$$r_n^2 = n\lambda R_0 + n^2\frac{\lambda^2}{4} \tag{12.16}$$

For not too high values of n, one can neglect the second term in Eq. (12.16), and the following result is then obtained:

$$r_n = \sqrt{n\lambda R_0} \tag{12.17}$$

We verify from Eq. (12.17) that, for a given observation point on the axis of the aperture, the *radii of the Fresnel zones are proportional to the square roots of integer numbers.*

Example 12.1

Plane waves with wavelength $\lambda = 500\,\text{nm}$ illuminate normally an opaque screen that has a circular aperture with radius $a = 1.0\,\text{mm}$. Determine the axial distances from the screen corresponding to the two farthest points with maximum irradiance.

Solution

From Eq. (12.17), we find that the farthest point with maximum irradiance corresponds to $n = 1$, in which case the aperture is completely filled by one Fresnel zone. The distance from that point to the aperture is given by

$$R_{01} = \frac{a^2}{\lambda} = \frac{\left(1 \times 10^{-3}\right)^2}{5 \times 10^{-7}} = 2\,\text{m}$$

The next point at which another maximum of irradiance occurs corresponds to $n = 3$, in which case the aperture is entirely filled by three Fresnel zones. The distance from that point to the aperture is given by

$$R_{03} = \frac{a^2}{3\lambda} = \frac{\left(1 \times 10^{-3}\right)^2}{3\left(5 \times 10^{-7}\right)} = 0.67\,\text{m}$$

12.2.3 Circular obstacles

In the case of an opaque obstacle with a circular shape, and assuming that it blocks the first (l) zones of Fresnel, the field at a point P_0 on its axis is given by

$$E = |E_{l+1}| - |E_{l+2}| + \cdots + |E_N| \tag{12.18}$$

As mentioned before, E_N approaches zero for large values of N. The sum in Eq. (12.18) can be calculated as done in Section 12.1 for the case of an unobstructed wave, with the result:

$$E \approx \frac{|E_{l+1}|}{2} \tag{12.19}$$

That is, there is a bright spot on the axis of the obstacle, which is a result unforeseen by geometrical optics. Such a spot is known as *Poisson's spot,* precisely in honour of the French scientist who considered the hypothesis of its existence absurd. However, Fresnel and Arago confirmed experimentally the existence of such a spot, whose irradiance is only slightly lower than that of the unobstructed wave.

12.3 Fresnel Zone Plates

The analysis in the previous section showed that the contributions of adjacent Fresnel zones tend to cancel each other out. This suggests that the irradiance at the observation point can be significantly increased by alternately blocking the Fresnel zones. For example, if the first 10 even zones are blocked, letting only the first 10 odd zones pass, we have that the field at the observation point will be. $E \approx 10E_1$ Taking into account that the field at that same point due to the unobstructed wave is equal to $E_1/2$, it turns out that this technique will, in this case, increase the irradiance at the observation point 400 times!

Considering the geometry of Fig. 12.3, one has that the radius y of an arbitrary zone is given by

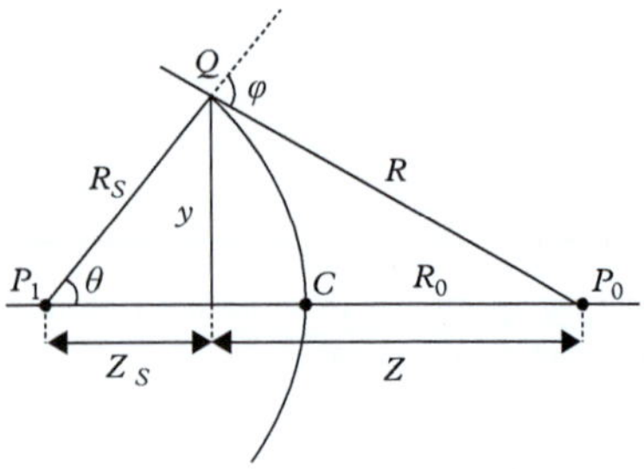

Figure 12.3 *Geometry for drawing a zone plate.*

$$R_s^2 - Z_s^2 = R^2 - Z^2 = y^2 \tag{12.20}$$

From Eq. (12.20), and assuming that θ is small enough, one can write:

$$R_s - Z_s = \frac{y^2}{R_s + Z_s} \approx \frac{y^2}{2Z_s} \tag{12.21a}$$

$$R - Z = \frac{y^2}{R + Z} \approx \frac{y^2}{2Z} \tag{12.21b}$$

If $y = r_n$, which corresponds to the radius of the Fresnel zone of order n, the difference in paths between the ray that goes from P_1 to Q and then to P_0 and the ray that goes from P_1 to P_0 along the axis is given by

$$(R_s + R) - (Z_s + Z) = \frac{n\lambda}{2} \tag{12.22}$$

Substituting the results of Eq. (12.21) into Eq. (12.22) and assuming $y = r_n$, one obtains:

$$\frac{1}{Z_s} + \frac{1}{Z} = \frac{n\lambda}{r_n^2} \tag{12.23}$$

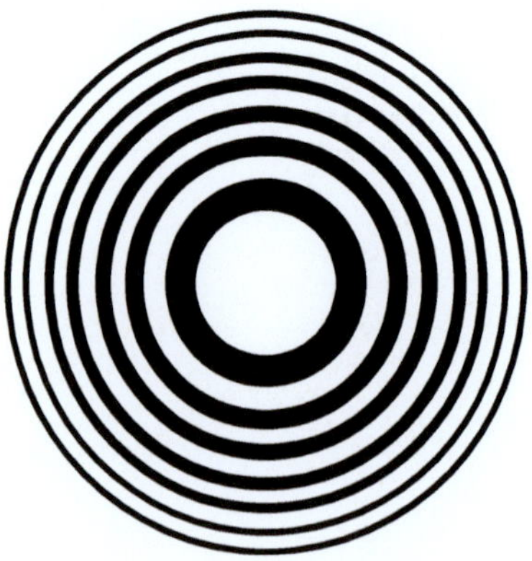

Figure 12.4 *Appearance of a zone plate with 14 Fresnel zones.*

One can verify from Eq. (12.23) that, once the positions of the source and observation point have been defined (Z_s and Z fixed), the radii, r_n, of the opaque zones used to block light are proportional to the square root of the integers, n. A device constructed according to this condition is called *a Fresnel zone plate*. Fig. 12.4 shows a zone plate with 14 Fresnel zones.

This aspect of Eq. (12.23) shows that a zone plate should operate similarly to a thin lens, having a focal length given by

$$f = \frac{r_n^2}{n\lambda} \tag{12.24}$$

Despite the similarity pointed out earlier, there are some significant differences between a Fresnel zone plate and a conventional lens. In fact, it should be noted that the operation of the zone plate is based on the phenomenon of light diffraction, whereas the operation of a conventional lens is based on the law of refraction. On the other hand, it can be seen from Eq. (12.24) that a zone plate is characterized by significant chromatic aberration.

Example 12.2

Light of wavelength $\lambda = 450$ nm illuminates a zone plate and an observation point is located at the focal distance $f = 20$ cm away on its central axis. Find (i) the radius of an aperture corresponding to the first zone relative to that point, (ii) the number of Fresnel zones contained in an aperture with a radius 10 times larger, and (iii) the ratio of irradiances observed in the two cases.

Solution

i) The first zone radius relative to the observation point is given from Eq. (12.17), considering $n = 1$ and $R_0 = f$: $r_1 = \sqrt{\lambda f} = 0.3$ mm.

ii) Considering an aperture with radius $a = 10r_1$, Eq. (12.17) gives $n = 10^2 = 100$ Fresnel zones.

iii) Half of the Fresnel zones found in (ii) are opaque. So the ratio of irradiances is $\frac{I}{I_0} = 50^2 = 2500$.

At the distance f given by Eq. (12.24), each ring of the zone plate corresponds exactly to a Fresnel zone of the plane wavefront, at which point a principal maximum of the irradiance distribution is observed. This distance f is called the first-order *focal length* of the zone plate. Actually, in contrast with conventional lenses, in the case of a zone plate there are other focal points of higher order, at distances $f/3$, $f/5$, $f/7, \ldots$ of the plate. The $f/3$ distance, for example, corresponds to the situation in which each ring of the zone plate includes three Fresnel zones. The fields due to two of these zones cancel each other out, leaving the contribution of the third zone. The field amplitude at such a point is reduced by 1/3, so that the irradiance is 1/9 of that at a distance f from the zone plate. The argument may, of course, be extended to an observation point at a distance $f/5$, when each plate zone includes five Fresnel zones, and the irradiance is 1/25 of that at a distance f.

Example 12.3

A collimated beam of light with wavelength λ illuminates a zone plate, and a maximum of irradiance is observed on the axis, 1 m behind the plate. Determine (i) the location of the three farthest focal points of this zone plate and (ii) the point where the light with the same wavelength will be focused, in case it diverges from a point situated 1.5 m in front of this plate.

Solution

i) The three farthest focal points are located at 1 m, 1/3 m, and 1/5 m from the zone plate.

ii) From Eq. (12.24), one has

$$\frac{1}{Z_s} + \frac{1}{Z} = \frac{1}{f}$$

where

$$\frac{1}{f} = \frac{n\lambda}{r_n^2}$$

In the case that the incident beam is collimated, one has $Z_s = \infty$, so $Z = f = 1$ m. On the other hand, if $Z_s = 1.5$ m, one has

$$\frac{1}{1.5} + \frac{1}{Z} = \frac{1}{1}$$

That is, the light is focused at a distance Z = 3 m from the zone plate.

12.4 Diffraction by a Rectangular Aperture

Let us consider the diagram in Fig. 12.5, where it is assumed that a rectangular aperture situated in the plane (x, y) is illuminated by a spherical wave $E_i = (E_0/R_s)\exp(i\mathbf{k}\cdot\mathbf{R}_s)$, originating from a point source S located in (x_s, y_s, Z_s). Considering the obliquity factor as a constant, one can calculate the field at the observation point P_0 based on Eq. (11.7), which takes the form

$$E_P = QE_0 \iint_{\Sigma} \frac{e^{i\mathbf{k}\cdot(\mathbf{R}+\mathbf{R}_s)}}{RR_s} dxdy \tag{12.25}$$

The geometry of Fig. 12.5 allows us to write the following expressions for the distances R and R_s:

$$\begin{aligned} R &= \sqrt{(x_p - x)^2 + (y_p - y)^2 + Z^2} \\ &\approx Z + \frac{(x_p - x)^2 + (y_p - y)^2}{2Z} \end{aligned} \tag{12.26a}$$

$$\begin{aligned} R_s &= \sqrt{(x - x_s)^2 + (y - y_s)^2 + Z_s^2} \\ &\approx Z_s + \frac{(x - x_s)^2 + (y - y_s)^2}{2Z_s} \end{aligned} \tag{12.26b}$$

One can approximate R and R_s in the denominator of Eq. (12.25) by Z and Z_s, respectively, assuming that the amplitudes of the spherical waves do not vary significantly over the area of integration. However, in the exponent of this equation, one must consider the approximations given by Eq. (12.26).

It is convenient to define two new parameters:

$$x_q = \frac{Z_s x_p + Z x_s}{D} \tag{12.27a}$$

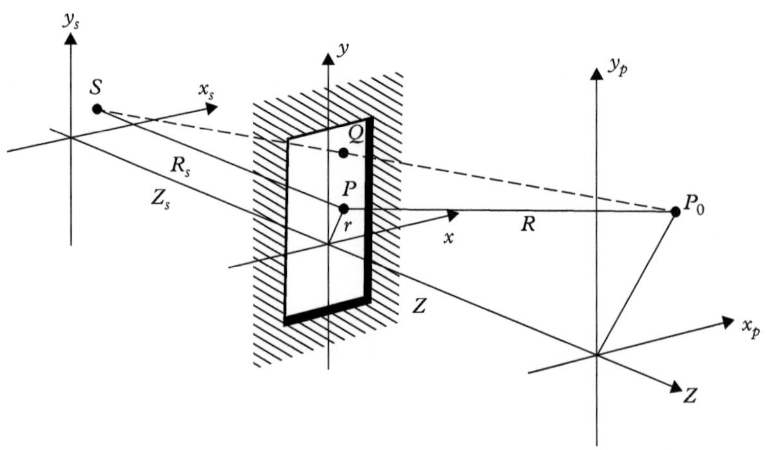

Figure 12.5 *Geometry for the analysis of Fresnel diffraction by a rectangular aperture.*

$$y_q = \frac{Z_s y_p + Z y_s}{D} \tag{12.27b}$$

where

$$D = Z + Z_s \tag{12.27c}$$

is the distance between the planes of the point source and the observation point P_0. It can be verified that the parameters x_q and y_q are the coordinates of the point Q, obtained by intersecting the plane of the aperture with the straight line passing through the source and observation points.

On the other hand, the distance between the source and the observation point P_0 is given approximately by

$$d \approx D + \frac{(x_p - x_s)^2 + (y_p - y_s)^2}{2D} \tag{12.28}$$

Using Eqs. (12.26)–(12.28), one can rewrite Eq. (12.25) in the form:

$$E_p = QE_0 \frac{e^{ikd}}{ZZ_s} \iint_\Sigma \exp\left\{\frac{ik}{2}\frac{D}{ZZ_s}\left[(x - x_q)^2 + (y - y_q)^2\right]\right\} dxdy \tag{12.29}$$

Introducing the variables,

$$u = \sqrt{\frac{2D}{\lambda ZZ_s}}(x - x_q) \tag{12.30a}$$

$$v = \sqrt{\frac{2D}{\lambda ZZ_s}}(y - y_q) \tag{12.30b}$$

one can write Eq. (12.29) in the form

$$E_P = \frac{iE_0}{2D} e^{ikd} \int_{u_1}^{u_2} e^{i\pi u^2/2} du \int_{v_1}^{v_2} e^{i\pi v^2/2} dv \tag{12.31}$$

Each of the integrals in Eq. (12.31) can be calculated on the basis of the so-called *Fresnel integrals*, $C(w)$ and $S(w)$, defined in the form:

$$C(w) = \int_0^\omega \cos\left[\frac{\pi s^2}{2}\right] ds \tag{12.32}$$

$$S(w) = \int_0^\omega \sin\left[\frac{\pi s^2}{2}\right] ds \tag{12.33}$$

where w represents u or v. In fact, we have:

$$\int_0^w e^{i\pi s^2/2} ds = C(w) + iS(w) \tag{12.34}$$

Table 12.1 *Fresnel integrals*

w	$C(w)$	$S(w)$	w	$C(w)$	$S(w)$
0.00	0.0000	0.0000	4.50	0.5261	0.4342
0.10	0.1000	0.0005	4.60	0.5673	0.5162
0.20	0.1999	0.0042	4.70	0.4914	0.5672
0.30	0.2994	0.0141	4.80	0.4338	0.4968
0.40	0.3975	0.0334	4.90	0.5002	0.4350
0.50	0.4923	0.0647	5.00	0.5637	0.4992
0.60	0.5811	0.1105	5.05	0.5450	0.5442
0.70	0.6597	0.1721	5.10	0.4998	0.5624
0.80	0.7230	0.2493	5.15	0.4553	0.5427
0.90	0.7648	0.3398	5.20	0.4389	0.4969
1.00	0.7799	0.4383	5.25	0.4610	0.4536
1.10	0.7638	0.5365	5.30	0.5078	0.4405
1.20	0.7154	0.6234	5.35	0.5490	0.4662
1.30	0.6386	0.6863	5.40	0.5573	0.5140
1.40	0.5431	0.7135	5.45	0.5269	0.5519
1.50	0.4453	0.6975	5.50	0.4784	0.5537
1.60	0.3655	0.6389	5.55	0.4456	0.5181
1.70	0.3238	0.5492	5.60	0.4517	0.4700
1.80	0.3336	0.4508	5.65	0.4926	0.4441
1.90	0.3944	0.3734	5.70	0.5385	0.4595
2.00	0.4882	0.3434	5.75	0.5551	0.5049
2.10	0.5815	0.3743	5.80	0.5298	0.5461
2.20	0.6363	0.4557	5.85	0.4819	0.5513
2.30	0.6266	0.5531	5.90	0.4486	0.5163
2.40	0.5550	0.6197	5.95	0.4566	0.4688
2.50	0.4574	0.6192	6.00	0.4995	0.4470
2.60	0.3890	0.5500	6.05	0.5424	0.4689
2.70	0.3925	0.4529	6.10	0.5495	0.5165

where I_f is the irradiance corresponding to an *unobstructed wavefront* at the observation point.

Fig. 12.6 shows the graph of $C(w)$ as a function of $S(w)$. This graph is known as the *Cornu spiral* in honour of Marie Alfred Cornu (1841–1902), who was the first person to use it in the calculation of Fresnel integrals. It should be noted that $C(w)$ and $S(w)$ are odd functions:

$$C(-w) = -C(w) \tag{12.38a}$$

$$S(-w) = -S(w) \tag{12.38b}$$

In practice, to use the Cornu spiral, one must begin by marking on this curve the positions of the boundaries w_1 e w_2 of the aperture. The result of each of the integrals in Eq. (12.31) is given by the length of the vector $\mathbf{V}_{12}$ that unites these two points, given by

$$\mathbf{V}_{12} = [C(w) + iS(w)]_{w_1}^{w_2} \tag{12.39}$$

Using this notation, one can rewrite Eq. (12.37) for the irradiance corresponding to a rectangular aperture in the form

$$I_P = \frac{I_f}{4} |\mathbf{V}_{12}(u)|^2 |\mathbf{V}_{12}(v)|^2 \tag{12.40}$$

If there is no obstruction of the wavefront along the direction of the *x-axis*, one has $u_1 = -\infty$ (corresponding to the point $(-0.5, -0.5)$) and $u_2 = \infty$ (corresponding to the point

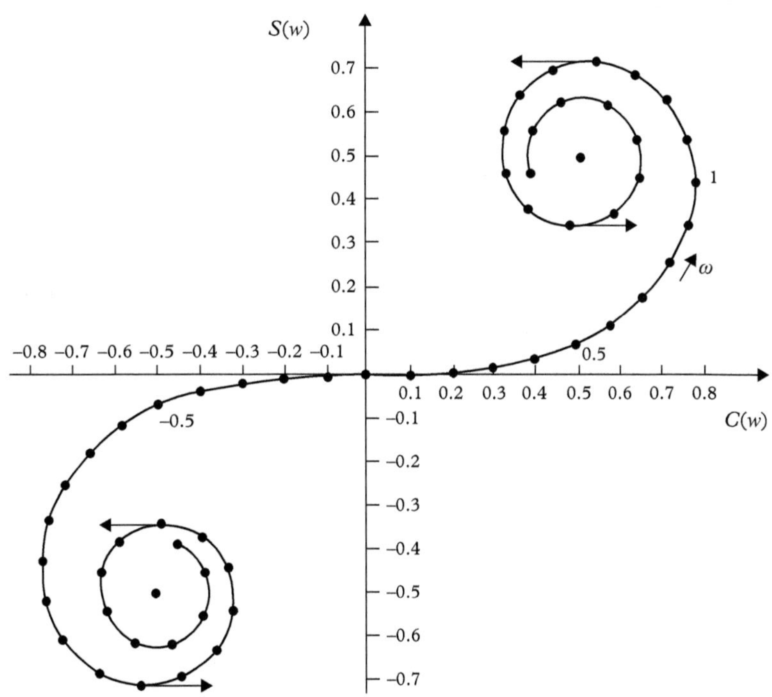

Figure 12.6 *Cornu spiral, obtained by arranging the values of $C(w)$ as a function of $S(w)$.*

w	C(w)	S(w)	w	C(w)	S(w)
2.80	0.4675	0.3915	6.15	0.5146	0.5496
2.90	0.5624	0.4101	6.20	0.4676	0.5398
3.00	0.6058	0.4963	6.25	0.4493	0.4954
3.10	0.5616	0.5818	6.30	0.4760	0.4555
3.20	0.4664	0.5933	6.35	0.5240	0.4560
3.30	0.4058	0.5192	6.40	0.5496	0.4965
3.40	0.4385	0.4296	6.45	0.5292	0.5398
3.50	0.5326	0.4152	6.50	0.4816	0.5454
3.60	0.5880	0.4923	6.55	0.4520	0.5078
3.70	0.5420	0.5750	6.60	0.4690	0.4631
3.80	0.4481	0.5656	6.65	0.5161	0.4549
3.90	0.4223	0.4752	6.70	0.5467	0.4915
4.00	0.4984	0.4204	6.75	0.5302	0.5362
4.10	0.5738	0.4758	6.80	0.4831	0.5436
4.20	0.5418	0.5633	6.85	0.4539	0.5060
4.30	0.4494	0.5540	6.90	0.4732	0.4624
4.40	0.4383	0.4622	6.95	0.5207	0.4591

The Fresnel integrals, given by Eqs. (12.23) and (12.33), have been calculated exhaustively, and their values are properly tabulated, as in Table 12.1. The use of these tables allows you to calculate easily the field given by Eq. (12.31), which may be presented in the form:

$$E_P = \frac{i}{2} E_f [C(u) + iS(u)]_{u_1}^{u_2} [C(v) + iS(v)]_{v_1}^{v_2} \tag{12.35}$$

In Eq. (12.35), E_f represents the field that would exist at the observation point if there were no obstacle and is given by

$$E_f = \frac{E_0}{d} e^{ikd} \tag{12.36}$$

The irradiance at the observation point is then given by

$$I_P = \frac{I_f}{4} \left\{ [C(u_2) - C(u_1)]^2 + [S(u_2) - S(u_1)]^2 \right\} \left\{ [C(v_2) - C(v_1)]^2 + [S(v_2) - S(v_1)]^2 \right\} \tag{12.37}$$

(0.5, 0.5)). The length of the vector connecting these two points is $\sqrt{2}$. Concerning the direction of the y-axis, an equal value is obtained, so the irradiance at the observation point, given by Eq. (12.40), is simply $I_P = I_f$, as expected.

12.4.1 Diffraction by a slit

If, instead of the rectangular aperture, there is a long slit parallel to the y-axis, it should be considered $y_1 \to -\infty$ and $y_2 \to \infty$, so $v_1 \to -\infty$ and $v_2 \to \infty$. Considering the Cornu spiral represented in Fig. 12.5, we have that the point corresponding to $v_1 \to -\infty$ has coordinates $(-0.5, -0.5)$, whereas the point corresponding to $v_2 \to \infty$ has coordinates (0.5, 0.5). The distance between these two points is $|\mathbf{V}_{12}(v)| = \sqrt{2}$, so the irradiance at the observation point in this case is given by Eq. (12.40) as

$$I_P = \frac{I_f}{2}|\mathbf{V}_{12}(u)|^2 = \frac{I_f}{2}\left\{[C(u_2) - C(u_1)]^2 + [S(u_2) - S(u_1)]^2\right\} \quad (12.41)$$

12.4.2 Diffraction by a semi-infinite opaque screen

The distribution of irradiance produced by a semi-infinite opaque screen can be described starting from the slit situation considered earlier. In fact, to have a screen in the form of an opaque semi-plane, it is enough to remove one of the semi-planes that define this slit. Assuming that one removes the semi-plane corresponding to the positive values of x, one has that $x_2 \to \infty$, so $u_2 \to \infty$. In this way, one has $C(u_2), S(u_2) \to \frac{1}{2}$, so the irradiance at a point beyond the semi-plane will be given by Eq. (12.41) as

$$I_P = \frac{I_f}{2}\left\{\left[\frac{1}{2} - C(u_1)\right]^2 + \left[\frac{1}{2} - S(u_1)\right]^2\right\} \quad (12.42)$$

If the observation point P_0 is at the level of the semi-plane edge ($P_0 \equiv P_b$ in Fig. 12.7(a)), one has $u_1 = 0$ and $C(0) = S(0) = 0$, so $I_P = 1/4$. As the point P_0 enters into the geometric shadow of the semi-plane ($P_0 \equiv P_s$), the (positive) value of u_1 increases and the length of the vector $\mathbf{V}_{12}(u)$ reduces, so the irradiance decreases monotonously. However, when the point P_0 goes out of the geometric shadow and moves away from the edge of the semi-plane ($P_0 \equiv P_l$), it turns out that the length of the vector $\mathbf{V}_{12}(u)$ begins to increase and subsequently acquires a damped oscillatory behaviour around $|\mathbf{V}_{12}(u)| = \sqrt{2}$, corresponding to the evolution of u_1 along the spiral centred at $(-0.5, -0.5)$. The variation of irradiance with u_1 is shown in Fig. 12.7(b).

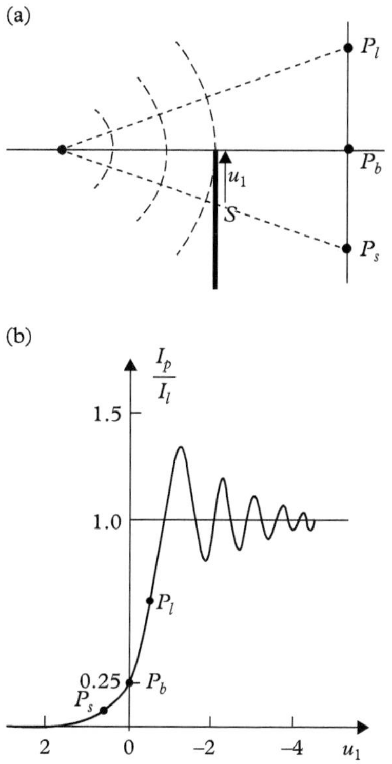

Figure 12.7 *(a) Schematic describing the Fresnel diffraction produced by an opaque semi-plane and (b) the variation of the irradiance.*

Example 12.4

Consider a semi-infinite opaque screen arranged vertically, such that its upper limit is horizontal. This screen is illuminated by a collimated, normally incident beam of light with a wavelength $\lambda = 500$ nm. Determine the irradiance at a point 4 m behind the screen and 2 mm below its edge level.

Solution

Since the beam is collimated, the wavefronts of the incident light are plane. In such a case, one has $Z_s = \infty$, so Eq. (12.30a) becomes:

$$u = (x - x_s)\sqrt{\frac{2}{\lambda Z}}$$

continued

Example 12.4 continued

Considering the problem data, one has

$$u_1 = (x_1 - x_s)\sqrt{\frac{2}{\lambda Z}} = 2 \times 10^{-3}\sqrt{\frac{2}{(500 \times 10^{-9})(4)}} = 2$$

Using Table 12.1, one has $C(2) = 0.4882$ and $S(2) = 0.3434$. Thus, based on Eq. (12.42), the irradiance at the point under consideration is given by

$$I_P = \frac{I_f}{2}\left\{[0.5 - 0.4882]^2 + [0.5 - 0.3434]^2\right\} = 0.012 I_f$$

12.5 Further Reading

Born, M., and E. Wolf, *Principles of Optics*, 6th Ed., Pergamon Press, 1980.
Cowley, J. M., *Diffraction Physics*, 3rd Ed., Elsevier, 1995.
Guenther, R., *Modern Optics*. John Wiley & Sons, 1990.
Heavens, O., and R. Ditchburn, *Insight into Optics*, John Wiley & Sons, 1991.
Hecht, E., *Optics*, 5th Ed., Pearson Education, 2016.
Klein, M., and T. Furtak, *Optics*, 2nd Ed., John Wiley & Sons, 1986.
Pedrotti, F., L. M. Pedrotti, and L. S. Pedrotti, *Introduction to Optics*, 3rd Ed., Cambridge University Press, 2018.
Ufimtsev, P., *Fundamentals of the Physical Theory of Diffraction*, 2nd Ed., John Wiley & Sons, 2014.

12.6 Problems

12.1 a) Distinguish between Fraunhofer and Fresnel diffraction regimes in terms of the experimental setup used.

b) In a diffraction experiment, an aperture with a diameter of 1 mm is illuminated normally by a plane wave of wavelength $\lambda = 500\,\text{nm}$. Determine whether Fraunhofer or Fresnel diffraction applies if the distance between the aperture and the screen is (i) 20 cm, (ii) 50 cm, and (iii) 6 m.

12.2 A plane wave ($\lambda = 600\,\text{nm}$) is normally incident on an opaque screen containing a circular aperture with a radius of 3 mm. Find the number of Fresnel zones contained in such an aperture when it is viewed from a position on the axis 50 cm away.

12.3 Let us consider an opaque screen containing a circular aperture 2 mm in diameter, which is normally illuminated by plane waves with wavelength $\lambda = 600\,\text{nm}$.

a) Indicate the farthest positions on the axis of the hole where it is possible to observe two maxima and two minima of irradiance.

b) Justify the existence of the two maxima in (a) and find the ratio between their irradiances.

12.4 A collimated beam of light with wavelength $\lambda = 500\,\text{nm}$ illuminates an opaque screen that has a circular aperture with an adjustable radius, r. One metre behind, and on the axis of the aperture, there is a detector. It turns out that the detected irradiance oscillates as the radius of the aperture increases from 0 to ∞.

a) Indicate the radius of the aperture for which the first maximum of irradiance is observed.

b) Obtain the value of such a radius corresponding to the first minimum.

c) Calculate the ratio of irradiances in case (a) and for $r = r_\infty$.

d) Suppose the screen is replaced by an opaque disc of radius equal to that found in (b). What is the irradiance detected in this case?

12.5 An opaque screen has an aperture with the shape indicated in Fig. 12.8, in which $\overline{BC} = 1\,\text{mm}$, $\overline{CD} = 1.41\,\text{mm}$, and $\overline{AD} = 3.14\,\text{mm}$. Considering the normal incidence on this aperture of plane waves of wavelength 500 nm, determine the amplitude of the light at a point on the axis at a distance of 2 m from the screen in terms of the amplitude that would exist there for an unobstructed wavefront.

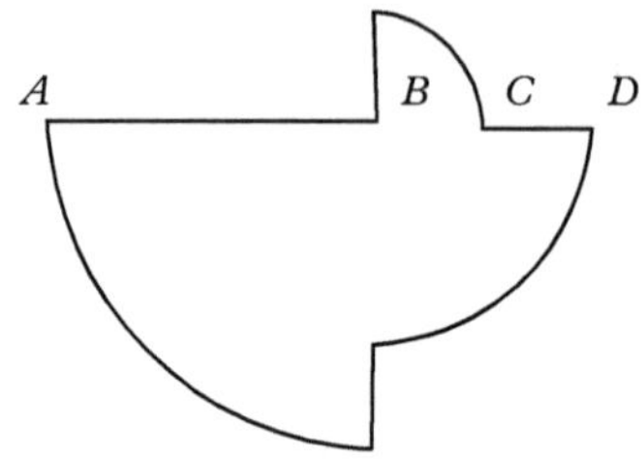

Figure 12.8 *An aperture in an opaque screen.*

12.6 a) Show that the area of a Fresnel zone is given by

$$A = \frac{R_s}{R_s + R_0}\pi R_0 \lambda$$

R_s being the radius of the primary spherical wavefront, and R_0 the distance from that wavefront to the point of observation.

12.7 Using the result of Problem 12.6, derive an expression for the number of Fresnel zones comprised in a circular aperture of radius (a), centred on the straight line passing through the point source of the primary wave and the observation point. Show that this expression can present itself in a form similar to that of the equation of a conventional lens:

$$\frac{1}{R_s} + \frac{1}{R_0} = \frac{1}{f}$$

and obtain an expression for parameter f.

12.8 Derive Eq. (12.23) from Eqs. (12.21) and (12.22). Describe the main properties of a Fresnel zone plate, comparing with a conventional converging thin lens.

12.9 A collimated beam of light with wavelength $\lambda = 550\,\text{nm}$ illuminates a zone plate of 3.0 cm in diameter. A bright spot is formed at a distance of 1.00 m from the zone plate. Determine how far the light is focused if it diverges from a point on the axis at a distance of 1.5 m in front of the zone plate.

12.10 Plane waves of wavelength $\lambda = 550$ illuminate normally a Fresnel zone plate, in which the first zone has a diameter of 0.4 mm. Find the first-order focal length of this zone plate and justify the existence of higher-order focal lengths.

12.11 Plane waves of wavelength 600 nm illuminate normally a Fresnel zone plate. Find (i) the first zone radius relative to a point 30 cm from the zone plate on the central axis and (ii) the two longest higher-order focal lengths for this zone plate.

12.12 A Fresnel zone plate consists of five zones. The first zone consists of a circular disc of radius r_1. The second is a concentric and transparent ring situated between r_1

and r_2, followed by an opaque ring between r_2 and r_3, another transparent ring between r_3 and r_4, and, finally, an opaque zone between r_4 and infinity. The values of these radii are in the ratio of the square roots of the respective indices. When the zone plate is illuminated by monochromatic plane waves of wavelength 500 nm, there is a more intense bright spot on the axis of the zone plate 1 metre behind it.

a) Find the value of the radius r_1.

b) Write the irradiance of this spot in terms of the irradiance of the incident wave.

c) Find the positions of the irradiance maxima along the axis.

12.13 Considering a semi-infinite opaque screen, determine the irradiance at the first maximum above the edge of the shadow.

12.14 A point source S, emitting light with wavelength $\lambda = 600\,\text{nm}$, is at a distance of 1.5 m from a flat, opaque screen containing a slit 0.4 mm wide. Using the Cornu spiral, calculate the irradiance at a point P located on the opposite side of the screen and 4.5 m from it. Suppose that the line $\overline{FP}$ is normal to the plane and passes through the centre of the slit.

12.15 An observation plane is situated at a distance $d = \overline{SP} = 6\,\text{m}$ from a point source S, which emits light with wavelength $\lambda = 500\,\text{nm}$. The point P results from the intersection between the observation plane and the normal that passes through the point source. Halfway between this source and the observation plane there is a semi-infinite opaque screen, whose upper boundary is rectilinear, horizontal, and tangent to the line $\overline{SP}$. Determine the irradiance (a) at point P, (b) 2 mm above point P, and (c) 2 mm below point P.

Holography

13

Holography is a technique invented by the British scientist Dennis Gabor in 1948, when he was seeking to improve the image quality provided by the electron microscope. The word 'holography' has its roots in the Greek term *holos*, which means totality. In fact, while in conventional photography only the irradiance of the light is recorded, in holography all the information carried by the wave coming from the object, namely its amplitude and phase, is recorded. This is achieved through the interference between the object wave and a reference wave. For this purpose, it is necessary to use coherent light so that there is a fixed phase relationship between these two waves. In order to reconstruct the three-dimensional image of the object, Gabor suggested that it would be enough to illuminate the hologram with a wave similar to the reference wave used during the recording process.

The idea of holography turned out not to be useful in the field of electron microscopy and remained largely ignored until the discovery of the laser in 1960. The interest of the scientific and artistic communities in this idea increased significantly with the development of the off-axis recording technique, proposed by Emmett Leith and Juris Upatnieks in 1962.

13.1 Photography and Holography

A conventional photograph records only the variation in irradiance corresponding to an object or scene. Since it is a two-dimensional record, it does not provide any information about the depth of the scene. When we look at a photograph, it sometimes seems that we have a perception of such depth. However, this usually happens only because the scene is already familiar to us.

In contrast to conventional photography, the hologram allows us to record the scene or the object with the sense of depth and parallax with which we can observe them in reality. This is possible because all the information of the wave coming from the scene, namely its amplitude and its phase, is recorded. When a hologram is illuminated under proper conditions, such a wave is completely reconstructed, and the observer can contemplate the original scene by looking through the hologram.

The characteristics of the image provided by a hologram are mainly due to the preservation of information related to the phase of the original wave. The recording of this information is important, since the phases of waves originating at different points of an object are also different. To record these phase differences, it is necessary to convert them into amplitude variations, which can be achieved through the interference of two waves: the wave coming from the object and another coherent wave, called the reference wave.

In conventional photography, there is a one-to-one relationship between object and image points, in the sense that all the light originating from a single point of the object is focused to a single point in the image. By contrast, in holography there is no one-to-one correspondence between object and wavefront points before reconstruction occurs. In other words, in a hologram one cannot observe any image of the object, since it records only a pattern of interference fringes resulting from the superposition of the object and reference waves. In fact, every object point illuminates the entire hologram, and each point of the hologram receives light from every point of the object.

Lasers, Optics, and Photonics. Mário F. S. Ferreira, Oxford University Press. © Mário F. S. Ferreira (2026).
DOI: 10.1093/oso/9780198952770.003.0013

13.2 The Holographic Process

Fig. 13.1 shows a typical set-up used for recording a hologram. The initial laser beam is divided into two mutually coherent beams. One of them goes directly to the registration plate and constitutes the so-called *reference wave*. The other beam illuminates the object whose image we wish to record, and the light it scatters constitutes the *object wave*. The object and reference waves interfere on the holographic plate, on which there is a light-sensitive medium that records the spatial distribution of irradiance of the interference pattern.

At a later time, the holographic plate can be illuminated by a replica of the reference wave, called the *reconstruction wave*. The interference pattern recorded on the plate diffracts this reconstruction wave, allowing the object wave to be reobtained.

The main aspects of the holographic process, namely those concerning the recording and reconstruction of the holographic image, will be analysed in more detail in the following sub-sections.

13.2.1 Holographic recording

Let us consider the holographic record of an object, made on a holographic plate that is located in the xy plane, where the object wave has a complex amplitude $E_O(x,y)$. If only this wave existed, the film would be darkened as a function of its irradiance, given by

$$I_O = |E_O|^2 = E_O E_O^* \tag{13.1}$$

In this case, the record would not contain any information concerning the phase of the object wave. However, in the presence of a reference wave with a complex amplitude $E_R(x,y)$, the resulting electric field at the point (x,y) of the holographic plate is given by

$$E(x,y) = E_O(x,y) + E_R(x,y) \tag{13.2}$$

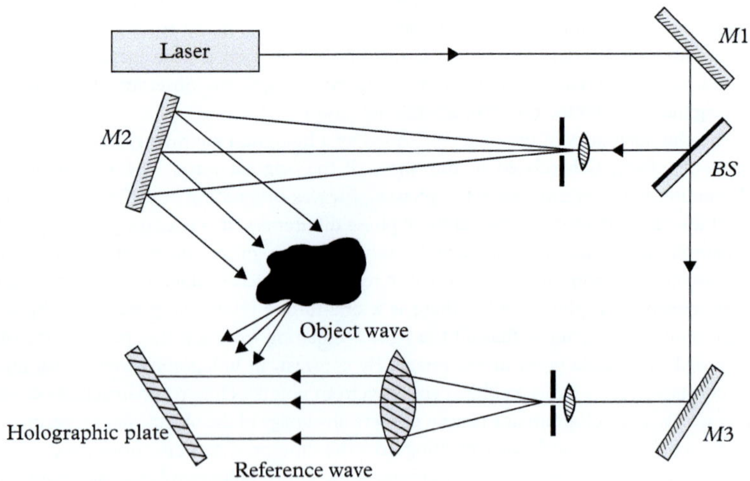

Figure 13.1 *Typical experimental set-up for recording a hologram. M1, M2, and M3: mirrors; BS: beam-splitter.*

Thus, the recorded irradiance is given by

$$I(x,y) = |E_O(x,y) + E_R(x,y)|^2 = E_R E_R^* + E_O E_O^* + E_R^* E_O + E_R E_O^* \qquad (13.3)$$

The pattern of irradiance given by Eq. (13.3) can be recorded using different methods. In the following, we consider the case of a recording medium provided by a photographic film, whose characteristics are given by the so-called Hurter–Driffield (H&D) curve, as shown in Fig. 13.2. This curve illustrates the dependence of emulsion optical density after recording, D, on the logarithm of exposure B (energy/area), given by

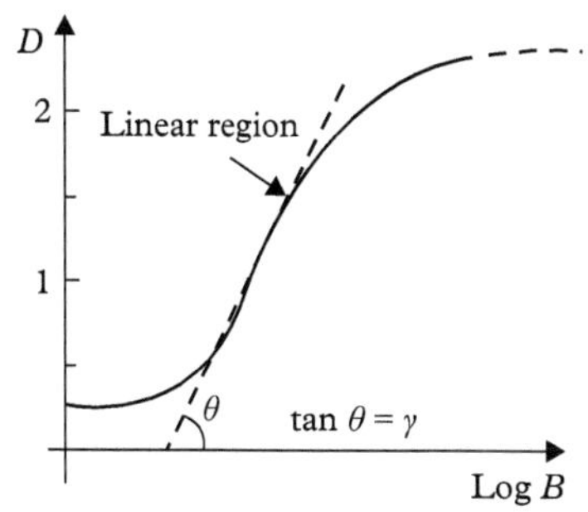

Figure 13.2 *Typical aspect of the H&D curve.*

$$B = It_B \qquad (13.4)$$

where I is the irradiance and t_B is the *exposure time*. The optical density is related to the amplitude transmission coefficient of the emulsion, τ, in the form:

$$D = \log \frac{1}{\tau^2} \qquad (13.5)$$

In the case of conventional photography, the developed film is made linear in terms of the recorded irradiance by adequately controlling the exposure time in order to operate in the linear region of the H&D curve. In this region, one can write the relation

$$D = \gamma_N \log B - D_0 = \gamma_N \log (It_B) - D_0 \qquad (13.6)$$

where γ_N is the so-called gamma of the film, given by the slope of the linear part of the H&D curve. The index N indicates that this record corresponds to the negative of the original signal. Illuminating this negative with a reconstruction wave of irradiance I_R, one has that the transmitted irradiance is given by

$$I_t = I_R \tau^2 = I_R 10^{-D} \qquad (13.7)$$

Substituting Eq. (13.6) into Eq. (13.7) allows one to obtain the following result:

$$I_t = C_N I_R I^{-\gamma_N} \qquad (13.8a)$$

where

$$C_N = 10^{D_0} t_B^{-\gamma_N} \qquad (13.8b)$$

Eq. (13.8) shows that the transmitted irradiance is not a linear reproduction of the original irradiance I. However, this condition can be achieved by realizing a positive version from the previous negative record, as is usually done in photography. Illuminating such a positive copy with a reconstruction wave $E_R(x,y)$, the transmitted irradiance is given by

$$I_{tP} = C_P I_R I^{\gamma_P \gamma_N} \qquad (13.9)$$

where γ_P is the gamma of the film used in the registration of the positive version. It follows from Eq. (13.9) that the relation between the transmitted and recorded irradiances is linear if the condition $\gamma_N \gamma_P = 1$ is satisfied.

In holography, the recording process must be such that the hologram provides a transmitted wave whose amplitude corresponds to a linear reproduction of the original

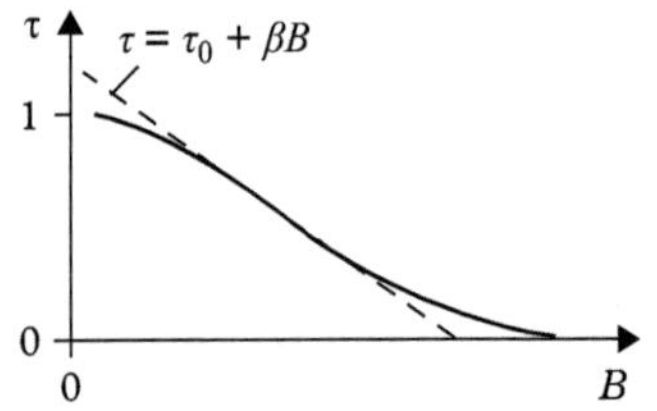

Figure 13.3 *Amplitude transmission coefficient, τ, of a holographic plate as a function of exposure, B.*

irradiance distribution resulting from the interference between the object and reference waves. From Eq. (13.9), it appears that in this case one should have

$$\gamma_N \gamma_P = 2 \tag{13.10}$$

The correct exposure for a linear holographic record corresponds to operating not in the linear region of the H&D curve, but rather in the linear region of the τ vs B curve, as depicted in Fig. 13.3. In this region, one has

$$\tau = \tau_0 + \beta B = \tau_0 + \beta I t_B \tag{13.11}$$

where τ_0 and β are constants.

Example 13.1

Consider the superposition of two plane light waves with the same amplitude and wavelength $\lambda = 632.8\,\text{nm}$. One of these waves (wave 1) propagates in the direction of the *z-axis*, whereas the other wave (wave 2) propagates in the *zx* plane, making an angle $\theta = 0.1$ rad with the *z-axis*. The interference pattern is recorded on a holographic plate located in the *xy* plane, which has a length $d = 10$ cm in the *x*-axis direction. Obtain the number of bright fringes recorded on the plate.

Solution

The fields of the two waves in the *xy* plane are given by

$$E_1 = E_0 \exp(i\mathbf{k}_1 \cdot \mathbf{r}) = E_0 \tag{13.12}$$

$$E_2 = E_0 \exp(i\mathbf{k}_2 \cdot \mathbf{r}) = E_0 \exp\left(i\frac{2\pi}{\lambda} x \,\text{sen}\,\theta\right) \tag{13.13}$$

The resulting field is obtained using the superposition principle:

$$E = E_1 + E_2 = E_0\left[1 + \exp\left(i\frac{2\pi}{\lambda} x \sin\theta\right)\right] \tag{13.14}$$

The irradiance distribution in the interference pattern is given by

$$I = EE^* = 2E_0^2\left[1 + \cos\left(\frac{2\pi}{\lambda} x \sin\theta\right)\right] \tag{13.15}$$

The spacing between the fringes corresponds to the spatial period of the irradiance distribution:

$$\Delta x = \frac{\lambda}{\sin\theta} \approx \frac{\lambda}{\theta} \tag{13.16}$$

Using the values given in this example, one has

$$\Delta x = \frac{632.8 \times 10^{-9}}{0.1} = 6.3\,\mu\text{m}$$

The number of fringes registered on the holographic plate is

$$N = \frac{l}{\Delta x} = 15873$$

13.2.2 Image reconstruction and localization

Let us consider that the recorded hologram is illuminated with a wave equal to the reference wave, now called the *reconstruction wave*. The field corresponding to the light transmitted by the hologram is given by

$$\begin{aligned} E_t(x,y) &= \tau(x,y)\,E_R \\ &= \tau_0 E_R + \beta t_B E_R\left(E_R E_R^* + E_O E_O^* + E_O E_R^* + E_R E_O^*\right) \end{aligned} \tag{13.17}$$

where Eqs. (13.11) and (13.3) were used.

The first three terms on the right-hand side of Eq. (13.17) correspond to an amplitude-modulated version of the reconstruction wave. Considering the hologram as a diffraction grating, these terms represent the undeflected light wave, i.e. the zero order of diffraction, whereas the fourth and fifth terms in Eq. (13.12) represent first-order diffracted waves. The fourth term corresponds to the object wave, amplitude-modulated by the factor $I_R = |E_R|^2 = E_R E_R^*$. This wave appears to come from the object. Hence, it diverges on emerging from the hologram as if coming from a *virtual image* located behind the hologram. The fifth term represents the object wave, modulated in amplitude and with *phase reversal*. Because of such phase reversal, originally diverging rays become converging and focus to form a *real image*, located in front of the hologram. Fig. 13.4 shows the location of both virtual and real images when the direction of propagation of the reference and reconstruction waves is perpendicular to the hologram.

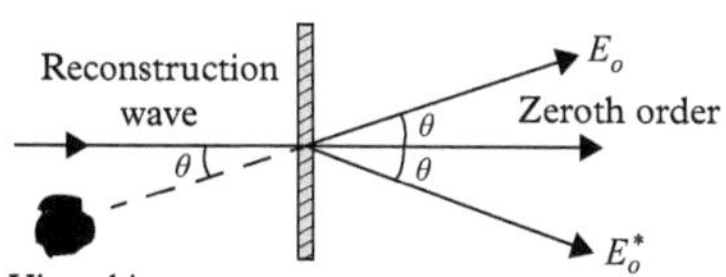

Figure 13.4 *Location of the images when both the reference and reconstruction waves are perpendicular to the hologram.*

Considering the characteristics of the conjugate image, it is generally of little interest. In fact, this image shows the object from the inside out, with the depth of the scene reversed along the axis of view. The points of the object closer to the observer are revealed to be farther apart in the conjugate image.

The fourth term on the right-hand side of Eq. (13.17) reproduces, at less than a constant factor, the object wave E_O. Looking through the hologram as if it were a window, one can see a three-dimensional image of the scene, with the objects situated exactly in their original positions. The parallax effects are easily observed, and one can see different aspects of the scene by moving slightly the head.

In order to obtain the object wave, E_O, under good conditions, one should try to avoid its spatial overlap with the other waves. In fact, the off-axis system represented in Fig. 13.1 can produce a hologram in which the two first-order diffracted waves are separated in direction from each other and from the zero-order wave.

Assuming that the object wave is a plane wave and that it is incident on the hologram at an angle θ to the normal, one can write:

$$E_O = \exp(ikx\sin\theta) \tag{13.18a}$$

$$E_O^* = \exp(-ikx\sin\theta) = \exp\left[ikx\sin(-\theta)\right] \tag{13.18b}$$

where a common factor $\exp(-i\omega t)$ has been omitted. The wave E_O^* leaves the hologram making an angle $-\theta$ with respect to the normal.

If the reference and reconstruction waves, E_R, are also plane waves incident at an angle γ, one has

$$E_R = \exp(ikx\sin\gamma) \tag{13.19a}$$

$$E_R^2 = \exp(ikx2\sin\gamma) \tag{13.19b}$$

Using Eqs. (13.18b) and (13.19b), the following result is obtained for $E_R^2 E_O^*$:

$$E_R^2 E_O^* = \exp[ikx(2\sin\gamma + \sin(-\theta))] = \exp(ikx\sin\phi) \tag{13.20}$$

where $\sin\phi = 2\sin\gamma + \sin(-\theta)$, ϕ is the angle that the wave corresponding to the conjugate image makes with the normal to the hologram. Increasing γ or θ, the conjugate image becomes more and more distorted until it disappears when $2\sin\gamma + \sin(-\theta) > 1$.

Example 13.2

When illuminated by a light wave, the holographic plate developed in Example 13.1 acts as a diffraction grating. Assuming that such a wave is similar to wave 1 of this example, show that the transmitted field can be considered as the sum of three waves. Comment on this result.

Solution

The amplitude transmission coefficient of the holographic plate is given by Eq. (13.11):

$$\tau = \tau_0 + \beta I t_B = \tau_0 + 2\beta t_B E_0^2\left[1 + \cos\left(\frac{2\pi}{\lambda}x\sin\theta\right)\right] \tag{13.21}$$

where τ_0, β, and t_B are constants, whereas the irradiance I has been substituted from Eq. (13.15). When illuminated by wave 1 of Example 13.1, the field transmitted by the holographic plate is given by

$$E_t = \tau E_1 = A\exp\left(\frac{2\pi}{\lambda}z\right) + B\exp\left[i\frac{2\pi}{\lambda}(x\sin\theta + z)\right] + B\exp\left[-i\frac{2\pi}{\lambda}(x\sin\theta - z)\right] \tag{13.22}$$

where

$$A = \tau_0 E_0 + 2\beta t_B E_0^3 \tag{13.23}$$

$$B = \beta t_B E_0^3 \tag{13.24}$$

The first term in Eq. (13.22) represents the undeflected wave transmitted through the holographic plate (zero-order diffraction), whereas the second and third terms represent the first-order diffracted waves, making angles θ and $-\theta$ with the z-axis, respectively. In particular, the third term represents the reconstructed wave 2 of Example 13.1.

13.3 Phase Holograms

In the most general case, the transmission coefficient of the holographic plate is a complex function and can be written in the form:

$$\tilde{\tau}(x,y) = \tau(x,y)\exp[i\varphi(x,y)] \tag{13.25}$$

Two cases can be considered in the recording of a hologram: the amplitude hologram ($\varphi = const.$) and the phase hologram ($\tau = const.$). In both cases, all the information contained in the object wave is recorded on the holographic plate.

The transmission coefficient phase of the holographic plate varies with exposure B, as shown in Fig. 13.5. In the linear zone of this curve, one can write:

$$\varphi(I) = \varphi_0 + \beta' I t_B \tag{13.26}$$

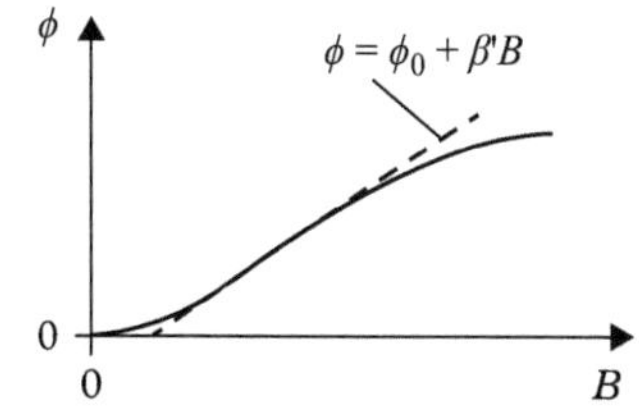

Figure 13.5 *Phase of the transmission coefficient of a holographic plate as a function of exposure B.*

In fact, the transmittance of the phase hologram varies in a non-linear fashion with the recorded irradiance I. However, if $\varphi(I) << \pi/2$, such transmittance can be approximated by a linear function of I:

$$\tilde{t} = \exp[i\varphi(I)] \approx 1 + i\varphi(I) = (1 + i\varphi_0) + i\beta' I t_B \tag{13.27}$$

When the phase hologram is illuminated by a reconstruction wave E_R, the transmitted field becomes:

$$\begin{aligned} E_t(x,y) &\approx (1 + i\varphi(I(x,y)))\,E_R \\ &= (1 + i\varphi_0)\,E_R + i\beta' t_B E_R\left(E_R E_R^* + E_O E_O^* + E_O E_R^* + E_R E_O^*\right) \end{aligned} \tag{13.28}$$

As in the case of the amplitude hologram, one can consider five terms in the result provided by Eq. (13.28). However, as a consequence of the additional approximation $\exp(i\varphi) \approx 1 + i\varphi$, it turns out that the non-linear dependence of the transmitted field on the recorded irradiance I is more significant in phase holograms than in amplitude holograms.

13.4 Holographic Configurations

An essential condition for the realization of a hologram is the existence of a sufficiently coherent light source. In fact, the object and reference waves must be mutually coherent so that their overlap results in a stable interference pattern. In this sense, the geometry of the set-up must be such that the paths of the two waves are similar. If the difference between these paths is greater than the coherence length of the light, the desired interference pattern on the photographic film will not form. The light used must also possess sufficient spatial coherence so that the beam is coherent across that portion of the wavefront needed to encompass the scene.

13.4.1 In-line holography

The simplest method of performing a hologram is to illuminate a holographic plate with a light beam expanded through the object, which should be sufficiently transparent for this purpose (Fig. 13.6). This method was originally used by Gabor and is undemanding with regard to the coherence of the light used. The object wave is formed by the light scattered by the object's particles, whereas the reference wave is made up of the light that passes between these particles without being affected by them.

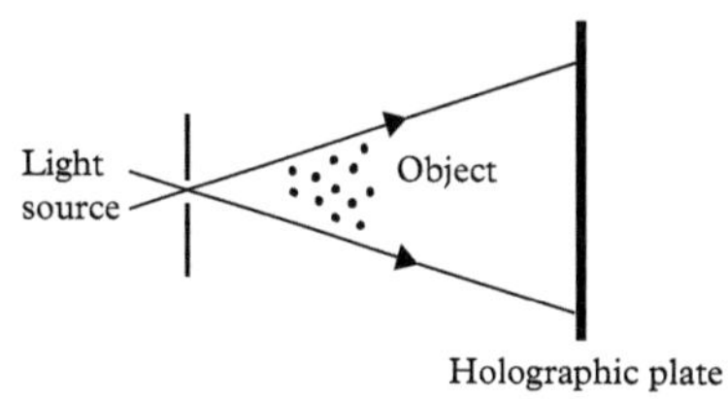

Figure 13.6 *Recording of an in-line hologram.*

In the case of a particle with very small dimensions, the scattered wave is practically spherical, and the interference pattern on the holographic plate consists of a series of concentric circular fringes. Fig. 13.7 illustrates the recording of such an interference pattern. The developed plate is called a *Gabor zone plate* and presents circular transmitting zones, whose transmittance is a sinusoidally varying function of the radius r. Fig. 17.7(b) illustrates such a zone plate. Each point of an extended object can be considered to produce a

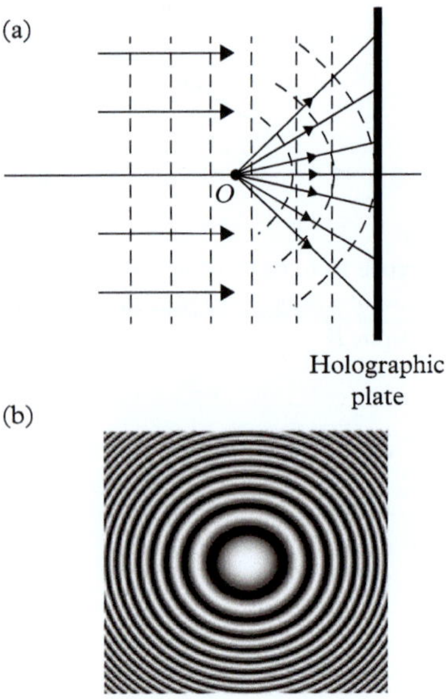

Figure 13.7 *(a) Recording of a Gabor zone plate and (b) appearance of such a plate.*

Gabor zone plate, and the resultant hologram corresponds to the overlapping of all these zone plates.

When illuminated with the same reference wave, the Gabor zone plate diffracts the light along directions that diverge from a point that is the *virtual image* of the original object point O. However, light is also diffracted in the direction of a second point on the exit side of the hologram, a point that is symmetrically placed with respect to the plate. This point corresponds to a *real image* of the original object point O that can be focused onto a screen.

13.4.2 Off-axis holography

The in-line configuration used by Gabor has the major drawback of the conjugate (real) image forming in the same direction as the virtual image. The method used to solve this problem is to make the object and reference waves propagate in different directions. This is the so-called off-axis technique, and it seems to be a trivial modification of the in-line configuration; however, it was about 10 years after Gabor's pioneering work before the new technique was suggested by Leith and Upatnieks.

Fig. 13.1 shows a typical assembly for performing off-axis holography. The laser light is divided into two beams, which are then expanded using microscopic lenses. The light scattered by the object interferes with the reference wave, and the resulting interference pattern is recorded on the holographic plate. Due to the limited coherence length of the source, it should be ensured that the optical paths of the object and reference waves are similar, thereby optimizing the contrast of the interference pattern on the holographic plate. When the holographic reconstruction is carried out using the same set-up, the object wave must be obstructed, or the beam-splitter must simply be removed.

The interference pattern must remain stationary during the exposure time in order for its recording to proceed satisfactorily. The quality of the hologram is greatly affected if the interference pattern moves even a small fraction of the spatial period of the fringes. It is therefore important that the positions of all the components of the set-up—mirrors, beam-splitters, the object itself, etc.—remain fixed during the exposure of the holographic plate. These requirements can be met easily if the holographic set-up is mounted on a sufficiently massive optical table supported by a set of pneumatic cylinders. The high mass of the table provides mechanical inertia, while the pneumatic system effectively insulates the table from vibrations that may affect the laboratory.

13.5 Classification of Holograms

Holograms can be classified according to the angle made between the directions of propagation of the object and reference waves.

13.5.1 Thin holograms

In most cases, photographic emulsions have a thickness between 5 μm and 16 μm. In this case, when the angle made between the object and reference waves is small ($\phi \leq 10°$), the spacing between the interference fringes is similar to the thickness of the emulsion. In this case, the recording medium can be considered two-dimensional, and the hologram thus obtained is called a *thin hologram*. When the angles of incidence of the object and reference waves, measured relative to the normal to the holographic surface, are symmetrical, the fringes recorded are perpendicular to that surface.

13.5.2 Thick holograms

When the angle between the object and reference waves is moderately large, falling within the range of 10° and 120°, the spacing between the fringes is less than the thickness of the recording medium, and the resulting hologram is called a *thick hologram*.

Typically, the spatial frequency of the interference pattern to be recorded is several hundred, and even on the order of a thousand lines per millimetre. Such fine structures cannot be recorded on common photographic films, which have a recording capacity limited to about 100 lines/mm. Special photographic emulsions such as the Eastman Kodak–type 649F and Agfa-Gevaert–type 8E70 are required for this purpose.

Example 13.3

A diffraction grating is to be produced using the holographic technique with light emitted by a helium–neon laser ($\lambda = 632.8\,\text{nm}$). Derive an expression for the spatial frequency as a function of the angles of incidence of the two interfering waves and calculate the angle between such waves in order that the grating has 500 lines/mm. What is the maximum possible spatial frequency of the grating?

Solution

A diffraction grating can be produced by recording the interference pattern of two plane waves on a holographic plate. The fields of the two plane waves incident on the plate can be represented by

$$E_1 = A_1 e^{i\phi_1}, \quad E_2 = A_2 e^{i\phi_2} \tag{13.29}$$

Suppose that the plate is in the xy plane and that the waves propagate in the yz plane; their phases can be written in the form

$$\phi_1 = k \sin\theta_1 y, \quad \phi_2 = k \sin\theta_2 y \tag{13.30}$$

where θ_1 and θ_2 are the angles that the wave vectors make with the plate normal (coincident with the z-axis).

The irradiance distribution on the holographic plate is given, up to a constant factor, by

$$I = A_1^2 + A_2^2 + 2\,|A_1|\,|A_2| \cos(\phi_1 - \phi_2) \tag{13.31}$$

There will be a bright fringe when $\cos(\phi_1 - \phi_2) = 1$, that is

$$\phi_1 - \phi_2 = \frac{2\pi}{\lambda} y [\sin\theta_1 - \sin\theta_2] = p2\pi \tag{13.32}$$

where p *is* an integer. From Eq. (13.32), one can obtain the position of the bright fringe of order p:

$$y_p = \frac{p\lambda}{\sin\theta_1 - \sin\theta_2} \tag{13.33}$$

The spacing between consecutive fringes is given by

$$\Delta y = y_{p+1} - y_p = \frac{\lambda}{\sin\theta_1 - \sin\theta_2} \tag{13.34}$$

If the normal to the plate bisects the angle (φ) formed between the two waves, one has $\theta_1 = -\theta_2 = \varphi/2$, so the spatial frequency $f = 1/\Delta y$ can be written as

$$f = \frac{2\sin(\varphi/2)}{\lambda} \tag{13.35}$$

continued

Example 13.3 continued

Considering a frequency of 500 lines/mm, one has

$$\varphi = 2\sin^{-1}\left(\frac{\lambda}{2}f\right) = 2\sin^{-1}\left(\frac{632.8\times10^{-9}}{2}\times5\times10^{5}\right) = 18.20° \tag{13.36}$$

So the two waves should make an angle of 9.1° relative to the normal.

The maximum frequency is obtained when $\sin(\varphi/2) = 1$ and is given by

$$f = \frac{2}{\lambda} = 3160,6/\text{mm} \tag{13.37}$$

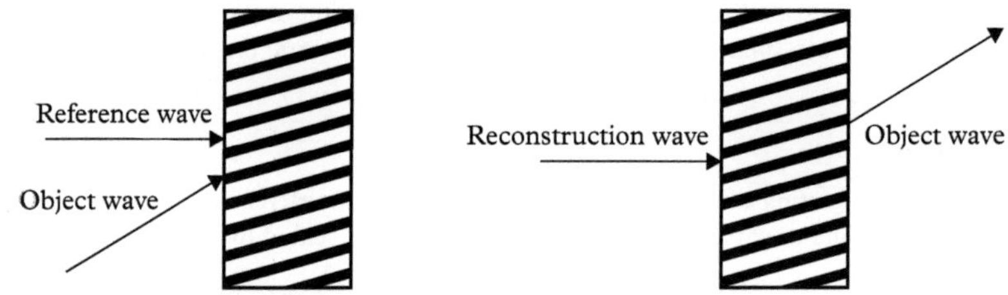

Figure 13.8 *Representation of the stages of (a) recording and (b) reconstruction of the object wave from a thick hologram.*

Fig. 13.8 illustrates (a) the recording and (b) the reconstruction of the object wave from a thick hologram. The theory used to describe light diffraction is based on Bragg's law. Assuming that the fringe spacing is d and that the angle between the plane of these fringes and the reconstruction wave is α (Fig. 13.9), the Bragg condition is given by

$$2d\sin\alpha = n\lambda, \quad n = 1, 2, \ldots \tag{13.38}$$

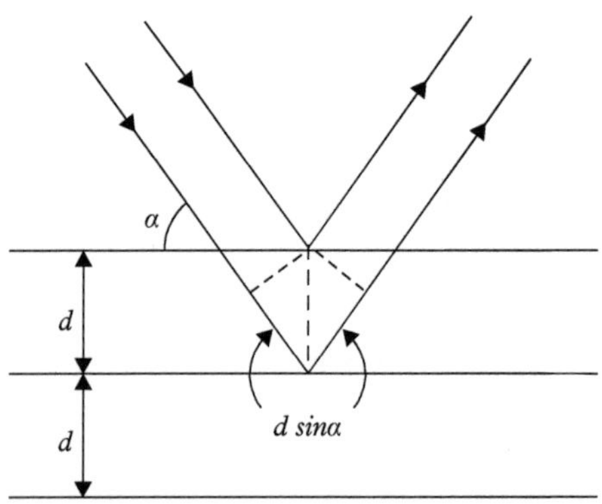

Figure 13.9 *Constructive interference of reflected waves as governed by the Bragg condition.*

The need to satisfy the Bragg condition makes the reconstruction of the holographic image from a thick hologram very sensitive to the angle that the reconstruction wave makes with the hologram. On the other hand, this angular dependence makes it possible, by using reference waves with different angles, to record different holograms in the same emulsion. From each hologram, it is possible to reconstruct the respective object wave, without this implying the reconstruction of the other object waves also recorded on the same plate. Eq. (13.38) shows that the wave diffracted by a thick hologram is also sensitive to the wavelength used for the reconstruction.

In both thick and thin holograms considered earlier, the observer looks through the hologram to see the holographic image. For this reason, they are called *transmission holograms*. This kind of hologram is obtained whenever the object and reference waves fall on the same side of the emulsion, as illustrated in Fig. 13.8(a).

13.5.3 Reflection holograms

One configuration that provides the maximum angle between the object and reference waves corresponds to the situation in which these waves illuminate opposite sides of the emulsion, as shown in Fig. 13.10(a). In this case, a *reflection hologram* is obtained. In the process of reconstruction, these holograms reflect light in the direction of the observer (Fig. 13.10(b)), who sees a virtual image behind the hologram, as if it were in a mirror.

The theory used to describe light diffracted by a reflection hologram is based on the Bragg condition given by Eq. (13.38). Assuming that one has $n = 1$ in this equation and considering a given angle of observation α, constructive interference between the reflected waves occurs only for a certain wavelength λ_α, without any disturbance of the waves with other wavelengths. That is, when illuminated by white light, this type of hologram works as a filter of wavelengths, allowing reconstruction of the image only for the wavelength λ_α. For this reason, this kind of hologram is also known as a white-light hologram.

Fig. 13.11 shows a set-up that allows the obtaining of white-light holograms. For recording, it is necessary to use a coherent light source (laser). The laser beam is expanded, collimated, and directed through the holographic plate to the object. The reference wave corresponds to the light that comes directly from the laser, whereas the object wave corresponds to the light that is scattered by the object.

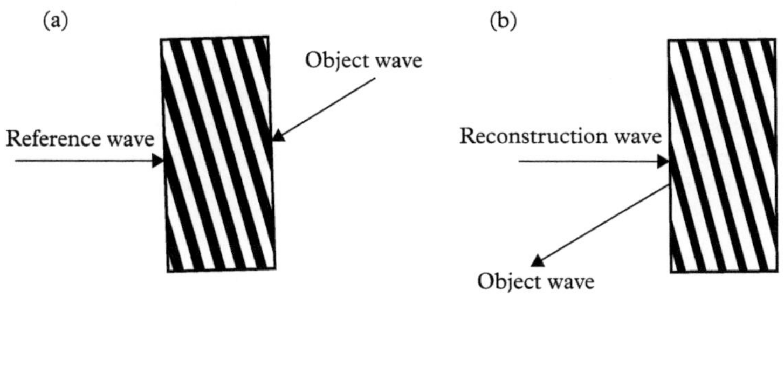

Figure 13.10 *Representation of the stages of (a) recording and (b) reconstruction of the object wave from a reflection hologram.*

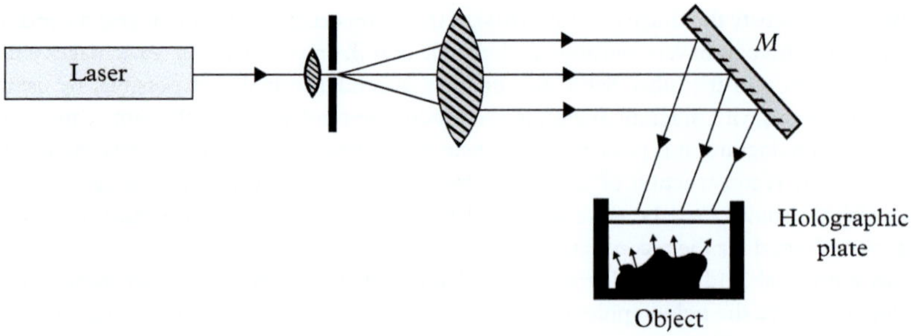

Figure 13.11 *Set-up for recording a white-light hologram.*

13.5.4 Rainbow holograms

If the holograms of Fig. 13.8 or 13.10 are viewed with a reference beam of a colour different from that used in their construction, it can be shown that the reconstructed images will appear at a different angle. If the reference/reconstruction beam used in the holograms of Fig. 13.8 or 13.10 is white light, the continuously displaced images due to different spectral regions of the light overlap and produce a coloured blur.

Rainbow holograms were invented in 1968 by Stephen A. Benton and are designed to be viewed under white-light illumination without the aforementioned coloured blur. In practice, the rainbow holography recording process uses a horizontal slit to eliminate vertical parallax in the output image, thus greatly reducing the confusion of images. In reconstruction, the hologram creates a clearer image when illuminated by white light.

The virtual image now appears coloured. The particular colour seen depends on the direction along which the hologram is viewed as the head is moved along a vertical line. The colour of the image sweeps through the colours of the rainbow, from red to blue, thereby giving rise to the name *rainbow hologram*. The holograms found on credit cards are examples of rainbow holograms.

13.6 Holographic Interferometry

One of the most important applications of holography lies in the area of interferometry. In fact, holography allows one to save the amplitude and phase of the object wave that can be used later for interferometric purposes. This technique also makes it possible to compare an object with itself at a later time.

Given its high sensitivity, holographic interferometry can be used to obtain important information regarding the structural characteristics of a given object by observing the movement of its surface when it is subjected to stress. Thus, this technique makes it possible to carry out a wide variety of non-destructive inspections whenever the parameter of interest is manifested through discontinuities in the displacement of the surface of that object. These discontinuities appear as an anomaly in a pattern of interference fringes that would, in principle, be regular, thus allowing one to identify the defective region.

13.6.1 Real-time technique

In this method, one begins by obtaining and processing the hologram of the object. Then, the hologram is illuminated with the reference wave, and the reconstructed image is made to interfere with the actual wave coming directly from the object. The interference fringes indicate the current deformation of the object relative to the situation in which its original image was recorded in the hologram. This deformation can be tracked in real time through this fringe system. In this method, the object generally cannot be removed from its position, as it becomes practically impossible to put it back in exactly the same situation with respect to the hologram. This technique is often applied to determine the maximum stress points on the subject as pressure is applied.

Let us consider a point P on the surface of the object, which undergoes a displacement d, of only a few wavelengths, to the point P' (see Fig. 13.12). In this case, the amplitudes of the waves E and E', coming from the hologram (reconstructed wave) and the object, respectively, can be considered equal in modulus, with only a phase difference $\Delta\psi = \psi_1 - \psi_2$ among them. One then has

$$E' = E\exp(i\Delta\psi) \tag{13.39}$$

The irradiance corresponding to the interference of the two waves is given by

$$I = (E+E')(E+E')^* = 4I_0\cos^2\frac{\Delta\Psi}{2} \tag{13.40}$$

where $I_0 = EE^*$.

Eq. (13.40) is similar to that obtained in Chapter 8 for the case of Young's experiment. The phase difference $\Delta\psi$ is determined by the difference in optical paths between the interfering waves. Considering the geometry of Fig. 13.11, this difference in optical paths is $s = u + v$, where

$$u = d\cos\theta, \quad v = d\cos\gamma$$

Thus, the phase difference is given by

$$\Delta\psi = k(u+v) = kd(\cos\theta + \cos\gamma) \tag{13.41}$$

Substituting Eq. (13.41) into Eq. (13.40), one obtains

$$I = 4I_0\cos^2\left(\frac{kd}{2}(\cos\gamma + \cos\theta)\right) \tag{13.42}$$

The angles γ and θ, as well as the wavenumber k, are known from the experimental configuration.

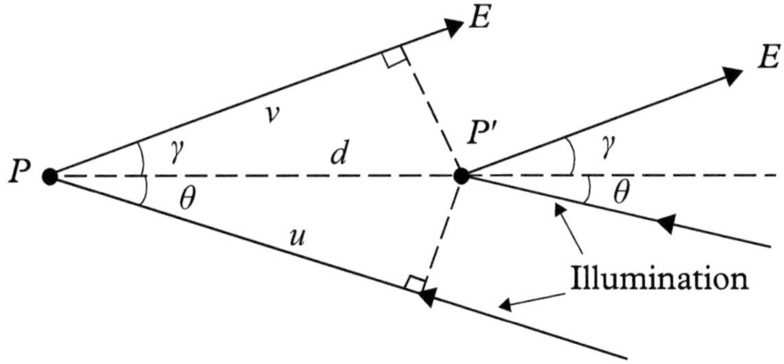

Figure 13.12 *Geometry for calculating the phase difference between fields E and E′, when the object point is shifted from P to P′.*

Example 13.4

In a holographic interferometry experiment, the surface to be analysed is illuminated perpendicularly by light with wavelength $\lambda = 500\,\text{nm}$. If the displacement of the surface is 4000 nm, find how many fringes will be counted when

a) the displacement is perpendicular to the surface;
b) the displacement makes an angle of 60° with the normal to the surface.

Solution

a) As seen in Chapter 9, when one of the mirrors of the Michelson interferometer undergoes a translation of $\lambda_0/2$, each fringe in the interference pattern shifts so that it takes the place of the adjacent fringe. In this way, a displacement Δd of the surface in the direction of the normal corresponds to a number, N, of fringes passing through a given reference position, given by

$$N = \frac{2\Delta d}{\lambda_0} \tag{13.43}$$

Using the values $\lambda_0 = 500\,\text{nm}$ and $\Delta d = 4000\,\text{nm}$, one obtains

$$N = \frac{2 \times 4 \times 10^{-6}}{5 \times 10^{-7}} = 16\,\text{fringes}$$

b) If the displacement makes an angle of 60° with the normal, one has

$$N = \frac{2\Delta d}{\lambda_0} \cos\theta = 16 \times \cos 60° = 8\,\text{fringes} \tag{13.44}$$

13.6.2 Double-exposure technique

In this method, the holographic plate is exposed twice: first with the undisturbed object and then again with the object under the effect of the disturbance. In this way, both wavefronts are recorded on the same holographic plate. When the hologram is illuminated with the reference wave, both waves are reconstructed and interfere, giving rise to a fringe pattern that reflects the displacement of the various points on the object. The theoretical description of this method is similar to that presented in the previous section for the real-time holography method.

13.6.3 Average-time technique

The method of average-time holography is more suitable in the case of objects subject to periodic vibratory motion. The exposure occurs over a relatively long time compared to the period of vibration. The final image can be considered as the result of the superposition of a large number of images, forming a pattern of standing waves. This pattern provides information on the relative vibrational amplitude as a function of position on the surface. The vibrational amplitudes of diffusely reflecting surfaces can be measured with high precision, and such measurements can be very useful for determining the modes of vibration of complex structures.

Let us consider the case of an object oscillating in a sinusoidal manner, with an amplitude d and an angular frequency ω. Based on the geometry of Fig. 13.10 for the case

$\gamma = \theta = 0$, one can write the object wave in the form:

$$E_O(t) = A\exp(ik2d\sin\omega t) \tag{13.45}$$

The exposure energy can be obtained in the form:

$$B = \int_0^{t_B} I(t)dt \tag{13.46}$$

where the resulting irradiance is given by

$$I(t) = (E_O(t) + E_R)(E_O(t) + E_R)^* \tag{13.47}$$

On the other hand, the amplitude transmission coefficient of the holographic emulsion in the linear zone of Fig. 13.3 can be written as follows:

$$\tau = \tau_0 + \beta B \tag{13.48}$$

Using the reference wave as a reconstruction wave E_R, we obtain the four terms referred to in Section 13.2, one of which corresponds to the direct image of the object, given by

$$\begin{aligned} E_d &= \beta|E_R|^2 \int_0^{t_B} E_O(t)\,dt \\ &= \beta|E_R|^2 \int_0^{t_B} A\exp(ik2d\sin\omega t)dt \end{aligned} \tag{13.49}$$

The integral in the right-hand member of Eq. (13.49) can be expressed in terms of the Bessel function of order zero, J_0, which is defined in the form:

$$J_0(\alpha) = \frac{1}{2\pi}\int_0^{2\pi} \exp(i\alpha\sin t)dt \tag{13.50}$$

Thus, the irradiance of the reconstructed object wave is given by

$$I_d = E_d E_d^* \propto J_0^2(2kd) \tag{13.51}$$

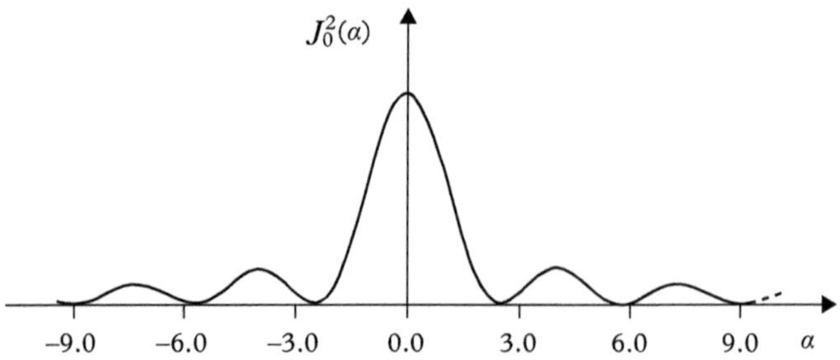

Figure 13.13 *Graphical representation of function $J_0^2(\alpha)$.*

Fig. 13.13 shows the graph of the function $J_0^2(\alpha)$. The function has zeros for $\alpha \approx 2.405, 5.52, \ldots$, which are not equidistant. On the other hand, the height of the higher-order maxima decreases with an increase in the phase difference, which corresponds to a decrease in the signal-to-noise ratio.

13.7 Other Applications of Holography

In addition to the holographic interferometry described in Section 13.6, holography provides a wide variety of other applications. Among them, we can mention the *recognition of patterns*. In this context, one may, for example, wish to examine a text in search of a given word. The light from this text passes through a hologram of the word sought in a suitable optical system. The presence of the word in question is indicated, at the place where it appears, by the appearance of a bright spot. The hologram thus functions as a filter that recognizes and transmits only the pattern of spatial frequencies similar to that recorded in it. This technique may eventually be used by robots, leading to the identification of certain objects and the consequent direction of their movement.

The recording, developing, and reconstruction processes occur sequentially in static holography, leading to the production of a permanent hologram. However, there exist some holographic materials that do not need the developing process and can record a hologram in a very short time. These real-time holograms can have some interesting applications, namely to perform phase-conjugate mirrors, image processing, and optical computing. In particular, *optical phase conjugation* allows the removal of the wavefront distortions that a light beam receives when passing through an aberrating medium by sending it back through the same aberrating medium with a conjugated phase.

Another very important application of holography lies in the *storage of information*. Given that the holographic technique allows one to reduce this information to dimensions of the order of the wavelength, it is estimated that a single hologram can store a huge amount of information. This amount can be increased by rotating the hologram and successively performing new exposures. In fact, instead of a conventional photographic plate, one can use for this purpose certain photosensitive crystals such as a lithium niobate crystal. Theoretically, it is assumed that all the information in a large library can be stored in one of these crystals with the dimensions of a sugar cube!

Another fascinating aspect of holographic information storage relates to its security. This characteristic has to do with the fact that any data are recorded in the entire volume of the hologram. Thus, damage to a part of the hologram does not prevent the further reconstruction of that data from the undamaged part. This damage will have an impact only at the level of the signal-to-noise ratio in the reconstructed image.

The waves used in the making of a hologram do not necessarily have to be electromagnetic in nature, but ultrasonic waves can also be used for this purpose. Given the ability of these waves to penetrate objects that are opaque to visible light, it is easily recognized that holograms made with these waves—the so-called *ultrasonic holograms*—could be quite useful. Ultrasonic holography could, for example, provide three-dimensional images of structures or cavities within a wide variety of opaque bodies, namely the human body.

Holography can also be used to realize different optical elements capable of performing as lenses, prisms, mirrors, gratings, etc. For example, a positive lens can be easily realized by recording the interference pattern produced by two spherical waves. Such holographic optical elements are generally lighter than the conventional optical components they replace, which becomes particularly important in some applications.

13.8 Further Reading

Blanche, P., *Field Guide to Holography*, SPIE Optical Engineering Press, 2014.

Colier, R., C. Burckhardt, and L. Lin, *Optical Holography*, Academic Press, 1971.

Guenther, R., *Modern Optics*, John Wiley & Sons, 1990.

Hariharan, P., *Basics of Holography*, Cambridge University Press, 2002.

Hecht, E., *Optics*, 5th Ed., Pearson Education, 2016.

Jones, R., C. Wykes, *Holographic and Speckle Interferometry*, Cambridge University Press, 1989.

Pedrotti, F., L. M. Pedrotti, and L. S. Pedrotti, *Introduction to Optics*, 3rd Ed., Cambridge University Press, 2018.

Toal, V., *Introduction to Holography*, CRC Press/Taylor & Francis, 2012.

Vest, C., *Holographic Interferometry*, John Wiley & Sons, 1979.

13.9 Problems

13.1 Describe the principle of holography by comparing it to photography. Comment on the importance of the coherence of the light used in holographic recording.

13.2 Starting from Eqs. (13.5) and (13.6), obtain Eqs. (13.8a) and (13.8b) for the transmitted irradiance when the holographic plate is illuminated by a reconstruction wave with irradiance I_R.

13.3 Show that, in order for the hologram to provide a transmitted wave whose amplitude is a linear reproduction of the recorded irradiance, the gamma parameters corresponding to the negative and positive copies must satisfy the condition $\gamma_N \gamma_P = 2$.

13.4 Consider a hologram located in the $z = 0$ plane, which records the interference pattern resulting from a reference plane wave, with irradiance I_R, propagating along the z-axis, and a spherical object wave originating at the point $(0, 0, -d)$. If a replica of the reference wave is used in the reconstruction stage, explain the significance of each of the components of the transmitted wave.

13.5 Assume that the positions of the object and the source of the reference beam are both located 100 cm from the recording medium and 10 cm apart. Determine the minimum resolution that the recording medium should have if the wavelength of the light is $\lambda = 500\,\text{nm}$. What kind of hologram will be recorded in this case?

13.6 Kodak Tri-X photographic film has a maximum resolution of about 50 lines/mm. What is the maximum allowed angle between the object and reference waves if this film is used for holographic registration, considering that the light has a wavelength $\lambda = 632.8\,\text{nm}$? Sketch a diagram of the set-up to be used for this purpose.

13.7 Show that the separation d of fringes in the formation of a holographic grating, as in Fig. 13.8, is given by $\lambda/2\sin\theta$, where λ is the wavelength of the beams in the film, and 2θ is the angle between the coherent beams in the film. Assume that the beams are incident symmetrically on the film's surface.

13.8 Considering the same case as Problem 13.7 and an angle of 120° between the beams, determine the number of fringes that are formed in a square holographic film of size 10 cm oriented perpendicularly to the fringes. Let the light wavelength be $\lambda = 500\,\text{nm}$ and the refractive index n of the emulsion be equal to 1.2.

13.9 A hologram is recorded in such a way that the object and reference waves strike the film at angles of 25° and 45°, respectively. Using the same reference wave during reconstruction, at what angle will the conjugate image form? For what value of the angle of incidence of the reconstruction wave does the conjugate image disappear?

13.10 A white-light hologram is recorded using light with wavelength $\lambda = 660\,\text{nm}$, such that the object and reference waves are both perpendicular to the holographic plate. During reconstruction, for what value of the angle of incidence does the image with a red colour appear ($\lambda = 500\,\text{nm}$)?

Anisotropy

14

Throughout the previous chapters, it was assumed that the light propagation medium had isotropic properties; that is, the refractive index did not depend on the direction of propagation in the medium. However, it is common to find in nature media with non-uniform physical properties. When a physical property of a material varies with the direction of propagation, the material is said to be *anisotropic*.

Anisotropic optical materials, also called *birefringent materials*, have a refractive index that varies with the direction of propagation in the material. When an object is observed through a birefringent material, two images of that object are seen, which move relative to each other when the material is rotated. The first recorded observation of this phenomenon dates back to 1669 and occurred when Erasmus Bartholinus (1625–1692) was studying the properties of calcite.

14.1 Index Ellipsoid

In Chapter 6, it was assumed that the medium through which electromagnetic waves propagated was characterized by a scalar quantity: the dielectric constant of the medium ε. On the other hand, the magnetic induction field **B** was related to the magnetic field **H** through another scalar quantity: the magnetic permeability of the medium μ. In this chapter, we will assume that the medium is also isotropic in terms of the magnetic field but not with regard to its electrical properties. In this case, the electric field vectors **E** and the electric displacement **D** are not, in general, parallel, and the components of these two vectors can be related in the form:

$$D_x = \epsilon_{xx} E_x + \epsilon_{xy} E_y + \varepsilon_{xz} E_z \tag{14.1a}$$

$$D_y = \varepsilon_{yx} E_x + \varepsilon_{yy} E_y + \varepsilon_{yz} E_z \tag{14.1b}$$

$$D_z = \varepsilon_{zx} E_x + \varepsilon_{zy} E_y + \varepsilon_{zz} E_z \tag{14.1c}$$

These relations can be written in a more compact form:

$$D_i = \sum_{j=1}^{3} \varepsilon_{ij} E_j \tag{14.2}$$

where, $i, j = 1$ for x, 2 for y e 3 for z.

The relations given in Eq. (14.1) may be presented in the following matrix form:

$$\begin{bmatrix} D_x \\ D_y \\ D_z \end{bmatrix} = \begin{bmatrix} \varepsilon_{xx} & \varepsilon_{xy} & \varepsilon_{xz} \\ \varepsilon_{yx} & \varepsilon_{yy} & \varepsilon_{yz} \\ \varepsilon_{zx} & \varepsilon_{zy} & \varepsilon_{zz} \end{bmatrix} \begin{bmatrix} E_x \\ E_y \\ E_z \end{bmatrix} \tag{14.3}$$

Lasers, Optics, and Photonics. Mário F. S. Ferreira, Oxford University Press. © Mário F. S. Ferreira (2026).
DOI: 10.1093/oso/9780198952770.003.0014

The matrix 3×3 in Eq. (14.3) corresponds to the so-called *dielectric tensor*. In general, the components ε_{ij} of this tensor are numbered nine. However, it can be shown that in the case of a non-absorbing medium, the dielectric tensor is real and symmetrical, so its independent components are reduced to six. Moreover, these components are reduced to only three by an appropriate choice of coordinate system. Written in this system of coordinates, whose axes are parallel to the so-called *principal axes*, the 3×3 matrix in Eq. (14.3) becomes diagonal, and its non-null elements are called the *principal values* of ε. In this case, one has the following relations:

$$D_x = \varepsilon_x E_x,\ D_y = \varepsilon_y E_y,\ D_z = \varepsilon_z E_z \tag{14.4}$$

The electric energy density is given by

$$U_E = \frac{1}{2}\mathbf{D} \cdot \mathbf{E} \tag{14.5}$$

Considering the coordinate system defined by the principal axes, this equation can be written in the form:

$$\frac{D_x^2}{2U_E\varepsilon_x} + \frac{D_y^2}{2U_E\varepsilon_y} + \frac{D_z^2}{2U_E\varepsilon_z} = 1 \tag{14.6}$$

Eq. (14.6) represents an ellipsoid, called the *index ellipsoid*, also known as the *optical indicatrix*. The principal semiaxes of this ellipsoid are equal to the square roots of the principal dielectric constants or, equivalently, to the principal *refractive indices*, n_j, defined in the form:

$$n_j \equiv \sqrt{\frac{\varepsilon_j}{\varepsilon_0}},\ j = x, y, z \tag{14.7}$$

Once the wave vector **k** is known, one can determine the vectors **D** and **E**. For this purpose, we must consider a plane perpendicular to **k** that goes through the origin. In the more general case, the curve obtained from the intersection of this plane and the ellipsoid is an ellipse whose semi-principal axes coincide with the two solutions for **D**, $\mathbf{D}_1$, and $\mathbf{D}_2$. However, for certain directions of **k**, this ellipse degenerates into a circumference. The directions of **k** for which this occurs are called *optical axes*. A crystal that has only one optical axis is called *uniaxial*, while one that has two optical axes is called *biaxial*.

14.2 Electromagnetic Waves in Anisotropic Media

In the absence of currents and free charges, Maxwell's four equations are given by

$$\nabla \cdot \mathbf{D} = 0 \tag{14.8}$$

$$\nabla \cdot \mathbf{B} = 0 \tag{14.9}$$

$$\nabla \times \mathbf{E} = -\frac{\partial \mathbf{B}}{\partial t} \tag{14.10}$$

$$\nabla \times \mathbf{H} = \frac{\partial \mathbf{D}}{\partial t} \tag{14.11}$$

Consider solutions of Eqs. (14.8)–(14.11) corresponding to plane waves, with a phase factor given by $\exp[i(\mathbf{k} \cdot \mathbf{r} - \omega t)]$. The wave vector $\mathbf{k}$ can be written in the form:

$$\mathbf{k} = \frac{\omega n}{c}\mathbf{s} \tag{14.12}$$

where $\mathbf{s}$ is a unit vector normal to the wavefront, and n is the refractive index. The following transformations can be used:

$$\frac{\partial}{\partial t} \rightarrow -i\omega \tag{14.13}$$

$$\nabla \rightarrow i\mathbf{k} \tag{14.14}$$

Using Eq. (14.13) and assuming that the medium is non-magnetic, so that $\mathbf{B} = \mu_0 \mathbf{H}$, the following relations are derived from Eqs. (14.10) and (14.11):

$$\nabla \times \mathbf{E} = i\omega\mu_0 \mathbf{H} \tag{14.15}$$

$$\nabla \times \mathbf{H} = -i\omega \mathbf{D} \tag{14.16}$$

Calculating the rotational of both members of Eq. (14.15) and using Eq. (14.16), one has

$$\nabla \times (\nabla \times \mathbf{E}) = \mu_0 \omega^2 \mathbf{D} \tag{14.17}$$

On the other hand, using Eqs. (14.12) and (14.14), one can write the left member of Eq. (14.17) in the form:

$$\begin{aligned} \nabla \times (\nabla \times \mathbf{E}) &= -\frac{\omega^2 n^2}{c^2}\mathbf{s} \times (\mathbf{s} \times \mathbf{E}) \\ &= -\frac{\omega^2 n^2}{c^2}[\mathbf{s}(\mathbf{s} \cdot \mathbf{E}) - \mathbf{E}] \end{aligned} \tag{14.18}$$

In obtaining the last result in Eq. (14.18), the following identity was used:

$$\mathbf{V}_1 \times (\mathbf{V}_2 \times \mathbf{V}_3) = \mathbf{V}_2(\mathbf{V}_1 \cdot \mathbf{V}_3) - (\mathbf{V}_1 \cdot \mathbf{V}_2)\mathbf{V}_3 \tag{14.19}$$

as well as the result $\mathbf{s} \cdot \mathbf{s} = 1$.

In contrast to what happens in an isotropic medium, the electric field $\mathbf{E}$ in an anisotropic medium is not perpendicular to the wave vector $\mathbf{k}$, for which one has $\mathbf{s} \cdot \mathbf{E} \neq 0$.

Example 14.1

Show that the electric displacement **D** is perpendicular to the wave vector **k**.

Solution

From Eqs. (14.17) and (14.18), the following relation is obtained between **D** and **E**:

$$\mathbf{D} = \varepsilon_0 n^2 [\mathbf{E} - \mathbf{s}(\mathbf{s} \cdot \mathbf{E})] \tag{14.20}$$

The components of the electric field parallel and perpendicular to **s** are given by

$$\mathbf{E}_{/\!/} = (\mathbf{E} \cdot \mathbf{s})\mathbf{s}, \;\; \mathbf{E}_{\perp} = \mathbf{E} - (\mathbf{E} \cdot \mathbf{s})\mathbf{s} \tag{14.21}$$

Thus, Eq. (14.20) can be written in the following form:

$$\mathbf{D} = \varepsilon_0 n^2 \mathbf{E}_{\perp} \tag{14.22}$$

It is thus verified that the electric displacement **D** is perpendicular to the wave vector **k**.

14.3 Fresnel Equation

Combining Eqs. (14.2) and (14.20) and considering the coordinate system defined by the principal axes, one obtains:

$$\varepsilon_j E_j = \varepsilon_0 n^2 \left[E_j - s_j (\mathbf{s} \cdot \mathbf{E})\right], \; j = x, y, z \tag{14.23}$$

From Eq. (14.23), three equations can be written in the form:

$$s_j E_j = \frac{n^2 s_j^2}{n^2 - n_j^2} (\mathbf{s} \cdot \mathbf{E}) \tag{14.24}$$

Adding the three equations given by (14.24) for $j = x, y, z$, one obtains:

$$(\mathbf{s} \cdot \mathbf{E}) = (\mathbf{s} \cdot \mathbf{E}) \sum_{j=x,y,z} \frac{n^2 s_j^2}{n^2 - n_j^2} \tag{14.25}$$

If $\mathbf{s} \cdot \mathbf{E} \neq 0$, this factor can be cancelled from both members of Eq. (14.25), and the following equation is obtained:

$$\sum_{j=x,y,z} \frac{s_j^2}{n^2 - n_j^2} = \frac{1}{n^2} \tag{14.26}$$

Eq. (14.26) is known as the *Fresnel equation* and makes it possible to determine the refractive index for a given direction of propagation.

Actually, Eq. (14.26) is a quadratic equation in n^2, presenting two positive roots: n_1^2 and n_2^2. Substituting these refractive index values into Eq. (14.23), two solutions are obtained for the electric field ($\mathbf{E}_1$ and $\mathbf{E}_2$) corresponding to the direction of propagation considered.

These solutions for the electric field can, in turn, be used in Eq. (14.20) to obtain two solutions for the electric displacement vector: $\mathbf{D}_1$ and $\mathbf{D}_2$.

Example 14.2

Show that the two electric displacement vectors, $\mathbf{D}_1$ and $\mathbf{D}_2$, corresponding to the two roots of Eq. (14.26), are perpendicular to each other.

Solution

Taking into account Eq. (14.2) and the fact that the dielectric tensor is symmetrical, one has

$$\begin{aligned}(\mathbf{E}_2 \cdot \mathbf{D}_1) &= \sum_i E_{2i} D_{1i} = \sum_{ij} E_{2i}\varepsilon_{ij}E_{1j} = \sum_{ij} E_{1j}\varepsilon_{ji}E_{2i} = \sum_j E_{1j}D_{2j} \\ &= (\mathbf{E}_1 \cdot \mathbf{D}_2)\end{aligned} \tag{14.27}$$

Since the electric field can be written as the sum of the parallel and perpendicular components to $\mathbf{s}$, we have from Eq. (14.27) that

$$((\mathbf{E}_{2/\!/} + \mathbf{E}_{2\perp}) \cdot \mathbf{D}_1) = ((\mathbf{E}_{1/\!/} + \mathbf{E}_{1\perp}) \cdot \mathbf{D}_2) \tag{14.28}$$

From Eq. (14.21), one has $(\mathbf{E}_{i/\!/} \cdot \mathbf{E}_{j\perp}) = 0$, whatever i and j may be. Combining this fact with Eq. (14.22), one concludes from Eq. (14.28) that

$$(n_1^2 - n_2^2)(\mathbf{E}_{2\perp} \cdot \mathbf{E}_{1\perp}) = 0 \tag{14.29}$$

If $n_1^2 \neq n_2^2$, Eq. (14.29) shows that

$$(\mathbf{E}_{1\perp} \cdot \mathbf{E}_{2\perp}) = 0 \tag{14.30}$$

From Eqs. (14.22) and (14.30) we conclude that the vectors $\mathbf{D}_1$ and $\mathbf{D}_2$ are perpendicular to each other. In the case $n_1^2 = n_2^2$, the directions of $\mathbf{D}_1$ and $\mathbf{D}_2$ are arbitrary but perpendicular to $\mathbf{s}$.

14.4 The Poynting Vector in Anisotropic Media

The Poynting vector, $\mathbf{S}$, indicates the direction and rate of energy flow and is given by

$$\mathbf{S} = \mathbf{E} \times \mathbf{H} \tag{14.31}$$

So the vector $\mathbf{S}$ is perpendicular to the magnetic field $\mathbf{H}$ and the electric field $\mathbf{E}$. From Eqs. (14.14) and (14.15), the following relation can be obtained between these two fields:

$$\mathbf{H} = \frac{1}{\omega\mu_0}(\mathbf{k} \times \mathbf{E}) = \frac{n}{c\mu_0}(\mathbf{s} \times \mathbf{E}) \tag{14.32}$$

Since the field $\mathbf{H}$ is normal to the plane containing the vectors $\mathbf{E}$, $\mathbf{D}$, and $\mathbf{s}$, the vector $\mathbf{S}$ must lie in this same plane. However, the vectors $\mathbf{s}$ and $\mathbf{S}$ are not generally parallel, making the same angle ϕ with each other as that between $\mathbf{E}$ and $\mathbf{D}$, as represented in Fig. 14.1.

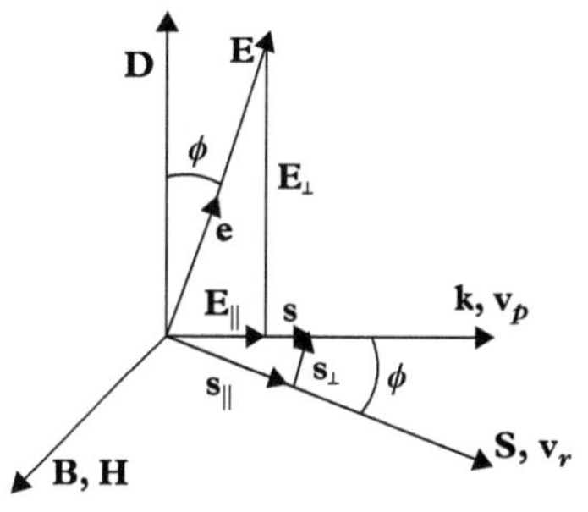

Figure 14.1 *Relationship between the various vectors referred to in the description of light propagation in an anisotropic medium.*

Using Eqs. (14.31) and (14.32) and the identity given by Eq. (14.19), the following result can be obtained for the Poynting vector:

$$\mathbf{S} = \frac{n}{c\mu_0}(\mathbf{E} \times (\mathbf{s} \times \mathbf{E})) = \frac{n}{c\mu_0}E^2 [\mathbf{s} - \mathbf{e}(\mathbf{s} \cdot \mathbf{e})] = \frac{n}{c\mu_0}E^2 \mathbf{s}_{\|} \tag{14.33}$$

where $\mathbf{e}$ is a unit vector in the direction of the electric field $\mathbf{E}$ and $\mathbf{s}_{\|}$ is the component of $\mathbf{s}$ parallel to $\mathbf{S}$. That is, in the case of an anisotropic medium, the direction of the Poynting vector, which defines the direction of the optical ray, does not generally coincide with the direction of the normal to the wavefront.

The so-called *optical ray velocity*, v_r, is a vector parallel to the Poynting vector and is given by

$$\mathbf{v}_r = \frac{\mathbf{S}}{U} \tag{14.34}$$

where U is the energy density of the electromagnetic field. On the other hand, the *phase velocity*, v_p, is the speed of propagation of surfaces with constant phase, $\mathbf{k} \cdot \mathbf{r} - \omega t =$ constant, and is given by

$$\mathbf{v}_p = \frac{\omega}{k}\mathbf{s} = \frac{c}{n}\mathbf{s} \tag{14.35}$$

The phase velocity is related to the optical ray velocity in the form:

$$v_p = v_r \frac{\mathbf{S} \cdot \mathbf{s}}{|\mathbf{S}|} = v_r \cos\phi \tag{14.36}$$

14.5 Index Surfaces

The *index surfaces* or *normal surfaces* are very useful geometric constructions that help in understanding the propagation of waves in an anisotropic medium. A given normal surface can be expressed as a function $n(\mathbf{s})$, representing in fact two surfaces for each vector $\mathbf{s}$, that correspond to the two values of the refractive index, n_1 and n_2. These values are obtained using Eq. (14.26). Starting from a given origin O, distances proportional to n_1 and n_2 are considered for each direction s. The set of all the points thus obtained constitutes the two index surfaces (see Fig. 14.2).

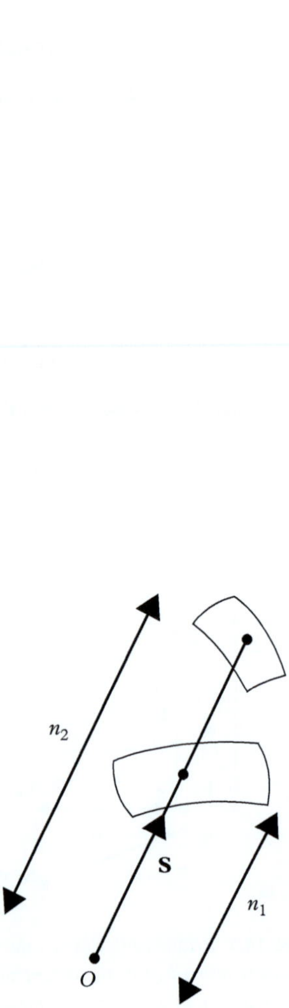

Figure 14.2 *Construction of index surfaces.*

In order to determine the shape of the index surfaces, consider the Fresnel equation given by Eq. (14.26) in the form:

$$s_x^2 n_x^2 (n^2 - n_y^2)(n^2 - n_z^2) + s_y^2 n_y^2 (n^2 - n_x^2)(n^2 - n_z^2) + s_z^2 n_z^2 (n^2 - n_x^2)(n^2 - n_y^2) = 0 \tag{14.37}$$

14.5.1 Uniaxial Crystals

In a uniaxial crystal, the dielectric tensor, when expressed in diagonal form, has two of its non-zero elements equal. If the axes are chosen conveniently, one has

$$n_x = n_y = n_o \tag{14.38}$$

$$n_z = n_e \tag{14.39}$$

where n_o and n_e are the ordinary and extraordinary refractive indices, respectively. In this case, the optical properties are invariant under rotation of the crystal around the *z-axis*, which is the optical axis, and the dielectric tensor takes the form:

$$\varepsilon = \varepsilon_0 \begin{bmatrix} n_o^2 & 0 & 0 \\ 0 & n_o^2 & 0 \\ 0 & 0 & n_e^2 \end{bmatrix} \tag{14.40}$$

Let us consider, without loss of generality, that the wave vector lies in the *xz* plane in a direction that makes an angle θ with the *z-axis*, such that

$$\mathbf{s} = (\sin\theta, 0, \cos\theta) \tag{14.41}$$

Substituting Eqs. (14.38), (14.39), and (14.41) into Eq. (14.37), one obtains

$$(n^2 - n_o^2)\left[n_o^2(n^2 - n_e^2)\sin^2\theta + n_e^2(n^2 - n_o^2)\cos^2\theta\right] = 0 \tag{14.42}$$

One of the solutions of this equation is given by

$$n^2 = n_o^2 \tag{14.43}$$

and corresponds to the ordinary wave $\mathbf{E}_o$. This wave is given by Eq. (14.23):

$$E_{ox} = E_{oz} = 0, \quad E_{oy} \neq 0 \tag{14.44}$$

In this case, $n(\mathbf{s})$ is a spherical surface of radius n_o, and the field $\mathbf{E}_o$ is perpendicular to both the wave vector $\mathbf{k}$ and the optical axis. On the other hand, the electric displacement vector $\mathbf{D}_o$ is collinear with $\mathbf{E}_o$.

The other solution of Eq. (14.42) is given by

$$\frac{1}{n^2} = \frac{\sin^2\theta}{n_e^2} + \frac{\cos^2\theta}{n_o^2} \tag{14.45}$$

and corresponds to the extraordinary wave $\mathbf{E}_e$, given by

$$E_{ey} = 0, \quad E_{ex} \neq 0, \quad E_{ez} \neq 0 \tag{14.46}$$

Eq. (14.45) represents the index surface for the extraordinary wave, which is an ellipsoid of revolution tangent to the spherical surface defined by Eq. (14.43). The points of contact between the two surfaces are located on the optical axis, for $\theta = 0, \pi$ rad, when $n = n_o$. For a direction of propagation in the *xy* plane ($\theta = \pi/2$ rad), the refractive index is $n = n_e$. Fig. 14.3 illustrates the index surfaces for the cases (a) of a negative uniaxial crystal ($n_e < n_o$) and (b) of a positive uniaxial crystal ($n_e > n_o$).

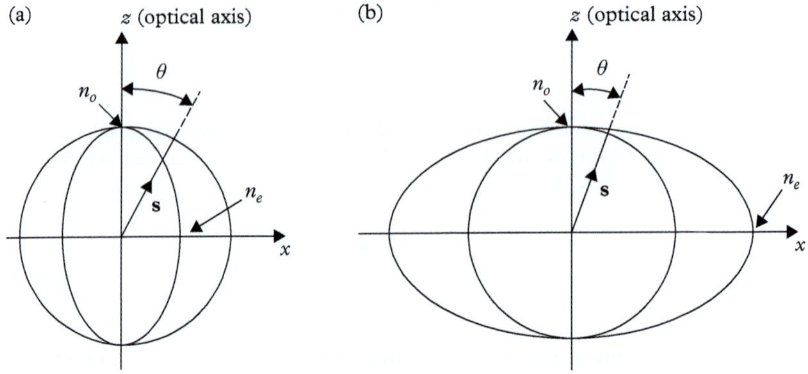

Figure 14.3 *Schematic of the index surfaces for cases (a) of a negative uniaxial crystal ($n_e < n_o$) and (b) a positive uniaxial crystal ($n_e > n_o$).*

14.5.2 Biaxial Crystals

In a biaxial crystal, the three main values of the dielectric tensor are distinct. In order to examine this case, the propagation vector must be considered to be situated in a plane perpendicular to one of the main axes of that tensor. For example, if that plane is the yz plane, we have $s_x = 0$ in Eq. (14.37). One of the solutions of this equation is

$$n^2 = n_x^2 \tag{14.47}$$

and corresponds to an electric field whose components can be obtained from Eq. (14.23):

$$E_y = E_z = 0, \quad E_x \neq 0 \tag{14.48}$$

The index surface in this case reduces to a circle in the yz plane. The electric displacement vector is collinear with the electric field vector and is given by

$$\mathbf{D} = \varepsilon_0 n_x^2 \mathbf{E} \tag{14.49}$$

It is also found that the velocity of the wave and the speed of the ray have the same direction, given by the vector **s**.

The other solution of Eq. (14.37) is given by

$$\frac{1}{n^2} = \frac{s_y^2}{n_z^2} + \frac{s_z^2}{n_y^2} \tag{14.50}$$

and corresponds to an electric field whose components are

$$E_x = 0, \quad E_y \neq 0, \quad E_z \neq 0 \tag{14.51}$$

In this case, when **s** has the direction of the *y-axis*, one has $n = n_z$, whereas, for **s** parallel to the z-axis, one has $n = n_y$. The index surface reduces to an ellipse in the yz plane, with

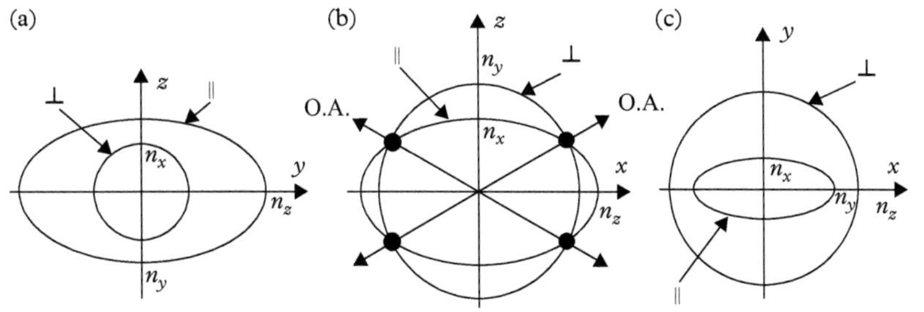

Figure 14.4 *Sections of the index surfaces for the case of a biaxial crystal, corresponding to the planes yz (a), xz (b), and xy (c). ⊥—polarization perpendicular to the plane of the figure; ||—polarization parallel to the plane of the figure. O.A.: optical axis.*

a semi-major axis n_z and a semi-minor axis n_y, which contains the circle corresponding to Eq. (14.47) if $n_x < n_y < n_z$.

When the vector **s** is perpendicular to the y and x axes, the results are similar to the previous ones. Fig. 14.4 shows schematically the sections of the index surfaces corresponding to the three planes of the coordinates. In the case of **s** being perpendicular to the y-axis, one of the index surfaces intersects the xz plane in a circle of radius n_y, whereas the other surface intersects the same plane in an ellipse given by:

$$\frac{1}{n^2} = \frac{s_z^2}{n_x^2} + \frac{s_x^2}{n_z^2} \tag{14.52}$$

Since it was assumed that $n_x < n_y < n_z$, this ellipse intersects the circumference of radius n_y at four points, as shown in Fig. 14.4(b). These points define the optical axes for the index surfaces, which are two in number in the case of a biaxial crystal.

14.6 Double Refraction

Consider a plane wave propagating in an isotropic medium and striking a plane boundary of separation from an anisotropic medium at an angle θ_i with the normal to that boundary. The incident and transmitted waves can be written in the form:

$$\mathbf{E}_i = \mathbf{e}_i E_i \exp\left[i\left(\frac{n\omega}{c}\mathbf{s}\cdot\mathbf{r} - \omega t\right)\right] \tag{14.53a}$$

$$\mathbf{E}_t = \mathbf{e}_t E_t \exp\left[i\left(\frac{n'\omega}{c}\mathbf{s}'\cdot\mathbf{r} - \omega t\right)\right] \tag{14.53b}$$

As a boundary condition, the phases of the two previous waves must be equal at the interface:

$$\frac{n\omega}{c}\mathbf{s}\cdot\mathbf{r} - \omega t = \frac{n'\omega}{c}\mathbf{s}'\cdot\mathbf{r} - \omega t \tag{14.54}$$

Since the frequency does not change on transmission through the interface, one has from Eq. (14.54) that

$$\mathbf{r}\cdot(n'\mathbf{s}' - n\mathbf{s}) = 0 \tag{14.55}$$

It can be assumed that the interface lies in the plane $z = 0$, where the vector **r** is also found. In this case, it follows from Eq. (14.55) that the vector $(n'\mathbf{s}' - n\mathbf{s})$ must be normal to the interface.

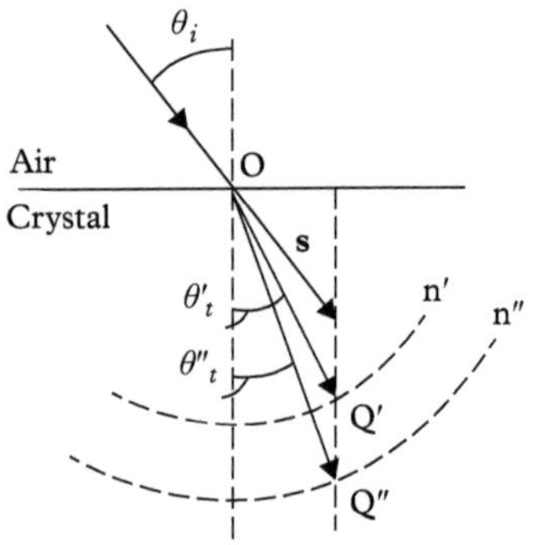

Figure 14.5 *Refraction at the boundary of an anisotropic medium.*

The index surfaces can be used to determine the vector **s**′ (Fig. 14.5). Starting from a given point O on the interface, one constructs the two index surfaces, as well as the vector **s**. The only vectors that satisfy Eq. (14.55) are the vectors of length OP′ and OP″, the ends of which lie on the two index surfaces, corresponding to n' and n''. The incident wave undergoes double refraction, giving rise to two transmitted waves in the directions OP′ and OP″.

Taking into account the two transmitted waves, one can write Eq. (14.55) in the form:

$$\mathbf{r}\cdot n\mathbf{s} = \mathbf{r}\cdot n'\mathbf{s}' = \mathbf{r}\cdot n''\mathbf{s}'' \tag{14.56}$$

That is, considering that the interface coincides with the plane $z = 0$, one has

$$n(xs_x + ys_y) = n'(xs'_x + ys'_y) = n''(xs''_x + ys''_y) \tag{14.57}$$

This equation must be satisfied for all values of x and y, so one has

$$\frac{s'_x}{s_x} = \frac{s'_y}{s_y} = \frac{s''_x}{s_x} = \frac{s''_y}{s_y} \tag{14.58}$$

From these relations, one can conclude that the propagation vectors of the two transmitted waves must be situated in the plane of incidence and that the angles θ_t' and θ_t'' in Fig 14.5 are given by

$$n\sin\theta_i = n'\sin\theta_t', \quad n\sin\theta_i = n''\sin\theta_t \tag{14.59}$$

The velocity of propagation does not depend on the direction for the index surface of spherical symmetry (ordinary wave) with refractive index n'. However, this is no longer the case for the orthogonal polarization, since the refractive index n'' is a function of the direction of propagation.

14.7 Controlled Anisotropy

The application of an electric field can make anisotropic a material that was previously isotropic or cause new optical axes to appear in already anisotropic materials. These effects can be understood at the atomic or molecular level, similarly to what happens in the case of materials with permanent anisotropy.

14.7.1 Pockels effect

The Pockels effect is an anisotropy induced in crystals that do not have a centre of symmetry, as is the case, for example, with lithium niobate. In a Pockels cell, the crystal is placed between two crossed polarizers and oriented in such a way that its optical axis aligns with the direction of the electric field (Fig. 14.6). The field is applied between two transparent electrodes placed at the ends of the crystal. In this situation, a fast axis and a slow axis appear in a plane normal to the field. The crystal of the Pockels cell can be rotated in such a way that the directions of these axes make an angle of 45° with the polarization direction of the incident field.

When the electric field E is increased, there is a variation in the birefringence of the crystal given by

$$\Delta n = n_o^3 gE \tag{14.60}$$

where n_o is the ordinary refractive index of the crystal in the absence of the electric field and g is the linear electro-optical constant, which will be seen in Chapter 20 to be related to the second-order non-linear susceptibility of the medium. As a consequence, the optical beam is either transmitted or blocked, depending on the phase difference between the ordinary and extraordinary waves. This phase difference is given by

$$\Delta\phi = \frac{2\pi}{\lambda_0} n_o^3 gEL = \frac{2\pi}{\lambda_0} n_o^3 gV \tag{14.61}$$

where $V=EL$ is the potential difference applied to the electrodes and L is the length of the crystal.

In the case where the Pockels cell functions as a half-wave plate, one has $\Delta\phi = \pi$ rad. The potential difference that is required to apply for this purpose is given by

$$V_{HW} = \frac{\lambda_0}{2gn_0^3} \tag{14.62}$$

In this case, the polarization plane of the emergent beam is rotated by 90° relative to the plane of polarization of the incident beam.

Variations in the applied voltage V in Fig. 14.6 modify the polarization state of the emergent light and, consequently, the transmitted irradiance. In fact, the polarizer/analyser pair transforms phase modulation into amplitude modulation, and a signal voltage superimposed on V is transformed into variations in light irradiance, as illustrated in Fig. 14.7. This explains why this device is known as a *Pockels electro-optic modulator*.

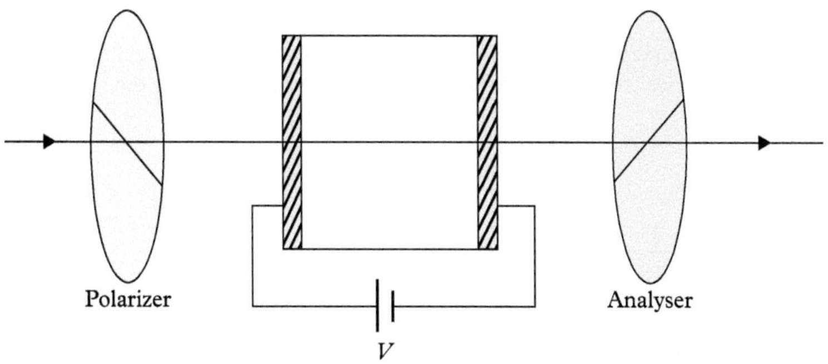

Figure 14.6 *Schematic representation of the Pockels cell.*

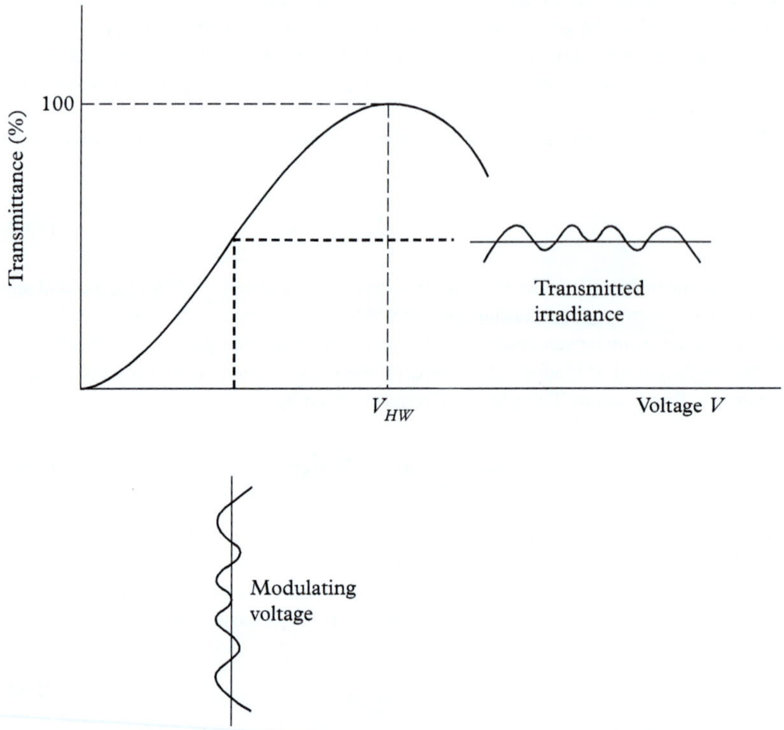

Figure 14.7 *Transmittance curve for the Pockels cell modulator.*

Besides light modulators, Pockels cells can also be used as ultrafast shutters and 'Q-switches' in lasers. In fact, they have been applied in a wide range of electro-optical systems for information processing.

Example 14.3

Calculate the potential difference V_{HW} in the case of a cell consisting of a crystal of potassium dihydrogenated phosphate (KDP), which has a refractive index $n_0 = 1.51$ and a linear electro-optical coefficient $g = 10.6$ pm/V at a wavelength $\lambda = 550$ nm.

Solution

Considering Eq. (14.62), one has

$$V_{HW} = \frac{\lambda_0}{2gn_0^3} = \frac{550 \times 10^{-9}}{2(10.6 \times 10^{-12})(1.51)^3} = 7.5 \text{ kV}$$

In the configuration of Fig. 14.6, the transmittance of the system is zero for $V = 0$ and is maximum for $V = V_{HW}$. When operating at these two voltage values, the Pockels cell behaves like an electro-optical switch.

14.7.2 Kerr effect

In the Kerr effect, birefringence is induced in the material by a transverse electric field (Fig. 14.7). The material is placed between two parallel electrodes and between two crossed polarizers, whose transmission axes make an angle of 45° with the electric field **E**. The difference between the extraordinary and ordinary refractive indices depends on the square of the electric field amplitude and is given by

$$\Delta n = \lambda K E^2 \tag{14.63}$$

where λ is the wavelength in the vacuum and K is the Kerr constant, which will be seen in Chapter 20 to be related to the third-order non-linear susceptibility of the medium. Thus, the state of polarization of the emergent beam depends on E^2.

The phase difference introduced between the ordinary and extraordinary components of the incident wave is given by

$$\Delta\phi = \frac{2\pi}{\lambda} d\Delta n = \frac{2\pi K d V^2}{L^2} \tag{14.64}$$

where $V = EL$ and L are the distances between the electrodes. Taking into account the system represented in Fig. 14.8, one can verify that the beam is blocked when $V = 0$ but is transmitted integrally when the Kerr cell functions as a half-wave plate ($\Delta\phi = \pi$rad). This case occurs for a potential difference $V = V_{HW}$ given by

$$V_{HW} = \frac{L}{\sqrt{2Kd}} \tag{14.65}$$

The Kerr cell can therefore be used as an electrically controlled light switch. However, the potential differences V_{HW} are generally superior to those given by Eq. (14.62) for the Pockels cell. Kerr cells are often used as Q-factor switches in pulsed lasers.

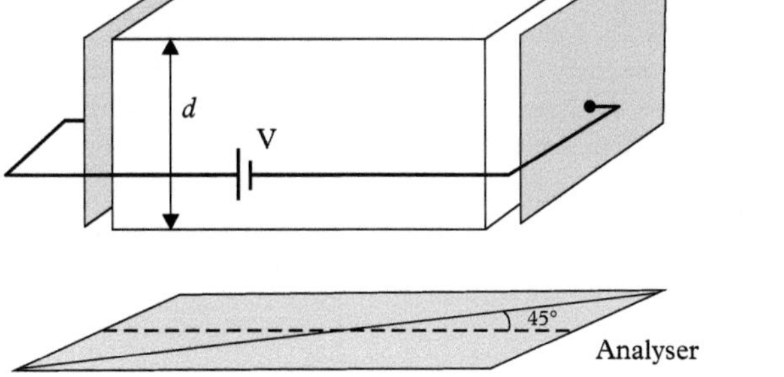

Figure 14.8 *Schematic of a Kerr cell.*

14.8 Further Reading

Azzam, R., and N. Bashara, *Ellipsometry and Polarized Light*, North Holland, 1977.
Born, M., and E. Wolf, *Principles of Optics*, 6th Ed., Pergamon Press, 1980.
Clarke, D., *Polarized Light and Optical Measurement*, Pergamon Press, 1971.
Guenther, R., *Modern Optics*, John Wiley & Sons, 1990.
Haussühl, S., *Physical Properties of Crystals*, John Wiley & Sons, 2007.
Nye, J. F., *Physical Properties of Crystals: Their Representation by Tensors and Matrices*, Oxford University Press, 1985.

14.9 Problems

14.1 Show that the Fresnel equation (Eq. (14.26)) is a quadratic equation and solve it so as to find the two positive roots n_1^2 and n_2^2.

14.2 Describe in detail how, from the normal surface $n(\mathbf{s})$ and given $\mathbf{s}$ and the magnitude of the electric field, one can find the vectors **D**, **E**, **H**, and **S**.

14.3 Demonstrate Eq. (14.36), which relates the speed of the ray to the speed of the wavefront.

14.4 Derive Eq. (14.37).

14.5 Assume a uniaxial crystal with $n_x = n_y = n_o$, $n_z = n_e$ (n_o and n_e are the refractive indices of ordinary and extraordinary refraction, respectively) and a wave propagating in the plane x, z making an angle θ with the z-axis (optical axis). Derive the expressions for the unit vector in the direction of wave propagation, as well as the vectors **D** and **E**.

14.6 Let us consider a uniaxial crystal in which the optical axis is in the direction of the *z-axis*. Determine, as a function of the refractive indices of the ordinary and extraordinary waves, the directions of propagation of the energy when the light wave propagates in the direction $\mathbf{s} = \left(\frac{1}{\sqrt{2}}, 0, \frac{1}{\sqrt{2}}\right)$.

14.7 A beam of light with a diameter of 1 mm and wavelength $\lambda = 600$ *nm* passes through a calcite crystal with a thickness of 1 cm at an angle of 30° to the optical axis. Determine the separation between the ordinary and extraordinary beams at the exit of the crystal.

14.8 Calculate the half-wave voltage, V_{HW}, of a Pockels cell made of ammonia dihydrogenated arsenate, considering a wavelength $\lambda_0 = 600$ nm, $g = 5.5$ pm/V and $n_o = 1.5$.

14.9 A Kerr cell is filled with water, whose Kerr constant is $5.23 \times 10^{-14}\mathrm{V}^{-2}$ m and is placed between two crossed polarizers. The electric field, E, is gradually increased, and the light transmitted by the analyser reaches a maximum value for $E = 5 \times 10^6$ V/m. Determine the thickness of the water traversed by the light.

14.10 Consider a nitrobenzene Kerr cell for which the Kerr constant is $K = 2.4 \times 10^{-12}\mathrm{V}^{-2}$ m at room temperature and $\lambda = 589$ nm. Find the half-wave voltage, assuming that $L = 1$ cm and $d = 3$ cm.

Optical Fibres

15

The first experiment in guided optics was performed by John Tyndall in 1854, when he demonstrated that light could be guided inside a water jet by successive total internal reflections. This jet of water functioned just like a primordial optical fibre. The first optical fibres of glass were manufactured as early as the twenties of the 20th century, but their use became more common only in the fifties, when the confinement characteristics of light were significantly improved by incorporating a cladding around the core. From that time until the seventies, optical fibres were used mainly in the field of medicine, guiding light over short distances. Their use in the field of telecommunications was not considered at all, given the high attenuation of these fibres (~ 1000 dB/ km).

In 1966, Kao and Hockham suggested that optical fibres could be used in communication systems if their losses were reduced below 20 dB/km. The limitations imposed by the attenuation of optical fibres began to be overcome in 1970, when it was possible to manufacture optical fibres with an attenuation of about 20 dB/km. Technological progress made it possible to reach, in 1979, an attenuation of only 0.2 dB/km in the spectral region of 1550 nm. The existence of fibres with such low attenuation was at the origin of a profound revolution that has been taking place since then in the field of optical communications.

15.1 Planar Waveguides

Analysing the propagation of light in a waveguide with the planar geometry is easier than in a waveguide with cylindrical geometry, such as an optical fibre. Besides elucidating the physical reasons for mode restriction and dispersion, this analysis also becomes particularly useful in the area of integrated optics and semiconductor devices.

Let us consider a three-layer planar waveguide, as represented in Fig. 15.1, with refractive indices such that $n_2 > n_3 > n_1$, where n_2 is the refractive index of the guiding layer (core). The rays propagating in this layer will be guided only if the angle of incidence at the boundaries is greater than the angle for total internal reflection, satisfying the condition:

$$\theta_2 > \text{sen}^{-1}(n_3/n_2) \tag{15.1}$$

Rays that satisfy Eq. (15.1) are associated with waves trapped in the region of index n_2, corresponding to the so-called *waveguide modes*. On the other hand, rays with angles of incidence such that $\theta_2 < \sin^{-1}(n_3/n_2)$ are transmitted to one or both of the adjacent layers and correspond to the *radiation modes* of the waveguide.

Considering their polarization, waves undergoing total internal reflection can be classified into two categories: transverse electric (TE) waves, presenting a longitudinal component of the magnetic field, and transverse magnetic (TM) waves, which have a longitudinal component of the electric field. Since the results are qualitatively the same for the two polarizations, we will consider only the TE waves in the following.

As seen in Chapter 2, the optical ray is parallel to the wave propagation vector **k**. The scalar components of this vector along the z-axis (direction of propagation) and along the

Lasers, Optics, and Photonics. Mário F. S. Ferreira, Oxford University Press. © Mário F. S. Ferreira (2026).
DOI: 10.1093/oso/9780198952770.003.0015

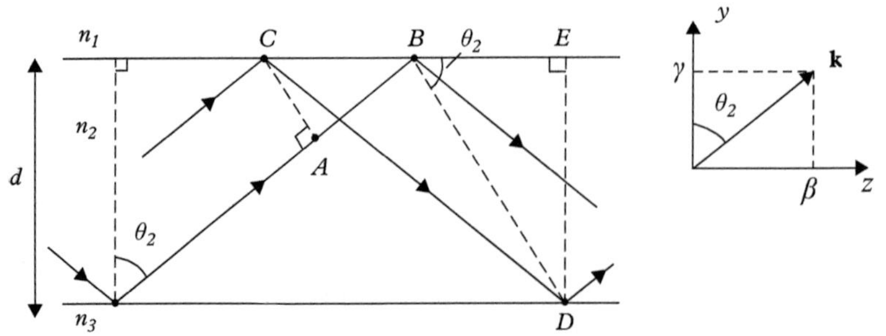

Figure 15.1 *Geometry for analysing the propagation of guided-wave modes in a planar waveguide.*

y-axis (perpendicular to the dielectric layers) are given, respectively, by

$$\beta = n_2 k_0 \sin\theta_2 \tag{15.2}$$

and

$$\gamma = n_2 k_0 \cos\theta_2 = \sqrt{n_2^2 k_0^2 - \beta^2} \tag{15.3}$$

where k_0 is the vacuum wave number. The β component corresponds to the effective propagation constant of a waveguide mode. In the case of a guided mode, this propagation constant must satisfy the condition (corresponding to Eq. (15.1)):

$$k_0 n_3 < \beta < k_0 n_2 \tag{15.4}$$

Using Fig. 15.1, let us consider a ray that travels from C to D and undergoes a reflection at each of these points. Meanwhile, a second ray goes from A to B and experiences no reflection. Since the wavefronts (represented by dashed lines) are surfaces of constant phase, the phase difference associated with the two propagation paths must be a multiple of 2π:

$$(n_2 k_0 \overline{CD} + \phi_1 + \phi_3) - n_2 k_0 \overline{AB} = 2p\pi \tag{15.5}$$

where p is an integer, ϕ_1 and ϕ_3 are phase shifts due to total reflections at the two boundaries, given by

$$\phi_{1,3} = -2\tan^{-1}\left(\frac{\sigma_{1,3}}{\gamma}\right) \tag{15.6}$$

where

$$\sigma_{1,3} = \sqrt{\beta^2 - (n_{1,3}k_0)^2} = \sqrt{\left(n_2^2 - n_{1,3}^2\right)k_0^2 - \gamma^2} \tag{15.7}$$

are the decay constants of the evanescent waves associated with total reflection at the guide boundaries. Such evanescent waves will be addressed in Section 15.2 concerning the case of step-index fibres.

Considering the geometry of Fig. 15.1, one has

$$\overline{BE} = d/\tan\theta_2 \tag{15.8a}$$

$$\overline{CE} = d\tan\theta_2 \tag{15.8b}$$

$$\overline{CD} = d/\cos\theta_2 \tag{15.8c}$$

$$\overline{AB} = \overline{CB}\sin\theta_2 \tag{15.8d}$$

where $\overline{CB} = \overline{CE} - \overline{BE}$. Introducing these results into Eq. (15.5), one obtains

$$2n_2k_0d\cos\theta_2 + \phi_1 + \phi_3 \equiv 2\gamma d + \phi_1 + \phi_3 = 2p\pi \tag{15.9}$$

Eq. (15.9) is the *dispersion relation* for the TE guided modes. This relation shows that the angle θ_2 can assume only certain discrete values, θ_p, determined by the integer values p. These angles θ_p correspond to the different modes of the waveguide.

From Eq. (15.1) one has

$$\cos\theta_p < \sqrt{1 - \left(\frac{n_3}{n_2}\right)^2} \tag{15.10}$$

Eqs. (15.9) and (15.10) show that there is a maximum number of guided modes, $p_{\max}$, given by

$$p_{\max} \leq \frac{V}{\pi} + \frac{\phi_1 + \phi_3}{2\pi} \tag{15.11}$$

where

$$V = k_0 d\sqrt{n_2^2 - n_3^2} \tag{15.12}$$

The parameter V given by Eq. (15.12) is known as the *normalized film thickness*, *normalized frequency*, or simply the *V number*. From Eqs. (15.11) and (15.12) one can conclude that the number of modes of the waveguide increases when the guiding layer thickness, d, or its refractive index, n_2, increases. The number of modes also increases when the refractive index of the substrate, n_3, or the light wavelength, λ, decreases.

In the case of a thick waveguide where $V >> 1$, the number of modes is well approximated by V/π. Actually, for each of these modes, there are two possible polarizations. If the different polarizations are considered to be different modes, then the total number of modes in the planar waveguide is $\sim 2V/\pi$.

Using Eqs. (15.6) and (15.7), Eq. (15.9) can be written in the form:

$$\gamma d - \tan^{-1}\left(\frac{\sigma_1}{\gamma}\right) - \tan^{-1}\left(\frac{\sigma_3}{\gamma}\right) = p\pi \tag{15.13}$$

or

$$\tan\gamma d = \frac{\gamma(\sigma_1 + \sigma_3)}{\gamma^2 - \sigma_1\sigma_3} \tag{15.14}$$

It can be verified that, for given values of the waveguide thickness and refractive indices, as well as the light wavelength, there is a maximum value of γd that satisfies Eq. (15.14), above which one of the decay constants given by Eq. (15.7) becomes imaginary. This situation occurs first with the constant γ_3, since we have assumed that $n_3 > n_1$. When $\sigma_3 = 0$, the angle θ_2 does not satisfy the condition for total reflection at the boundary between layers 2 and 3. At this point, the guided modes disappear and substrate modes are produced. From Eq. (15.7) one obtains the maximum value of γd, which is given by

$$\gamma d = k_0 d\sqrt{n_2^2 - n_3^2} = V_c \tag{15.15}$$

where V_c is the cut-off value of the normalized film thickness. Considering that $\sigma_3 = 0$ and using Eqs. (15.14) and (15.15), we also obtain the following expression for the cut-off condition:

$$\tan V_c = \tan\gamma d = \frac{\sigma_1}{\gamma} = \sqrt{\frac{n_3^2 - n_1^2}{n_2^2 - n_3^2}} \tag{15.16}$$

For a given guide thickness d, there is a minimum wavelength that will propagate in the guide.

On the other hand, there is a minimum value of γd, $\gamma d = \pi/2$, below which there are no solutions of Eq. (15.14) in an asymmetric guide. This means that the asymmetric guide can only support modes in a band of wavelengths defined by $\pi/2 \leq \gamma d \leq V_c$.

Example 15.1

Show that, in the case of a symmetric waveguide, there is not a maximum value, but only a minimum value for the wavelength of the guided modes.

Solution

In the case of a symmetric waveguide, the refractive indices in media 1 and 3, as well as the decay constants, are equal ($\sigma_1 = \sigma_3 = \sigma$). The dispersion relation is given by Eq. (15.9) with $\phi_1 = \phi_2 = -2\tan^{-1}\left(\frac{\sigma}{\gamma}\right)$, and can be written in the form:

$$\tan\left[\frac{1}{2}(\gamma d - p\pi)\right] = \frac{\sigma}{\gamma} \tag{15.17}$$

In this case, Eq. (15.16) shows that there is not a minimum value for γd, since one has $V_c = 0$ for $n_3 = n_1$. The situation $\gamma d = 0$ corresponds to an angle $\theta_2 = \pi/2$, when the ray propagates along the z-axis. On the other hand, the maximum value of γd occurs when the decay constant σ becomes zero, corresponding to the absence of total internal reflection. As a consequence, a symmetric guide does not impose any maximum value but only a minimum value for the wavelength of the guided modes.

15.2 Step-Index Fibres

In its simplest configuration, the optical fibre is constituted by a core with uniform refractive index, surrounded by a cladding with a refractive index that is also uniform but slightly inferior to that of the core. Due to the radial dependence profile of the refractive index, these fibres are known as *step-index fibres*. Fig. 15.2 shows a schematic of the refractive-index variation in this case.

Fig. 15.3 illustrates the ray propagation in a step-index fibre. Considering the geometry associated with ray (1), the refraction of the ray entering the fibre core is described by the equation:

$$n_0 \sin\theta_i = n_1 \sin\theta_t \tag{15.18}$$

where n_1 and n_0 are the refractive indices of the fibre core and of the medium at the fibre entrance, respectively. The ray experiences total reflection at the core/cladding boundary if the angle of incidence at this boundary is greater than the critical angle given by

$$\sin\phi_c = \frac{n_c}{n_1} \tag{15.19}$$

where n_c is the cladding refractive index. All rays with $\phi > \phi_c$ remain confined to the core and correspond to the guided modes of the fibre.

One can use Eqs. (15.18) and (15.19) to determine the maximum angle of incidence at the fibre input in order to have a guided ray. Introducing $\theta_t = \pi/2 - \phi_c$ in Eq. (15.18) and using Eq. (15.19), we obtain:

$$n_0 \sin\theta_{i,\max} = n_1 \cos\phi_c = \left(n_1^2 - n_c^2\right)^{1/2} \tag{15.20}$$

Light incident on the fibre core with entrance angles within the range $0 < \theta_i < \theta_{i,\max}$ is accepted into guided modes, whereas light incident outside of this range goes into unguided modes. In three dimensions, this corresponds to an acceptance cone of half-angle $\theta_{i,\max}$, as illustrated in Fig. 15.4. Using the principle of reversibility mentioned in Section 2.4, the acceptance cone describes at the same time the exit cone of light at the fibre exit.

The result given by Eq. (15.20) is used to define the *numerical aperture* (*NA*) of the optical fibre, given by

$$NA = \left(n_1^2 - n_c^2\right)^{1/2} \tag{15.21}$$

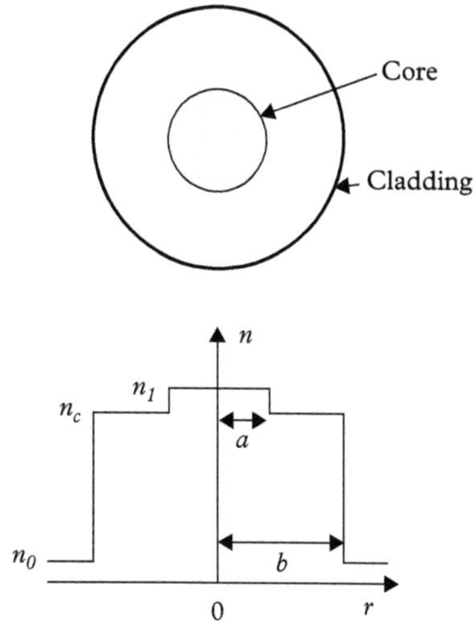

Figure 15.2 *Schematic representation of the refractive-index profile in the case of a step-index fibre.*

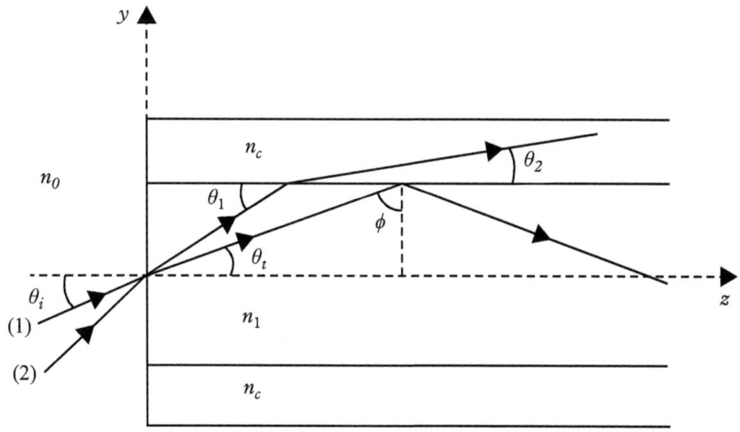

Figure 15.3 *Propagation of (1) a guided ray and (2) an unguided ray in a step-index fibre.*

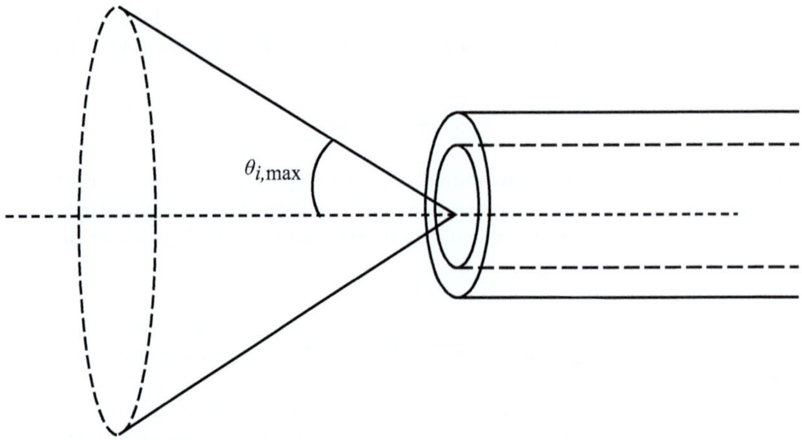

Figure 15.4 *Schematic representation of the acceptance and exit cones of a fibre.*

The NA is a measure of the light-gathering capacity of the fibre. Considering the typical values $n_1 = 1.48$ and $n_c = 1.46$, the NA is $NA = 0.242$, and the maximum angle of incidence at its entrance is $\theta_{i,\max} \approx 14°$ assuming we are coupling from air ($n_0 = 1$).

In the case $n_1 \approx n_c$, the NA can be approximated by

$$NA = n_1(2\Delta)^{1/2} \tag{15.22}$$

where

$$\Delta = \frac{n_1 - n_c}{n_1} \tag{15.23}$$

is the fractional index difference. Eq. (15.22) shows that the NA increases when Δ increases. However, as will be seen in Chapter 16, an increase of Δ also determines an increase in the intermodal dispersion, which is undesirable from the viewpoint of optical fibre communication systems.

In the previous discussion we have neglected the fact that in addition to meridional rays there can also be the so-called skew rays, which have a helical trajectory and are transmitted without passing through the fibre axis. These rays are illustrated in Fig. 15.5 and are accepted at the fibre entrance at larger axial angles than meridional rays. Actually, skew rays tend to propagate only in the annular region near the outer surface of the core and provide some increase in the fibre light-gathering ability. However, for most communication design purposes, the expressions given by Eqs. (15.21) and (15.22) for the NA are considered adequate.

15.2.1 Evanescent waves

Considering the geometry associated with ray (2) in Fig. 15.3 and using Snell's law, one has

$$\sin\theta_2 = \sqrt{1 - \cos^2\theta_2} = \sqrt{1 - \frac{n_1^2}{n_c^2}\cos^2\theta_1} \tag{15.24}$$

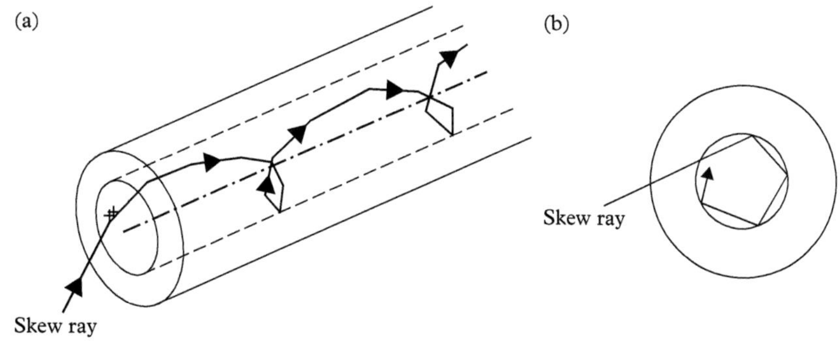

Figure 15.5 *(a) Skew ray path down the fibre; (b) cross-sectional view of the fibre.*

When the angle of incidence at the core–cladding boundary exceeds the critical value for total reflection, the radicand in Eq. (15.24) becomes negative and one can write:

$$\sin\theta_2 = \sqrt{1 - \frac{n_1^2}{n_c^2}\cos^2\theta_1} = \pm i\delta \tag{15.25}$$

In this case, the amplitude of the transmitted wave in the cladding can be written in the form:

$$E_t = A\exp\left[ik_2\left(z\cos\theta_2 + y\sin\theta_2\right)\right] \tag{15.26}$$

$$= A\exp\left[ik_2\frac{n_1}{n_c}z\cos\theta_1\right]\exp\left(-\delta k_2 y\right) \tag{15.26}$$

Eq. (15.26) shows that the amplitude of the transmitted wave decreases exponentially along the y-axis. For this reason, it is called an *evanescent wave*. The *penetration depth*, ξ, of this evanescent wave is given by the inverse of δk_2:

$$\xi = \left[k_2^2\left(\frac{n_1^2}{n_c^2}\cos^2\theta_1 - 1\right)\right]^{-1/2} \tag{15.27}$$

Near the critical angle and considering a wavelength $\lambda = 1.3\,\mu\text{m}$, one has $\xi \approx 10\,\mu\text{m}$. This result shows that the cladding must be sufficiently thick in order for the amplitude of the evanescent wave to be practically zero at its outer boundary.

15.3 Optical Fibre Modes

Geometrical optics provides only an approximate description of the guiding properties of optical fibres, which is reasonable when the core radius is much larger than the light wavelength. However, a more rigorous approach to describing the propagation of light in fibres is given by electromagnetic theory. This approach starts with Maxwell's equations presented in Chapter 6.

The fibre modes can be classified as guided modes, leaky modes, and radiation modes (Snyder and Love, 1983). Since the transmission of information in actual optical fibre communication systems makes use of the guided modes only, the following discussion will be restricted to them.

By introducing the Fourier transform of the electric field in the form:

$$\tilde{\mathbf{E}}(\mathbf{r},\omega) = \frac{1}{2\pi}\int_{-\infty}^{\infty} \mathbf{E}(\mathbf{r},t)\exp(i\omega t)\,dt \tag{15.28}$$

the wave Eq. (6.13) becomes

$$\nabla^2\tilde{\mathbf{E}} + n^2(\omega)k_0^2\tilde{\mathbf{E}} = 0 \tag{15.29}$$

where

$$k_0 = \frac{\omega}{c} \tag{15.30}$$

is the free-space wave number.

Taking advantage of the cylindrical symmetry of fibres, Eq. (15.29) is written in the cylindrical coordinates r, ϕ and z as

$$\frac{\partial^2\tilde{\mathbf{E}}}{\partial r^2} + \frac{1}{r}\frac{\partial\tilde{\mathbf{E}}}{\partial r} + \frac{1}{r^2}\frac{\partial^2\tilde{\mathbf{E}}}{\partial\phi^2} + \frac{\partial^2\tilde{\mathbf{E}}}{\partial z^2} + n^2k_0^2\tilde{\mathbf{E}} = 0 \tag{15.31}$$

where $n = n_1$ for $r \leq a$ and $n = n_c$ for $r > a$. A similar equation can be written for the magnetic field $\tilde{\mathbf{H}}$. Since $\mathbf{E}$ and $\tilde{\mathbf{H}}$ satisfy Maxwell's Eqs. (6.1)–(6.4), it follows that, of the six components, only two are independent. $\tilde{E}_z$ and $\tilde{H}_z$ are usually chosen as the independent components, and both satisfy Eq. (15.31). Writing $\tilde{E}_z$ as

$$E_z(r,\phi,z) = F(r)\Phi(\phi)Z(z) \tag{15.32}$$

and substituting into the wave Eq. (15.31), we obtain the solutions $Z(z) = \exp(i\beta z)$ and $\Phi(\phi) = \exp(im\phi)$, where β is the propagation constant and m is an integer. On the other hand, $F(r)$ satisfies the following ordinary differential equation:

$$\frac{\partial^2 F}{\partial r^2} + \frac{1}{r}\frac{\partial F}{\partial r} + \left(n^2k_0^2 - \beta^2 - \frac{m^2}{r^2}\right)F = 0 \tag{15.33}$$

Eq. (15.33) is a form of Bessel's equation. Its solution is given in terms of Bessel functions and assumes the form

$$F(r) = \begin{cases} AJ_m(pr) + A'Y_m(pr); & r \leq a \\ CK_m(qr) + C'I_m(qr); & r > a \end{cases} \tag{15.34}$$

where A, A', C, and C' are constants; J_m and Y_m are ordinary Bessel functions of the first and second kind of order m, respectively, whereas K_m and I_m are modified Bessel functions. The parameters p and q are given by:

$$p = \sqrt{n_1^2k_0^2 - \beta^2} \tag{15.35}$$

$$q = \sqrt{\beta^2 - n_c^2k_0^2} \tag{15.36}$$

and satisfy the important relation

$$p^2 + q^2 = \left(n_1^2 - n_c^2\right) k_0^2 \tag{15.37}$$

As $Y_m(pr)$ has a singularity at $r = 0$, $A' = 0$ for a physically meaningful solution. In the cladding region ($r > a$), the solution $F(r)$ must decay exponentially for large values of r, which happens only if $C' = 0$. As a result, the general solution given by Eq. (15.32) becomes

$$E_z = \begin{cases} AJ_m(pr)\exp(im\phi)\exp(i\beta z); & r \le a \\ CK_m(qr)\exp(im\phi)\exp(i\beta z); & r > a \end{cases} \tag{15.38}$$

Similarly, the solution for H_z can be written in the form

$$H_z = \begin{cases} BJ_m(pr)\exp(im\phi)\exp(i\beta z); & r \le a \\ DK_m(qr)\exp(im\phi)\exp(i\beta z); & r > a \end{cases} \tag{15.39}$$

Maxwell's equations are used to find general expressions for all transverse components in terms of the first derivatives of the z components. In cylindrical coordinates, we find

$$E_r = \frac{i}{p^2}\left(\beta\frac{\partial E_z}{\partial r} + \omega\mu\frac{1}{r}\frac{\partial H_z}{\partial \phi}\right) \tag{15.40}$$

$$E_\phi = \frac{i}{p^2}\left(\frac{\beta}{r}\frac{\partial E_z}{\partial \phi} - \omega\mu\frac{\partial H_z}{\partial r}\right) \tag{15.41}$$

$$H_r = \frac{i}{p^2}\left(\beta\frac{\partial H_z}{\partial r} - \omega\varepsilon\frac{1}{r}\frac{\partial E_z}{\partial \phi}\right) \tag{15.42}$$

$$H_\phi = \frac{i}{p^2}\left(\frac{\beta}{r}\frac{\partial H_z}{\partial \phi} + \omega\varepsilon\frac{\partial E_z}{\partial r}\right) \tag{15.43}$$

These equations are also valid in the cladding region if p^2 is replaced by $-q^2$.

The four constants A, B, C, and D in Eqs. (15.38) and (15.39) are determined using the requirement that all field components that are tangent to the core–cladding boundary at $r = a$ be continuous across it. All ϕ and z components apply, producing four equations in all. A nontrivial solution for A, B, C, and D is obtained only if the determinant of the coefficient matrix vanishes. This condition provides the following eigenvalue equation:

$$\left[\frac{J_m'(pa)}{pJ_m(pa)} + \frac{K'_m(qa)}{qK_m(qa)}\right]\left[\frac{n_1^2}{n_c^2}\frac{J'_m(pa)}{pJ_m(pa)} + \frac{K'_m(qa)}{qK_m(qa)}\right] = \frac{m^2}{a^2}\left(\frac{1}{p^2} + \frac{1}{q^2}\right)\left(\frac{n_1^2}{n_c^2} + \frac{1}{p^2} + \frac{1}{q^2}\right) \tag{15.44}$$

where a prime indicates differentiation with respect to the argument.

The solutions of the eigenvalue equation determine the propagation constant β for the fibre modes. For each integer value of m, there are several solutions β_{mn} ($n = 1, 2, \ldots$). Each value of β_{mn} corresponds to one specific mode of propagation whose field distribution is given by Eqs. (15.38)–(15.43). For $m = 0$, the fibre modes are denoted by TE_{0n} and TM_{0n}, since they correspond to transverse-electric ($E_z = 0$) and transverse-magnetic ($H_z = 0$) modes of propagation. However, for $m > 0$, fibre modes become hybrid, since all six components of the electromagnetic field are nonzero. These hybrid modes are denoted by HE_{mn} or EH_{mn}, due to an early system in which the contributions of H_z and E_z to a given transverse field component at some reference position are compared. If, for example, E_z makes the larger contribution, then the mode is designated HE.

Using the weak-guidance approximation, stating that $n_1 \approx n_c$, the eigenvalue equation (15.40) can be simplified and written in the form

$$\frac{J_m(pa)}{pJ_m(pa)} + \frac{K'_m(qa)}{qK_m(qa)} = \pm \frac{m}{a}\left(\frac{1}{p^2} + \frac{1}{q^2}\right) \tag{15.45}$$

Eq. (15.45) with a positive right-hand side is the eigenvalue equation for EH_{mn} modes, whereas with a negative right-hand side it is the equation for HE_{mn} modes. The case in which the right-hand side is zero applies to TE and TM modes.

Each fibre mode has an effective refractive index $n_{eff} = \beta / k_0$, known as the *effective mode index*, whose value is such that $n_1 > n_{eff} > n_c$. The mode is said to reach cut-off when $n_{eff} = n_c$, which, from Eq. (15.36), corresponds to $q = 0$. It is useful to introduce a *normalized frequency* or *V-number* parameter, similar to that given by Eq. (15.12) for planar waveguides, except that the waveguide thickness d is replaced by the fibre core radius a:

$$V = k_0 a\left(n_1^2 - n_c^2\right)^{1/2} \tag{15.46}$$

The V-number parameter embodies both the fibre structural parameters and the optical wavelength. It is also useful to introduce a normalized propagation constant b in the form

$$b = \frac{n_{eff} - n_c}{n_1 - n_c} \tag{15.47}$$

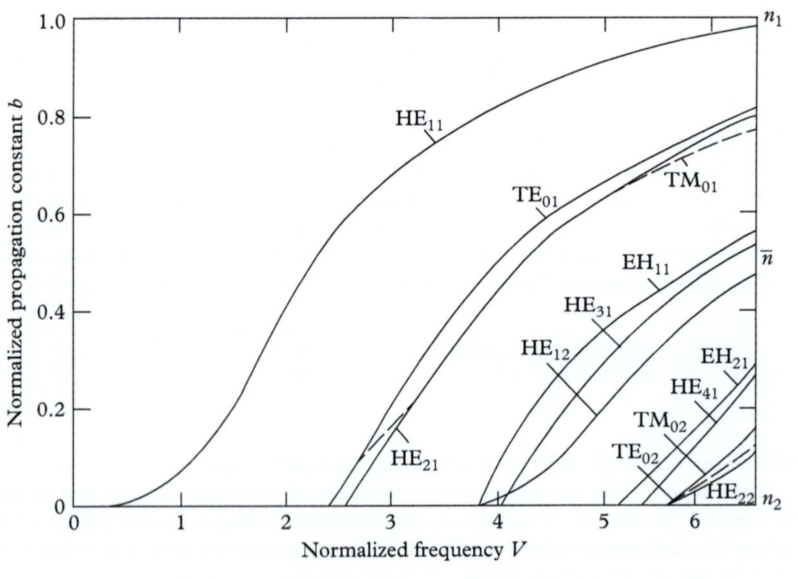

Figure 15.6 *Normalized propagation constant b against normalized frequency V for some low-order fibre modes. The right axis shows the effective mode index n_{eff}. (Reprinted with permission from Keck, 1981 © Elsevier.)*

Fig. 15.6 represents b against the V parameter for some low-order fibre modes obtained by solving Eq. (15.40).

As seen in Section 15.1, the number of guided modes in a one-dimensional planar waveguide is given by $\sim 2V/\pi$ for large values of V. In a two-dimensional planar waveguide, there is a similar result for the orthogonal direction, and the total number of guided modes is given by the product of these two numbers, or $\sim (2V/\pi)^2$. In the case of an optical fibre, it can be shown that the number of guided modes is approximately given by $V^2/2$ for a multimode fibre with a large value of V.

15.3.1 Single-mode fibres

The single-mode condition is given by the value of V at which the TE_{01} and TM_{01} modes reach cut-off. From the eigenvalue Eq. (15.44) or (15.45), it can be easily verified that the cut-off condition for both modes is given by $J_0(V) = 0$. The smallest value of V for which this condition is satisfied is 2.405. As seen from Fig. 15.6, for $V<2.405$ the fibre supports only the HE_{11} mode, also referred to as the *fundamental mode*. This mode has no cut-off and is always supported by a fibre.

Considering Eq. (15.46), the condition for single-mode propagation is seen to depend on fibre structural parameters and the optical wavelength. In principle, either the core radius or the NA could be reduced to achieve single-mode operation. However, making the NA too small restricts the acceptance angle for incident light and reduces source-to-fibre coupling efficiency. The dependence on wavelength implies that any fibre will be single mode for a sufficiently long wavelength of light. A typical fibre with $n_1 = 1.45$, $\Delta = 3 \times 10^{-3}$, and $a = 4$ μm becomes single mode for $\lambda > 1.2$ μm. The wavelength at which the fibre just becomes single mode is termed the *cut-off wavelength*, λ_c, defined from the relation:

$$2.405 = V_c = k_0 a \left(n_1^2 - n_c^2\right)^{1/2} \tag{15.48}$$

where

$$k_0 = 2\pi/\lambda_c \tag{15.49}$$

In the weak-guidance approximation, the axial components E_z and H_z of the fundamental mode are negligible, and it is approximately linearly polarized. Assuming a linear polarization along the x-axis, the electric field for the HE_{11} mode is given by Marcuse (1977):

$$E_x = \begin{cases} E_0 J_0(pr)\exp(i\beta z); & r \le a \\ E_0\left[J_0(pa)/K_0(qa)\right]K_0(qr)\exp(i\beta z); & r > a \end{cases} \tag{15.50}$$

The same fibre supports another mode linearly polarized along the y-axis. The two orthogonally polarized modes of a single-mode fibre are degenerate (i.e. they have the same propagation constant) under ideal conditions. However, in practice, a small deviation from the cylindrical geometry or a slight anisotropy of the material determines a breaking of the mode degeneracy. In such a case, the propagation constant β turns out to be slightly different for the modes polarized along the two orthogonal directions. This is the origin of the polarization-mode dispersion phenomenon, which will be discussed in Chapter 16.

Due to fluctuations in the fibre geometry or in the material anisotropy, the fibre birefringence does not remain constant along its length. As a consequence, when linearly polarized light is launched at the input, it generally acquires an arbitrary polarization during its propagation along the fibre. However, a proper design of the fibre can intentionally introduce significant birefringence in order to make possible the propagation of light with a constant polarization. Fibres obtained in this way are called *polarization-maintaining fibres*.

Gaussian Approximation for the Fundamental Mode Field

The field distribution for the fundamental mode given by Eq. (15.50) can be approximated by a Gaussian distribution having a width w, known as the *mode-field radius*. The Gaussian field polarized along the x-axis is given by

$$E_{gx} = E_{g0} \exp\left(-\frac{r^2}{w^2}\right) \exp(i\beta z) \tag{15.51}$$

The mode-field radius w is determined by fitting the exact distribution to the Gaussian function, and it is given approximately by the empirical formula:

$$\frac{w}{a} \approx 0.65 + \frac{1.619}{V^{3/2}} + \frac{2.879}{V^6} \tag{15.52}$$

This formula approximates the exact result to better than 1% accuracy for $1.2 < V < 2.4$. The mode-field radius w is used to define the effective core area as

$$A_{eff} = \pi w^2 \tag{15.53}$$

The parameter A_{eff} is important, since it indicates how tightly light is confined to the core.

Eq. (15.52) shows that the ratio w/a becomes relatively large for small values of the fibre parameter V. In this case, the mode is referred to as *weakly guided*, a significant part of its energy being contained in the cladding region rather than in the fibre core. Such behaviour can be useful for the operation of some devices, namely fibre sensors and fibre couplers, but it is detrimental in the area of optical communications. For best confinement of the mode, the parameter V is often chosen to be not much below the cut-off value of 2.405.

Example 15.2

Consider an optical fibre where the core has a refractive index $n_1 = 1.465$ and the cladding has a refractive index $n_c = 1.460$. Determine (i) the maximum value of the core radius for having a single-mode fibre and (ii) the number of guided modes when that radius is $a = 16\ \mu\text{m}$. Consider that light has a wavelength $\lambda = 1.25\ \mu\text{m}$.

Solution

i) The maximum value of the radius that guarantees a single-mode regime for the fibre can be calculated from the condition $V < 2.4048$, obtaining

$$a < \frac{2.4048}{k_0\sqrt{n_1^2 - n_c^2}} = \frac{2.4048}{2\pi\sqrt{(1.465)^2 - (1.46)^2}}\left(1.25 \times 10^{-6}\right) = 3.95\ \mu\text{m}$$

Example 15.2 continued

ii) The V-number of the fibre whose core has a radius $a = 16\ \mu m$ is

$$V = k_0 a\sqrt{n_1^2 - n_c^2} = \frac{2\pi}{1.25 \times 10^{-6}} 16 \times 10^{-6}\sqrt{(1.465)^2 - (1.46)^2} = 9.73$$

So the number of guided modes is

$$n_{\max} = \frac{1}{2}V^2 = 47$$

15.4 Fibre Attenuation

Attenuation is one important characteristic of an optical fibre, since it determines the repeater spacing in a fibre transmission system. The lower the attenuation, the greater will be the repeater spacing required and the lower will be the cost of that system.

There are three attenuation mechanisms in an optical fibre: absorption, scattering, and bending loss. The absorption mechanism corresponds to the loss of propagating photons in the fibre material, which are generally converted into heat. In the scattering process, the photon does not disappear but changes its direction. Finally, the bending loss is related to the failure to satisfy the total internal reflection requirement at the boundary between the core and the cladding.

Representing by P_0 the power launched at the input of a fibre of length L, it is straightforward to show that the output power is given by the so-called *Beer's law*:

$$P_t = P_0 \exp(-\alpha L) \tag{15.54}$$

where α is the attenuation constant. Usually, the fibre attenuation is given in dB/ km, using the relation:

$$\alpha_{dB} = -\frac{10}{L}\log\left(\frac{P_t}{P_0}\right) = 4.343\alpha \tag{15.55}$$

For example, if the output power is half the input power, then the loss is $10\log(2) \approx 3\text{dB}$.

Pure silica glass presents an attenuation below 1 dB/km for wavelengths in the range 0.8–1.8 μm. Outside this range, the attenuation increases rapidly. Fig. 15.7 shows a plot of the spectral attenuation in a $GeO_2 - SiO_2$ silica fibre. The attenuation presents minimum values around 1.3 and 1.55 μm, achieving in the latter case a value of 0.2 dB/km. The two bands around the above wavelengths correspond to the windows currently used in fibre communication systems.

Material absorption and Rayleigh scattering are the dominant contributions to the residual attenuation of fused silica. Pure silica absorbs in the ultraviolet region, as well as in the infrared region beyond 2 μm. However, even a relatively small quantity of impurities can result in significant absorption in the range 0.5–2 μm. The most difficult extrinsic loss contributor to remove is the hydroxyl group, OH, which enters the glass through water vapour and whose main absorption peak occurs at 2.73 μm. The peak in Fig. 15.7 near 1.37 μm corresponds to the second harmonic of this value.

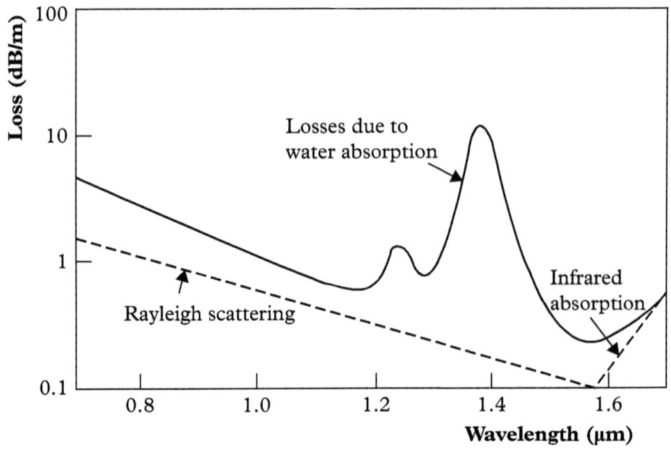

Figure 15.7 *Typical spectral attenuation profile of a* $GeO_2 - SiO_2$ *fibre.*

Rayleigh dispersion is a fundamental effect and results from random fluctuations of the material density. As a consequence, there are local fluctuations of the refractive index, which determine light scattering in all directions. This effect varies with λ^{-4}, becoming more significant for short wavelengths. The minimum of attenuation near 1.55 μm is mainly due to Rayleigh scattering.

As mentioned in Section 15.2, near the critical angle the evanescent wave extends about $\xi \approx 10\ \mu m$ into the cladding. In these circumstances, considering that a significant fraction of the transmitted energy is carried by this evanescent wave, the quality of the cladding glass must be as high as that of the core in order to minimize the losses.

15.5 Specialty Optical Fibres

Optical fibres used in communication systems are appropriately designed to transmit modulated optical signals at specific wavelengths. The fibre design is optimized for low loss and modal properties, taking into account transmission distance and bandwidth requirements. These characteristics are generally accomplished in a standard step-index fibre, made of silica and presenting common values of parameters such as the core size and NA.

Specialty optical fibres refer to optical fibres that have at least one special property, distinguishing them from *standard fibres*. Specialty optical fibres are designed and manufactured in order to realize certain special functions other than signal transmission, namely as amplifiers, sensors, lasers, filters, ring resonators, etc. Considering such a variety of applications, there are many different types of specialty fibres, with variations in glass composition, core and cladding structures, geometric properties, coatings, and specially tailored optical performance characteristics.

One type of specialty optical fibre corresponds to the so-called *active fibres*, which are glass fibres generally doped with laser-active rare-earth ions. These ions can absorb pump light which excites them into some metastable levels, allowing for light amplification via stimulated emission. In this way, active fibres provide a gain medium with a particularly high gain efficiency, being used to realize fibre amplifiers and lasers. Some of these devices are described in Chapter 22. The operational wavelengths of rare-earth-doped fibre amplifiers are determined by the emission spectra of the rare-earth ions, moderately modified by the host material in which they are embedded. The technologically most important rare-earth-doped fibres are erbium-doped fibres and ytterbium-doped fibres. Erbium-doped fibre amplifiers (EDFAs) have attracted particular attention because they operate in the wavelength region near 1.55 μm, the main window of current communication systems.

Another type of specialty optical fibres uses non-standard materials instead of silica. For example, there are *fluoride fibres*, which can transmit in longer (infrared) wavelength regions in comparison with silica optical fibres. This becomes particularly important considering that the optical communications area is actually facing exponential growth in bandwidth transmission with the introduction of higher-bandwidth applications. Low-loss fluoride fibres are indeed excellent candidates for applications in the 2.4–5.5 μm spectral region.

Plastic optical fibres (POFs) are one of the non-silica-glass-based specialty optical fibres, which are generally used for optical communication and data transmission in automobiles. POFs are also commonly used in local area networks (LANs) as data lines. Major research in POFs has been concerned with reducing their relatively high transmission loss and increasing transmission bandwidth.

Radiation-resistant fibres are optical fibres made from materials that are substantially less affected by radiation than other types of fibres. They are used, for example, in space satellites, airplanes, high-energy physics experiments, nuclear plants, and medical facilities involving X-rays or proton radiation. Special materials may also be used for fibre coatings and jackets, which provide the possibility of withstanding much higher temperatures than standard fibre coatings.

Single-crystal fibres (SCFs) are made of monocrystalline material rather than amorphous material, as in standard silica fibres. SCFs allow only very limited bending and do not properly contain a fibre core, since the inclusion of a cladding layer with a slightly lower refractive index proves to be very difficult to realize. Moreover, multimode waveguiding can be achieved only due to the strong refractive-index contrast at the crystal–air interface. Single-crystal fibres can achieve much higher doping concentrations of laser-active ions and are suitable for supporting high power levels, since they usually have large effective mode areas.

Polarization-maintaining fibres were already mentioned earlier and can be used to maintain a linear polarization state of light over arbitrarily long propagation distances. In polarization-maintaining single-mode fibres, the fibre symmetry is broken by integrating stress elements into the fibre cladding, thus introducing significant birefringence in the fibre. The light is then guided in two perpendicular states of polarization with different propagation constants—the fast and the slow axes. The linear polarization of light coupled into one of these axes is maintained during propagation. Different types of polarization-maintaining fibres are designed depending on the geometry of the stress elements.

For some applications, namely high-power fibre lasers, it is desirable to use *large mode area* (LMA) *fibres*. For a given optical power, LMAs imply reduced optical intensities. Therefore, such fibres effectively exhibit weaker non-linearities and a higher damage threshold, which makes them suitable for, e.g., the amplification of intense pulses in fibre amplifiers. Whereas standard single-mode fibres have an effective mode area below 100 μm^2, large mode area fibres can reach values of hundreds or even thousands of μm^2.

In contrast to LMA fibres, *highly non-linear fibres* (HNLFs) are designed in order to enhance optical non-linearities. This characteristic is needed in certain applications in the area of all-optical signal processing, namely pulse compression and regeneration, wavelength conversion, supercontinuum generation, etc. Using standard optical fibres, a length of several kilometres is usually required for these applications due to their relatively small non-linear parameter. However, the required fibre length can be reduced significantly by using microstructured optical fibres (MOFs), which contain air holes and are therefore also called *holey fibres*. MOFs having either a solid core or a hollow core are discussed in the next section and can present significantly enhanced non-linear characteristics. A further improvement consists in employing materials with a non-linear refractive index higher than that of the silica glass, namely lead silicate, tellurite, bismuth glasses, and chalcogenide glasses. Using such highly non-linear optical fibres, the required fibre length for all-optical signal processing can be reduced to the order of centimetres. Further discussion of highly non-linear optical fibres will be presented in Chapter 25 concerning their application to supercontinuum generation.

15.6 Microstructured Optical Fibres

MOFs represent a class of specialty optical fibres that are characterized by the fact that the silica cladding presents an array of embedded air holes. They are also referred to as *photonic crystal fibres* (PCFs), since they were first realized in 1996 in the form of a photonic crystal cladding with a periodic array of air holes (Knight et al., 1996).

MOFs can be divided into two main types. One type has a solid core, in which the light is guided mainly due to total internal reflection, since the core has a higher refractive index than the cladding. In such fibres, the air holes in the cladding provide an effective reduction of its refractive index below that of the silica core. This helps to concentrate the mode field in a very small area, which is particularly the case for MOFs with small-scale cladding features and large air-filling fractions.

Fig. 15.8 illustrates a solid-core (SC) MOF with a hexagonal pattern of holes in the cladding. This kind of fibre is characterized by two main structural parameters: the air-hole diameter d and the hole-to-hole spacing Λ.

The number of guided modes in a conventional fibre is determined by the normalized frequency V, given by Eq. (15.46), which depends on the wavelength. In conventional fibres, the single-mode condition $V<2.405$ is only satisfied if the wavelength is in the infrared region. For lower values of the wavelength, the fibre becomes multimode. However, in a MOF the normalized frequency is given by a slightly different expression from Eq. (15.46), with a replaced by Λ and n_c replaced by the effective cladding index of the microstructured cladding. Such an effective refractive index depends significantly on the ratios d/Λ and λ/Λ and asymptotically approaches the core index in the short-wavelength limit. As a consequence, both the effective index difference and the NA go to zero in this limit.

Surprisingly, in a MOF it is verified that the behaviour of the cladding effective index cancels out the wavelength dependence of the normalized frequency V, such that it approaches a constant value as $\Lambda/\lambda \rightarrow \infty$. As a consequence, a MOF shows single-mode behaviour at all wavelengths if $d/\Lambda < 0.43$; such a fibre is known as the *endlessly single-mode* (ESM) fibre.

The second type of MOFs presents a central region containing air and is usually called hollow-core (HC) MOFs. Compared with SC-MOFs, HC-MOFs can offer a much higher power delivery capability, with a considerably higher damage threshold and significantly lower dispersion and non-linearity.

HC-MOFs can guide light inside the hollow core based on two different mechanisms. One of them is based on the formation of photonic bandgaps (PBGs) in the periodic surrounding holey dielectric cladding region. The PBG prevents light at well-prescribed

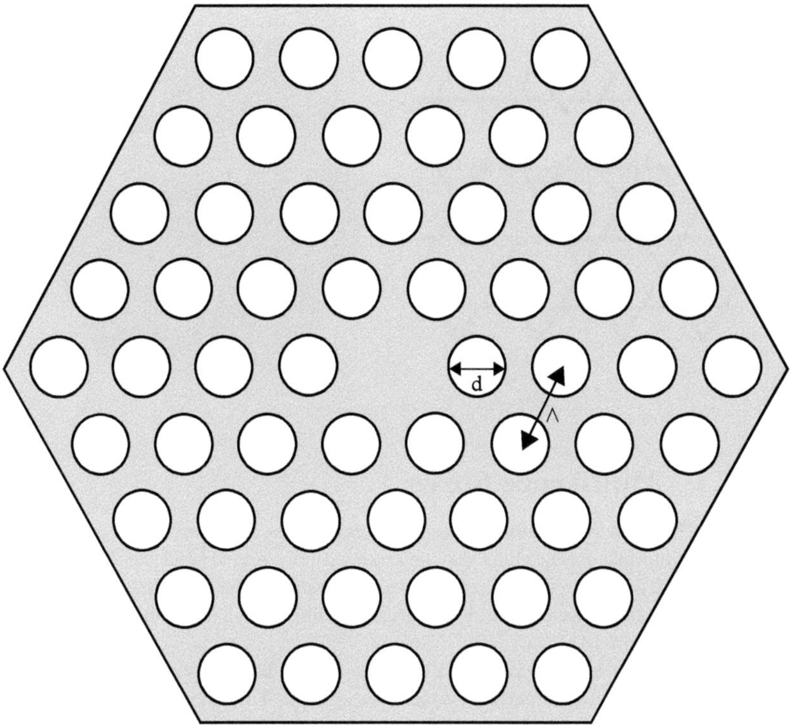

Figure 15.8 *Schematic of a solid-core microstructured optical fibre with a hexagonal pattern of holes.*

wavelengths and angles of incidence from propagating into the cladding and hence allows tight optical confinement in a suitably designed low-refractive-index central defect.

The main drawback of PBG MOFs is their intrinsically narrow transmission bandwidth, determined by the bandgaps, which excludes their implementation in a large number of applications in ultrafast non-linear optics requiring broadband guidance or guidance in the visible and UV. Moreover, PBG MOFs present a relatively low damage threshold due to the significant overlap between the air-guided field and the cladding glass.

A different type of optical guidance mechanism in HC-MOFs requires neither a higher-index core material nor a cladding PBG. The guidance mechanism in this case is called inhibited coupling (IC) and was first experimentally observed with the introduction of kagome-lattice HC-PCF. Coupling between the core and the cladding modes in this kind of fibres is inhibited for two reasons. First, there is a strong transverse phase mismatch between them. Second, there is a weak spatial overlap between the core mode and the cladding modes due to good confinement of the optical field in the core and strong localization of the cladding modes in the cladding glass web. The possibility of filling the fibre core with a suitable liquid or gas and of conveniently adjusting its pressure also offers a new degree of freedom over conventional fibres. Fig. 15.9 shows (a) the crystal lattice and (b) the spatial distribution of the optical field for a kagome MOF with a core diameter d = 40.0 μm and a strut thickness of t = 0.1 μm, filled with helium at a pressure p = 10 bar and pump wavelength λ = 790 nm. The dispersion characteristics of kagome MOFs filled with noble gases will be discussed in Chapter 16 of this book.

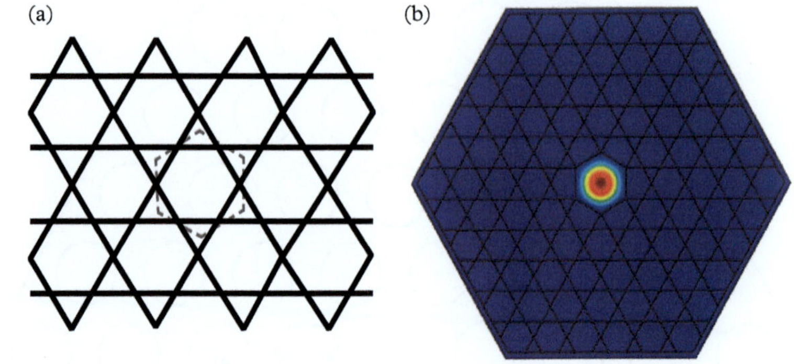

Figure 15.9 *(a) Schematic of a kagome crystal lattice and (b) optical field of the fundamental mode of a kagome MOF with a core diameter* $d = 40.0\ \mu\text{m}$ *and strut thickness* $t = 0.1\ \mu\text{m}$*, filled with helium at a pressure* $p = 10$ *bar and pump wavelength* $\lambda = 790$ *nm.*

15.6.1 Multi-core fibres

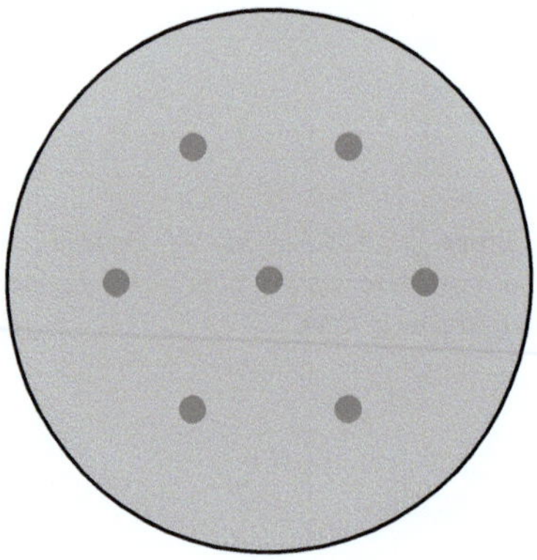

Figure 15.10 *A seven-core fibre with a hexagonal lattice structure.*

Instead of having a single fibre core, as in a standard fibre, there are some specialty fibres containing multiple cores, which may be arranged on a ring around the fibre axis or on some 2D grid. Such fibres are called *multi-core fibres* and can be realized with all-glass fibre technology or alternatively as MOFs. A seven-core fibre with a hexagonal lattice structure is shown in Fig. 15.10.

In principle, each of the cores in a multi-core fibre can act as a separate waveguide, so that light can independently propagate along them. However, some mode coupling between the cores can occur if the distance between them is so small that the corresponding mode fields have a significant spatial overlap. Such mode coupling can be useful for some applications. However, it is clearly undesirable in optical fibre communication systems. In this case, mode coupling can be minimized by using sufficiently large spacings between the cores. A distance of the order of 30–40 μm between each pair of cores may be sufficient to avoid significant coupling.

15.7 Optical Fibre Communications

Considering optical communications in a broad sense, we can say that the use of light for communication purposes dates back to the beginning of humanity. In fact, hand signals, waving of the arms, and fire and smoke signalling are basic means of optical communication. The use of mirrors and, more recently, signalling lamps, flags, and other semaphore devices have been successful in conveying a single piece of information. This technique was extended further when Claude Chappe constructed the first optical telegraph in 1791, which was put into service between Paris and Lille in July 1794. This system included repeater stations every few kilometres and allowed messages to be sent between the two cities in just six minutes—corresponding to twice the speed of sound. By 1830, the network had expanded throughout Europe. However, the use of light was replaced by electricity with the advent of telegraphy during the 1830s. In 1880, the inventor of the telephone, Alexander Graham Bell, introduced what he called the *photophone*, which allowed information to be transmitted through air on a beam of modulated sunlight. The lack of suitable light sources and the problem that light transmission in the atmosphere is restricted to line of sight and is severely affected by several disturbances limited the investigation of optical communication in the early part of the 20th century.

The invention of the laser in 1960 stimulated a renewed interest in optical communications. The availability of such a powerful coherent light source, together with the low divergence of the laser beam, made free-space optical transmission a practical possibility. However, the aforementioned constraints of light transmission in the atmosphere restricted these systems to relatively short-distance applications. As an alternative to light transmission in air, Kao and Hockham suggested in 1966 using optical fibres as a transmission medium. The main obstacle to realizing this idea was the losses of the glass fibres at that time, which were of the order of 1000 dB/km. However, as early as 1970 it was possible to have fibres with an attenuation of 20 dB/km, which was reduced to 0.2 dB/km in 1979.

The possibility of being used for optical communications gave a tremendous impetus to the development of fibre optics. In fact, the advantages of optical fibre communications over conventional two-wire, coaxial cable, or microwave waveguide systems are impressive. On the one hand, using light waves instead of microwaves or radio waves in communication systems is particularly attractive, since the information-carrying capacity of the optical carrier wave is much greater. On the other hand, compared with copper wires, fibre-optic cables provide greater communications capacity with lower loss, besides being more flexible and having a smaller size and weight. Additionally, in contrast with metallic conduction techniques, optical fibres are free from electromagnetic and radio-frequency interference. Finally, optical fibres offer the potential for a low-cost communication line, since the natural resources needed to make them are not scarce—mostly the sand found on beaches.

Fig. 15.11 gives a schematic overview of the basic components of a fibre-optic communications system. At the input end of the communication system, the information to be transmitted, in the form of audio, video, or computer data, is converted by some type of transducer from an electrical signal to an optical one. After transmission along the optical fibre to the receiving location, the optical signal is reconverted into an electrical signal by a photodetector. The electrical signal from the photodetector is then decoded and converted back into a replica of the original source data. Using a number of repeaters to regenerate the optical signal along the transmission fibre may be necessary.

Source data in a fibre-optic communication system can be in analogue or digital form and modulate the light source either directly or indirectly by passing the light through an external modulator. One may choose to modulate the amplitude, the frequency, or the phase of the lightwave. In the case of analogue modulation, the three modulation choices are known as amplitude modulation (AM), frequency modulation (FM), and phase modulation (PM). The same modulation techniques can be applied in the digital case and are called amplitude-shift keying (ASK), frequency-shift keying (FSK), and phase-shift keying (PSK), depending on whether the amplitude, frequency, or phase of the carrier wave is

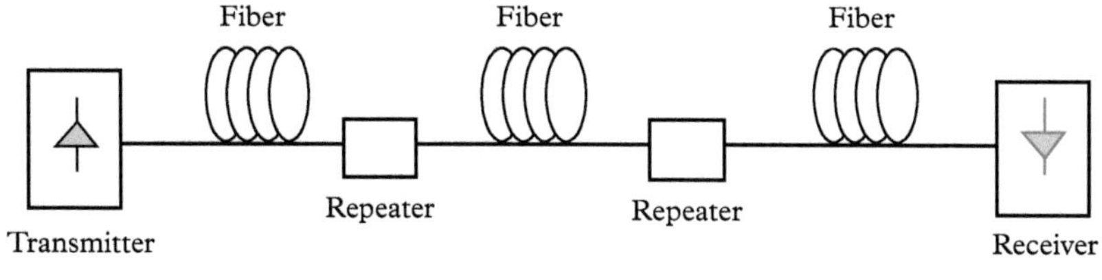

Figure 15.11 *Overview of the basic components of an optical fibre communication system.*

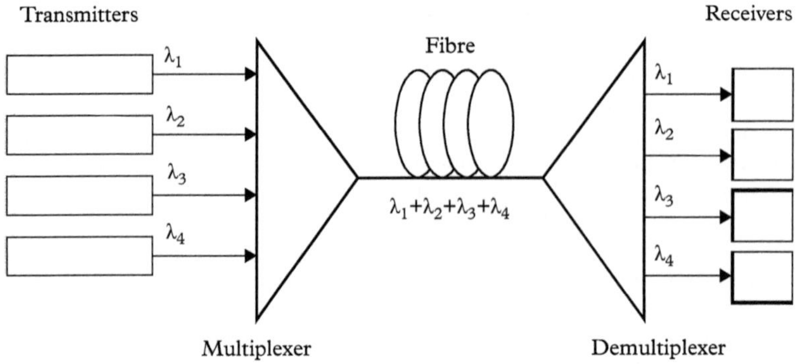

Figure 15.12 *Schematic of a wavelength-division multiplexing optical fibre communication system.*

shifted between the two levels of a binary digital signal. The light source is typically either a light-emitting diode (LED) or a semiconductor laser (SL).

Commonly, the transmission fibre is step-index, either single mode or multimode. An optical fibre allows the transmission of many distinct information channels, each at a slightly different wavelength. The technique is known as wavelength-division multiplexing (WDM), and it has provided a significant expansion of the capacity of optical fibre communication systems. A schematic of the WDM technique is shown in Fig. 15.12.

As the optical signal propagates along the optical fibre, it suffers both *attenuation* (loss of power) and *distortion* (change in shape) due to several mechanisms. If the fibre is very long, it may be necessary to amplify the signal at several positions along the fibre. Until 1995, this task was realized using optoelectronic repeaters, devices in which the optical signal is first converted into electrical form using a receiver and then regenerated using a transmitter. Such regenerators become quite complex and expensive for WDM systems because they require the demultiplexing of individual channels. This task is accomplished by using optical fibre amplifiers, in which the weak light signal is directly amplified within a fibre without requiring conversion to the electrical domain. The amplification is realized based either on the process of stimulated emission or on some non-linear mechanism, as will be discussed in Chapter 22. At the remote end of the fibre, light can be detected with a *p-i-n* photodiode or avalanche photodiode (APD), which reconverts the optical signal into an electrical signal. The signal processing at this stage involves filtering and amplification and, possibly, digital-to-analogue conversion.

Considering the transmission distances, optical fibre networks can be divided into three broad categories. LANs are used to connect different places in a very restricted area, covering distances of $\sim 0.1 - 2$ km. *Metropolitan area networks* (MANs) are used to connect different parts of a city or different regions of a suburban area, encompassing transmission distances that can be as great as ~200 km. Finally, *wide area networks* (WANs) correspond to transmission distances greater than 200 km and can connect cities, countries, or continents.

Several generations of optical fibre transmission systems have been developed during the last four decades, corresponding to the different windows provided by fibre attenuation. The *first generation* was developed in the period 1977–1979 and operated around 850 nm, a spectral region known as the *first window* for fibre transmission. This wavelength was favoured by the availability at that time of very economical gallium arsenide semiconductor lasers. The transmission rate was 45 Mb/s, and the spacing between repeaters was on the order of 10 km.

The *second generation* operated at a wavelength around 1300 nm (the *second window*), where fibre losses were less than 1 dB/km. This wavelength was convenient because the fibre's dispersion is particularly low there, as will be seen in the next chapter. This fact led to a major development effort of both InGaAsP semiconductor lasers and detectors operating at 1.3 μm. The use of single-mode fibres allowed the overcoming of the problems arising from intermodal dispersion and provided the possibility of implementing by 1987 commercial systems operating at transmission rates up to 1.7 Gb/s with repeaters spaced 50 km apart.

The *third generation* operates around 1550 nm (the *third window*), a spectral region where fibres made of fused silica have their global loss minimum (~ 0.2 dB/ km), as seen in Section 15.4. However, the high dispersion in this spectral region was a main problem. This obstacle can be overcome by using dispersion-shifted fibres, which have a minimum dispersion close to 1550 nm, and/or by using semiconductor lasers operating in a single longitudinal mode. The combination of these two conditions allowed the installation of third-generation commercial communication systems in 1990, operating at 2.5 Gb/s.

Fourth-generation communication systems make use of optical amplification instead of electronic repeaters, as well as WDM technology, which increases the transmission rate. Optical amplification was realized using EDFAs, which became commercially available in 1990. These amplifiers operate only in the so-called telecommunications C-band, corresponding to the range 1530–1570 nm.

The *fifth generation* of fibre-optic communication systems makes use of other bands adjacent to the C-band, both at lower wavelengths—the S-band—and at upper wavelengths—the L-band. Due to the limited operating range of EDFAs, new amplification techniques, namely Raman amplification, were developed. On the other hand, in order to increase the spectral efficiency of WDM systems, advanced modulation formats were implemented. In 2010, 64 Tb/s transmission was achieved over 320 km using 640 WDM channels spaced at 12.5 GHz, filling both the C- and L-bands. The successive results reveal a tenfold increase in bit rate every four years, which corresponds to the so-called *optical Moore's law*.

15.8 Optical Fibre Devices

Besides the area of optical communications, optical fibres are also used to realize many important devices and systems, namely optical sensors, fibre amplifiers and lasers, couplers, etc. This section provides a brief description of two kinds of devices using standard optical fibres, namely couplers and sensors. Fibre amplifiers and lasers will be discussed in Chapter 22.

15.8.1 Optical fibre couplers

Among the most important components of an optical system is the fibre-optic directional coupler, depicted in Fig. 15.13.

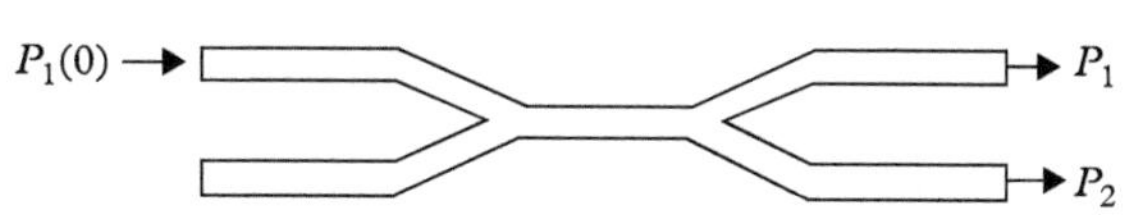

Figure 15.13 *Schematic representation of a fibre-optic coupler.*

The operation of the fibre-optic directional coupler is based on the fact that the guided mode field extends beyond the core–cladding interface. Thus, when the nuclei of two fibres are arranged in parallel and are sufficiently close to each other to provide the overlap of their modal fields, there is a periodic transfer of power between the two fibres. This transfer is incomplete in the more general situation where the propagation constants of the modes of the individual fibres are different. However, complete power transfer between the two fibres can be achieved when their propagation constants are equal. In this case, designating by $P_1(0)$ the power in fibre 1 at the beginning of the coupling region ($z = 0$), the powers in the two fibres at a position $Z > 0$ in that region are given by

$$P_1(z) = P_1(0)\cos^2(\mu z) \quad (15.56)$$

$$P_2(z) = P_1(0)\sin^2(\mu z) \quad (15.57)$$

In the previous equations, μ represents the so-called *coupling coefficient* and is a measure of the magnitude of the interaction between the two fibres. The coupling coefficient depends on the parameters of the fibres, the separation between their nuclei, and the wavelength of the propagated light.

From Eqs. (15.56) and (15.57), it can be seen that for

$$z = \frac{p\pi}{\mu}, \qquad p = 0, 1, 2, 3, \ldots \quad (15.58)$$

one has $P_1(z) = P_1(0)$ e $P_2(0) = 0$. That is, all the power is in the input fibre. On the other hand, for

$$z = (2p+1)\frac{\pi}{2\mu}, \qquad p = 0, 1, 2, 3, \ldots \quad (15.59)$$

one has $P_1(z) = 0, P_2(z) = P_1(0)$, so the power is entirely in the other fibre. The minimum distance at the end of which all the power is transferred from the input fibre to the other fibre is given by

$$z_a = \frac{\pi}{2\mu} \quad (15.60)$$

This distance is known as the *coupling distance*, and its value is typically only a few millimetres.

One of the most important applications of the directional coupler is its operation as a power divider. It can be seen from Eqs. (15.56) and (15.57) that, if the interaction distance of the linear coupler is $L = \pi/4\mu$, then the powers in each of the fibres at the coupler output are

$$P_1(L) = P_2(L) = \frac{1}{2}P_1(0) \quad (15.61)$$

In other words, the directional coupler works in this case as a 50:50 beam-splitter. However, given that the coupling constant μ varies with the wavelength, this result is valid, strictly speaking, only for a certain wavelength.

The variation of the coupling constant with the wavelength is the basis of another very important application of directional couplers in the area of WDM/demultiplexing.

To understand this possibility, consider a directional coupler of length L, consisting of identical fibres and with coupling constants μ_1 and μ_2, corresponding to wavelengths λ_1 and λ_2, respectively, such that

$$\mu_1 L = p\pi \tag{15.62a}$$

and

$$\mu_2 L = (2p+1)\frac{\pi}{2} \tag{15.62b}$$

In these circumstances, suppose that light is launched with these two wavelengths simultaneously into one of the coupler fibres, such that

$$P_1(0) = P_1(\lambda_1, 0) + P_1(\lambda_2, 0) \tag{15.63}$$

In this case, we have, for light with wavelength λ_1, that

$$P_2(\lambda_1, L) = P_1(\lambda_1, 0)\sin^2(\mu_1 L) = 0 \tag{15.64}$$

While, for light with wavelength λ_2, one has

$$P_2(\lambda_2, L) = P_1(\lambda_2, 0)\sin^2(\mu_2 L) = P_1(\lambda_2, 0) \tag{15.65}$$

That is, light with wavelength λ_1 will come out through the input fibre, while light with wavelength λ_2 will come out through the other fibre, as illustrated in Fig. 15.14.

15.8.2 Optical fibre sensors

Optical fibres can also be useful as sensors for the measurement of various physical and chemical quantities: pressure, temperature, electric and magnetic fields, current, rotation, acceleration, displacement, chemical concentration, pH, etc. Fibre-optic sensors have some characteristics that make them particularly attractive: they are immune to electromagnetic interference, they are very versatile, they can provide spatially distributed measurements, they have a very short response time, and information from several sensors can be combined and transmitted over long distances through the optical fibres themselves, etc. Actually, many specialty fibre types are designed for sensing systems. The breadth of sensing applications is a key reason why there are so many types of specialty fibres.

In a typical situation, the light cast on an optical fibre is guided to the desired region. There, some of the properties of the light (intensity, phase, polarization state, or frequency) are modulated by the quantity to be measured, after which the modulated light is sent through the same or another fibre for detection and processing of information. When modulation is performed on the light that propagates in the fibre, one has an *intrinsic* fibre sensor. In the case of this modulation being performed when light travels between two fibres, one has an *extrinsic* fibre sensor. In general, the use of single-mode fibres allows the sensor to have a much higher sensitivity than that provided by multimode fibres.

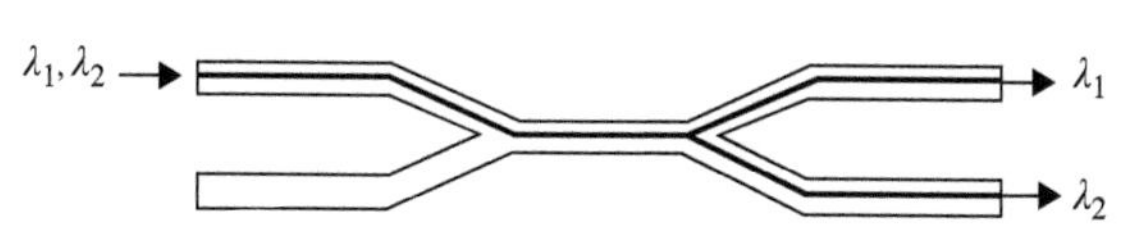

Figure 15.14 *Schematic representation of a fibre-optic coupler used as a wavelength-division demultiplexer.*

A particularly useful feature of fibre-optic sensors is that they can, if required, provide distributed sensing over distances of up to one metre. In contrast, highly localized measurements can be provided by integrating miniaturized sensing elements with the tip of the fibre. These can be implemented by various micro- and nano-fabrication technologies, such that they do not exceed the microscopic boundary of the fibre tip, allowing for such applications as insertion into blood vessels via a hypodermic needle.

Extrinsic fibre-optic sensors use an optical fibre cable, normally a multimode one, to transmit modulated light from either a non-fibre optical sensor or an electronic sensor connected to an optical transmitter. A major benefit of extrinsic sensors is their ability to reach otherwise inaccessible places. They can be used to measure temperature, vibration, rotation, displacement, velocity, acceleration, torque, and torsion. The *fibre-optic gyroscope* (FOG), which was already described in Chapter 9, exploits the Sagnac effect to detect mechanical rotation.

Fig. 15.15 shows a schematic representation of an optical microphone, which is an example of an extrinsic fibre sensor. The microphone membrane is made to vibrate by a sound wave. On the opposite side, this membrane is illuminated by laser light coming from an optical fibre. Another fibre receives the reflected light. Due to the vibration of the membrane, the direction of the reflected light varies, thus determining a modulation of the intensity received by the second fibre. In practice, the dimensions of this sensor can be greatly reduced, allowing it to be used in a wide variety of situations.

The interferometric Mach–Zehnder sensor, depicted in Fig. 15.16, is an example of an intrinsic all-fibre sensor, which has particularly high sensitivity. The light coming from a laser passes through a directional coupler (DC) that acts as a 50:50 beam-splitter. The resulting two beams are guided by two single-mode fibres, which make up the two arms of the Mach–Zehnder interferometer. These two fibres connect to the inputs of a second coupler, identical to the previous one, which allows the overlapping of the two beams and the subsequent detection and processing of the resulting field. One of the arms of the interferometer is made up of the fibre that is sensitive to the quantity to be measured, while the other arm is made up of a fibre that is intended to remain immune to external disturbances. When the quantity to be measured acts on the sensitive fibre, the phase of the light that propagates in it is altered. Since the phase in the other fibre is not affected, the phase difference between the two beams when they overlap at the second coupler will generally determine different values of the powers at the two outputs of that coupler. The measurement of these powers makes it possible to characterize the quantity in question.

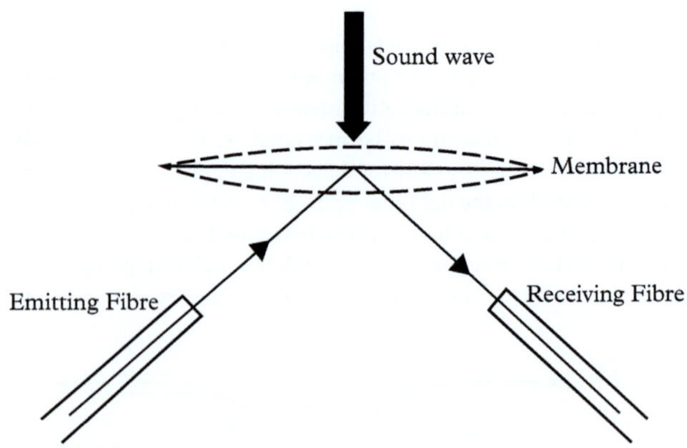

Figure 15.15 *Schematic representation of an optical microphone.*

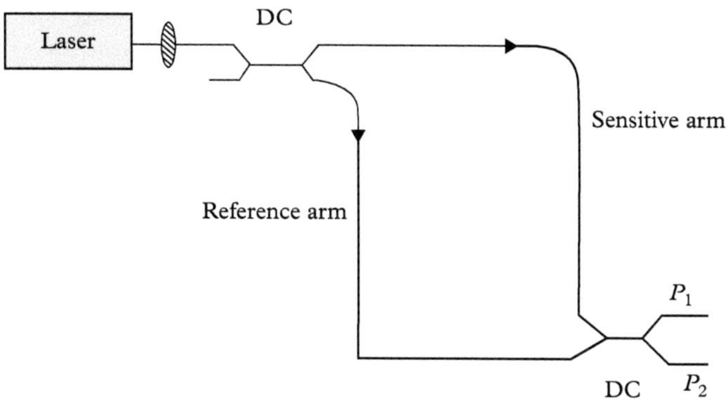

Figure 15.16 *Schematic representation of the Mach–Zehnder interferometric fibre sensor.*

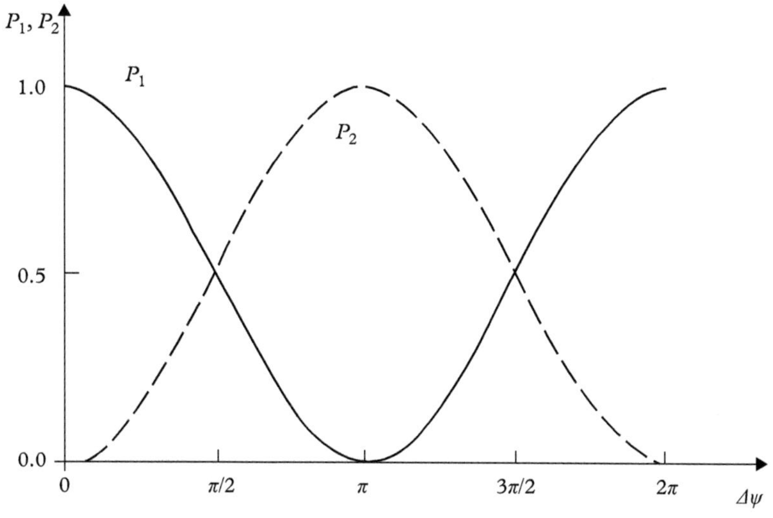

Figure 15.17 *Powers at the output of the Mach–Zehnder interferometer as a function of phase difference.*

Indicating by $\Delta\psi$ the phase difference between the two beams at the input of the second coupler, the powers at the output of this coupler are given by

$$P_1 = P_0 \cos^2 \frac{\Delta\psi}{2} \tag{15.66}$$

$$P_2 = P_0 \sin^2 \frac{\Delta\psi}{2} \tag{15.67}$$

where P_0 is the initial power. If $\Delta\psi = 0$, all the power comes out of fibre 1, while if $\Delta\psi = \pi$, all the power comes out through fibre 2. For other phase difference values, the power is divided between the two outputs. This behaviour is illustrated in Fig. 15.17.

Given that the variations of the phase difference $\Delta\psi$ between the two beams caused by the external disturbance are usually quite small (on the order of milli- or microradians), it is found that if the sensor operates close to an extreme (minimum or maximum) of the powers P_1 and P_2, the resulting modulation of this power will be minimal. The operating point that provides maximum sensitivity corresponds to a phase difference $\Delta\psi = (2p+1)\frac{\pi}{2}$, and it is usually referred to as a *quadrature point.*

Example 15.3

Using a Mach–Zehnder interferometric sensor, it is possible to measure phase variations of the order of 10^{-6} rad. Determine the corresponding variation of fibre length, assuming that the light used has a wavelength $\lambda_0 = 1.0\ \mu$m.

Solution

A phase variation $\Delta\psi$ corresponds to a change in the length of the fibre given by

$$\Delta l = \frac{\Delta\psi}{\left(\frac{2\pi}{\lambda_0}n\right)}$$

Considering a phase change of 10^{-6} rad and assuming that the refractive index of the fibre core is $n = 1.5$, one has

$$\Delta l = \frac{10^{-6}}{\left(\frac{(2)(3.14)}{10^{-6}}(1.5)\right)} \approx 10^{-13}\ \text{m}$$

This result reflects the remarkable sensitivity of the Mach–Zehnder fibre sensor.

15.9 Further Reading

Ferreira, M. F. (Ed), *Photonic Crystal Fibers—Advances in Research and Applications*, NOVA Science Publishers, 2022.

Ferreira, M. F., and M. C. Paul (Eds), *Optical Fiber Technology and Applications: Recent Advances*. Institute of Physics Publishing (IOP), 2021.

Ferreira, M. F., and M. Paul (Eds), *Specialty Optical Fibers—Materials, Fabrication Technology and Applications*, Elsevier, 2024.

Ghatak, A. G., and K. Thyagarathan, *An Introduction to Fiber Optics*, Cambridge University Press, 1998.

Kaiser, A., *Optical Fibre Communications*, 2nd Ed., McGraw-Hill, 1991.

Keck, D. B., 'Optical Fibers', in *Fundamentals of Optical Fiber Communications*, M. K. Barnoski (Ed.) Academic Press, 1981.

Knight, J., T. Birks, P. Russell, and D. Atkin, 'All-Silica Single-Mode Optical Fiber with Photonic Crystal Cladding', *Optics Letters*, **21**, 1547, 1996.

Marcuse, D., 'Loss Analysis of Single-Mode Fiber Splices', *Bell System Technical Journal*, **56**, 703, 1977.

Mitschke, F., *Fiber Optics—Physics and Technology*, Springer, 2009.

Palais, J. C., *Fiber Optic Communications*, Prentice Hall, Englewood Cliffs, 1988.

Senior, J. M., *Optical Fibre Communications: Principles and Practice*, 2nd Ed., Prentice Hall, 1992.

Snyder, A. W., and J. D. Love, *Optical Waveguide Theory*, Chapman & Hall, 1983.

Quimby, R. S., *Photonics and Lasers—An Introduction*, John Wiley & Sons, 2006.

15.10 Problems

15.1 Starting from Eqs. (15.14) and (15.15), derive the cut-off condition for a planar waveguide, as given by Eq. (15.16).

15.2 Consider a symmetric slab waveguide of AlGaAs for which $n_1 = 1.3.60$ and $n_2 = 3.55$. Find the number of modes that can propagate in this waveguide if it has a thickness $d = 10\lambda$.

15.3 A glass waveguide (refractive index $n = 1.5$) has a rectangular section of size d and is bent as shown in Fig. 15.18. The waveguide curvature has a radius R. A beam of collimated light falls perpendicularly on the input end A. Determine the minimum value of the ratio R/d for which all the light entering the glass through surface A emerges through surface B.

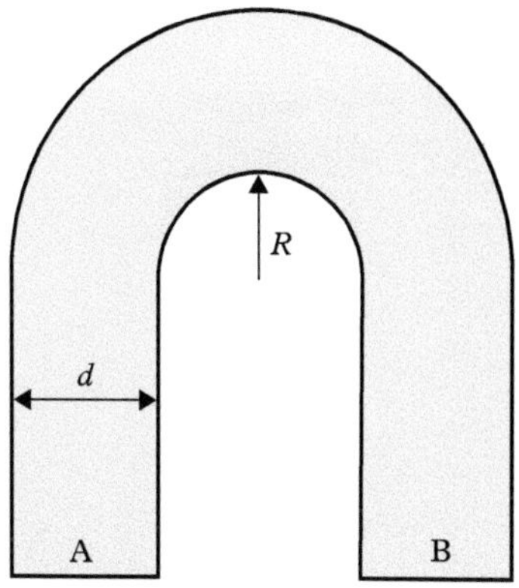

Figure 15.18 *Glass waveguide with a rectangular section of size d and radius of curvature R.*

15.4 The refractive-index difference between the core and the cladding of an optical fibre is 1%. Assuming that the fibre numerical aperture is $NA = 0.18$, calculate the core refractive index.

15.5 Show how Eq. (15.22) follows from Eq. (15.21) under the approximation $n_1 \approx n_c$.

15.6 Consider a step-index fibre whose core has a radius $a = 20$ mm and a refractive index $n_1 = 1.475$, whereas the cladding refractive index is $n_c = 1.460$. Obtain the maximum angle between a guided ray and the fibre axis.

15.7 The cut-off wavelength for a step-index fibre is 1400 nm. How many modes can propagate at a signal wavelength of (a) 1550 nm and (b) 750 nm?

15.8 Considering that the fibre of Problem 15.6 has a length of 5 km, determine (a) the number of reflections experienced by a guided ray propagating according to the angle obtained in that problem and (b) the total attenuation in dB/km, assuming that there is a power attenuation of 0.01% at each reflection.

15.9 A signal of power 4 μW is launched inside an optical fibre 200 m long. Considering that the power arriving at the fibre exit is only 1 μW, find the attenuation constant of the fibre in dB/km.

15.10 An optical fibre cable 4 km long is made up of three 1-km lengths spliced together. Each length has a 3-dB loss, and each splice contributes a 1-dB loss. Find the output signal power if a signal of power 4 mW is launched inside the cable.

15.11 Show that the actual distance a ray travels following the critical angle for total internal reflection during one skip distance L is given by $L/\sin\phi_c$, where $\sin\phi_c = \frac{n_c}{n_1}$

15.12 Using Maxwell's equations, obtain Eqs. (15.40)–(15.43) for the transverse components of the electric and magnetic fields in terms of the longitudinal components E_z and H_z.

15.13 An optical fibre has core index $n_1 = 1.495$, cladding index $n_c = 1.485$, and core diameter 50 μm. Light of free-space wavelength 1.3 μm is coupled into this fibre. Determine the V parameter and the maximum number of modes that can propagate in such a fibre.

15.14 Assuming a fibre with a core refractive index $n_1 = 1.5$ and a free-space wavelength of 1300 nm, find the core diameter such that only one mode can propagate in it.

15.15 Find the mode field diameter, considering a standard telecommunications fibre designed for operation at 1550 nm, having a core diameter 8.4 μm, a core index

$n_1 = 1.49$, and a fractional index difference of 0.004. For what range of wavelengths will the fibre be single mode?

15.16 For the fibre of Problem 15.13 and considering a wavelength of 1550 nm, determine the radial distance from the fibre axis at which the light irradiance is reduced to 5% of the peak (on-axis) value.

15.17 Explain how a directional coupler can be used for wavelength-division multiplexing/demultiplexing.

Dispersion in Optical Fibres

16

In digital optical fibre communication systems, information to be sent is first coded in the form of pulses, which are then transmitted through the fibre to the receiver, where the information is decoded. *Dispersion* refers to the phenomenon of pulse broadening in time as it propagates through the fibre, which becomes an important limiting factor, especially in high bit rate and long-haul transmission systems. When each pulse broadens and overlaps with its neighbours, they eventually become indistinguishable at the receiver input. The effect is known as *inter-symbol interference* and causes an increasing number of errors in the digital optical channel as it becomes more pronounced.

In multimode fibres, the dispersion is primarily due to the different propagation speeds of the various modes, which is known as *intermodal dispersion.* This kind of dispersion can be reduced using *graded-index fibres,* whose cores are designed so that different modes have a more nearly equal transit time down the fibre. A more radical approach is to eliminate all modes but one, that is, to use single-mode fibres. However, although these fibres have no intermodal dispersion, they have other sources of dispersion, namely *material dispersion*, *waveguide dispersion*, and *polarization mode dispersion.* The dispersion that occurs in single-mode fibres is termed *intramodal dispersion.*

16.1 Intermodal Dispersion

Light propagates along a multimode optical fibre in one or more waveguide modes, each mode being characterized by a different propagation angle ϕ, as represented in Fig. 16.1. The information in these modes propagates down the waveguide at the group velocity. The time it takes for a light pulse to propagate a distance L down the waveguide will then depend on angle ϕ.

To analyse the consequences of such dependence, let us consider the schematic representation in Fig. 16.1. Different guided rays in this figure have different trajectories and different travel times along the fibre. The shortest path occurs for $\theta_i = 0$ and is just equal to the fibre length L. The longest path occurs for an angle $\theta_i = \theta_{i,\max}$, given by Eq. (15.20), and it has a length $L/\sin\phi_c$, where $\sin\phi_c = \dfrac{n_c}{n_1}$ is given by Eq. (15.19). Considering that the

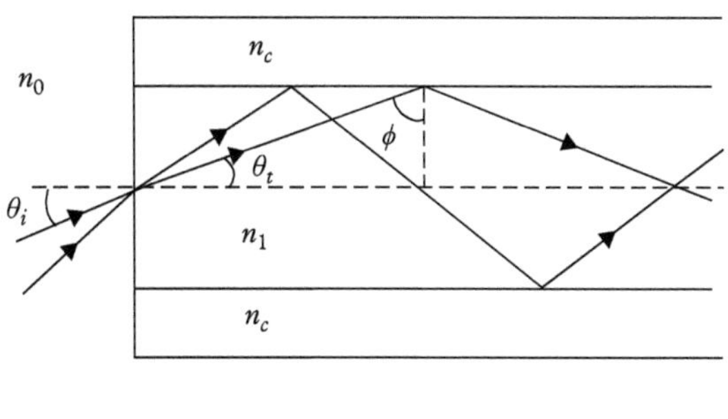

Figure 16.1 *Trajectories of different guided rays in a step-index fibre.*

Lasers, Optics, and Photonics. Mário F. S. Ferreira, Oxford University Press. © Mário F. S. Ferreira (2026).
DOI: 10.1093/oso/9780198952770.003.0016

two rays have the same velocity of propagation $v = c/n_1$, the time delay between them is given by

$$\Delta t = \frac{n_1}{c}\left(\frac{L}{\sin\phi_c} - L\right) \tag{16.1}$$

Assuming $n_1 \approx n_c$, the time delay can be written in the form

$$\Delta t \approx \frac{L}{c}\frac{n_1^2}{n_c}\Delta \tag{16.2}$$

where

$$\Delta = \frac{n_1 - n_c}{n_1} \tag{16.3}$$

is the fractional index difference between the core and cladding. The time delay given by Eq. (16.1) is a measure of the broadening of a pulse launched at the fibre input.

Example 16.1

Determine the temporal pulse spread per unit length for a fibre having a core index $n_1 = 1.46$ and a cladding index $n_c = 1.45$.

Solution

Using Eq. (16.1), the temporal pulse spread per unit length is given by

$$\frac{\Delta t}{L} = \frac{n_1}{c}\frac{n_1 - n_c}{n_c} = \frac{1.46}{3 \times 10^8}\frac{1.46 - 1.45}{1.45} = 34\,\text{ns/km}$$

So the pulse broadens by 34 ns in each km of fibre.

In an optical fibre communication system, the time delay Δt must be less than the bit slot $t_B = 1/B$, where B is the bit rate. This imposes a limit on the *bit rate–distance product*, *BL*, which is frequently used to measure the capacity of an optical communication system. Using Eq. (16.1), one can obtain the following condition:

$$BL < \frac{c}{n_1}\left(\frac{\sin\phi_c}{1 - \sin\phi_c}\right) \tag{16.4}$$

Eq. (16.4) shows clearly the convenience of using a fibre with a low numerical aperture, in order to achieve a value for the bit rate–distance product that is sufficiently high. Typical values of intermodal dispersion are ~ 50 ns/km, which limit the useful propagation range for a 100 Mb/s signal to ~ 100 m.

Example 16.2

Calculate the maximum transmission rate imposed by intermodal dispersion for a communication system using optical fibres with a length $L = 10$ km and refractive indices

i) $$n_1 = 1.5 \text{ e } n_c = 1;$$

continued

Example 16.2 continued

ii) $n_1 = 1.5$ e $n_c = 1.497.$

Compare and comment on the obtained results.

Solution

From Eq. (16.4) one has

$$B < \frac{1}{\Delta t} = \frac{c}{n_1 L}\left(\frac{\sin\phi_c}{1-\sin\phi_c}\right) \tag{16.5}$$

Considering the two suggested cases, we obtain the following results for the maximum transmission rate imposed by intermodal dispersion:

i) $B < 0.04\,\text{Mb/s}$

ii) $B < 10\,\text{Mb/s}$

One can verify that introducing a cladding with a refractive index close to that of the core allows a significant increase in the transmission rate in spite of reducing the numerical aperture of the fibre.

16.2 Graded-Index Fibres

Graded-index fibres were developed in order to minimize the effects of intermodal dispersion. In such fibres, the refractive index of the core is not uniform but decreases from a maximum value (n_1) at its axis to a minimum value (n_c) at the core–cladding boundary. Fig. 16.2 illustrates the profile of the refractive index for this type of fibre.

The refractive index of a graded-index fibre can be described by the expression

$$n(r) = \begin{cases} n_1\left[1-\Delta(r/a)^{\alpha}\right] & r < a \\ n_1(1-\Delta) = n_c & r \geq a \end{cases} \tag{16.6}$$

where a is the core radius, Δ is given by Eq. (16.3), and α is a constant. Fig. 16.3 shows the refractive index variation corresponding to a power-law profile given by Eq. (16.6). The case $\alpha = 2$ corresponds to a parabolic profile of the refractive index. A step-index fibre corresponds to the limit $\alpha \to \infty$.

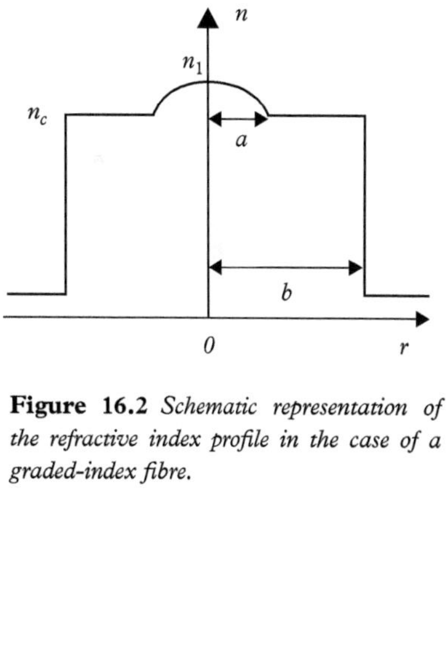

Figure 16.2 *Schematic representation of the refractive index profile in the case of a graded-index fibre.*

In order to describe the propagation of a ray in a graded-index fibre, let us consider the eikonal equation derived in Chapter 6:

$$\nabla L = n\hat{\mathbf{s}} \tag{16.7}$$

where

$$\hat{\mathbf{s}} = \frac{d\mathbf{R}}{|d\mathbf{R}|} \tag{16.8}$$

Figure 16.3 *Refractive index variation corresponding to a power-law profile given by Eq. (16.6). For $\alpha = 2$, the profile is parabolic, whereas for $\alpha \to \infty$, it becomes rectangular (step-index profile).*

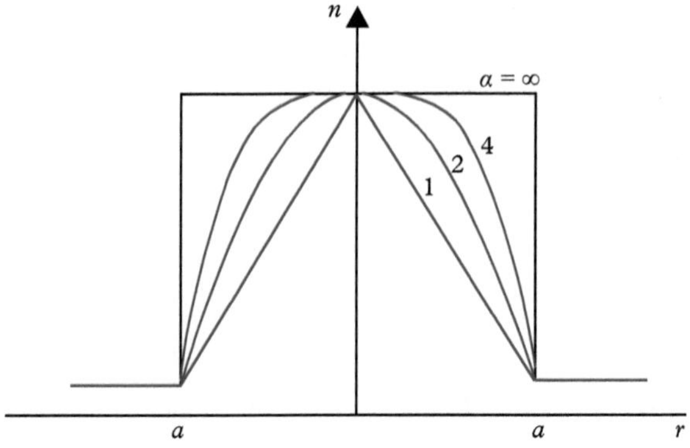

and **R** is the position vector. Writing this vector in cylindrical coordinates and performing the differentiation, one has

$$d\mathbf{R} = dr\hat{\mathbf{u}}_r + rd\phi\hat{\mathbf{u}}_\phi + dz\hat{\mathbf{u}}_z \tag{16.9}$$

Using cylindrical coordinates, Eq. (16.7) gives:

$$|\nabla L|^2 = \left(\frac{\partial L}{\partial r}\right)^2 + \left(\frac{1}{r}\frac{\partial L}{\partial \phi}\right)^2 + \left(\frac{\partial L}{\partial z}\right)^2 = n^2(r) \tag{16.10}$$

In this case, the eikonal can be written in the form:

$$L(r,\phi,z) = \int \left(n^2 - \frac{c_2^2}{r^2} - c_1^2\right)^{1/2} dr + c_2\phi + c_1 z + c_0 \tag{16.11}$$

where c_0, c_1, and c_2 are constants.

From Eqs. (16.7)–(16.11), one can derive the following equations:

$$n\frac{dr}{dR} = \frac{\partial L}{\partial r} = \sqrt{n^2 - \frac{c_2^2}{r^2} - c_1^2} \tag{16.12}$$

$$nr^2\frac{d\phi}{dR} = \frac{\partial L}{\partial \phi} = c_2 \tag{16.13}$$

$$n\frac{dz}{dR} = \frac{\partial L}{\partial z} = c_1 \tag{16.14}$$

In the case of a parabolic profile for the refractive index, assuming that the point of incidence of the ray at the fibre entrance coincides with the origin of the reference system

$(z = 0, r = 0, \phi = 0)$ and that the ray direction makes an angle θ_0 with the z-axis, the constants c_2 and c_1 are given by

$$c_2 = 0,\ c_1 = n_1 \cos\theta_0 \tag{16.15}$$

It can be verified that, in this case, the ray remains in the meridian plane $\phi = 0$.

In order to obtain the ray trajectory, one can divide each member of Eq. (16.12) by the corresponding member of Eq. (16.14), introduce Eq. (16.6) with $\alpha = 2$, and substitute the constants c_2 and c_1 using Eq. (16.15). The following equation is then obtained:

$$\frac{dr}{dz} = \frac{\sqrt{\sin^2\theta_0 - 2\Delta(r/a)^2}}{\cos\theta_0} \tag{16.16}$$

The integration of Eq. (16.16) provides the result:

$$r = h \sin\theta_0 \sin\left(\frac{z}{h\cos\theta_0}\right) \tag{16.17}$$

where $h = a/\sqrt{2\Delta}$. Eq. (16.17) shows that the propagating ray has a sinusoidal trajectory with an amplitude $A = h\sin\theta_0$ and a spatial period $z_p = 2\pi h \cos\theta_0$.

When the initial conditions of the incident ray are not those assumed earlier, the solution for the ray trajectory is more complex than that given by Eq. (16.17). In general, the guided ray is not restricted to the meridian plane, and its trajectory is helical, as seen in Chapter 15.

Fig. 16.4 illustrates the evolution of three meridional guided rays, corresponding to three values of the angle θ_0. The ray propagating along the fibre axis has a shorter path, but its velocity is reduced due to the higher refractive index. On the other hand, a greater value of θ_0 determines a longer path for the guided ray. However, in such circumstances, the ray experiences a lower refractive index in regions far from the fibre axis, where it travels with a greater velocity. As a consequence, a significant reduction of the intermodal dispersion can be expected when using graded-index fibres with an adequate refractive index profile.

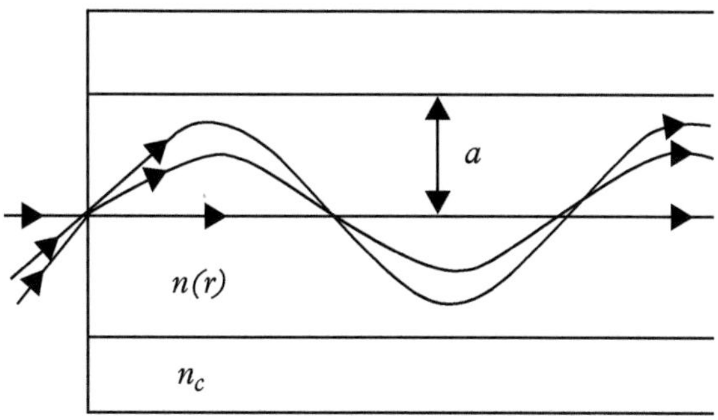

Figure 16.4 *Trajectories of meridional guided rays in a graded-index fibre.*

16.3 Chromatic Dispersion

A medium exhibits *chromatic dispersion* or *group velocity dispersion* (GVD) if the propagation constant β of a wave within it varies non-linearly with frequency. As a consequence, signal distortion occurs as the frequency components of the signal propagate with different velocities. In an optical waveguide, this effect arises from two mechanisms: (i) refractive index variations with wavelength (material dispersion) and (ii) waveguide-related effects (waveguide dispersion).

Let us consider a pulse described by a function $\psi(0,t)$ at $z = 0$. This pulse can be represented as a superposition of harmonic waves:

$$\psi(0,t) = \int_{-\infty}^{\infty} A(\omega)\, e^{-i\omega t} d\omega \tag{16.18}$$

After propagation along a distance z, the spectral component of frequency ω experiences a phase variation $\beta(\omega)z$, where $\beta(\omega)$ is the propagation constant at this frequency. The total field is then given by

$$\psi(z,t) = \int_{-\infty}^{\infty} A(\omega)\, e^{i(\beta(\omega)z - \omega t)} d\omega \tag{16.19}$$

The propagation constant $\beta(\omega)$ is related to the refractive index $n(\omega)$ in the form

$$\beta(\omega) = \frac{\omega n(\omega)}{c} \tag{16.20}$$

One can expand $\beta(\omega)$ in a Taylor series about the carrier frequency ω_0:

$$\beta(\omega) = \beta_0 + \frac{d\beta}{d\omega}(\omega - \omega_0) + \frac{1}{2}\frac{d^2\beta}{d\omega^2}(\omega - \omega_0)^2 + \ldots \tag{16.21}$$

where $\beta_0 = \beta(\omega_0)$,

$$\frac{d\beta}{d\omega} \equiv \beta_1 = \frac{1}{c}\left[n + \omega\frac{dn}{d\omega}\right] = \frac{n_g}{c} = \frac{1}{v_g} \tag{16.22}$$

and

$$\frac{d^2\beta}{d\omega^2} \equiv \beta_2 = -\frac{dv_g/d\omega}{v_g^2} \tag{16.23}$$

In Eq. (16.22), $n_g = n + \omega(dn/d\omega)$ and $v_g = c/n_g$ represent the group refractive index and the group velocity, respectively. The parameter β_2 characterizes the GVD. It determines how much an optical pulse would broaden on propagation inside the fibre.

In the presence of GVD, the information carried by different frequency components of $\psi(0,t)$ propagates at different speeds and thus arrives at different times. The relative delay

Δt_D of arrival time of information at frequencies ω_1 and ω_2, after propagating a distance z, is given by

$$\Delta t_D = \frac{z}{v_g(\omega_1)} - \frac{z}{v_g(\omega_2)} = \frac{\frac{dv_g}{d\omega}(\omega_2 - \omega_1)\, z}{v_g^2} \tag{16.24}$$

Using Eq. (16.23), one has

$$\Delta t_D = \beta_2 z \Delta\omega \tag{16.25}$$

where $\Delta\omega = \omega_1 - \omega_2$. Eq. (16.25) shows that the difference in the arrival time of information is proportional to the GVD, β_2, the difference of the frequency components, $\Delta\omega$, and the propagation distance, z. One can verify that, if $\beta_2 < 0$ (anomalous dispersion regime), higher frequency components of information arrive first, whereas for $\beta_2 > 0$ (normal dispersion regime) the reverse occurs. If the information at different frequency components arrives at different times, the information may be lost. This problem becomes more serious when $\Delta\omega$ is large, as occurs in the case of ultrashort pulses.

In practice, the group dispersion is commonly characterized by a *group delay parameter*, *D*, defined as the delay of arrival time, in ps, for two wavelength components separated by 1 nm over a distance of 1 km. From Eq. (16.25), the group delay may be described by

$$\Delta t_D = \beta_2 c\left(\frac{2\pi}{\lambda_1} - \frac{2\pi}{\lambda_2}\right) z = -\beta_2 c \frac{2\pi}{\lambda^2}\Delta\lambda z = Dz\Delta\lambda \tag{16.26}$$

The parameter D has units of ps/nm/km and is related to β_2 through

$$D = -\frac{2\pi c\beta_2}{\lambda^2}. \tag{16.27}$$

For a typical single-mode fibre, one has $D = 0\,(16\,\text{ps}/(\text{nm}-\text{km}))$ at $\lambda = 1.31\,\mu\text{m}$ $(1.55\,\mu\text{m})$. This indicates that two wavelength components of a pulse separated by 1 nm arrive with a time delay on the order of a few ps when they propagate over a distance of 1 km in a typical fibre.

The effect of dispersion on the bit rate B can be estimated by using the criterion $B\Delta t_D < 1$ in a manner similar to that used in Section 16.1. Using Δt_D from Eq. (16.26), this condition becomes

$$BL|D|\Delta\lambda < 1. \tag{16.28}$$

For a semiconductor laser, the spectral width $\Delta\lambda$ is 2–4 nm even when the laser operates in several longitudinal modes. The *BL* product of such lightwave systems can exceed $100\,(\text{Gb/s}) - \text{km}$. However, the same *BL* product can exceed 1 (Tb/s)-km when single-mode semiconductor lasers are used to reduce $\Delta\lambda$ below 1 nm.

The fibre chromatic dispersion is given by the sum of the material dispersion, D_M, and waveguide dispersion, D_w. The material dispersion D_M is related to the rate of variation of the group index with wavelength, $D_M = c^{-1}\left(dn_g/d\lambda\right)$. In the case of fused silica, $dn_g/d\lambda = 0$ and consequently $D_M = 0$ at $\lambda = \lambda_{ZD} = 1.276\,\mu\text{m}$, which is referred to as the material zero-dispersion wavelength (ZDW).

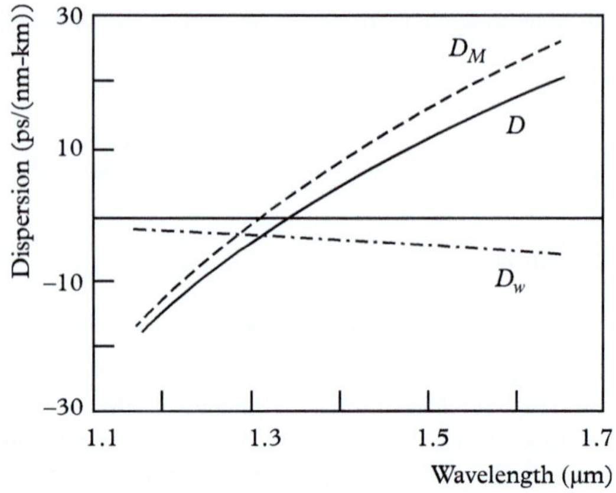

Figure 16.5 *Relative contributions of material dispersion D_M and waveguide dispersion D_w, as well as total dispersion $D = D_M + D_w$ for a typical single-mode fibre.*

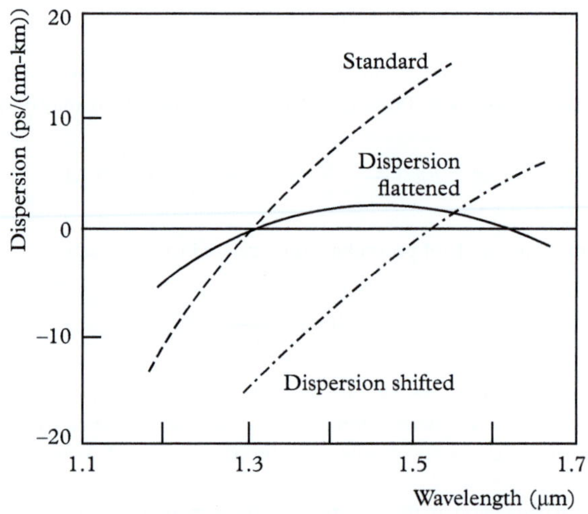

Figure 16.6 *Fibre dispersion as a function of wavelength for several kinds of commercial fibres.*

The waveguide dispersion D_w depends on the fibre parameters, such as the core radius and the index difference between the core and the cladding, and is negative in the range 0–1.6 μm. Fig. 16.5 shows D_M, D_w, and their sum $D = D_M + D_w$ for a typical single-mode fibre. The main effect of waveguide dispersion in a standard single-mode fibre is to shift λ_{ZD} so that the total dispersion is zero near 1.31 μm.

It is possible to design the fibre such that the ZDW λ_{ZD} is shifted to the vicinity of 1.55 μm. Such fibres are referred to as *dispersion-shifted fibres* (DSFs). It is also possible to design the fibre such that the total dispersion D is relatively small over a wide wavelength range extending from 1.3 to 1.6 μm. Such fibres are called *dispersion-flattened fibres*. Fig. 16.6 illustrates the fibre dispersion as a function of wavelength for several kinds of commercial fibres.

16.3.1 Unchirped input pulses

Let us consider the case of an input pulse with a Gaussian profile:

$$E(0,t) = E_0 \exp\left\{-\frac{1}{2}\left(\frac{t}{t_0}\right)^2\right\} \exp\{-i\omega_o t\} \tag{16.29}$$

where E_0 is the peak amplitude, ω_o is the carrier frequency, and the parameter t_0 represents the half width at the 1/e-intensity point. It is related to the full width at half maximum (FWHM) by the relation

$$t_{FWHM} = 2(\ln 2)^{1/2} t_0 \approx 1.665 t_0 \tag{16.30}$$

The amplitude spectrum of this pulse is obtained by calculating its Fourier transform:

$$E(0,\omega) = \frac{1}{2\pi}\int_{-\infty}^{\infty} E_0 \exp\left\{-\frac{1}{2}\left(\frac{t}{t_0}\right)^2\right\} \exp\{i(\omega-\omega_0)t\}dt = \frac{E_0 t_0}{\sqrt{2\pi}} \exp\left\{-\frac{t_0^2(\omega-\omega_o)^2}{2}\right\} \tag{16.31}$$

Each spectral component of (16.31) is propagated over a distance z in the medium with propagation constant $\beta(\omega)$ and acquires a phase variation $\beta(\omega)z$. Thus, the pulse at z is found through the inverse Fourier transform of $E(0,\omega)$ after propagation:

$$E(z,t) = \int_{-\infty}^{\infty} E(0,\omega) \exp\{i[\beta(\omega)z - \omega t]\}\, d\omega \tag{16.32}$$

Using the expansion of the propagation constant given by Eq. (16.21), and retaining terms up to second order in $(\omega - \omega_0)$, one obtains:

$$E(z,t) = \frac{E_0}{(1+\sigma^2)^{1/4}} exp\left\{-\frac{(t - z/v_g)^2}{2t_p^2(z)}\right\} \exp\{i[\beta_0 z - \phi(z,t)]\} \tag{16.33}$$

where

$$\phi(z,t) = \omega_o t + \kappa\left(t - \frac{z}{v_g}\right)^2 - \frac{1}{2}\tan^{-1}(\sigma) \tag{16.34}$$

$$\kappa = \frac{\sigma}{2(1+\sigma^2)t_o^2} \tag{16.35}$$

$$\sigma = \frac{\beta_2 z}{t_o^2} \tag{16.36}$$

$$t_p^2(z) = t_o^2(1+\sigma^2) \tag{16.37}$$

Eq. (16.37) shows that the half width of the pulse power $t_p(z)$ increases with the propagation distance by an amount that is governed by the GVD coefficient β_2. The *dispersion length*, L_D, is defined as the distance over which a Gaussian pulse broadens to $\sqrt{2}$ times its initial width t_0. Considering Eqs. (16.35) and (16.36), it can be seen that

$$L_D = \frac{t_0^2}{|\beta_2|} \tag{16.38}$$

The broadening effect due to fibre dispersion is illustrated in Fig. 16.7, which shows the Gaussian pulse profile at the input (dashed curve) and after propagation over a distance $z = 3L_D$ (full curve).

It is apparent from Eqs. (16.36) and (16.37) that an optimum initial pulse width exists for which the output width is minimized. By substituting (16.36) into (16.37) and differentiating with respect to t_0, one finds that the optimum initial pulse width is given by $t_{opt} = \sqrt{\beta_2 z}$.

Eq. (16.34) shows that the phase of the pulse varies with the square of time. As a consequence, the instantaneous frequency also varies with time, and the pulse is said to be *chirped*. The instantaneous frequency is found by taking the time derivative of the phase:

$$\omega(t) = \omega_0 + 2\kappa\left(t - \frac{z}{v_g}\right) \tag{16.39}$$

This result shows that in the normal dispersion regime ($\kappa > 0$), the instantaneous frequency at a fixed position z increases with time (positive linear chirp). As a consequence, the leading edge of the pulse is red-shifted, whereas the trailing edge is blue-shifted. On the other hand, in the anomalous dispersion regime ($\kappa < 0$) the instantaneous frequency decreases with time (negative chirp). In this case, the leading edge of the pulse is blue-shifted, and the trailing edge becomes red-shifted. Fig. 16.8 illustrates the pulse chirping in the normal and anomalous dispersion regimes.

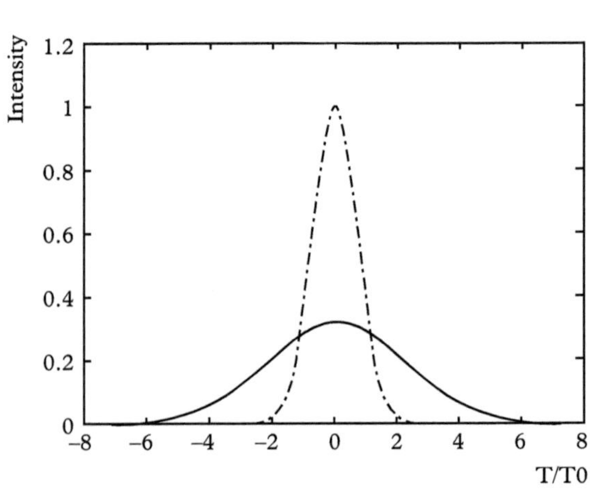

Figure 16.7 *Broadening of a Gaussian pulse due to fibre dispersion. The dashed curve corresponds to the initial pulse profile, whereas the solid curve gives the pulse profile after propagation over a distance $z = 3L_D$.*

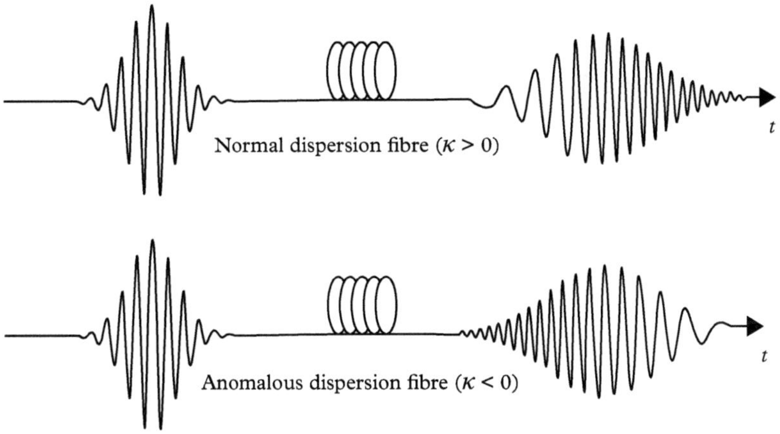

Figure 16.8 *Illustration of the pulse chirp in the cases of anomalous and normal dispersion.*

The rate of frequency change with time over the pulse envelope is found by differentiating (16.39) with respect to time:

$$\frac{\partial \omega}{\partial t} = 2\kappa = \frac{\sigma}{t_p^2} \tag{16.40}$$

where we identify the *dimensionless linear chirp parameter* σ.

From Eq. (16.33), one finds that the peak intensity of the broadened pulse is

$$|E'_0|^2 = \frac{|E_0|^2}{(1+\sigma^2)^{1/2}} \tag{16.41}$$

From Eqs. (16.33) and (16.41), one sees that the pulse energy (proportional to $|E'_0|^2 t_p$) is independent of position, as expected.

16.3.2 Chirped input pulses

When a pulse having an initial chirp is input into a dispersive medium, the pulse is either further broadened or compressed, depending on the sign of the dispersion. Let us consider a chirped Gaussian input pulse of width t_0 and chirp C_0, whose electric field is given by

$$E(0,t) = E_0 \exp\left\{-\frac{1+iC_0}{2}\left(\frac{t}{t_0}\right)^2\right\}\exp(-i\omega_0 t) \tag{16.42}$$

The pulse spectrum is given by the Fourier transform of (16.42), which is

$$E(0,\omega) = \frac{E_0 t_0}{\sqrt{2\pi(1+iC_0)}}\exp\left\{-\frac{t_0^2(\omega-\omega_o)^2}{2(1+iC_0)}\right\} \tag{16.43}$$

Using Eq. (16.43), and considering that the propagation constant is given by the expansion (16.21) limited to second order in $(\omega - \omega_0)$, the output pulse field is obtained by calculating the inverse Fourier transform, as indicated by Eq. (16.32). The result can be presented in the form:

$$E(z,t) = \tilde{E}_0 \exp\left\{-\frac{(1+iC)(t-z/v_g)^2}{2t_p^2(z)}\right\} \exp\{i[\beta_0 z - \omega_0 t]\} \tag{16.44}$$

where

$$\tilde{E}_0 = \frac{E_0 t_0 [1 + i\sigma(1 - iC_0)]^{1/2}}{t_p} \tag{16.45}$$

$$C = C_0 + \sigma(1 + C_0^2) \tag{16.46}$$

$$t_p^2 = t_0^2\left[(1 + C_0\sigma)^2 + \sigma^2\right] \tag{16.47}$$

The quantity $(1 + C^2)/t_p^2$ corresponds to the spectral width of the pulse, and it can be shown (Problem 16.8), from Eqs. (16.46) and (16.47), that it remains constant and equal to its initial value $(1 + C_0^2)/t_0^2$. On the other hand, it can be seen from Eqs. (16.36) and (16.47) that the input pulse broadens monotonously during propagation if C_0 and β_2 are of the same sign. However, if these parameters are of opposite sign, the pulse compresses until a given location $z_{\min}$, after which it starts to broaden. This behaviour is illustrated in Fig. 16.9 for the case $\beta_2 < 0$.

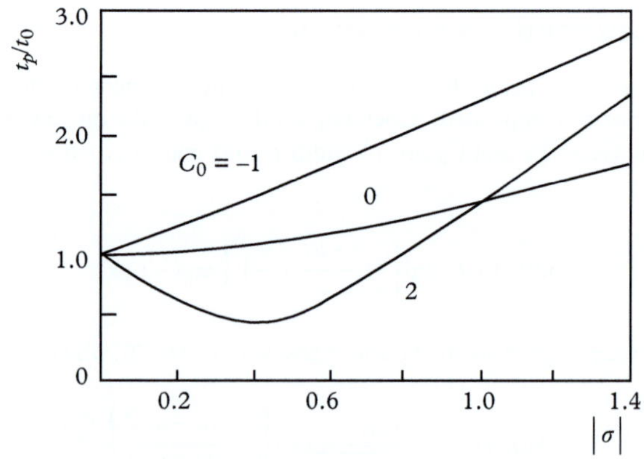

Figure 16.9 *Evolution of the pulse width with the normalized distance $|\sigma|$ for three values of the chirp parameter C_0.*

Example 16.3

Assuming the case $C_0\beta_2 < 0$, derive expressions for the minimum pulse width and the fibre length at which such a minimum occurs.

Solution

The location z_{min} of maximum compression is found from Eq. (16.46) by setting $C = 0$:

$$z_{min} = \left|\frac{C_0}{\beta_2}\right| \frac{t_0^2}{1 + C_0^2} \quad (16.48)$$

Substituting Eq. (16.48) into Eq. (16.47), the following result is obtained for the minimum pulse width:

$$t_{min} = \frac{t_0}{\sqrt{1 + C_0^2}} \quad (16.49)$$

16.3.3 Dispersion compensation

As mentioned earlier, standard single-mode fibres have a ZDW of 1.31 μm. On the other hand, most of the installed communication systems using these fibres operate at 1.55 μm, where the fibre loss is minimum but the typical dispersion is +16 ps/km-nm. In the presence of this dispersion, the optical pulses will suffer from broadening, and thus, before they completely overlap, they have to be processed. Such a task can be realized optically using the concept of *dispersion compensation.*

When pulses at 1.55 μm propagate through a conventional fibre, they experience anomalous dispersion and their leading edges are blue-shifted, whereas the trailing edges are red-shifted. If such pulses are then propagated through an appropriate length of a fibre having normal dispersion ($D < 0$), they can be compressed to their original width. Such a fibre is called a *dispersion-compensating fibre* (DCF). If the transmission and compensating fibres have dispersions D_1 and D_2 and lengths L_1 and L_2, respectively, the requirement for complete dispersion compensation at a given wavelength λ is

$$D_2 L_2 = -D_1 L_1 \quad (16.50)$$

Dispersion compensation using DCFs can be realized in three schemes, as illustrated in Fig. 16.10. In the first configuration, known as *precompensation,* the dispersion accumulated over the entire link is compensated at the transmitter end. In the second configuration, known as *postcompensation,* a DCF of appropriate length is placed at the receiver end. In the third configuration, known as *periodic compensation,* dispersion is compensated periodically along the link.

One handicap of using a DCF for dispersion compensation is the added loss associated with the increased fibre span. Concerning this limitation, a useful figure of merit is the ratio of the DCF dispersion magnitude to its loss in dB/km, expressed in units of ps/nm/dB. Clearly, a high value of this parameter is desired. Another disadvantage is related to non-linear effects, which may degrade the signal over the long length of the fibre if its intensity is relatively high. This problem arises because the effective core area of a DCF is usually much smaller than that of a conventional single-mode fibre.

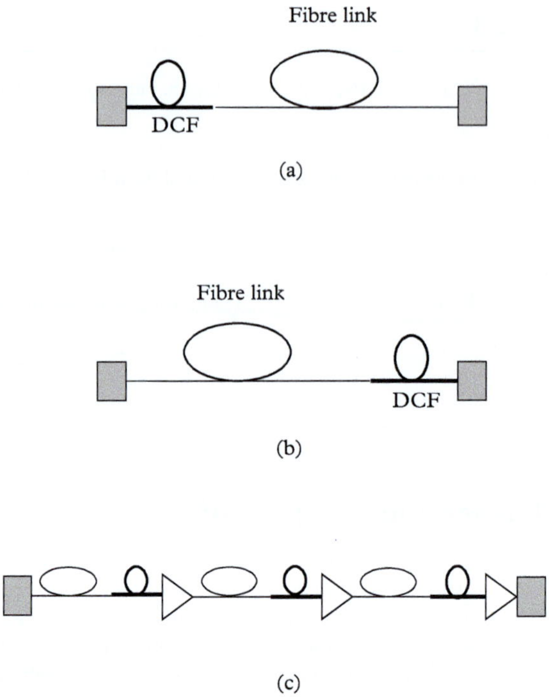

Figure 16.10 *Schematic of three dispersion-compensation schemes: (a) precompensation, (b) postcompensation, and (c) periodic compensation.*

Example 16.4

Let us consider a multi-channel system, in which there are a number of wavelengths propagating simultaneously through the fibre. Assuming that Eq. (16.50) is satisfied for a channel of wavelength λ, find the condition to be satisfied for dispersion compensation to occur at a nearby wavelength $\lambda + \Delta\lambda$.

Solution

Assuming that Eq. (16.50) is satisfied at wavelength λ, the residual dispersion at a nearby wavelength $\lambda + \Delta\lambda$ is given by

$$\text{Dispersion} = D_1(\lambda + \Delta\lambda) L_1 + D_2(\lambda + \Delta\lambda) L_2 \approx D_1' L_1 \Delta\lambda \left(1 - \frac{D_2'/D_2}{D_1'/D_1}\right) \tag{16.51}$$

where the prime denotes differentiation with respect to wavelength. Thus, for dispersion compensation over a band of wavelengths, the relative dispersion slopes D'/D of the two fibres must be equal:

$$\frac{D_2'}{D_2} = \frac{D_1'}{D_1} \tag{16.52}$$

If the condition (16.52) is not satisfied, the dispersion may accumulate and reach unacceptable levels in multi-channel systems.

16.4 Chromatic Dispersion of Microstructured Optical Fibres

The dispersive properties of microstructured optical fibres (MOFs) are quite sensitive to their structural parameters, such as the air-hole diameter d and the hole-to-hole spacing Λ. By changing each of these parameters, the waveguide dispersion and, consequently, the overall dispersive properties can be adjusted conveniently. This possibility is illustrated in Fig. 16.11, which shows the dispersion characteristics of a solid-core MOF with a hexagonal pattern of holes (see Fig. 15.8) spaced by (a) $\Lambda = 3\,\mu$m, (b) $\Lambda = 2$ µm, and (c) $\Lambda = 1$ µm, at $\lambda = 1.55$ µm for different values of the hole-pitch ratio.

In a solid-core MOF, as the holes become larger, the core becomes increasingly isolated until it resembles an isolated strand of silica glass, also known as a jacketed air-suspended rod (JASR). The core diameter of a MOF is given by $d_{core} = 2\Lambda - d$. As observed for standard fibres, MOFs with larger cores exhibit semi-infinite anomalous dispersion above the

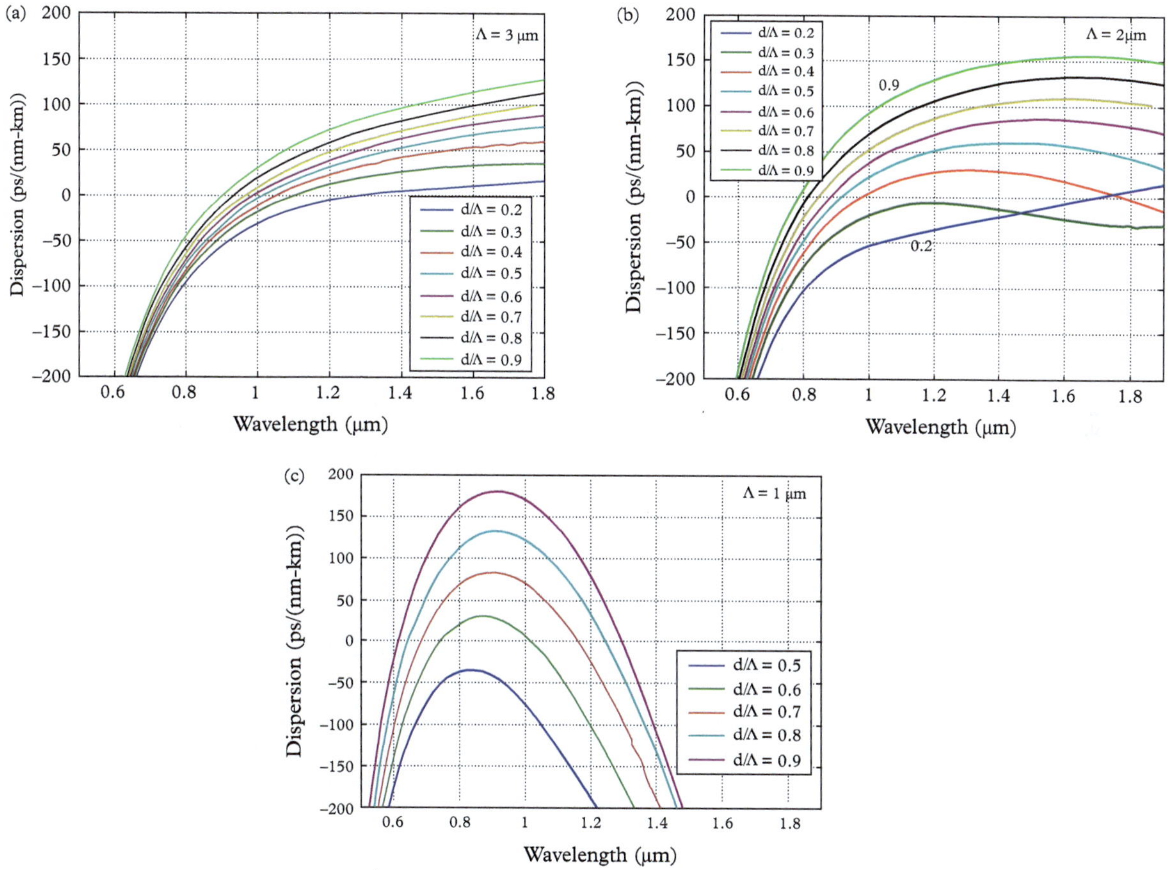

Figure 16.11 *Dispersion properties of a microstructured fibre with a hexagonal pattern of holes spaced by (a) $\Lambda = 3$ µm, (b) $\Lambda = 2$ µm, and (c) $\Lambda = 1$ µm at $\lambda = 1.55$ µm for different values of the hole-pitch ratio.*

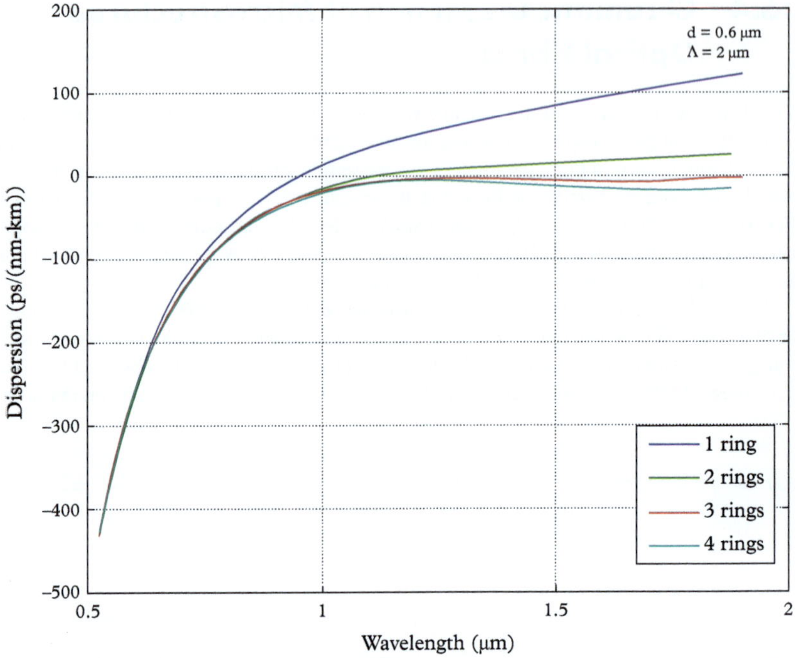

Figure 16.12 *Dispersion properties of a microstructured fibre with a hexagonal pattern of holes spaced by* $\Lambda = 2\,\mu m$ *and hole diameter* $d = 0.6\,\mu m$ *at* $\lambda = 1.55\,\mu m$ *for different numbers of hole rings around the fibre core.*

ZDW. By decreasing the core size, the ZDW tends to shift to a shorter wavelength, leading to anomalous dispersion at near-infrared and visible wavelengths. When the core size is decreased further, a second ZDW arises on the longer-wavelength side such that the GVD is anomalous in the spectral window between the two ZDWs and normal outside it. This situation can be observed in some curves of Figs. 16.11(b) and 16.11(c).

A nearly zero dispersion and flattened behaviour is observed in some cases of Fig. 16.11. This characteristic can be further controlled by adjusting the number of hole rings around the core of a MOF, as illustrated in Fig. 16.12 for the case $\Lambda = 2\,\mu m$ and $d = 0.6\,\mu m$. The availability of fibres with nearly zero dispersion is particularly important for supercontinuum generation and other telecom applications making use of non-linear effects.

16.4.1 Hollow-core kagome fibres

As already mentioned in Chapter 15, hollow-core (HC) MOFs offer the additional possibility of filling the fibre core with a suitable liquid or gas. The case of a kagome HC-MOF filled with a noble gas will be considered in the following. The effective modal refractive index, n_{eff}, of kagome HC-MOFs can be approximated by that of a glass capillary, which is given by Im et al. (2009):

$$n_{eff} = n_{gas} - \frac{1}{2}\left(\frac{u_{01}\lambda}{\pi d_f}\right)^2 \tag{16.53}$$

where d_f is the core diameter, u_{01} is the first zero of the J_0 Bessel function, and λ is the light wavelength. In Eq. (16.53), n_{gas} is the refractive index of the core-filling gas, which depends on temperature and can be obtained from the following *Sellmeier formula* (Börzsönyi et al. 2008):

$$n_{gas} = \sqrt{1 + \frac{p}{p_0}\left(\frac{b_1\lambda^2}{\lambda^2 - c_1} + \frac{b_2\lambda^2}{\lambda^2 - c_2}\right)} \qquad (16.54)$$

In Eq. (16.54), p is the gas pressure, p_0 = 1 bar, and the coefficients b_i and c_i depend on the particular noble gas (Table 16.1).

The propagation constant β of an optical pulse is related to the effective modal refractive index as given by Eq. (16.20).

The kagome MOF provides ultra-broadband guidance at low loss levels and, when evacuated, exhibits weak anomalous GVD over the entire transmission window. However, when filled with a noble gas, the normal GVD of the gas can be balanced against the anomalous GVD of the fibre, allowing the ZDW to be tuned across the ultraviolet, visible, and near-infrared spectral regions. This can be observed from Fig. 16.13, which shows (a) the dependence of the GVD parameter $\beta_2 = d^2\beta/d\omega^2$ on the wavelength for a kagome MOF with a 40 μm core diameter filled with helium at various pressures and (b) the GVD parameter for helium, argon, and xenon at 10 bar. At pumping wavelengths lower than the ZDW, the dispersion regime of these fibres is normal, whereas at pumping wavelengths higher than λ_{ZD} the dispersion regime is anomalous. Therefore, we can change the dispersion profile and the position of the ZDW by adjusting the pressure or by changing the filling gas. Compared with a solid-core MOF, we observe that the dispersion magnitude is significantly smaller in a kagome HC-MOF, which means that ultrafast pulses broaden much more slowly in this case.

Fig. 16.14 shows the ZDW as a function of pressure for the fundamental mode of a kagome MOF with a core diameter of 20 μm, filled with three different gases: xenon, argon, and helium. As the pressure is raised from 1 to 150 bar, the ZDW tunes smoothly from ~400 to 1400 nm in the case of xenon, whereas it varies from 300 to 950 nm for the argon-filled core. Considering a pump wavelength of 790 nm, the dispersion in this case is anomalous (normal) below (above) 65.8 bar. However, in the case of a helium-filled core, and considering the same pump wavelength, the dispersion is anomalous for any pressure in the range 1–150 bar.

Table 16.1 *Sellmeier coefficients for various gases (values obtained from Börzsönyi et al. 2008).*

	argon	helium	Xenon
b1	*b1=20332.29∗1e-8*	*4977.77∗1e-8*	*103701.61∗1e-8*
c1	*c1=206.12∗1e-6*	*28.54∗1e-6*	*12.75∗1e-6*
b2	*b2=34458.31∗1e-8*	*1856.94∗1e-8*	*31228.61∗1e-8*
c2	*c2=8.066∗1e-3*	*7.760∗1e-3*	*0.561∗1e-3*

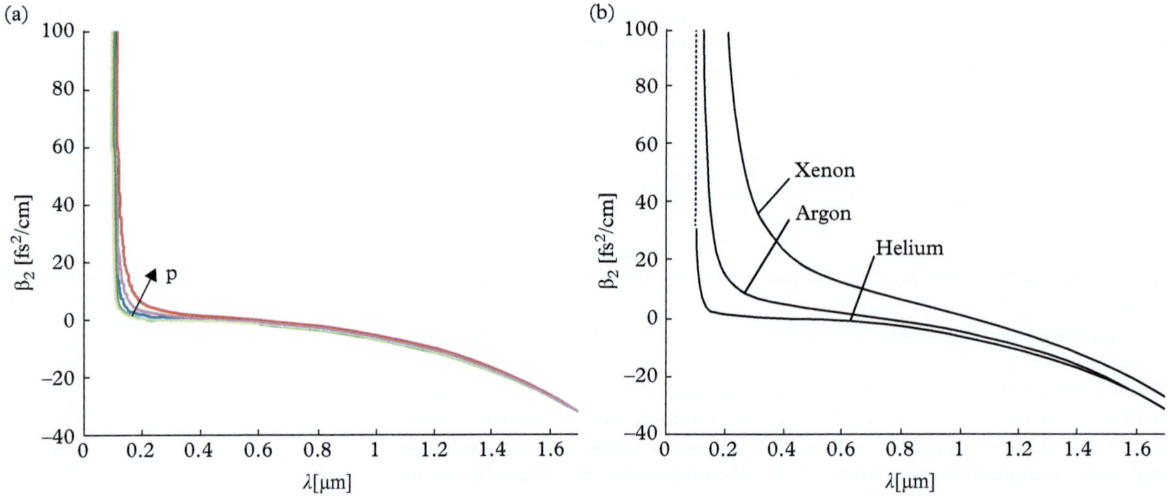

Figure 16.13 *(a) Group velocity dispersion for the helium-filled kagome MOF for p = 10.0, 20.0, 40.0, and 80.0 bar; (b) group velocity dispersion at p = 10 bar for helium, argon, and xenon.*

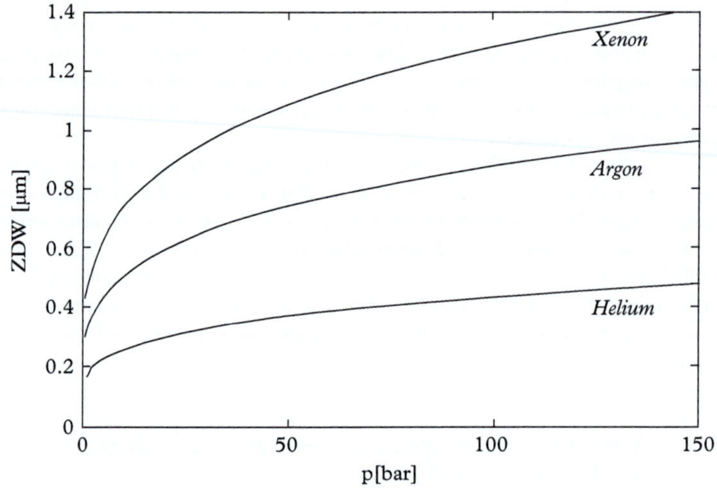

Figure 16.14 *Zero-dispersion wavelength (ZDW) as a function of pressure for the fundamental mode of a kagome MOF with a core diameter of 20 μm, filled with three different gases: xenon, argon, and helium.*

16.5 Polarization Mode Dispersion

Another pulse-broadening mechanism that can occur in single-mode fibres, besides chromatic dispersion, is related to *polarization mode dispersion* (PMD). PMD originates in the optical fibre birefringence, which causes a difference in the delays corresponding to different polarizations of the propagating pulses. When this difference approaches a

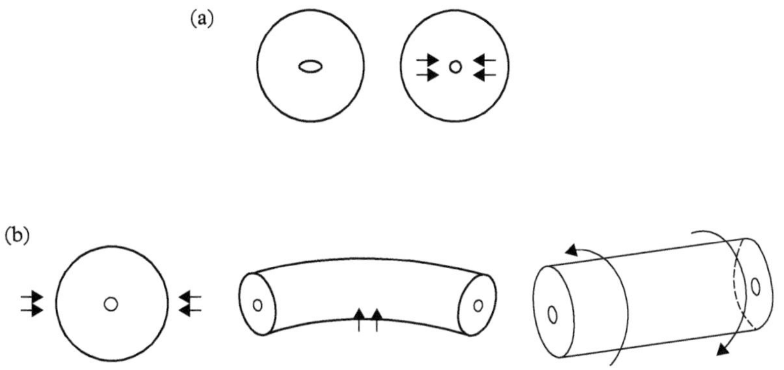

Figure 16.15 *Different kinds of birefringence mechanisms that can be found in single-mode fibres: (a) intrinsic mechanisms and (b) extrinsic mechanisms.*

significant fraction of the bit period, pulse distortion and system penalties occur. Limitations due to PMD become particularly significant in high-speed, long-haul optical fibre transmission systems.

The mechanisms inducing birefringence in optical fibres can be divided into two main categories: intrinsic and extrinsic. Intrinsic mechanisms are related to the manufacturing process and constitute a permanent feature of the fibre. The non-circular geometry of the core and a non-symmetrical stress field in the glass around the core are two examples, as represented in Fig. 16.15(a). Birefringence can also be created in a fibre affected by external forces during handling or cabling. Fig. 16.15(b) shows three examples of the so-called extrinsic mechanisms: lateral stress, bending, and twisting. Due to the extrinsic mechanisms, fibre birefringence changes randomly over time, reflecting the environmental conditions. This effect is responsible for the stochastic behaviour of PMD.

16.5.1 Fibre birefringence and the intrinsic PMD

In a birefringent fibre, the effective mode index varies continuously with the field orientation angle in the transverse plane. The directions that correspond to the maximum and minimum mode indices are orthogonal and define the *principal axes* of the fibre. Let us assume that these principal axes coincide with the x and y axes. A linearly polarized field along x or y has a propagation constant $\beta_x = n_x\omega/c$ or $\beta_y = n_y\omega/c$, where n_x and n_y are the effective mode indices associated with x and y polarizations, respectively. The fibre *birefringence* is given by

$$\Delta\beta = \beta_x - \beta_y = \frac{\omega}{c}\Delta n_{eff} \tag{16.55}$$

where $\Delta n_{eff} = n_x - n_y$. Typically, Δn_{eff} range between 10^{-7} and 10^{-5}, which is much smaller than the index difference between the core and the cladding $\left(\sim 3 \times 10^{-3}\right)$.

Let us consider an input field given by

$$\mathbf{E}(z) = \exp(i\beta_x z)\left[E_{0x}\hat{\mathbf{i}} + E_{0y}\exp(-i\Delta\beta z)\hat{\mathbf{j}}\right] \tag{16.56}$$

where E_{0x} and E_{0y} are the input field amplitudes along the principal axes. Eq. (16.56) describes a wave whose polarization state transforms from linear to elliptical and then returns to the original linear state when $\Delta\beta z = 2\pi$. The spatial period of this evolution process, known as the *beat length*, is given by

$$L_b = \frac{2\pi}{\Delta\beta} = \frac{\lambda}{\Delta n_{eff}} \tag{16.57}$$

Eqs. (16.55) and (16.57) show that both the birefringence and the beat distance depend on the frequency. Therefore, the output polarization state also varies in a cyclic manner with frequency for a fixed fibre length. Changes in frequency or distance thus produce equivalent effects on the polarization state.

A different behaviour occurs when the input field is polarized along one of the two principal axes. In this case, the polarization remains along the given principal axis for the entire span, even if the frequency changes. This behaviour provides the basis for the definition of what are called the *input principal polarization states* of the fibre. Such states correspond to the input polarizations that lead to output polarization states that are *invariant with frequency*.

The difference in the local propagation constants indicated by Eq. (16.55) is usually accompanied by a difference in the local group velocities for the two polarization modes. This differential group velocity is described by a differential group delay (DGD), $\Delta\tau$, per unit length between the two modes, often referred to as *intrinsic polarization mode dispersion* (PMD_i), given by

$$\text{PMD}_i = \frac{\Delta\tau}{L} = \frac{d}{d\omega}(\Delta\beta) = \frac{1}{c}\left(\Delta n_{eff} + \omega\frac{d\Delta n_{eff}}{d\omega}\right) \tag{16.58}$$

The quantity PMD_i is often expressed in units of picoseconds per kilometre of fibre length. The linear length dependence of DGD applies only when the birefringence can be considered uniform, as in a short fibre. Fig. 16.16 illustrates the effect of PMD when a signal is introduced into such a short fibre with equal components along the two principal axes. DGD can be interpreted as the effective pulse spread corresponding to the group delay difference between the two orthogonally polarized components.

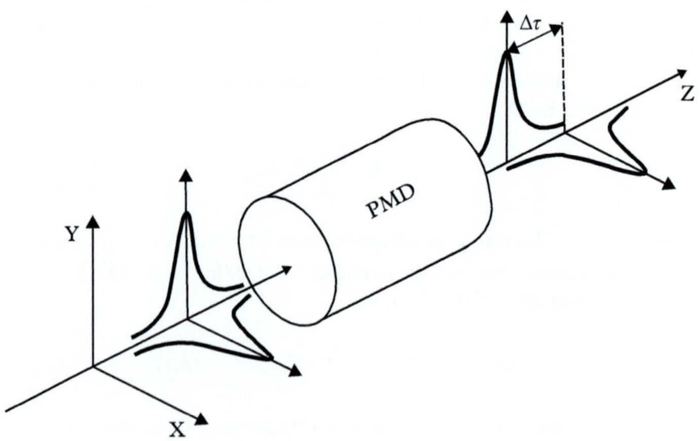

Figure 16.16 *Effect of PMD when a signal is introduced into a short fibre with equal components along the two principal axes.*

From Eq. (16.58), and ignoring the dispersion of Δn_{eff}, one can see that the DGD for a single beat length is given by

$$\Delta\tau_b = L_b \frac{\Delta n_{eff}}{c} = \frac{\lambda}{c} = \frac{1}{f} \tag{16.59}$$

which corresponds to an optical cycle. At $\lambda = 1.55\,\mu\text{m}$, one has $\Delta\tau_b = 5.2\,\text{fs}$.

The equivalence between changes in frequency and distance in affecting polarization provides a method of measuring PMD. Over a distance L, the two principal polarizations establish a phase difference given by

$$\Delta\phi = \Delta\beta L = \frac{\omega}{c}\Delta n_{eff} L \tag{16.60}$$

On the other hand, the change in frequency $\Delta\omega$ producing an additional phase shift of 2π is such that

$$\Delta\phi + 2\pi = \frac{\omega + \Delta\omega}{c}\left(\Delta n_{eff} + \Delta\omega \frac{dn_{eff}}{d\omega}\right) L \tag{16.61}$$

Carrying out the product in Eq. (16.61), neglecting the term of the order of $(\Delta\omega)^2$ and subtracting (16.60), one can obtain the result

$$\Delta\tau \approx \frac{2\pi}{\Delta\omega} \tag{16.62}$$

where Eq. (16.58) was also used. Eq. (16.62) provides a direct relation between the measurement of polarization evolution in the frequency domain and the time-domain effects of PMD.

16.5.2 PMD in long fibre spans

The assumption of constant birefringence and polarization axis orientation is no longer valid in the case of long fibre spans. In fact, a long span is likely to possess a very complicated progression of features along its length that will modify the polarization of light. As a consequence, even if the light is linearly polarized along a principal axis of the fibre at the input, it couples into the orthogonal polarization during its progress along the fibre. Long fibres are often modelled as a concatenation of a large number of segments, as shown schematically in Fig. 16.17. Both the degree of birefringence and the orientation of the principal axes remain constant in each segment but change randomly from segment to segment. In practice, each fibre segment is treated as a phase plate with different birefringence characteristics.

Due to mode coupling, the birefringence of each section may either add to or subtract from the total birefringence. As a result, PMD in long fibre spans does not accumulate

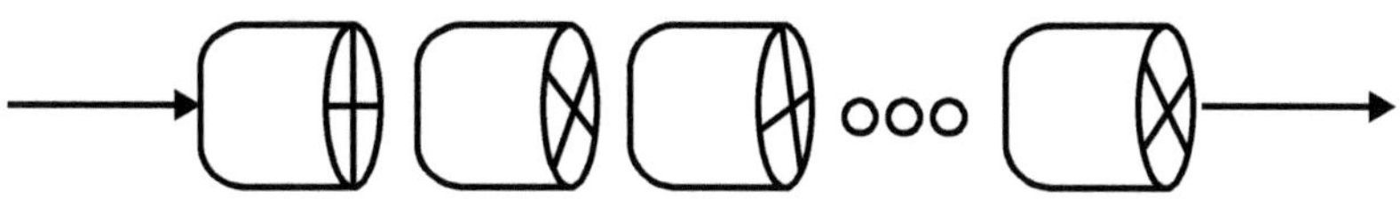

Figure 16.17 *Schematic of the technique used for calculating the PMD effects in long fibres.*

linearly with fibre length but accumulates in a random-walk-like process that leads to a square-root-of-length dependence.

Since most of the perturbations that act on a fibre depend on the temperature, the transmission properties also vary with ambient temperature. This manifests, for example, as a random, time-dependent drift of the state of polarization at the output of a fibre. Therefore, a statistical approach to PMD must be adopted.

To distinguish between the short- and long-length regimes, it is important to consider a parameter called the *correlation length*, L_C. One considers the evolution of the polarizations as a function of length in an ensemble of fibres with statistically equivalent perturbations. While the input polarization is fixed (e.g. along the x axis), it is equally probable to observe any polarization state at large lengths. The correlation length is defined as the length at which the average power in the orthogonal polarization mode, $\langle P_y \rangle$, is within $1/e^2$ of the power in the starting mode $\langle P_x \rangle$, i.e.

$$\frac{\langle P_y \rangle - \langle P_x \rangle}{P_{tot}} = \frac{1}{e^2} \tag{16.63}$$

where $P_{tot} = \langle P_x \rangle + \langle P_y \rangle$. The value of L_c can vary over a wide range from 1 m to 1 km for different fibres, typical values being 10 m. When the fibre transmission distance L satisfies $L << L_c$, the fibre is in the short-length regime, whereas the case $L >> L_c$ corresponds to the long-length regime.

It can be shown that the mean square DGD of the fibre for arbitrary values of the fibre length is given by Wai and Menyuk (1996):

$$\langle \Delta\tau^2 \rangle = 2\left(\Delta\tau_b \frac{L_c}{L_b}\right)^2 \left(\frac{L}{L_c} + e^{-L/L_c} - 1\right) \tag{16.64}$$

For $L << L_c$, the root mean square of the DGD simplifies to

$$\sqrt{\langle \Delta\tau^2 \rangle} = \Delta\tau_{rms} = \Delta\tau_b L/L_b \tag{16.65}$$

On the other hand, for $L >> L_c$ Eq. (16.64) gives

$$\Delta\tau_{rms} = (\Delta\tau_b/L_b)\sqrt{2LL_c} \tag{16.66}$$

reflecting the length dependence discussed earlier.

Eqs. (16.64)–(16.66) give the mean (and most likely) DGD values that would be observed. However, in practice, significant variations of DGD are observed about the mean, and so the form of this distribution is important. When $L >> L_c$, it was found that the PMD has a Maxwellian probability distribution given by Wai and Menyuk (1996):

$$p(\Delta\tau) = \frac{8}{\pi^2 \langle \Delta\tau \rangle}\left(\frac{2\Delta\tau}{\langle \Delta\tau \rangle}\right)^2 e^{-\left(\frac{2\Delta\tau}{\langle \Delta\tau \rangle}\right)^2 \frac{1}{\pi}} \qquad (\Delta\tau > 0) \tag{16.67}$$

where $\langle \Delta\tau \rangle$ is the mean DGD, which is related to $\Delta\tau_{rms}$ by

$$\langle \Delta\tau \rangle = \sqrt{\frac{8}{3\pi}}\Delta\tau_{rms} \tag{16.68}$$

The propagation of a pulse through a long span is very complicated due to the random sequence of perturbations along the length that produce local changes in the polarization. However, a surprising feature of PMD is that even for long fibres there are two *input principal states of polarization* (PSPs). These two principal states are orthogonal and are defined as those input polarizations for which the output states of polarization (SOPs) are independent of frequency to first order, i.e. over a small frequency range. The latter states, known as *output principal states of polarization*, are also orthogonal and are not necessarily the same as the input states. Therefore, if a pulse having a sufficiently narrow frequency spectrum is coupled into the span in either one of the input PSPs, the output pulse (appearing in the associated output PSP) experiences no broadening or change of shape and presents a group delay τ_i $(i = 1, 2)$. The differential group delay between the PSPs is $\Delta\tau = \tau_1 - \tau_2$.

Using the principal states model, PMD can be characterized by the PMD vector, $\mathbf{\Omega}$, defined in the three-dimensional Stokes space, around which the SOP, $\mathbf{s}$, rotates when the carrier frequency is changed, such that $d\mathbf{s}/d\omega = \mathbf{\Omega} \times \mathbf{s}$. The direction of the vector $\mathbf{\Omega}$ defines an axis whose two intercepts with the surface of the Poincaré sphere correspond to the two output PSPs. The magnitude of the PMD vector is the DGD, $\Delta\tau$.

16.6 Fibre Bragg Gratings

As mentioned in Section 16.3, the use of dispersion compensation fibres in a transmission link suffers from high insertion losses because of their relatively long lengths. They also enhance the impact of non-linear effects within a long-haul communication system because of their reduced effective core area. Both of these issues can be avoided to a large extent by using fibre-based Bragg gratings for dispersion compensation.

A fibre Bragg grating (FBG) is obtained when the refractive index varies periodically along the fibre length, as represented in Fig. 16.18. The realization of such index gratings is possible due to the photosensitivity of optical fibres, which permanently change their optical properties when subjected to intense ultraviolet radiation. Fibre gratings can be designed to operate over a wide range of wavelengths extending from the ultraviolet to the infrared region. Index changes occur only in the fibre core and are typically of the order of $\sim 10^{-4}$ in the 1550 nm spectral region.

The holographic technique is now routinely used to fabricate such fibre gratings with a controllable period. This technique uses two optical beams, obtained from the same laser and making an angle 2θ between them, which are made to interfere at the exposed core of an optical fibre. The grating period, Λ, is determined by that angle and by the wavelength of the ultraviolet radiation, λ_{uv}, according to the relation

$$\Lambda = \frac{\lambda_{uv}}{2\sin\theta} \tag{16.69}$$

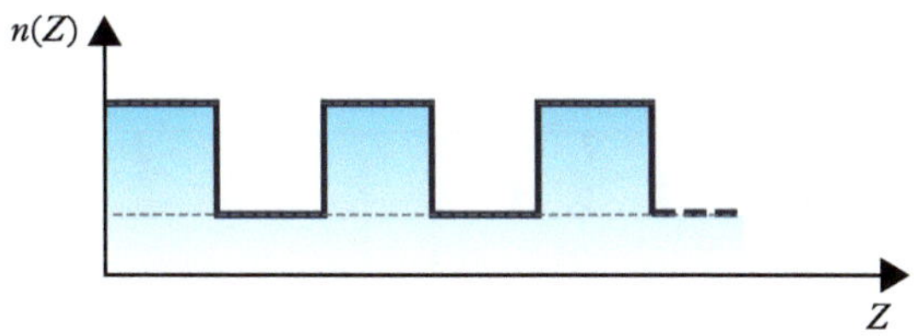

Figure 16.18 *Periodic variation of refractive index along the fibre grating.*

Taking into account the frequency dependence, in addition to its periodic variation along the fibre length, the refractive index of a FBG can be written in the form

$$n(\omega, z) = \bar{n}(\omega) + \Delta n_g(z) \tag{16.70}$$

where $\bar{n}(\omega)$ is the average refractive index and $\Delta n_g(z)$ describes the periodic variation of the grating refractive index. In the following, we will consider a sinusoidal grating of the form

$$\Delta n_g = n_a \cos(2\pi z/\Lambda) \tag{16.71}$$

where n_a is the refractive index modulation depth.

A fibre grating acts as a reflector for a given wavelength of light λ, such that

$$\lambda \equiv \lambda_B = 2\bar{n}\Lambda \tag{16.72}$$

The Bragg wavenumber and the Bragg frequency are given by $\beta_B = \pi/\Lambda$ and $\omega_B = \pi c/(\bar{n}\Lambda)$, respectively. The Eq. (16.72) is known as the *Bragg-resonance condition*, and it ensures that the phase of the reflected light adds up constructively, producing a strong reflection of the light wave.

The periodic nature of the index variations couples the forward- and backward-propagating waves at wavelengths close to the Bragg wavelength and, as a result, provides frequency-dependent reflectivity to the signal over a bandwidth determined by the grating strength (Fig. 16.19). In order to study wave propagation along the fibre grating, we solve the wave equation in the frequency domain, as given by Eq. (15.29). In this case, the field $\tilde{E}$ includes both forward- and backward-propagating waves and can be written in the form

$$\tilde{E}(r, \omega) = F(r)\left[\tilde{A}_f(z, \omega)\exp(i\beta_B z) + \tilde{A}_b(z, \omega)\exp(-i\beta_B z)\right] \tag{16.73}$$

where $F(r)$ describes the transverse modal distribution in a single-mode fibre and β_B is the Bragg wave number. Using Eqs. (16.69)–(16.73) and the slowly varying approximation for the amplitudes $\tilde{A}_f$ and $\tilde{A}_b$, the following coupled-mode equations are obtained:

$$\frac{d\tilde{A}_f}{dz} = i\Delta(\omega)\tilde{A}_f + i\kappa_g\tilde{A}_b \tag{16.74}$$

$$-\frac{d\tilde{A}_b}{dz} = i\Delta(\omega)\tilde{A}_b + i\kappa_g\tilde{A}_f \tag{16.75}$$

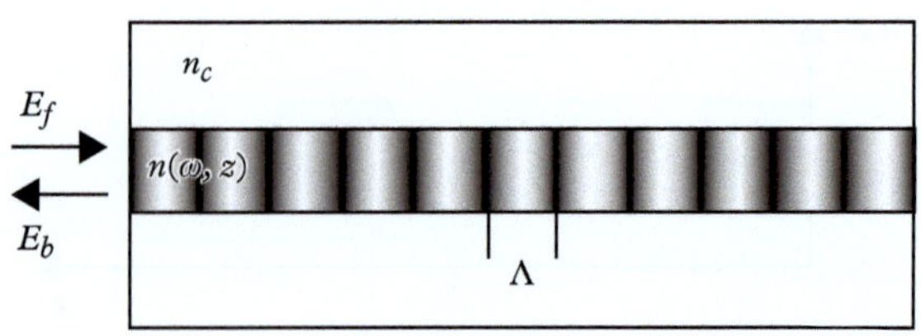

Figure 16.19 *The periodic nature of index variations couples the forward- and backward-propagating waves and provides frequency-dependent reflectivity.*

where

$$\Delta = (\bar{n}/c)(\omega - \omega_B) \equiv \beta(\omega) - \beta_B \tag{16.76}$$

is the detuning from the Bragg frequency ω_B and

$$\kappa_g = \frac{\pi n_a}{\lambda_B} \tag{16.77}$$

is the coupling coefficient.

From Eqs. (16.74) and (16.75), one can obtain the following result for the reflection coefficient of a FBG of length L_g:

$$r_g = \frac{A_b(0)}{A_f(0)} = \frac{i\kappa_g \sin(QL_g)}{Q\cos(QL_g) - i\Delta\sin(QL_g)} \tag{16.78}$$

where Q satisfies the linear dispersion relation

$$\Delta = \pm\sqrt{\kappa_g^2 + Q^2} \tag{16.79}$$

The reflectivity $R = |r_g|^2$ is illustrated in Fig. 16.20 as a function of detuning Δ for $\kappa_g L = 1, 2, 4$. One can see that the central lobe grows in amplitude with $\kappa_g L$ and then broadens and saturates. Maximum reflectivity occurs at the centre of the stop band and can be obtained from Eqs. (16.78) and (16.79) by $\Delta = 0$:

$$R_{\max} = |r_g|^2 = \tanh^2 \kappa_g L \tag{16.80}$$

One has $R_{\max} = 0.93$ for $\kappa_g L = 2$ and $R_{\max} \approx 1$ for $\kappa_g L \geq 3$. The bandwidth of the reflection spectrum, which is defined as the wavelength spacing between the two reflection minima on either side of the central peak, is approximately given by

$$\Delta\lambda = \frac{\lambda_B^2}{\pi n_{eff} L}\left(\kappa_g^2 L^2 + \pi^2\right)^{1/2} \tag{16.81}$$

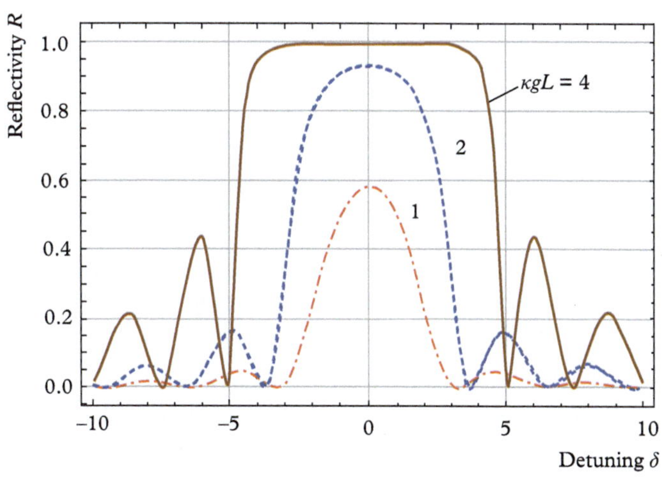

Figure 16.20 *Reflectivity $R = |r_g|^2$ as a function of detuning Δ for $\kappa_g L = 1, 2, 4$.*

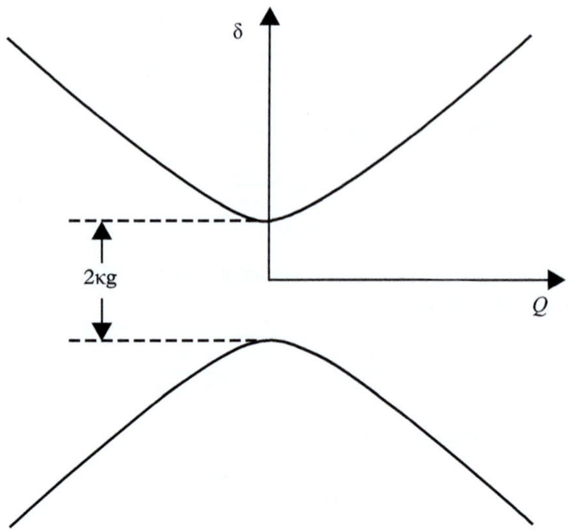

Figure 16.21 *Dispersion relation of a uniform fibre Bragg grating.*

Fig. 16.21 illustrates the relation between Q and Δ given by Eq. (16.79). The most interesting feature in this figure is the existence of a *photonic bandgap* for detunings Δ such that $-\kappa_g < \Delta < \kappa_g$. Light with frequencies within this bandgap cannot propagate and is reflected. Frequencies outside the gap, however, can propagate but at velocities that are significantly reduced in comparison with the speed of light in the uniform medium. According to Fig. 16.20, the group velocity vanishes at the band edge and asymptotically approaches $c/\bar{n}$ far from the Bragg resonance.

Outside but close to the stop-band edges, the grating provides large chromatic dispersion. Noting that the effective propagation constant of the forward- and backward-propagating waves is $\beta_e = \beta_B \pm Q$, where the choice of sign depends on the sign of δ, the following result can be obtained for the group velocity inside the grating:

$$V_G = \left[\frac{dQ}{dw}\right]^{-1} = \pm v_g\sqrt{1-x^2} \tag{16.82}$$

where $x = \kappa_g/\Delta$. On the other hand, one can obtain the following result for the GVD parameter of a fibre grating:

$$\beta_2^g = \frac{d^2Q}{d\omega^2} = -\frac{1}{v_g^2}\frac{x^2}{\Delta(1-x^2)^{3/2}} \tag{16.83}$$

The grating-induced GVD depends on the sign of the detuning δ. The GVD is anomalous $(\beta_2^g < 0)$ on the high-frequency side of the stop band and becomes normal $(\beta_2^g > 0)$ on the low-frequency side. Since an FBG acts as a dispersive delay line, it can be used for dispersion compensation in fibreoptic transmission systems. In particular, the normal GVD on the low-frequency side of the stop band can be used to compensate the dispersion of standard fibres near 1550 nm $(\beta_2 \sim -20\,\text{ps}^2/\text{km})$. Since β_2^g can exceed $1000\,\text{ps}^2/\text{cm}$, a single 2-cm-long grating can compensate for dispersion accumulated over 100 km of fibre.

FBGs have also been used more recently in high-power fibre lasers. In the case of a simple fibre laser, the FBGs can be used as the high reflector and output coupler to form the laser cavity, thus providing a monolithic, all-fibre device. The gain for the laser is provided by a length of rare-earth-doped optical fibre. The major advantage of these all-fibre systems, in which the free-space mirrors are replaced with a pair of FBGs, is the elimination of realignment during the life of the system, since the FBGs are spliced directly to the doped fibre and never need adjusting.

An FBG can also be used in place of the bulk-grating pair in a grating-fibre pulse compression system, providing a compact all-fibre device. Both the compression factor and the quality of the compressed pulse can be significantly enhanced by using an FBG in which the grating period varies linearly with position—the so-called *chirped fibre grating*. In this case, the grating reflects different frequency components of the pulse at different points along its length.

16.7 Further Reading

Börzsönyi, A., Z. Heiner, M. Kalashnikov, A. Kovács, and K. Osvay, Dispersion Measurement of Optical Components Using White-Light Interferometry. *Appl. Opt.*, 47, no. 25, 4856–4863, 2008.

Ferreira, M. F. (Ed), *Photonic Crystal Fibers—Advances in Research and Applications*, NOVA Science Publishers, 2022.

Ferreira, M. F., and M. C. Paul (Eds), *Optical Fiber Technology and Applications: Recent Advances*. Institute of Physics Publishing (IOP), 2021.

Ferreira, M. F., and M. Paul (Eds), *Specialty Optical Fibers—Materials, Fabrication Technology and Applications*, Elsevier, 2024.

Ghatak, A., and K. Thyagarathan, *An Introduction to Fiber Optics*, Cambridge University Press, 1998.

Im, S., A. Husakou, and J. Herrmann, Supercontinuum Generation in a Gas-Filled Photonic Crystal Fiber. *Opt. Express*, 17, no. 16, 13050–13058, 2009.

Kaiser, A., *Optical Fibre Communications*, 2nd Ed., McGraw-Hill, 1991.

Palais, J. C., *Fiber Optic Communications*, Prentice Hall, Englewood Cliff, 1988.

Senior, J. M., *Optical Fiber Communications*, Pearson Education, 2009.

Quimby, R., *Photonics and Lasers—An Introduction*, John Wiley & Sons, 2006.

Wai, P. K., and C. R. Menyuk, Polarization Mode Dispersion, Decorrelation, and Diffusion in Optical Fibers with Randomly Varying Birefringence. *J. Lightwave Technol.* 14, no. 2, 148–157, 1996.

16.8 Problems

16.1 Determine (a) the length and (b) the transit time for the longest and shortest trajectories in a step-index fibre of length 2 km, having a core index of 1.46 and a cladding index of 1.45.

16.2 Calculate the time delay between an axial ray and one that enters a 1-km-long fibre at an angle of 10°. The core index is 1.47.

16.3 Find the numerical aperture of a multimode fibre with a 50-μm core diameter and designed to limit the intermodal dispersion to 10 ns/km.

16.4 Obtain the maximum transmission rate determined by the intermodal dispersion for a fibre with the characteristics specified in Problem 16.2 and a length $L = 20$ km.

16.5 Unchirped Gaussian pulses of 100-ps width (FWHM) are launched into a single-mode fibre at a wavelength of 1.55 μm. Calculate their FWHM after 50 km if the fibre has a dispersion of 16 ps/(km-nm).

16.6 Pulses at two discrete wavelengths, 850 and 860 nm, and with a 2 nm spectral width are coupled simultaneously into a long step-index, single-mode fibre with core radius 4 m. Find the time delay between the two pulses at the fibre end.

16.7 Using Eq. (16.37) for the evolution of a Gaussian pulse width, calculate the distance z after which the pulse width becomes $t_1(z) = 20t_0$, with $t_0 = 5$ ps. Consider that the fibre GVD is $\beta_2 = 20\,\text{ps}^2/\text{km}$

16.8 Using Eqs. (16.46) and (16.47), show that the quantity $\left(1 + C^2\right)/t_p^2$ remains constant and equal to its initial value $\left(1 + C_0^2\right)/t_0^2$.

16.9 10-ns (FWHM) pulses are transmitted in a communication system over a 10-km single-mode fibre, which has a dispersion parameter $D = -80\text{ps}/(\text{km} - \text{nm})$. Determine the maximum bit rate if the light source emits at 0.88 μm and has a spectral FWHM of 30 nm.

16.10 Determine the dispersion-limited maximum fibre length of a 1.55-μm optical communication system operating at 5 Gb/s using Gaussian pulses of width 100 ps (FWHM). Assume that $\beta_2 = -20\text{ps}^2/\text{km}$ and that the pulses are (i) unchirped and (ii) chirped such that $C = -6$.

16.11 A fibre link uses standard silica telecommunications fibre with a core radius of 4 m. Chromatic dispersion is periodically compensated using a 1-km-long dispersion compensation fibre (DCF), with a dispersion parameter $Dc = -300\text{ps}/(\text{nm} - \text{km})$ at the operating wavelength of 1550 nm. Determine the distance required between the DCF insertions along the fibre link.

16.12 Explain the origin of polarization mode dispersion (PMD) in optical fibres and the reason why it leads to pulse broadening.

16.13 Estimate the intrinsic PMD at 1550 nm of a polarization-maintaining fibre that has a beat length of 4 mm.

16.14 Solve Eqs. (16.74) and (16.75) and show that the transfer function of a Bragg grating is indeed given by Eq. (16.78).

16.15 Use the dispersion relation of a Bragg grating given by Eq. (16.79) to show that the second-order dispersion parameter of the grating is given by Eq. (16.83).

Laser Fundamentals

17

When the laser was first discovered, it was often said to be 'a solution in search of a problem'. In fact, it was a novel device with interesting properties, but its usefulness was not clear at that time. Today, however, the situation is quite different, and the laser is used in various applications in our day-to-day lives, from reading barcodes to CD players, and in the field of technology, from optical communications to materials processing or medicine. It has also become an important tool in the research that is done in virtually all areas of science.

The process of *stimulated emission* is the basis of the operation of the laser and was demonstrated theoretically by Albert Einstein as early as 1917. However, it was only in the early fifties of the 20th century, following the work of Charles H. Townes (USA), Alexander M. Prokhorov and Nikolai G. Basov (former Soviet Union), that a microwave device that amplified radiation through the process of stimulated emission was put into operation. This device was called MASER, an acronym for the English expression 'Microwave Amplification by Stimulated Emission of Radiation'. In 1958, Townes and Arthur L. Schawlow published a paper discussing the extension of the maser principles to the optical region of the electromagnetic spectrum. The operation of the so-called optical maser, or laser, was first achieved by Theodore H. Maiman in May 1960. For this purpose, he used a small synthetic crystal of pink ruby, with faces parallel to each other and mirrored, installed on the axis of a helical gaseous discharge lamp. Since then, a wide variety of lasers has been developed.

Any type of laser consists of three essential elements (see Fig. 17.1): (1) an *active medium*, i.e. a set of atoms, molecules or ions (hereinafter referred to generically as 'atoms') capable of emitting optical radiation and providing optical gain; (2) a *pumping source*

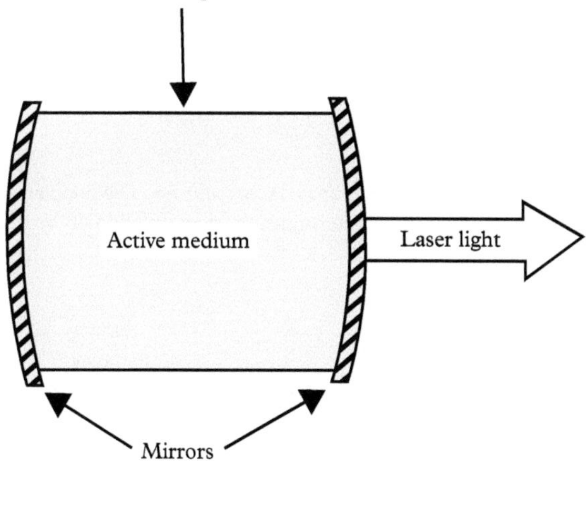

Figure 17.1 *Schematic of the essential elements of a laser: the active medium, the pumping source, and the resonant cavity.*

Lasers, Optics, and Photonics. Mário F. S. Ferreira, Oxford University Press. © Mário F. S. Ferreira (2026).
DOI: 10.1093/oso/9780198952770.003.0017

capable of exciting the atoms of the active medium; and (3) a *resonant cavity*, consisting of two mirrors in which the beam of radiation is successively reflected, forcing it to pass a large number of times through the active medium. In the following sections, each of these aspects will be dealt with separately.

17.1 The Active Medium—Absorption and Emission of Radiation

According to quantum theory, the interaction of light with matter takes place in discrete packets of energy or quanta, called photons. Furthermore, atoms exist only in certain discrete energy states such that the absorption and emission of light cause them to make a transition from one discrete energy state to another.

Let us consider an active medium whose atoms are characterized by two energy states : the ground state, with energy E_0, and an excited state, with energy E_1 ($E_1 > E_0$). In the presence of radiation with a frequency

$$\nu_{10} = \frac{E_1 - E_0}{h} \tag{17.1}$$

where $h = 6.626 \times 10^{-34}$ J.s is Planck's constant, the transition of some atoms from the ground state to the excited state occurs. This process is called *stimulated absorption*.

Once in the excited state, an atom can emit a photon of energy ν_{10} and return to the ground state. This process can occur without any external influence, in which case it is called *spontaneous* emission, or it can be induced by a pre-existing photon with that same frequency, a situation which we refer to as *stimulated emission*. In the first case, the random nature of the spontaneous emission process, in which light is emitted by electronic transitions from a large number of atoms, gives rise to incoherent radiation. In the second case, the emitted photon has the same characteristics as the stimulating photon, namely the same phase and polarization. Therefore, in contrast to spontaneous emission, this process provides coherent radiation. Fig. 17.2 illustrates the three transition processes referred to.

The rates of the three transition processes mentioned earlier are related to each other. In the case of spontaneous emission, the respective rate is proportional to the number of atoms, N_1, which are in the excited state of energy E_1, and we can write

$$\left(\frac{dN_0}{dt}\right)_{sp} = -\left(\frac{dN_1}{dt}\right)_{sp} = A_{10}N_1 \tag{17.2}$$

where A_{10} is called *Einstein's coefficient for spontaneous emission*. The population N_1 decreases with a time constant $\tau = 1/A_{10}$, which determines the depopulation of the level with energy

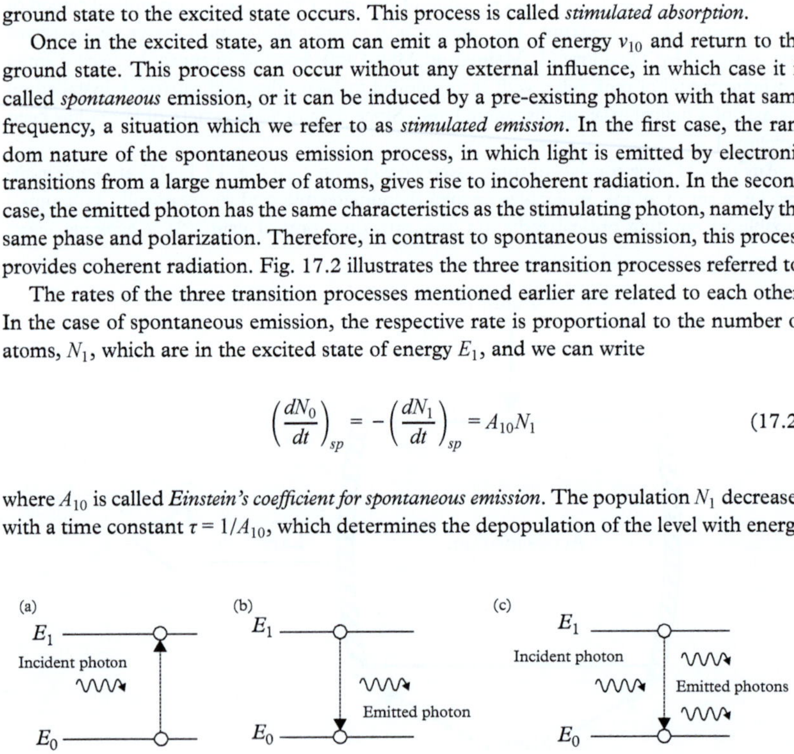

Figure 17.2 *Diagrams representing the transition processes in an active medium: (a) stimulated absorption, (b) spontaneous emission, and (c) stimulated emission.*

E_1 at a rate N_1/τ, whereas the population of the level with energy E_0 increases at the same rate. The constant τ is known as the *lifetime of spontaneous emission*.

Transitions concerning stimulated absorption or emission occur only in the presence of stimulant radiation. In the case of stimulated absorption, Einstein postulated that its transition rate would be proportional to the number of atoms in the ground state, N_0, and to the spectral energy density of the incident field $\rho(\nu_{10})$ (with units of energy per unit volume per unit frequency range, J.m^{-3}Hz^{-1}) with frequency ν_{10}:

$$-\left(\frac{dN_0}{dt}\right)_{ab} = \left(\frac{dN_1}{dt}\right)_{ab} = B_{01}N_0\rho(\nu_{10}) \tag{17.3}$$

In Eq. (17.3), the proportionality constant, B_{01}, is the so-called *Einstein coefficient for stimulated absorption*. Similarly, the rate of transition corresponding to the stimulated-emission process is proportional to the number of atoms in the excited state, N_1, and to the spectral energy density of the incident field $\rho(\nu_{10})$, and is given by

$$\left(\frac{dN_0}{dt}\right)_{st} = -\left(\frac{dN_1}{dt}\right)_{st} = B_{10}N_1\rho(\nu_{10}) \tag{17.4}$$

where B_{10} is the so-called *Einstein coefficient for stimulated emission*.

In a situation of thermal equilibrium, there is a balance between the upward and downward transition rates, which provides the condition

$$A_{10}N_1 + B_{10}N_1\rho(\nu_{10}) = B_{01}N_0\rho(\nu_{10}) \tag{17.5}$$

The populations N_1 and N_0 are related through the *Boltzmann distribution*:

$$\begin{aligned}\frac{N_1}{N_0} &= \exp\left(-\frac{E_1 - E_0}{k_B T}\right)\\ &= \exp\left(-\frac{h\nu_{10}}{k_B T}\right)\end{aligned} \tag{17.6}$$

where T is the absolute temperature and k_B is the Boltzmann constant.

From Eqs. (17.5) and (17.6) it follows that

$$\begin{aligned}\rho(\nu_{10}) &= \frac{A_{10}}{B_{01}(N_0/N_1) - B_{10}}\\ &= \frac{A_{10}}{B_{01}\exp(h\nu_{10}/k_B T) - B_{10}}\end{aligned} \tag{17.7}$$

On the other hand, since the atomic system is in thermal equilibrium, it produces a radiation density that is identical to black-body radiation. Under these conditions, the spectral energy density $\rho(\nu)$ is given by *Planck's radiation law*:

$$\rho(\nu_{10}) = \frac{8\pi h\nu_{10}^3}{c^3}\frac{1}{\exp(h\nu_{10}/k_B T) - 1} \tag{17.8}$$

Eqs. (17.7) and (17.8) lead to the equality:

$$\frac{A_{10}}{B_{01}\exp(h\nu_{10}/k_B T) - B_{10}} = \frac{8\pi h\nu_{10}^3}{c^3}\frac{1}{\exp(h\nu_{10}/k_B T) - 1} \tag{17.9}$$

Eq. (17.9) is satisfied when

$$A_{10} = \frac{8\pi h\nu_{10}^3}{c^3} B_{10} \tag{17.10}$$

and

$$B_{01} = B_{10} \tag{17.11}$$

Eqs. (17.10) and (17.11) are called the *Einstein relations*. They can be generalized to an arbitrary pair of atomic energy states, E_i and E_j $(E_j > E_i)$, by substituting E_0 with E_i and E_1 with E_j.

The ratio of the stimulated-emission rate to the spontaneous emission rate in a situation of thermal equilibrium is given by $\rho(\nu_{10}) B_{10}/A_{10}$. Considering an incandescent lamp with an average wavelength of 0.5 μm operating at a temperature of 1000 K, and using Eqs. (17.8) and (17.19), it can be found that such a ratio is $\sim 3 \times 10^{-13}$. This result shows that the dominant mechanism in the origin of radiation emitted by ordinary optical sources in the visible spectrum is spontaneous emission, proving that these sources are incoherent.

From the previous discussion, two important ideas for successfully achieving a coherent optical source and the amplification of a light beam stand out. The first one has to do with the existence of the *stimulated-emission* process, which provides the amplification of light. The second refers to the need to achieve *population* inversion between the two energy levels of the atoms that constitute the active medium so that the effect of stimulated emission is dominant over the opposite effect due to stimulated absorption.

Example 17.1

Assume that the Einstein coefficient for spontaneous emission in a transition corresponds to a wavelength $\lambda = 600\,\text{nm}$ and $A_{10} = 10^5/\text{s}$.

a) What is the corresponding Einstein coefficient for stimulated emission?

b) What must be the irradiance within the cavity for the stimulated-emission rate to be three times higher than the spontaneous emission rate?

Solution

a) The Einstein coefficient for stimulated emission is obtained from the Einstein coefficient for spontaneous emission using Eq. (17.10), which can be rewritten in the form:

$$B_{10} = \frac{\lambda^3}{8\pi h} A_{10}$$

Substituting the values of the parameters, one has

$$B_{10} = 1.3 \times 10^{18}\ \text{J}^{-1}\text{m}^3\text{s}^{-2}.$$

b) The spectral energy density, $\rho(\nu)$, is related to the irradiance in the form

$$\rho(\nu) = \frac{I(\nu)}{c} \tag{17.12}$$

Example 17.1 continued

Using Eqs. (17.2), (17.4) and (17.12), one finds that the rate of stimulated emission will be three times higher than the rate of spontaneous emission if

$$B_{10}\frac{I}{c} = 3A_{10}$$

Substituting the values of B_{10}, A_{10} and c, one obtains

$$I = 6.9 \times 10^{-5}\text{Wm}^{-2}\text{Hz}^{-1}$$

17.2 Absorption and Gain Coefficients

Let us consider a collimated beam of monochromatic light, with frequency $\nu_{ij} = (E_j - E_i)/h$, which propagates in the direction of the *z-axis* and passes through a homogeneous active medium. The variation of the irradiance between two planes situated at z and $z + \Delta z$ is given by

$$\begin{aligned}\Delta I(z) &= I(z+\Delta z) - I(z)\\ &= -\alpha I(z)\,\Delta z\end{aligned} \tag{17.13}$$

where α is the so-called *absorption coefficient*, and the minus sign indicates a reduction of the irradiance determined by absorption in the material.

Eq. (17.13) can be written in the form of a differential equation:

$$\frac{dI(z)}{dz} = -\alpha I(z) \tag{17.14}$$

The solution of Eq. (17.13) is given by

$$I(z) = I(0)\exp(-\alpha z) \tag{17.15}$$

where $I(0)$ is the irradiance at the entrance of the active medium. This result is known as *Beer's law*, and it shows that the irradiance decreases exponentially when the beam propagates in a medium in which the absorption effect is dominant.

If the number of stimulated emissions exceeds the number of absorptions, the beam irradiance will increase rather than decrease during its propagation through the active medium. The number of stimulated emissions depends on the spectral energy density, $\rho(\nu)$, which is related to the irradiance as in Eq. (17.12). In the following, we will consider that the beam consists of perfectly monochromatic radiation, so the dependence of irradiance on frequency is ignored.

The number of photons, p_{ij}, gained by the beam every second when passing through the active medium can be written in terms of the irradiance at z and Δz, in the form:

$$\begin{aligned}\frac{dp_{ij}}{dt} &= [I(z+\Delta z) - I(z)]\,\frac{A}{h\nu_{ij}}\\ &= \frac{dI(z)}{dz}\Delta z\frac{A}{h\nu_{ij}}\end{aligned} \tag{17.16}$$

where A is the area of the beam section, and it is assumed that Δz is small enough.

On the other hand, the rate of change of the number of photons can be expressed in terms of the stimulated emission and absorption rates:

$$\begin{aligned}\frac{dp_{ij}}{dt} &= N_j\frac{I(z)}{c}B_{ji} - N_i\frac{I(z)}{c}B_{ij}\\ &= B_{ij}\left(N_j - N_i\right)\frac{I(z)}{c}\end{aligned} \tag{17.17}$$

where Eq. (17.11) was used, and N_i and N_j are the number of atoms in the lower and upper energy states, respectively, in the volume element $A\Delta z$. When writing Eq. (17.17), the attenuation of the beam caused by dispersion in the medium, as well as the effects of spontaneous emission, has been neglected.

From Eqs. (17.16) and (17.17), one can write:

$$\begin{aligned}\frac{dI(z)}{dz} &= B_{ij}\left(N_j - N_i\right)\frac{I(z)}{c}\frac{h\nu_{ij}}{\Delta zA}\\ &= B_{ij}\left(n_j - n_i\right)\frac{h\nu_{ij}}{c}I(z)\end{aligned} \tag{17.18}$$

where $n_i = N_i/A\Delta z$ and $n_j = N_j/A\Delta z$ represent the number of atoms per unit volume, or *population density*, of the two energy states.

Integrating Eq. (17.18), one obtains

$$I(z) = I(0)\exp(gz) \tag{17.19}$$

where g is the so-called *gain coefficient*, given by

$$\begin{aligned}g &= B_{ij}(n_j - n_i)\frac{h\nu_{ij}}{c}\\ &= \left(n_j - n_i\right)\frac{c^2 A_{ij}}{8\pi\nu_{ij}^2}\end{aligned} \tag{17.20}$$

One can see from Eq. (17.20) that g is positive and that the irradiance increases with distance if $n_j > n_i$. The amplification of light provided by the stimulated-emission process is the basis for the operation of lasers.

The result obtained earlier shows that the amplification of light by stimulated emission is possible only when the population of the upper energy level exceeds the population of the lower level, a condition known as *population inversion*. This can be achieved by properly pumping the active medium. The fact that the gain coefficient given by Eq. (17.20) is inversely proportional to the square of the frequency ν_{ij} indicates that it will be more difficult to build a laser emitting light in the ultraviolet region than in the infrared region.

17.3 Population Inversion

Considering an arbitrary pair of energy states E_i and E_j $E_j\left(E_j > E_i\right)$, Eq. (17.6) takes the form:

$$\frac{N_j}{N_i} = \exp\left(-\frac{E_j - E_i}{k_B T}\right) = \exp\left(-\frac{\Delta E_{ji}}{k_B T}\right) \tag{17.21}$$

where $\Delta E_{ji} = E_j - E_i$. If $k_B T << \Delta E_{ji}$ the ratio given by Eq. (17.21) is quite small, and there are practically no atoms in the excited state. On the other hand, for sufficiently high temperatures, in which case $k_B T >> \Delta E_{ji}$, the population at the energy level E_j approaches the population at the level E_i. In any case, the population at the energy level E_j can never be higher than the population at the level E_i in a situation of thermal equilibrium. To achieve population inversion, atoms must be excited, or pumped, using an external source of energy.

Example 17.2

Consider an atomic system in which the ground state has energy E_0 and the first excited state has energy E_1. The difference between these energies corresponds to a wavelength $\lambda = 600\,\text{nm}$.

a) Determine the ratio $\frac{N_1}{N_0}$ between the populations of the two levels at a temperature $T = 300\,\text{K}$.

b) At what temperature does one achieve the ratio $\frac{N_1}{N_0} = \frac{1}{2}$?

Solution

a) The difference between the two energy levels is

$$E_1 - E_0 = h\nu = h\frac{c}{\lambda}$$

$$= \frac{6.626 \times 10^{-34} \times 3 \times 10^8}{600 \times 10^{-9}} \approx 3.31 \times 10^{-19}\text{J}$$

On the other hand, at a temperature T = 300 K, one has

$$k_B T = 1.38 \times 10^{-23} \times 300 \approx 4.14 \times 10^{-21}\text{J}$$

So the ratio between the populations of the two levels at this temperature is

$$\frac{N_1}{N_0} = e^{-80} \approx 10^{-35}.$$

The aforementioned result shows that it will be quite unlikely to find an atom in the excited state at a temperature of 300 K.

b) Considering $\frac{N_1}{N_0} = \frac{1}{2}$ in Eq. (17.6) and calculating the logarithm of both members, one has

$$T = \frac{E_1 - E_0}{k_B (\ln 2)} = \frac{3.31 \times 10^{-19}}{\left(1.38 \times 10^{-23}\right)(\ln 2)} = 34,600\,\text{K}$$

This value is much higher than the temperature at the surface of the sun. It becomes clear that, in order to achieve population inversion between the two levels, some form of energy pumping must be used.

Pumping of the active medium of a laser can be optical, electrical, chemical or thermal. In the case of gaseous lasers, the most common pumping method is an electrical discharge. In some lasers of this type, the free electrons generated in the process of discharge collide with and directly excite the atoms of the active medium. In other gaseous lasers, excitation occurs through inelastic collisions between atoms. In this case, a mixture of two gases is used, such that the two species of atoms have identical excited states. An example is the He–Ne laser, in which neon atoms are excited by energy transfer provided by helium atoms that are in a metastable state. Helium atoms receive this energy from the free electrons that collide with them.

Population inversion may be obtained in systems with three- or four-energy levels. In order to achieve such a condition, both systems must display a central metastable state in which the atoms spend an unusually long time. It is from this metastable level that stimulated emission or lasing takes place.

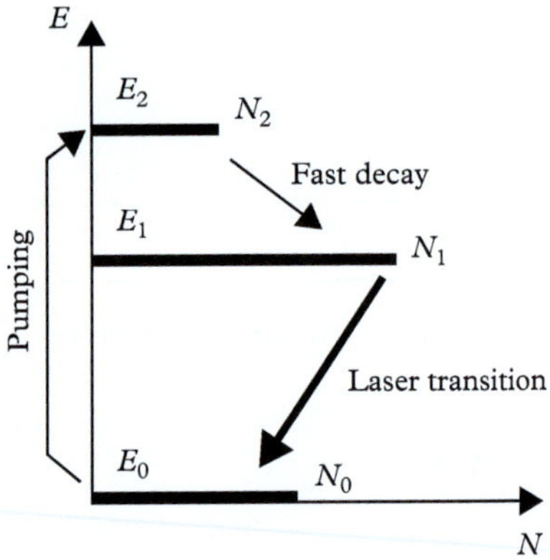

Figure 17.3 *Population distribution in the case of a three-level laser system.*

Fig. 17.3 illustrates a laser pumping scheme making use of three energy levels. When the atoms of the active medium are excited by the pumping source, a part of them passes from the ground state to the state with higher energy, E_2, from where they decay to the metastable state of energy E_1. Population inversion is achieved when the population of the latter state is higher than that of the ground state. For this situation to be possible, it is necessary that the transition from the level E_2 to the level E_1 be relatively fast.

In the case of a three-level system, relatively high pumping powers are required, since the lower energy state of the laser transition is the ground state and more than half of the atoms at this level must be pumped into the excited state in order to achieve population inversion. However, the pumping power can be greatly reduced in the case of a four-level system, as illustrated in Fig. 17.4. In this case, the ground-state atoms are excited by pumping to the highest energy level, E_3, from which they decay to the metastable state of energy E_2. If the transitions from E_3 to E_2 and from E_1 to E_0 are fast compared with the transition from E_2 to E_1, population inversion in this latter transition can be achieved with modest values of pumping power.

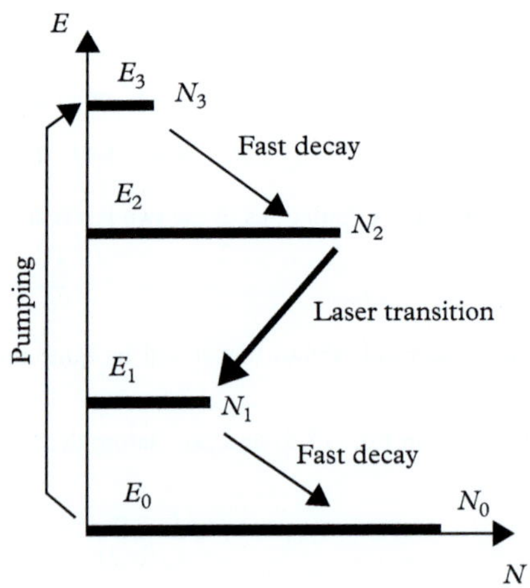

Figure 17.4 *Population distribution in the case of a four-level laser system.*

The pumping schemes of some types of lasers are much more complex than the three- or four-level schemes mentioned earlier. However, these schemes provide excellent models for a large number of real lasers.

17.4 Resonant Optical Cavity

The amplification provided by the active medium when light passes through it only once is in general relatively low. However, this limitation can be overcome by using mirrors, with which the effective length of the amplifier medium becomes a high multiple of the laser cavity length. The mirrors provide optical feedback and form a *resonant optical cavity* that withstands electromagnetic oscillations with the frequency of the laser transition. The oscillation originates from the spontaneous emission of radiation, which is subsequently amplified until a certain steady state is reached. A stable output is obtained at saturation when the optical gain is exactly matched by the losses incurred in the active medium. In the stationary regime, all the additional energy provided by the stimulated-emission process serves to feed the output power of the laser.

In many cases, the resonant cavity is constituted by spherical mirrors, since they generally provide a more stable configuration for the confinement of light compared with other types of mirrors. Fig. 17.5 illustrates the case of a resonant cavity formed by two concave spherical mirrors with radii R_1 and R_2, which are centred on the *z-axis* and separated by a distance *d*. A plane mirror corresponds to the case $R = \infty$.

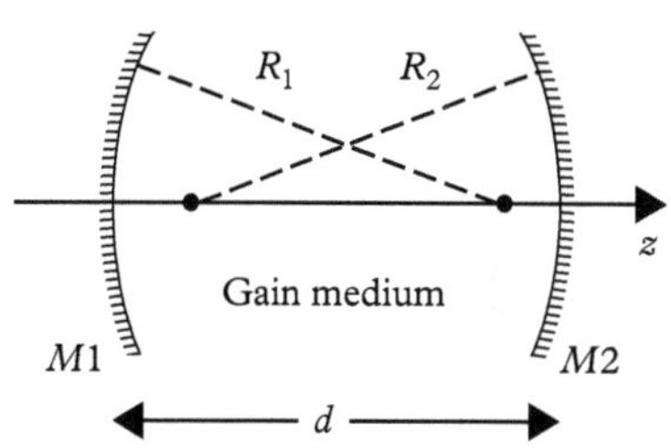

Figure 17.5 *Schematic of a resonant cavity formed by two concave spherical mirrors (M1 and M2).*

17.4.1 Laser oscillation modes

The two mirrors of a laser form a resonant cavity in which light can be stored by multiple reflections between them. The light in such a cavity consists of counterpropagating electromagnetic waves that form stationary waves, presenting nodes at the mirrors. This situation is illustrated in Fig. 17.6. As illustrated in this figure, the requirement that nodes exist at the mirror positions restricts the wavelengths that can be supported by the cavity to discrete values. If the distance between the cavity mirrors is *d*, the cavity will support stationary waves with free-space wavelengths λ satisfying the condition:

$$d = p\frac{\lambda}{2n},\ p\,\text{integer} \tag{17.22}$$

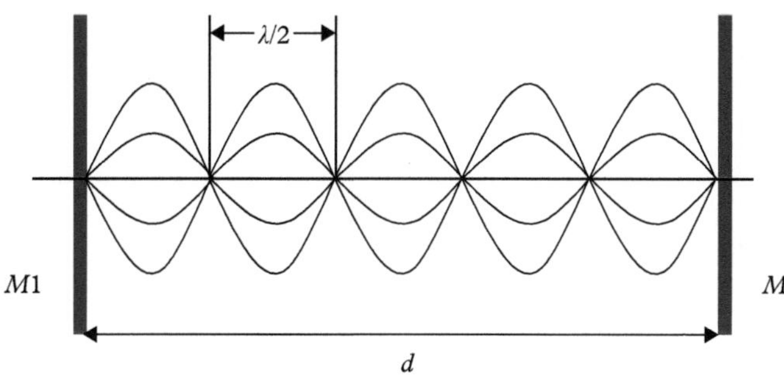

Figure 17.6 *Standing wave in a laser cavity with mirror spacing d.*

where n is the refractive index of the gain medium. Since p can be any integer, there are several wavelengths within the laser transition line that satisfy the previous condition. The fields corresponding to these different wavelengths are called *longitudinal*, or *axial*, *modes*. The frequencies of these modes are given by

$$\nu = \frac{c}{\lambda} = \frac{c}{(2dn/p)} = p\frac{c}{2dn} \tag{17.23}$$

The difference in frequency between neighbouring longitudinal modes is given by

$$\Delta\nu = \nu_{p+1} - \nu_p = \frac{c}{2dn} \tag{17.24}$$

If one considers, for example, a spacing between the mirrors d = 30 cm and n = 1, one obtains for the separation between the modal frequencies the result $\Delta\nu$ = 500 MHz. Given that the width of the laser transition lines is usually much greater than 500 MHz, it is possible for the laser to emit light at several discrete frequencies separated from each other by $c/(2d)$.

Although a large number of longitudinal modes may be generated within the laser cavity, the spectral output from the device will only include those modes contained within the spectral width of the gain curve. This situation is illustrated in Fig. 17.7. A laser device emitting light corresponding to several longitudinal modes is said to be multimode.

Resonant modes transverse to the direction of propagation can also occur in an optical cavity. These modes are designated similarly to the transverse modes in a waveguide by TEM*lm* where the integers l and m indicate the number of transverse modes. In the case of the TEM00 mode, all parts of the propagating wavefront are in phase. In contrast, the wavefronts of higher-order transverse modes present phase reversals, which produce different mode patterns. Higher-order transverse modes may be eliminated by a suitable narrowing of the laser cavity.

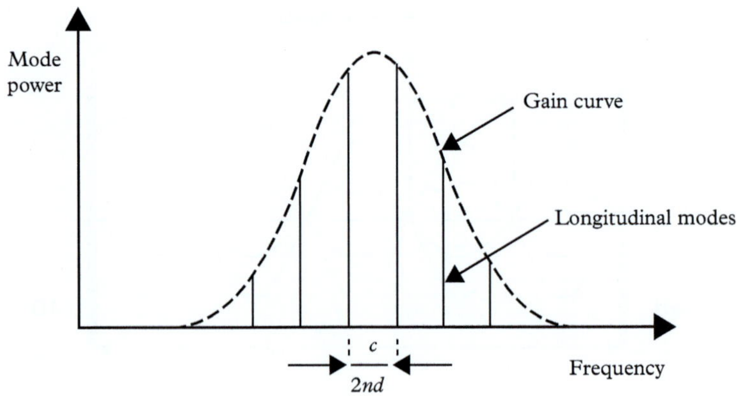

Figure 17.7 *Longitudinal modes in the laser output.*

Example 17.3

A helium–neon laser emitting at 632.8 nm has a resonant cavity formed by two mirrors spaced 0.5 m apart. Determine the number of longitudinal modes that contribute to the laser emission, considering that the spectral width of the spontaneous emission line of Ne at 632.8 nm is 0.0016 nm.

Solution

According to Eq. (17.24) and considering $n = 1$, the frequency separation between successive longitudinal modes of the laser is

$$\Delta\nu = \frac{c}{2d} = \frac{3 \times 10^8}{2\,(0.5)} = 3 \times 10^8\ \text{Hz}$$

On the other hand, the spectral width of the Ne line at 632.8 nm in the frequency domain is

$$\Delta\nu' = c\frac{\Delta\lambda}{\lambda^2} = 3 \times 10^8 \frac{1.6 \times 10^{-12}}{\left(632.8 \times 10^{-9}\right)^2} = 12 \times 10^8\ \text{Hz}$$

Thus, one has

$$\frac{\Delta\nu'}{\Delta\nu} = 4$$

So the laser will operate with four longitudinal modes.

17.4.2 Threshold oscillation condition

Let us consider that the active medium completely fills the region between the two mirrors, $M1$ and $M2$, of the laser and that the pumping is uniform. When radiation travels from the mirror $M1$ to the mirror $M2$, the irradiance increases from its initial value I_0 up to a value I given by

$$I = I_0 \exp\{(g - \alpha_{\text{int}})\,d\} \tag{17.25}$$

where d is the separation between the mirrors, g is the gain coefficient, and α_{int} are the losses due to absorption and scattering in the active medium, as well as absorption, scattering, and diffraction at the mirrors. After reflection at the mirror $M2$, the ratio of the reflected irradiance to the initial irradiance I_0 is given by $r_2^2 \exp\{(g - \alpha_{\text{int}})\,d\}$, r_2 being the reflection coefficient in amplitude of the mirror. After one pass through the active medium and a reflection at the mirror E_1, the ratio between the final and initial irradiances is given by

$$\frac{I_f}{I_0} \equiv G = r_1^2 r_2^2 \exp\{2\,(g - \alpha_{\text{int}})\,d\} \tag{17.26}$$

where r_1 is the amplitude reflection coefficient of the mirror $M1$.

The parameter G, given by Eq. (17.26), represents the net gain in power in one pass through the laser cavity. If $G > 1$, there is amplification of the radiation with a frequency equal to the resonance frequency of the laser, which corresponds to an increase in the oscillations of the cavity. On the other hand, if $G < 1$, the oscillations inside the cavity eventually disappear. The *threshold oscillation condition* of the laser is given by $G = 1$. Only if this condition is achieved for a certain time does the output power of the laser acquire a stationary value.

Making $G = 1$ in Eq. (17.26), the following result can be obtained for the threshold value of the gain coefficient, g_{th}:

$$g_{th} = \alpha_{\text{int}} + \frac{1}{d}\ln\left(\frac{1}{r_1 r_2}\right) = \alpha_{\text{int}} + \alpha_{mir} \tag{17.27}$$

where

$$\alpha_{mir} = \frac{1}{d}\ln\left(\frac{1}{r_1 r_2}\right) \tag{17.28}$$

It turns out from Eq. (17.27) that, at steady level, the threshold gain coefficient must be equal to the sum of the absorption losses in the material (α_{int}) and the losses corresponding to the emission of light to the outside (α_{mir}). Stabilization of the gain coefficient at its threshold value is commonly referred to as *gain saturation*.

17.4.3 Confinement condition

One can use the matrix method, presented in Chapter 2, to determine the conditions for the confinement of light rays in the resonant cavity. In fact, a resonant cavity is an example of a periodic optical system, given that light rays travel through the same system repeatedly.

If x_p and γ_p represent the position and the angle of the ray, respectively, after p turns in the cavity, the respective values after $p + 1$ turns are given by the following matrix relation:

$$\begin{bmatrix} x_{p+1} \\ \gamma_{p+1} \end{bmatrix} = \begin{bmatrix} A & B \\ C & D \end{bmatrix} \begin{bmatrix} x_p \\ \gamma_p \end{bmatrix} \tag{17.29}$$

where

$$\begin{bmatrix} A & B \\ C & D \end{bmatrix} = \begin{bmatrix} 1 & 0 \\ 2/R_1 & 1 \end{bmatrix} \begin{bmatrix} 1 & d \\ 0 & 1 \end{bmatrix} \begin{bmatrix} 1 & 0 \\ 2/R_2 & 1 \end{bmatrix} \begin{bmatrix} 1 & d \\ 0 & 1 \end{bmatrix} \tag{17.30}$$

represents the transfer matrix for one turn in the resonant cavity, obtained from the matrices representing the following steps (from right to left): propagation of distance d, reflection at the spherical mirror with radius R_2, again propagation of distance d, and reflection at the spherical mirror with radius R_1.

Multiplying the matrices on the right-hand side of Eq. (17.30), one obtains the following expressions for the elements of the ABCD matrix:

$$A = 1 + \frac{2d}{R_2} \tag{17.31a}$$

$$B = 2d + \frac{2d^2}{R_2} \tag{17.31b}$$

$$C = \frac{2}{R_1} + \frac{2}{R_2} + \frac{4d}{R_1 R_2} \tag{17.31c}$$

$$D = \frac{4d}{R_1} + \frac{2d}{R_2} + \frac{4d^2}{R_1 R_2} + 1 \tag{17.31d}$$

In this case, one has

$$\begin{vmatrix} A & B \\ C & D \end{vmatrix} = AD - BC = 1 \tag{17.32}$$

and the solution of Eq. (17.29) is given, as seen in Chapter 2, by

$$x_p = x_{\max} \sin\,(p\varphi + \varphi_0) \tag{17.33a}$$

$$\varphi = \cos^{-1} q \tag{17.33b}$$

$$q = \frac{A+D}{2} = 2\left(1 + \frac{d}{R_1}\right)\left(1 + \frac{d}{R_2}\right) - 1 \tag{17.33c}$$

The solution given by Eq. (17.33) is harmonic (and therefore bounded) if φ is real, which happens if $-1 \le q \le 1$. Introducing the so-called *g parameters*, given by

$$g_1 = 1 + \frac{d}{R_1} \quad \text{and} \quad g_2 = 1 + \frac{d}{R_2} \tag{17.34}$$

This condition can be written in the form:

$$0 \le g_1 g_2 \le 1 \tag{17.35}$$

which constitutes the so-called *confinement condition*. If this condition is not satisfied, the resonant cavity is said to be *unstable*. In the case where one has any of the equalities in Eq. (17.35), the cavity is *conditionally stable*.

In the case of a symmetrical resonator, one has $g_1 = g_2 = g$ and the condition given by Eq. (17.35) reduces to $g \le 1$, which is equivalent to $2R \ge d$. If $g = -1$, one has the case of a *concentric cavity*, in which case the centres of curvature of the two mirrors coincide at the middle of the cavity. On the other hand, if $g = 0$, one has $R = -d$, and the cavity is termed *confocal*, since the focal points of the two mirrors coincide.

17.5 Pumping Rate and Laser Dynamics

Fig. 17.8 shows a simplified scheme of the dominant transitions for the case of a four-level laser system, analogous to that represented in Fig. 17.4. The pumping rate R represents the number of atoms that pass each second from the fundamental level to the highest energy level.

The rates of change of the populations in each of the four levels are given by the following differential equations:

$$\frac{dN_3}{dt} = R - N_3 A_{32} \tag{17.36}$$

$$\frac{dN_2}{dt} = N_3 A_{32} + N_1 B_{21} \frac{I}{c} - N_2 B_{21} \frac{I}{c} - N_2 A_{21} \tag{17.37}$$

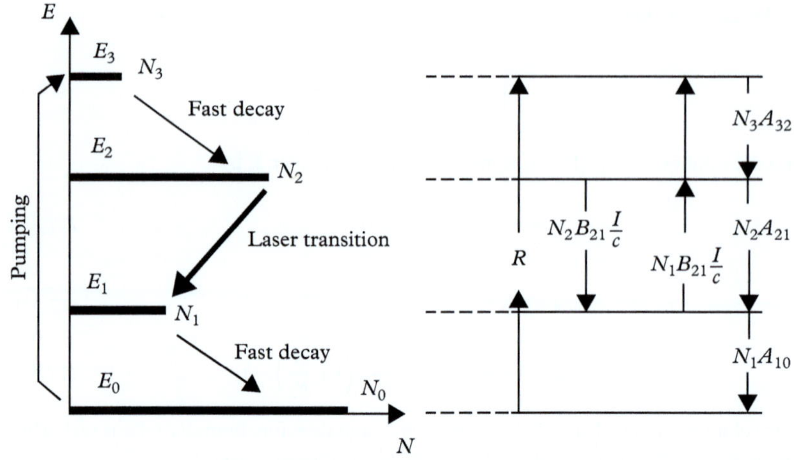

Figure 17.8 *Schematic representation of the dominant transition processes in a four-level laser system.*

$$\frac{dN_1}{dt} = N_2 B_{21}\frac{I}{c} + N_2 A_{21} - N_1 B_{21}\frac{I}{c} - N_1 A_{10} \tag{17.38}$$

$$\frac{dN_0}{dt} = N_1 A_{10} - R \tag{17.39}$$

where A_{32}, A_{21}, A_{10} and B_{21} are the Einstein coefficients for the transitions between levels 0, 1, 2, and 3. At steady state, one has

$$\frac{dN_3}{dt} = \frac{dN_2}{dt} = \frac{dN_1}{dt} = \frac{dN_0}{dt} = 0 \tag{17.40}$$

In this case, one can obtain from Eqs. (17.36–17.39) the following expression for the population difference $N_2 - N_1$:

$$N_2 - N_1 = \frac{R(A_{10} - A_{21})}{A_{10}\left[A_{21} + B_{21}\left(\frac{I}{c}\right)\right]} \tag{17.41}$$

It turns out from Eq. (17.41) that only for $A_{21} < A_{10}$ the numerator is positive, and the population inversion can be achieved.

Below threshold, the stimulated emission from the upper level of the laser transition is small, given that the irradiance in the cavity, I, is also quite low. Under these circumstances, Eq. (17.41) can be approximated by

$$N_2 - N_1 \approx R\left(\frac{1}{A_{21}} - \frac{1}{A_{10}}\right) \tag{17.42}$$

It is verified from Eq. (17.42) that the population difference increases linearly with the pumping rate.

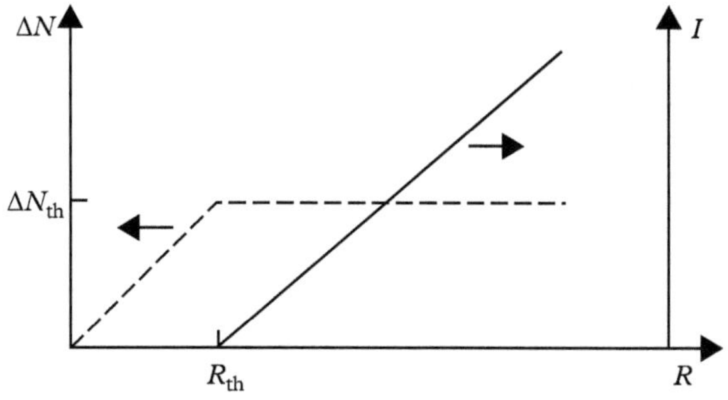

Figure 17.9 *Difference in population (left axis) and irradiance within the cavity (right axis) as a function of the pumping rate.*

At threshold, the irradiance I is still relatively small, so the population difference in this case, ΔN_{th}, is given by Eq. (17.42) with $R = R_h$. The threshold value of the pumping rate, R_h, is then given by

$$R_{th} = \Delta N_{th} \frac{A_{21}A_{10}}{A_{10} - A_{21}} \tag{17.43}$$

At steady state, the population difference does not exceed its value at the threshold, ΔN_{th}. It should be noted that the gain coefficient g given by Eq. (17.20) is proportional to the difference in populations and that its threshold value is determined by the sum of the material absorption losses and the losses corresponding to the emission of light to the outside, as shown in Eq. (17.27). Thus, for pumping rates above threshold, Eq. (17.41) gives:

$$\Delta N_{th} = \frac{R\left(1 - \frac{A_{21}}{A_{10}}\right)}{A_{21} + B_{21}\left(\frac{I}{c}\right)}, \quad \text{for } R > R_h \tag{17.44}$$

Using Eqs. (17.43) and (17.44), one can write the irradiance in the cavity as a function of the pumping rate R as follows:

$$I = \left[\frac{R - R_l}{R_l}\right] \frac{c}{B_{21}} A_{21} \tag{17.45}$$

It can be seen from this result that the irradiance in the cavity increases linearly once the threshold condition is exceeded. Fig. 17.9 illustrates the dependence of ΔN and I on the pumping rate R.

17.6 Line Broadening

Spectral line broadening is a consequence of fluctuations in the energy difference between the energy levels involved in laser emission. Such fluctuations lead to the transformation of sharp, discrete spectral lines corresponding to transitions between these energy states into broader lines. Depending on the nature of the fluctuation, there can be two types of broadening: homogeneous broadening and inhomogeneous broadening.

17.6.1 Homogeneous broadening

Homogeneous broadening occurs if all the atoms of the active medium experience the same external conditions, such as temperature or pressure. These uniform external perturbations cause the energy levels of the atoms to be affected in a similar way, leading to identical transition energies. As a consequence, the spectral lines corresponding to the emitted frequencies are broadened by the same amount, resulting in a symmetric lineshape.

An important homogeneous broadening mechanism is called *natural* or *lifetime broadening*. This mechanism is due to photon emission, and the resulting linewidth is referred to as the *natural linewidth*, since it occurs naturally for an isolated atom. However, there are other processes that contribute to the homogeneous linewidth. In a gas, for example, collisions between atoms occur at a rate that is proportional to the gas pressure, resulting in a *pressure-broadening* contribution to the transition linewidth.

The homogeneous broadening phenomenon can be described by a Lorentzian lineshape, which exhibits a symmetric, bell-shaped curve that extends infinitely along the frequency axis (Fig. 17.10). The *Lorentzian* line-shape function can be written in the form:

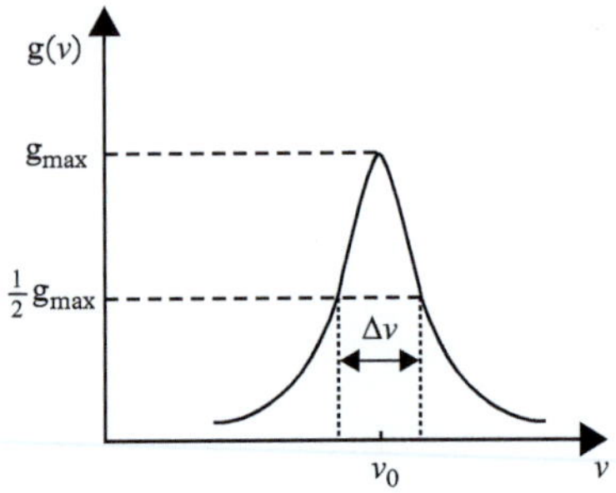

Figure 17.10 *Illustration of a Lorentzian line shape for homogeneous broadening.*

$$g(\nu) = \frac{1}{\pi} \frac{\Delta\nu/2}{(\nu - \nu_0)^2 + (\Delta\nu/2)^2} \tag{17.46}$$

where ν_0 is the centre frequency of the atomic transition, and $\Delta\nu$ is the full width at half maximum (FWHM), as illustrated in Fig. 17.10. The maximum value of the line-shape function occurs at the line-centre frequency ν_0 and is given by

$$g_{\max} = g(\nu_0) = \frac{2}{\pi\Delta\nu} \tag{17.47}$$

This inverse relation between $g_{\max}$ and $\Delta\nu$ shows that transitions with a narrow linewidth are expected to have a higher gain.

17.6.2 Inhomogeneous broadening

Inhomogeneous broadening refers to the broadening of a spectral line caused by different atoms emitting radiation at different frequencies. The differences in frequencies cause the spectral lines to broaden unevenly, leading to an irregular or asymmetric shape. In this case, the lineshape is due to the superposition of many homogeneous line-shape components with different centre frequencies, as illustrated in Fig. 17.11. Inhomogeneous broadening is often observed in dilute gases, where the atoms or molecules are well separated and do not experience frequent collisions. In such media, the Doppler effect causes the spectral lines to broaden. The Doppler effect arises due to the random motion, causing them to emit radiation at slightly different frequencies due to the relative motion between the atom and the observer

The physical processes that give rise to inhomogeneous broadening are random in nature, and the lineshape therefore represents the statistical distribution of a randomly varying centre frequency. The line-shape function resulting from these processes is well described by a *Gaussian* function:

$$g(\nu) = a \exp\left(-b(\nu - \nu_0)^2\right) \tag{17.48}$$

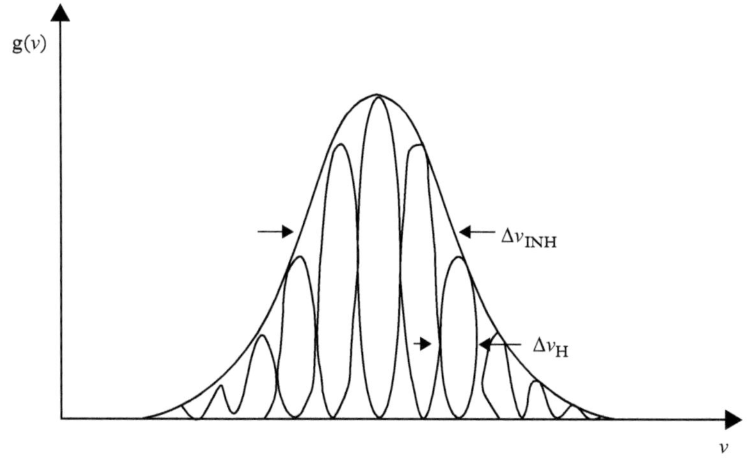

Figure 17.11 *An inhomogeneously broadened lineshape is the superposition of many homogeneous lineshape components with different centre frequencies.*

where ν_0 is the average centre frequency, and a and b are constants.

Eq. (17.48) is applicable when the homogeneous linewidth $\Delta\nu_H$ is much smaller than the inhomogeneous linewidth $\Delta\nu_{INH}$, whereas Eq. (17.46) is valid in the opposite limit, $\Delta\nu_H >> \Delta\nu_{INH}$. In the intermediate case in which $\Delta\nu_H \sim \Delta\nu_{INH}$, the lineshape is given by the *Voigt profile*, which is the convolution of these two functions.

17.7 Pulsed Operation

The most common lasers, such as the He–Ne laser and semiconductor lasers, produce light in the form of a continuous beam. However, some types of lasers are designed to produce very short, sometimes high-power, pulses. These ultra-short pulses can have many applications. For example, several very fast processes in atoms, molecules, and some materials can be excited and investigated using these pulses. They can also be used for pumping X-ray lasers, as well as for studying plasmas with very high temperatures and densities.

A simple means of producing optical pulses from a laser would be to gain-switch the system, turning the source of laser energy on and off. Using this technique, however, determines a complex exchange of energy between the field and the atomic population, which makes it difficult to control the pulse characteristics. Two more effective techniques for producing ultra-short light pulses, namely *Q-switching* and *mode-locking*, are described in the subsequent section.

17.7.1 Q-switching

Under normal conditions, the laser resonant cavity has relatively low losses: it is said that the cavity has a high-quality factor Q. However, laser action can be inhibited by reducing this Q factor, for example by misaligning one of the mirrors. During this reduction of the Q factor, a large population inversion grows in the gain medium, but laser oscillation is inhibited because the small-signal gain coefficient is less than its threshold value. In this situation, the cavity losses are deliberately and suddenly reduced (the Q factor of the cavity being switched to a high value), which causes a consequent reduction in the threshold gain. As a result, the radiation in the cavity increases rapidly and all the available energy is emitted in a single pulse. The sudden variation of the losses in a laser cavity is known as *Q-switching*. The large irradiance in the cavity induces a high stimulated-emission rate, which causes a rapid depopulation of the excited state until the gain is reduced below its threshold value and the laser action ceases. The quality factor Q is then switched back to a low value, allowing the population inversion to grow once again to a large value, and the cycle repeats. Q-switching was independently discovered and demonstrated in 1961 or 1962 by R. W. Hellwarth and F. J. McClung at Hughes Research Laboratories using electrically switched Kerr-cell shutters in a ruby laser.

The *average power* output from a Q-switched system is roughly the same as the CW power of the same system. Q-switching simply redistributes the energy so that it comes out in large bursts separated by periods of nearly zero energy output.

The duration of a pulse obtained by Q-switching is on the order of the cavity lifetime or photon lifetime, given by

$$\tau_c = \frac{2d}{c(1 - R_1 R_2)} \tag{17.49}$$

where d is the length of the resonant cavity, n *is* the refractive index of the active medium, c is the speed of light in vacuum, and R_1, R_2 are the reflectances of the mirrors. Typically, the pulse width in a Q-switched system is on the order of 1 microsecond or less, whereas the *pulse repetition time* is on the order of 1 ms. As a result, the *peak power* in a Q-switched pulse can be a factor of 1000 or more larger than the CW output power from the same laser system.

The Q-switching effect can be achieved using different techniques, either active or passive. Active Q-switching uses an externally controlled variable attenuator. This may be performed by a mechanical device, such as a shutter, a chopper wheel, or a spinning mirror or prism placed inside the cavity. It can also be realized using a modulator, such as an acousto-optic, a magneto-optic, or an electro-optic device, namely Pockels or Kerr cells. Modulators generally allow a faster transition from low to high Q and provide better control. Passive Q-switching is performed using a saturable absorber, a material whose transmission increases when the intensity of light exceeds some threshold. In the low-Q state, the population inversion grows, which leads to an increase in spontaneous emission. Q-switching occurs when the irradiance due to spontaneous emission reaches a value large enough to saturate the absorber and the cavity loss is suddenly reduced.

17.7.2 Mode-locking

A laser typically oscillates in several longitudinal modes. The variation in time of the power emitted by the laser depends on the frequencies, phases, and relative amplitudes of these modes. In an ordinary laser, these characteristics vary with time, so the emitted power fluctuates randomly. However, if the phases of the various modes are fixed relative to each other, the emitted light has repetitive characteristics, and the laser is said to operate in a *mode-locking regime*.

The mode-locking regime can be achieved by varying the losses in the cavity. These losses can be induced using the same methods mentioned earlier in relation to the Q-switching technique, provided that the loss-modulation gate can be opened and closed over a time period shorter than a cavity round-trip time. The process is initiated by spontaneous emission into the different cavity modes in such a way that the fields in these modes interfere constructively at the gate when it is open. Field components not phase-locked to form constructive interference at the gate position are blocked by the closed gate.

In order to achieve pulses of high power and short duration, it is necessary to have a relatively high number of oscillating modes. Suppose one has N modes with the same amplitude, with a constant relative phase, and with a spacing between them $\delta\nu = c/2d$, where d is the length of the resonant cavity. The total electric field is then given by

$$E(t) = E_0 \sum_{n=1}^{N} \exp(-i2\pi\nu_n t) = E_0 \exp(-i2\pi\nu t)\frac{\sin(N\pi ct/2d)}{\sin(\pi ct/2d)} \tag{17.50}$$

where $\nu_n = \nu + n\delta\nu$. The corresponding irradiance is given by

$$I(t) = I_0 \frac{\sin^2(N\pi ct/2d)}{\sin^2(\pi ct/2d)} \tag{17.51}$$

where $I_0 = E_0^2$. Thus, the power at the positions of equal phase, where the fields constructively interfere, is given by

$$P_p = N^2 P_0 \tag{17.52}$$

where P_0 is the power of a single longitudinal mode. The output of the laser is constituted by a train of pulses evenly spaced in time with a period

$$T = \frac{2d}{c} = \frac{1}{\delta\nu} \tag{17.53}$$

which corresponds to the time of one turn of a pulse in the cavity. The width of this pulse, τ, is approximately given by the inverse of the width of the laser emission spectrum:

$$\tau = \frac{1}{\left(\frac{Nc}{2d}\right)} = \frac{2d}{Nc} \tag{17.54}$$

From Eqs. (17.53) and (17.54) it appears that the ratio of the time spacing to the width of the pulses, T/τ, is approximately equal to the number of modes, N. Therefore, we verify that to obtain fairly short pulses using the mode-locking technique, one must have a relatively high number of oscillating modes.

Example 17.4

Consider a laser whose cavity has a length d = 1.5 m and whose gain curve has a width $\Delta\omega = 6 \times 10^{12}\text{s}^{-1}$. Considering that this laser operates in the mode-locking regime, obtain the spacing between the longitudinal modes of the laser, the irradiance as a function of time, the peak value of the irradiance, the duration of each pulse, and the separation between consecutive pulses.

continued

Example 17.4 continued

Solution

The spacing between the longitudinal modes is given by

$$\delta\nu = \frac{c}{2d} = 10^8\ \mathrm{s}^{-1}$$

The irradiance is given from Eq. (15.20) by

$$I(t) = I_0 \frac{\sin^2(N\pi ct/2d)}{\sin^2(\pi ct/2d)}$$

where

$$N = Int\left(\frac{\Delta\omega}{2\pi\delta\nu}\right) = 9549$$

The pulse duration is

$$\tau = \frac{2d}{Nc} = 1.05 \times 10^{-12}\ \mathrm{s}$$

The separation between consecutive pulses is given by

$$T = \frac{1}{\delta\nu} = 10^{-8}\ \mathrm{s}$$

The peak irradiance can be obtained by assuming $t = 0$:

$$I(0) = I_0 N^2 = 9.12 \times 10^7\, I_0$$

17.8 Some Types of Lasers

Actually, there is a great variety of lasers using gaseous, liquid, or solid active media. Some of the most representative lasers are briefly described in the next section, and some of their applications are also mentioned. Semiconductor lasers, which today constitute the most economically important class of lasers, will be covered in more detail in the next chapter, whereas fibre lasers will be discussed in Chapter 21.

17.8.1 The first laser—ruby laser

As mentioned at the beginning of this chapter, the first working laser made by Theodore H. Maiman in May 1960 was a laser whose active medium consisted of a small synthetic crystal of pink ruby, that is, a crystal of Al_2O_3 with about 0.05% of Cr_2O_3. The crystal had faces parallel to each other, perpendicular to the axis and partially mirrored (forming a resonant cavity), being installed on the axis of a helical gaseous discharge lamp (pumping source). Ruby is one of the few solid-state lasers that produce light in the visible range of the spectrum. Typical ruby laser pulse lengths are on the order of a millisecond.

The ruby laser is an example of a laser with three levels. When discharging in the pumping lamp, there is the excitation of many Cr^{3+} ions for absorption bands. The excited ions decay through non-radioactive transitions to a pair of metastable intermediate levels, where they remain for several milliseconds until they decay to the ground state, emitting a red fluorescent radiation centred at 694.3 nm. Ruby lasers are today used as pulsed sources of coherent and high-intensity radiation, in interferometry, plasma analysis, holography, etc.

17.8.2 The helium–neon laser

Seven months after the discovery of the ruby laser by Maiman, Ali Javan, W. R. Bennett, and D. R. Herriott showed the possibility of building a laser with continuous emission using, as an active medium, a gas at low pressure, consisting of a mixture of helium and neon. The He–Ne laser is currently the most popular laser, operating almost always in visible light and providing a few milliwatts of continuous power.

A typical configuration of the He–Ne laser is shown in Fig. 17.12. The optical cavity is provided by two spherical mirrors, which are coated with multilayer dielectric films in order to achieve high reflectance at the desired wavelength. One mirror has a very high reflectivity, while the other (output mirror) has a transmission that optimizes the laser output. Windows at the ends of the glass tube are set at Brewster's angle in order to obtain the maximum possible transparency. This results in linearly polarized laser light because the windows are highly transparent to one preferred direction of polarization, as discussed in Section 6.6.

The glass tube with two metal electrodes inside is evacuated and then backfilled with a low-pressure mixture of helium and neon gas. An electrical discharge is produced within the glass tube by applying a high voltage between the anode and the cathode. The fast-moving electrons originating in such a discharge collide with He atoms, raising them to an excited state, a process termed *electron-impact excitation*. When one of the metastable helium atoms, which has a high average lifetime, collides with a neon atom in the ground state, there is a high probability of energy transfer to this atom, exciting it to one of the 3s or 2s levels, while the helium atom returns to the ground state (Fig. 17.13). Under suitable discharge conditions, population inversion can be achieved with respect to transitions between the levels 3s and 3p (emission at $3.39\,\mu m$), 3s and 2p ($632.8\,nm$), and 2s and 2p ($1.15\,\mu m$). The optimum value of the total pressure is about 1 torr, and the most favourable ratio of helium to neon is found to be about 7:1.

He–Ne lasers with a tube length of 10–20 cm have typical output powers of a few mW. The efficiency is quite low, with electrical-to-optical conversion efficiencies of $\sim 0.02\%$. In comparison, a semiconductor laser operating in the same wavelength range can have efficiencies above 10% and requires a lower-voltage power supply.

In spite of the above limitations, He–Ne lasers are still widely used, mainly because they provide an inherently circular beam, with a beam divergence that is close to being diffraction limited. In contrast, the semiconductor laser beam is inherently asymmetrical.

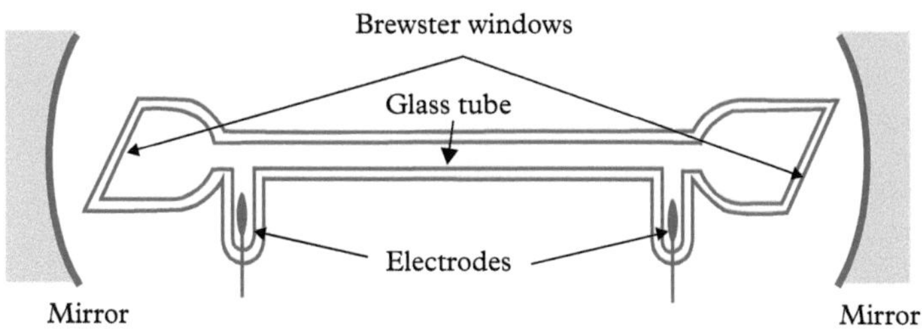

Figure 17.12 *Schematic of the He–Ne laser.*

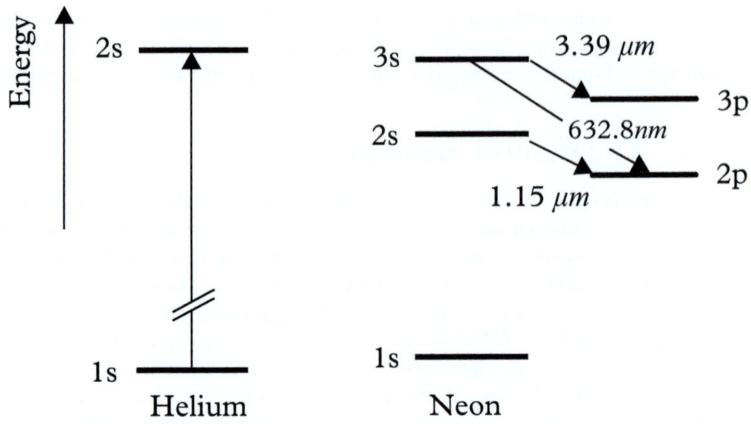

Figure 17.13 *Helium–Neon laser energy levels.*

Another advantage of the He–Ne laser is that the coherence length is relatively long—typically 10–20 cm for a laser with multiple longitudinal modes and hundreds of metres for a laser with a single longitudinal mode. This makes the He – Ne laser ideal for holography and interferometry applications.

Example 17.5

Assume a helium–neon laser emitting light at wavelength $\lambda = 632.8\,\text{nm}$. Considering that the 2p level of the neon atom has an energy of $15.2 \times 10^{-19}\,\text{J}$, determine the pumping energy that must be used.

Solution

The pumping energy must be at least equal to the energy difference between the ground level and the metastable state 3s, $\Delta E_{3s,0}$. This difference in energy is given by

$$\Delta E_{3s,0} = \Delta E_{2p,0} + \Delta E_{3s,2p}$$

where $\Delta E_{2p,0}$ is the energy of the 2p level relative to the fundamental level, and $\Delta E_{3s,2p}$ is the difference between the energies of the 3s and 2p levels. This difference is given by

$$\Delta E_{3s,2p} = E_{3s} - E_{2p} = h\nu = \frac{hc}{\lambda}$$

$$= \frac{\left(6.6 \times 10^{-34}\right)\left(3 \times 10^{8}\right)}{6.32.8 \times 10^{-9}} = 3.12 \times 10^{-19}\text{J}$$

Therefore, one has

$$\Delta E_{3s,0} = \Delta E_{2p,0} + \Delta E_{3s,2p}$$

$$= 15.2 \times 10^{-19} + 3.12 \times 10^{-19} = 18.32 \times 10^{-19}\,\text{J}$$

17.8.3 Other types of lasers

Since Maiman discovered the ruby laser, it has been found that many other materials are likely to provide the same effect, emitting energy from the ultraviolet to the submillimetre region (0.774 mm for the HCN laser). All gaseous elements can be used as active media in a laser, as can many metals and molecules with two or three atoms. Among other examples, laser phenomena have been observed in systems using trivalent rare-earth ions inserted into solid glass arrays and yttrium and aluminium crystalline garnets (YAG, from 'yttrium and aluminium garnet').

Among *gaseous lasers* and in addition to the He–Ne laser, the argon laser is well known, emitting in the green, blue-green, and violet regions. Gas atomic lasers are typically Doppler-broadened and have low efficiencies (less than 1%). Another important gas laser is the CO_2 laser, emitting in the infrared at 10.6 μm. The CO_2 laser is one of the most versatile and efficient laser systems, capable of CW power outputs ranging from mW to MW, with efficiencies ranging from 10% to 40%. These gaseous lasers are typically pumped by an electric discharge.

An *excimer* laser typically uses a combination of a noble gas (argon, krypton, or xenon) and a reactive gas (fluorine or chlorine). Rare-gas atoms have filled outer shells and so tend not to bond to other atoms. However, an excited rare-gas atom has a single electron in its outer shell and so can form an ionic bond with a halogen atom. Under the appropriate conditions of electrical stimulation and high pressure, a pseudo-molecule called an excimer is created, which can exist only in an energized state and can give rise to laser light in the ultraviolet range. Excimer lasers are commonly used in various applications, including eye surgery, semiconductor manufacturing, and micromachining.

The first *liquid laser* operated in 1963 and emitted at 613.1 nm, its active medium consisting of an alcoholic solution of europium benzoyl acetonate. Subsequently, it was possible to observe the laser phenomenon in a wide variety of dye solutions so that today one can build a laser at almost any frequency in the visible spectrum. Dye lasers have very large gain bandwidths (50–100 nm) and are usually tunable in a fairly extensive wavelength band. Such a broad gain bandwidth also enables the production of extremely short (ps-width) mode-locked pulses.

In addition to the lasers already mentioned, one could also mention chemical lasers, which are pumped by the energy released in a chemical reaction; vapour lasers; free-electron lasers; X-ray lasers; and many others with particular properties, generating ultra-short pulses or with a great stability in frequency. The latter are particularly useful in high-resolution spectroscopy.

Solid-state lasers are a class of lasers in which the lasing atomic species is doped into a transparent host material. The most common solid-state laser is the Nd:YAG laser, in which the dopant, neodymium, typically replaces a small fraction (1%) of the yttrium ions in the host crystal structure of YAG, since the two ions are of similar size. It is the neodymium ion that provides the lasing activity in the crystal. Nd:YAG lasers are optically pumped using a flashtube or semiconductor lasers and typically emit light with a wavelength of 1064 nm. They can operate in both pulsed and continuous modes. Pulsed Nd:YAG lasers are typically operated in the so-called Q-switching mode, with pulse durations of 10 to 25 nanoseconds and output powers of 250 megawatts. The high-intensity pulses may be efficiently frequency-doubled to generate laser light at 532 nm, or higher harmonics at 355, 266, and 213 nm.

17.9 Further Reading

Hecht, E., *Optics*, Pearson Education, 5th Ed., 2016.
Miloni, P., and J. H. Eberly, *Lasers*, John Wiley & Sons, 1988.
O'shea, D. C., *Introduction to Lasers and Their Applications*, Addison-Wesley, 2nd Ed., 1978.
Pedrotti, F., L. M. Pedrotti, and L. S. Pedrotti, *Introduction to Optics*, Cambridge University Press, 3rd Ed., 2018.
Siegman, A. E., *Lasers*, Oxford University Press, 1986.
Svelto, O., *Principles of Lasers*, Plenum Press, 1998.
Yariv, A., *Optical Electronics in Modern Communications*, Oxford University Press, 5th Ed., 1997.

17.10 Problems

17.1 Consider the equation

$$\frac{dN_1}{dt} = -A_{10}N_1$$

for the decay of the population of the energy level E_1 due to spontaneous emission. Show that the initial population N_{10} decreases to a value N_{10}/e after a time $\tau = 1/A_{10}$. The constant τ is known as the *lifetime of spontaneous emission*.

17.2 Explain why it will be more difficult to make a laser that emits light in the ultraviolet region than in the infrared region. Base your explanation on the ratio A_{10}/B_{10} and the meaning of the coefficients A_{10} e B_{10}.

17.3 Considering a system in thermal equilibrium and a wavelength $\lambda = 10\,\mu\text{m}$, determine the temperature at which the rates of spontaneous emission and stimulated emission become equal.

17.4 An atom has two energy levels with a transition wavelength $\lambda = 591.6\,\text{nm}$. At what temperature are 10% of the atoms in the higher energy state? Assume that all atoms are in one or the other of these two energy states.

17.5 Considering two mirrors with radii of curvature $R_1 = -100\text{cm}$ and $R_2 = -200\,\text{cm}$, determine for which separation values between them the configuration is (i) stable, (ii) unstable, and (iii) conditionally stable.

17.6 Assuming that the central wavelength of a He–Ne laser is $\lambda = 632.8\,\text{nm}$, what is the corresponding frequency? If the separation between the mirrors of the cavity is 50 cm, what is the value of the integer q corresponding to the mode closest to the centre of the line? If the width of the gain curve of the He–Ne laser is 1 GHz, what is the number of longitudinal modes for this cavity?

17.7 A helium–neon laser emitting at 632.8 nm has a resonant cavity formed by two plane mirrors spaced by 0.5 m.

a) Determine the number of longitudinal modes that contribute to the laser emission, knowing that the spectral width of the spontaneous emission line of Ne at 632.8 nm is 0.0016 nm.

b) Obtain the coherence length of this laser and compare it with that of a stabilized Nd–YAG laser operating in a single longitudinal mode, with spectral width $\Delta\nu = 30\,\text{kHz}$.

17.8 The ends of a ruby crystal ($n = 1.76$) are cut and polished to form the mirrors of a laser, which emits light with a wavelength $\lambda = 694$ nm. Considering that the crystal has a length of 12 cm, determine the number of longitudinal modes and their frequency separation.

17.9 Considering that the ruby crystal in Problem 17.8 has a diameter of 6 mm, discuss the importance of diffraction in the calculation of energy losses in the cavity of this laser.

17.10 Derive the results in Eqs. (17.31a–d) for the elements of the ABCD matrix corresponding to the laser resonant cavity, as well as Eq. (17.33c) for the parameter q.

17.11 Considering the rate Eqs. (17.36)–(17.39) and assuming the steady-state conditions, derive the result given by Eq. (17.41) for the population difference $N_2 - N_1$.

17.12 Consider a laser system with four levels, in which the pumping rate is proportional to the difference in the populations of the lowest and highest energy levels, $R = (N_2 - N_1)\,W$, where W is the transition probability. Considering that the total population of the four levels is N_t, and that the lifetimes of levels 3 and 1 are very short, obtain an expression for the inversion of the population $N_2 - N_1$.

17.13 Discuss the pumping scheme with three levels and compare it with the case of four levels. Considering that the total population is N_t and that the lifetime of level 2 is very short, obtain (i) an expression for the difference in populations $N_1 - N_0$ in the steady state and (ii) the minimum pumping rate that ensures population inversion.

17.14 A Nd:YAG laser with a CW output power of 5 W is Q-switched with a pulse repetition rate of 2 kHz and a pulse width of 1 μs. Find (a) the peak power and (b) the energy of each Q-switched pulse.

17.15 The resonant cavity of a He–Ne laser has mirrors with reflectances $R_1 = 0.99$ and $R_2 = 0.98$. Find the minimum length of such a cavity in order to have a small-signal gain coefficient of 0.001/cm.

18 Semiconductor Lasers

Semiconductor lasers (SLs) were invented in 1962, but their use became common only in the seventies of the last century, after achieving continuous operation at room temperature. The development they have experienced since then has been such that they are today of all types of lasers, the most numerous and the most important from an economic perspective.

The applications of SLs are currently very varied, namely in fibre optic communication systems, barcode readers, laser printers, CD players, etc. In the case of fibre optic communication systems, the preference for SLs is related to their small dimensions (typically less than 1 mm), emission area compatible with the dimensions of the fibre core, high efficiency, adequate wavelength of light emitted, and possibility of direct modulation at very high frequencies.

As with most lasers, SLs rely on the stimulated emission of photons to produce a coherent beam of light. However, the processes of pumping and generating photons are distinct from those of other types of lasers, as will be seen in the following sections.

18.1 Carrier Distribution in a Semiconductor

In the solid state, the spacing between the atoms that make up matter is on the order of only a few angstroms. Since the electron cloud surrounding the nucleus of each of these atoms extends into a volume of similar dimensions, there is an alteration of the discrete energy levels characteristic of the isolated atom, resulting in the formation of *energy bands* in the solid. The upper energy band that is completely filled by electrons at 0 K is known as the *valence band*, whereas the empty band immediately above is called the *conduction band*. These two bands are separated by the so-called *forbidden energy gap*, or *bandgap*, whose value, E_g is indicative of some important properties, both optical and electrical, of the solid.

Semiconductors are materials whose electrical conductivity values lie between those of a good conductor and those of an insulator. Most semiconductors correspond to elements of groups II to VI of the Periodic Table, presenting bandgaps in the order of 0.25–2.5 eV. In the case that the bandgap is much larger, the solid is an insulator. In the opposite case, when there is no bandgap, the solid corresponds to a conductor.

At absolute zero temperature, the conduction band has no electrons and the material is a perfect insulator. When the temperature increases, a certain number of electrons may pass into the conduction band, leaving an equal number of empty hole states in the valence band. These holes can be considered, for many purposes, as positively charged particles that can move between the remaining electrons. If an external electric field is applied, electrons and holes move in opposite directions, with both types of particles contributing to the conductivity of the semiconductor. Both electrons and holes are commonly referred to as *carriers*.

Lasers, Optics, and Photonics. Mário F. S. Ferreira, Oxford University Press.
DOI: 10.1093/oso/9780198952770.003.0018

For a given absolute temperature T, the probabilities of occupation of the different energy states by electrons are given by the Fermi–Dirac distribution:

$$f(E) = \frac{1}{1 + e^{-(E - E_f)/k_B T}} \tag{18.1}$$

where k_B is the Boltzmann constant and E_f is a constant known as the *Fermi level* or *Fermi energy*. The Fermi energy represents the energy for which the probability of electron occupancy is 0.5. In a pure semiconductor, the number of electrons in the conduction band is equal to the number of holes in the valence band, and the Fermi level lies approximately halfway between the upper limit of the valence band and the lower limit of the conduction band (see Fig. 18.1). Since $f(E)$ is the probability that the energy level E is occupied, $1 - f(E)$ is the probability that it is empty, i.e., that it is occupied by a hole. At $T = 0°$ K. $f(E) = 1$ for $E \leq E_f$ and $f(E) = 0$ for $E > E_f$ (see Fig 18.1(c)); there are then no electrons in the conduction band and no holes in the valence band.

The energy bands of a semiconductor can be altered by introducing atoms of other elements (impurities) into the crystal lattice. This process is called *doping*. For example, silicon sits in group IV of the Periodic Table, with each of its atoms having four electrons in the outer shell. The atom of an element of group V has five electrons in the outer shell, so that if the impurity consists of these atoms, there will be one more electron than those necessary for the establishment of covalent bonds with the silicon atoms. This additional electron is easily lost to the conduction band, so it is said that the impurity is of the *donor* type. On the other hand, in the case of the impurity being made up of an element of group III, it functions as an *acceptor*, since each of its atoms has only three electrons in the last layer, contributing a hole to the valence band.

A dopant-free semiconductor is sometimes called an *intrinsic* semiconductor, while once doped it is called an *extrinsic semiconductor*. When it is doped with a donor, it is said to be an *n-type semiconductor*. On the other hand, when it is doped with an acceptor, it is said to be a *p-type semiconductor*.

In an n-type semiconductor, there are a greater number of electrons in the conduction band than holes in the valence band, so the Fermi level is closer to the boundary of the

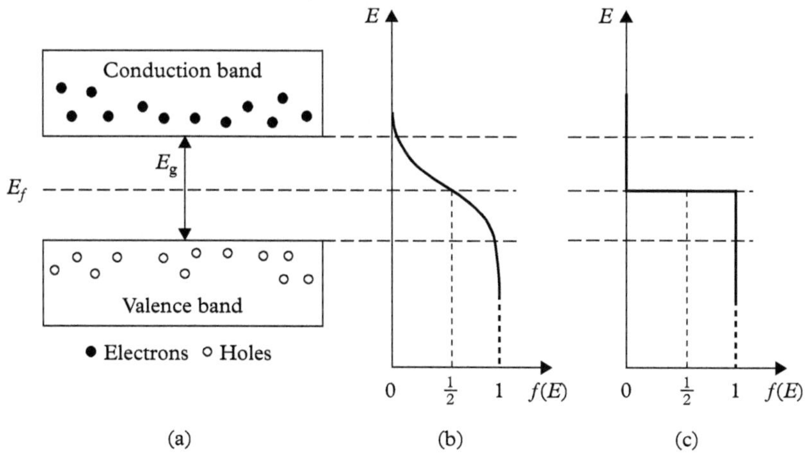

Figure 18.1 *(a) Energy band structure of an intrinsic semiconductor and carrier population above absolute zero. (b) The Fermi–Dirac distribution corresponding to case (a) and at $T = 0°K$ (c). In case (c), $f(E) = 1$ for $E \leq E_f$, and $f(E) = 0$ for $E > E_f$; there are then no electrons in the conduction band and no holes in the valence band.*

conduction band. On the other hand, in a *p*-type semiconductor, there are a greater number of holes in the valence band than electrons in the conduction band, thus placing the Fermi level closer to the boundary of the valence band.

The previous description is valid only for a semiconductor in thermal equilibrium, and it is not generally applicable when this equilibrium is disturbed. However, there are situations in which the electrons in the conduction band are in thermal equilibrium with each other, as are the holes in the valence band, although there is no thermal equilibrium between the electrons and the holes. These situations are known as *quasi-equilibrium* situations and happen when the relaxation times for transitions within each of the bands are much shorter than the relaxation time between the two bands.

In quasi-equilibrium situations, it is appropriate to use distinct Fermi functions for the conduction band and for the valence band. The respective Fermi levels, E_{fc} and E_{fv}, are known as *quasi-Fermi levels*. The probabilities of occupation of the different energy states in the conduction band and in the valence band are given, respectively, by

$$f_c(E) = \frac{1}{1 + e^{-(E - E_{fc})/k_B T}} \tag{18.2}$$

and

$$f_v(E) = \frac{1}{1 + e^{-(E - E_{fv})/k_B T}} \tag{18.3}$$

18.2 Absorption and Emission in a Semiconductor

In the semiconductor band model, it is assumed that the electrons with energy E move through the semiconductor material with a wave vector $\mathbf{k}$. The energy of the electron is related to the modulus of the wave vector in the form

$$E = \frac{\hbar k^2}{2m^*} \tag{18.4}$$

where m^* is the effective mass of the electron and $\hbar = h/2\pi$, h being Planck's constant. This parabolic relationship between E and k has important consequences for electron transitions in the semiconductor. Semiconductors in which the minimum of the conduction band and the maximum of the valence band occur for the same $\mathbf{k}$ are known as direct-recombination or *direct-bandgap* semiconductors (Fig. 18.2(a)); those in which this does not occur are referred to as semiconductors with *indirect recombination* or an *indirect bandgap* (Fig. 18.2(b)).

Figure 18.2 – *$E - k$ diagrams for semiconductor cases with (a) a direct bandgap and (b) an indirect bandgap.*

As in the case of an atomic system, one can distinguish three distinct processes involving an electron, a hole, and a photon in a semiconductor: absorption, spontaneous emission, and stimulated emission. In the latter case, the presence of a photon with adequate energy induces the recombination of an electron in the conduction band with a hole in the valence band, resulting in the emission of a photon with the same characteristics as the original photon.

If one considers that the states of the conduction band and the valence band involved in the electron transition have energies E_2 and E_1, respectively, the law of conservation of energy requires that the photon involved in this transition has an energy

$$h\nu = E_2 - E_1 \tag{18.5}$$

On the other hand, if $\mathbf{k}_1$ and $\mathbf{k}_2$ represent the wave vectors associated with the electrons in these two states, the law of conservation of momentum requires that

$$\hbar\mathbf{k}_1 + \hbar\mathbf{k}_\nu = \hbar\mathbf{k}_2 \tag{18.6}$$

where $\mathbf{k}_\nu = 2\pi/\lambda$ is the wave vector associated with the photon.

Since k_ν is 2–3 orders of magnitude lower than the wave vectors associated with the electrons, one has

$$\Delta k = k_2 - k_1 \approx 0 \tag{18.7}$$

This condition is known as the *k-selection rule* and has the consequence that the transitions between the conduction and valence bands are mainly 'vertical transitions' in the $E-k$ diagram.

In addition to vertical transitions, one can verify the occurrence of transitions in which the initial and final moments of the electron are different, provided that certain quantum units—called *phonons*—related to the state of vibration of the crystal lattice also participate in these transitions. The probability of this type of transition occurring is, however, much lower than that of vertical transitions. For this reason, the semiconductors used in practically all light sources have a direct bandgap.

18.3 Optical Amplification in a Semiconductor

If E_1 and E_2 are the energies of the electron states in the valence band and the conduction band, respectively, the probability of the absorption of a photon with energy $h\nu = E_2 - E_1$ in a quasi-equilibrium situation is given by

$$P_{ab}(\nu) = f_v(E_1)[1 - f_c(E_2)] \tag{18.8}$$

where f_c and f_v represent the distribution functions in the conduction and valence bands, respectively, given by Eqs. (18.2) and (18.3). It should be noted that if $f_v(E_1)$ represents the probability of occupying a state in the valence band with energy E_1, $[1 - f_c(E_2)]$ gives the probability of a state in the conduction band with energy E_2 being unoccupied.

The probability of the emission of a photon with the same energy $h\nu = E_2 - E_1$ is given by

$$P_{em}(\nu) = f_c(E_2)[1 - f_v(E_1)] \tag{18.9}$$

The emission process will dominate over the absorption process if the condition is met:

$$P_{em}(\nu) > P_{ab}(\nu) \tag{18.10}$$

Using Eqs. (18.2), (18.3), (14.8), and (18.9), Eq. (18.10) results in the condition:

$$E_{fc} - E_{fv} > h\nu \tag{18.11}$$

Since it must be $h\nu \geq E_g$, it follows that

$$E_{fc} - E_{fv} > E_g \tag{18.12}$$

Eq. (18.12) represents the necessary condition for optical amplification in a semiconductor medium, and it shows that the separation between the Fermi levels must exceed the bandgap energy for population inversion to occur. In thermal equilibrium, the two Fermi levels coincide. They can be separated only by pumping energy into the semiconductor from an external energy source.

18.4 The $p - n$ Junction

A semiconductor diode is a junction formed between an *n*-type doped semiconductor and a *p*-type doped semiconductor. Due to the difference in concentration of electrons and holes in the two regions, there is electron diffusion from the *n* region to the *p* region and diffusion of holes in the opposite direction. This diffusion and the recombination of electrons and holes produce a thin region free of mobile carriers at the junction, across which a potential barrier, V_0, is established. This potential barrier prevents the flow of carriers from one region to another if no externally applied voltage is applied. Parts (a) and (b) of Fig. 18.3 represent the *p*-type and *n*-type doped semiconductors, respectively, whereas part (c) represents the *p–n* junction. The Fermi level for the *p*- and *n*-type semiconductors is the same in the equilibrium state, as shown in Fig. 18.3(c).

The application of an external positive voltage to the *p*-type region with respect to the *n*-type determines a reduction of both the depletion region width and the resulting potential barrier. In this case, the diode is said to be forward biased, producing an increased electron flux from region *n* to region *p* and holes in the opposite direction. The corresponding electrical current increases exponentially with the applied voltage, in this form.

$$I = I_s\,[\exp(qV/k_B) - 1] \tag{18.13}$$

where q is the carrier charge, V is the applied voltage, and I_s is the saturation current. The application of a potential voltage with *inverse polarization* also causes a small current

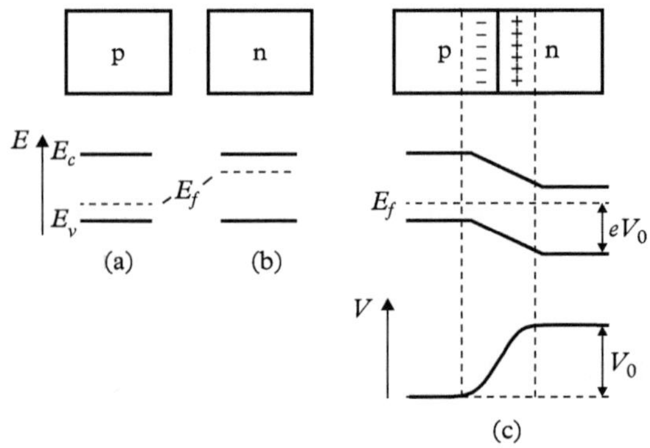

Figure 18.3 *Diagrams illustrating the development of the potential difference across the p–n junction.*

to appear in the opposite direction. Fig. 18.4 illustrates the typical dependence of this current on the voltage applied to the semiconductor diode.

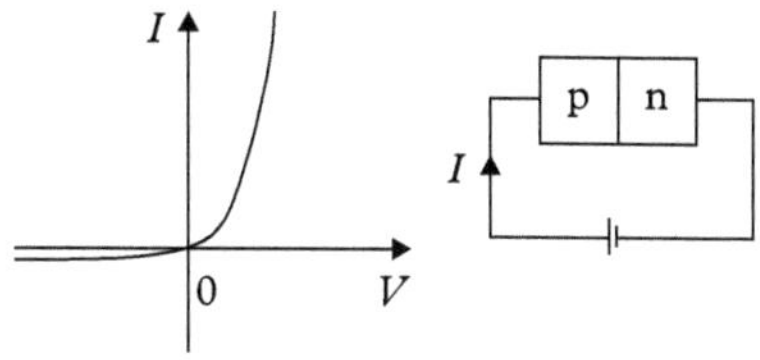

Figure 18.4 *Characteristic voltage–current curve for a semiconductor diode.*

18.5 Light-Emitting Diodes

As mentioned earlier, the application of an external potential difference in the forward direction to a *p–n* junction gives rise to a current through the junction: the electrons on the n side of the junction are injected into the p region, while the holes from the p region are injected into the n region. In this process, there is recombination of these two types of carriers, giving rise to optical radiation through the process of spontaneous emission. This phenomenon is called *injection electroluminescence* and is the basis of the operation of *light-emitting diodes* (LEDs). Given its origin in the spontaneous emission process, the light emitted from an LED is incoherent, with a relatively wide spectral width and large angular spread.

Besides radiative recombination giving rise to the creation of a photon, electrons and holes can also recombine non-radiatively. Non-radiative recombination mechanisms include recombination at defects and Auger recombination. In the Auger recombination process, the energy released during electron–hole recombination is given to another electron or hole as kinetic energy.

All non-radiative processes are harmful, as they reduce the number of electron–hole pairs that emit light. Their effect can be incorporated through the *internal quantum efficiency*:

$$\eta_{int} = \frac{R_{rr}}{R_{tot}} = \frac{R_{rr}}{R_{rr} + R_{nr}} \tag{18.14}$$

where R_{rr} and R_{nr} are the radiative and non-radiative recombination rates, respectively. We define the *carrier lifetime*, τ_c, as the total recombination time of charged carriers in the absence of stimulated emission, such that

$$R_{sp} + R_{nr} = \frac{N_d}{\tau_c} \tag{18.15}$$

where N_d is the carrier density.

At a given current I, the carrier-injection rate is I/q, where q is the carrier charge. The rate of electron–hole pair recombination through radiative and non-radiative processes is the same. Given the internal quantum efficiency η_{int}, the internal optical power is given by

$$P_{\text{int}} = hv\,(\eta_{\text{int}} I/q) \tag{18.16}$$

where hv is the photon energy. If η_{ext} is the fraction of photons escaping from the device, the emitted power is given by

$$P_{ext} = \eta_{ext} P_{\text{int}} = \eta_{ext}\eta_{\text{int}} hv\,(I/q) \tag{18.17}$$

The quantity η_{ext} is called the *external quantum efficiency*. This emitted power increases linearly with the injection current, as given by Eq. (18.17). However, for high injection

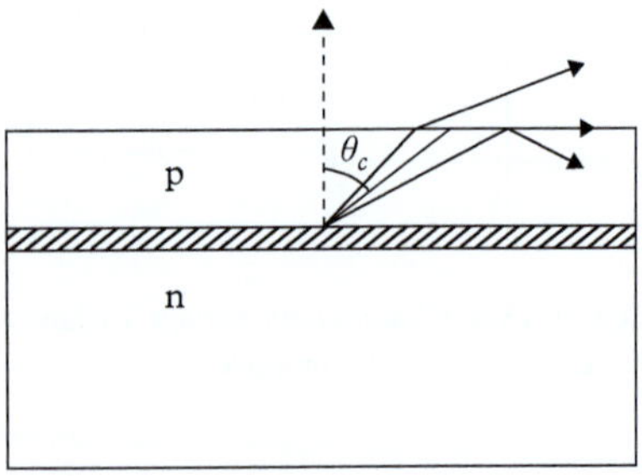

Figure 18.5 *Total internal reflection at the output of a LED.*

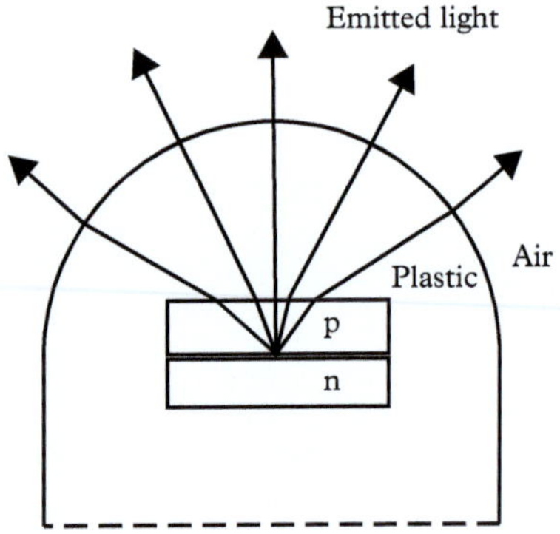

Figure 18.6 *Encapsulating the p–n junction in a plastic dome structure reduces the total internal reflection losses.*

currents, a saturation of the output power is observed at values in the order of a few milliwatts.

The external quantum efficiency η_{ext} can be calculated by considering the internal absorption and reflection at the air–semiconductor interface. As seen in Fig. 18.5, only light emitted within a cone defined by the critical angle for total internal reflection escapes from the LED surface. Typical values are η_{ext}~1.5%, indicating that only a small fraction of the internal power becomes useful output power. A further loss in useful power occurs when the emitted light is coupled into an optical fibre. Because of the incoherent nature of the emitted light, a LED acts as a *Lambertian source*, with an angular distribution $S(\theta) = So \cos\theta$, where So is the intensity in the direction $\theta = 0$. The coupling efficiency for such a source scales with the square of the numerical aperture of the fibre. In practice, only a few per cent of the emitted power is coupled into the fibre (100 μW or less).

In order to improve the external quantum efficiency, the *p–n* junction can be encapsulated in a plastic dome structure, as shown in Fig. 18.6. The critical angle for total internal reflection at the semiconductor–plastic interface is larger than that of a semiconductor–air interface, thus increasing the fraction of the internal power which becomes useful output power. Using this design, the Fresnel losses at the dome–air interface are also minimized because all emitted rays pass through at normal incidence.

A measure of the LED performance is the *total quantum efficiency* η_{tot}, given by the ratio of the emitted optical power to the applied electrical power $P = VI$. One has

$$\eta_{tot} = \frac{P_{ext}}{VI} = \eta_{ext}\eta_{\text{int}}\,(hv/qV) \tag{18.18}$$

Typically, one has $hv \approx qV$ and $\eta_{tot} = \eta_{\text{int}}\,\eta_{ext}$.

The spectrum of light emitted by a LED is usually quite wide (on the order of 30–80 nm). This fact determines a high dispersion when a LED source is used in fibre optic communication systems. On the other hand, the fact that the spontaneous emission is random and occurs in all directions means that the light emitted by a LED presents a relatively large divergence angle This characteristic makes the coupling of light to single-mode optical fibres relatively inefficient.

The LED structures can be classified as *surface-emitting* or *edge-emitting*, depending on whether the LED emits light from a surface that is parallel to the junction plane or from the edge of the junction region. The beam divergence of edge-emitting LEDs differs from that of surface-emitting LEDs because of waveguiding in the plane perpendicular to the junction. Surface-emitting LEDs operate as a Lambertian source, and the resulting beam divergence has a FWHM of about 120°. In contrast, edge-emitting LEDs have a divergence of about 30° in the direction perpendicular to the junction plane.

18.6 Semiconductor Lasers

In general, it is not possible to satisfy condition (18.12) when the *p–n* junction is formed from moderately doped *n*-type and *p*-type semiconductors. However, such a condition can be satisfied if both semiconductors are strongly doped (~ 0.1%) and the applied potential difference is also sufficiently strong, such that the injected carrier density becomes high enough. This is the basis of the operation of *semiconductor lasers* (SLs).

SLs light through stimulated emission. As a consequence, they are not only capable of emitting high powers (~100 mW), but also have other advantages related to the coherent nature of the emitted light. The relatively narrow angular spread of the output beam compared with LEDs permits high coupling efficiency (~50%) into single-mode fibres. The relatively narrow spectral width of the emitted light allows operation at high bit rates (~ 40 Gb/s), since fibre dispersion becomes less critical.

In its simplest version, the semiconductor laser uses the same semiconductor on both sides of the *p–n* junction and is therefore called *a homojunction laser*. A problem with the homojunction is that electron–hole recombination occurs over a relatively wide region (~ $1 - 10\ \mu$m), which makes it difficult to achieve high carrier densities. This problem can be solved by sandwiching a thin layer between the *p*-type and *n*-type layers, such that the bandgap of the sandwiched layer is smaller than that of the adjacent layers. Such junctions are called *heterojunctions*.

Currently, the basic configuration used in the vast majority of SLs is based on a *double heterojunction*. Typical materials for this purpose are GaAs and AlGaAs. To obtain a double heterojunction, a thin layer of a given semiconductor material (e.g., GaAs) is inserted between two layers of semiconductor material with a larger bandgap (e.g., AlGaAs). Fig. 18.7(a) presents a schematic representation of a laser with a double heterojunction.

A heterojunction has three important advantages over a homojunction. First, it keeps the carriers in a limited region where their recombination becomes more likely. Second, the active central region has a higher refractive index than the adjacent layers (Fig. 18.7(b)), thus helping to confine the optical field and increasing the efficiency of stimulated emission. Finally, the absorption of light in the regions adjacent to the active region is negligible, so that losses are minimized. These three effects cause the threshold current for the operation of the heterojunction laser to be much lower than that of the homojunction laser, allowing its operation in the continuous regime at room temperature.

The characteristics of a double heterojunction laser are significantly altered when the thickness of the active layer becomes comparable to the de Broglie wavelength associated with the electron or hole, given by Eq. (1.2). In a double heterojunction, the electrons and the holes are confined to the central region, where the separation between bands E_g is lower than that of the adjacent regions. That is, electrons and holes are confined to a potential well, and their energy levels are dependent on the dimensions of that well. SLs based on these structures are called *quantum well lasers* and are characterized by a fairly low threshold current and high gain.

In SLs, the faces at the ends of the crystal are cut parallel to each other and conveniently polished to form a resonant cavity. The large difference between the refractive indices in the semiconductor ($n \approx 3.5$) and in air ($n \approx 1.0$) causes the interface between the two media to present a reflectance of about 30%, which is sufficient to constitute the

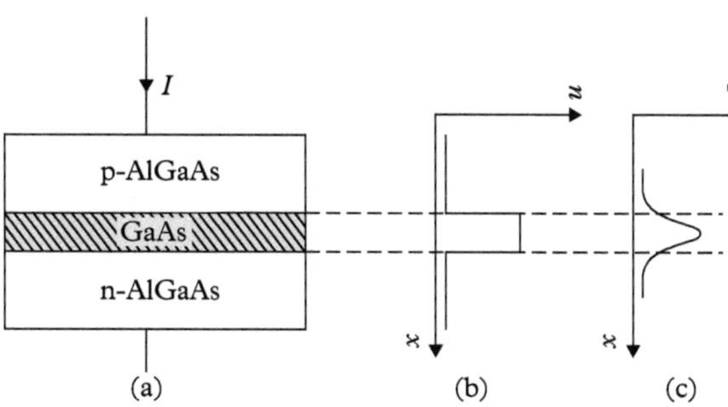

Figure 18.7 *(a) Diagram of a semiconductor laser with a double heterojunction, in which the active layer of GaAs is sandwiched between the p and n regions of AlGaAs; (b) refractive index profile at the heterojunction and (c) optical power distribution.*

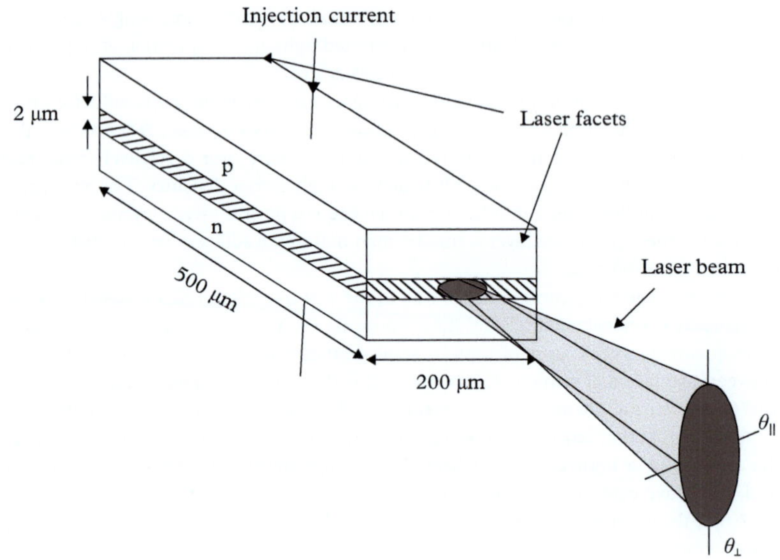

Figure 18.8 *Schematic representation and typical dimensions of a semiconductor laser.*

cavity. Fig. 18.8 shows a schematic representation and the typical dimensions of a semiconductor laser. Pumping of the semiconductor laser is carried out through the injection current *I*.

The condition of population inversion is achieved when the injected carrier density in the active layer exceeds a given value, N_T, known as the transparency value. In this case, an input signal propagating inside the active layer will be amplified as exp(gz), where g is the gain coefficient.

The peak value of the gain coefficient g_p increases almost linearly with the carrier density, N_d, and can be written as

$$g_p = \sigma_g(N_d - N_T) \tag{18.19}$$

where σ_g is the differential gain.

The recombination time of the carriers due to stimulated emission depends on the density of photons within the cavity and is on the order of 10 ps. As for the lifetime of photons, it corresponds to the average time that a photon takes inside the laser cavity before being absorbed or emitted to the outside and is given by

$$\tau_p = \frac{n}{cg_{th}} \tag{18.20}$$

where g_{th} is the threshold value of the gain coefficient, given by Eq. (17.27). Considering typical values for the parameters involved, one has $\tau_p \approx 2$ps.

Example 18.1

Calculate the threshold gain coefficient g_{th} and the lifetime of photons τ_p of a semiconductor laser with length $d = 300\ \mu\text{m}$, reflectance $R_1 = R_2 = 0.30$, and attenuation $\alpha_{\text{int}} = 10\ \text{cm}^{-1}$.

Solution

Using Eq. (17.27), we have:

$$g_{th} = \alpha_{\text{int}} + \frac{1}{2d}\ln\left(\frac{1}{R_1R_2}\right) = 50\ \text{cm}^{-1}$$

The lifetime of photons is given by Eq. (18.19):

$$\tau_p = \frac{3.5}{(3\times 10^8)(5\times 10^3)} = 2.3\ \text{ps}$$

18.7 Materials for Semiconductor Lasers

The energy of the photons emitted by a semiconductor laser is close to the bandgap energy, E_g. In a GaAs laser, one has $E_g = 1.42\ \text{eV}$ and the wavelength of the emitted light is about 904 nm. The addition of Al to the active layer allows one to increase the bandgap and consequently decrease the wavelength of the emitted light. In this way, it is possible to build lasers based on AlGaAs/GaAs alloys emitting light in the band of 780–880 nm. These lasers are widely used in various day-to-day applications, namely in CD players, laser printers, and communication systems operating in the region of 850 nm.

Most of the SLs currently in use are made up of a combination of elements from the third and fifth groups of the Periodic Table and are therefore called III–V lasers. This is the case for lasers based on the ternary alloys AlGaAs and InGaAs or the quaternary alloy InGaAsP. Lasers based on the latter alloy are particularly useful in the field of optical fibre communications, since they can emit light with wavelengths in the region of 1300 nm and 1550 nm, in which the attenuation of optical fibres is minimal.

The compound $In_{1-x}Ga_xAs_yP_{1-y}$ is formed from the binary III–V compound InP, replacing an x fraction of In atoms by Ga, which is another element of group III, and a y fraction of P by As, which is another element of group V. One can choose the x and y fractions in such a way that the bandgap corresponds to the desired wavelength. Further, $In_{1-x}Ga_xAs_yP_{1-y}$ is lattice matched to InP when $x \approx 0.45y$.

Lasers based on the combination of elements from the second and sixth groups of the Periodic Table, called II–VI lasers, provide light in the blue-green region. The continuous emission of blue light at 410 nm is possible, for example, with an InGaN laser.

18.8 Longitudinal Modes in a Semiconductor Laser

The most common SLs—known as *Fabry–Perot* (FP) *lasers*—emit light not just at one but at various wavelengths. These wavelengths correspond to the longitudinal modes of laser oscillation and are determined by the resonance condition of the cavity, given by

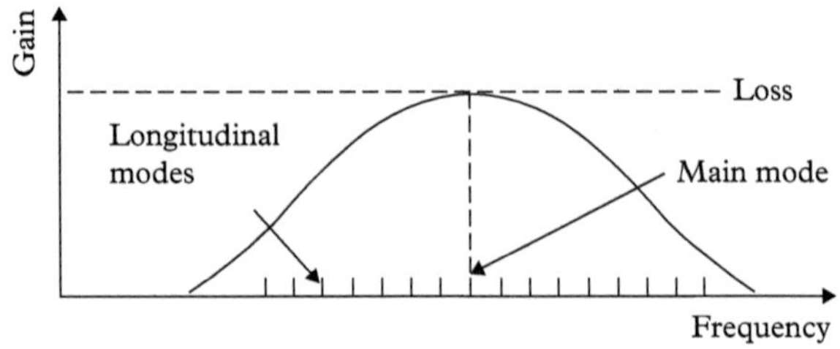

Figure 18.9 *Diagram of the gain curve and longitudinal modes of a semiconductor laser.*

$$\nu = \frac{c}{2n(\nu)d}p, \quad p = 1, 2, 3, ... \tag{18.20}$$

where d is the length of the cavity and $n(\nu)$ is the refractive index of the semiconductor material.

The spacing $\Delta\nu$ between adjacent oscillation frequencies can be obtained from Eq. (18.20) in the form:

$$\Delta\nu = \frac{c}{2nd}\left(1 + \frac{\nu}{n}\frac{dn}{d\nu}\right)^{-1} \tag{18.21}$$

In obtaining Eq. (18.21), an expansion of the refractive index is given by

$$n(\nu + \Delta\nu) = n(\nu) + \frac{dn}{d\nu}\Delta\nu \tag{18.22}$$

If typical values are considered, namely $\frac{\nu}{n}\frac{dn}{d\nu} \approx 0.4$, $n = 3.6$ and $d = 300\,\mu$m, one has

$$\Delta\nu \approx 100\,\text{GHz} \tag{18.23}$$

The gain spectrum of SLs is wide enough (bandwidth ~ 10 THz) that many longitudinal modes of the FP cavity experience gain simultaneously (Fig. 18.9). The mode closest to the gain peak becomes the main mode. However, since the gain difference of the neighbouring modes is extremely small, some of these modes also carry a significant portion of the laser power. Such lasers are called *multimode SLs*.

Example 18.2

When the injection current of a semiconductor laser is modulated, the refractive index of the cavity varies due to carrier injection. Assuming that the relative variation of the refractive index is $\frac{\Delta n}{n} = 10^{-6}$, calculate the corresponding variation in wavelength and frequency. Consider that the laser emits light with a wavelength $\lambda = 1.55\,\mu$m.

Solution

From Eq. (18.20), one has

$$\nu = \frac{c}{\lambda} = \frac{c}{2nd}p, \quad p = 1, 2, 3, ...$$

continued

Example 18.2 continued

The variation of the wavelength, $\Delta\lambda$, corresponding to the variation Δn of the refractive index is then

$$\Delta\lambda = \lambda\frac{\Delta n}{n} = \left(1.55 \times 10^{-6}\right)\left(10^{-6}\right) = 1.55 \times 10^{-12}\,\text{nm}$$

The corresponding frequency variation is

$$\Delta\nu = \frac{c}{\lambda^2}\Delta\lambda = \frac{3 \times 10^8}{\left(1.55 \times 10^{-6}\right)^2}\left(1.55 \times 10^{-12}\right) = 193.5\,\text{MHz}$$

18.9 Single-Mode Semiconductor Lasers

The fact that SLs have a wide emission spectrum is undesirable when they are used as light sources in fibre optic communication systems, since it results in a large dispersion of pulses. To minimize this dispersion, the laser should oscillate in a single longitudinal mode (SLM). The performance of an SLM laser is characterized by the mode-suppression ratio (MSR):

$$MSR = \frac{P_{mm}}{P_{sm}} \tag{18.24}$$

where P_{mm} and P_{sm} are the powers of the main mode and of the side mode. The MSR should exceed 1000 (or 30 dB) for a good single-mode laser.

An effective method of achieving operation in a SLM is the introduction of components or mechanisms into the laser cavity that cause losses for all longitudinal modes except one of them. Thus, only one longitudinal mode—the one for which the gain exceeds the losses—can satisfy the oscillation condition. One of the techniques that allows to achieve single-mode operation makes use of external cavities, corresponding to the so-called *external cavity lasers* (ECLs). An example of an ECL is shown in Fig. 18.10(a), in which the light from a semiconductor laser is coupled to an external diffraction grating. It is necessary to reduce the natural reflectivity of the cleaved facet facing the grating to make such coupling sufficiently strong. In the presence of the external grating, the effective reflectivity of the laser facet facing it becomes wavelength-dependent and leads to low losses for certain wavelengths. The dominant mode will be the longitudinal mode that has the lowest cavity loss and is closest to the gain peak. Besides providing SLM operation, using an external diffraction grating also gives a narrow linewidth at a desired wavelength. Spectral linewidths as narrow as 10 kHz have been reported with such devices (Wyatt and Devlin, 1983). An additional feature of this ECL is the possibility of tuning the wavelength of such a mode (about 90 nm) simply by rotating the grating (Severin and Shaw, 1985). However, the dimensions and non-monolithic nature of the ECL can affect its mechanical stability in some practical applications.

A monolithic version of the external cavity laser is provided by the distributed Bragg reflector (DBR) laser that uses a Bragg grating at one or both ends of the active region, as illustrated in Fig. 18.10(b). The reflectivity of Bragg grating is maximum for a wavelength, λ_B, satisfying the Bragg condition:

$$2n_{eff}\Lambda = m\lambda \tag{18.25}$$

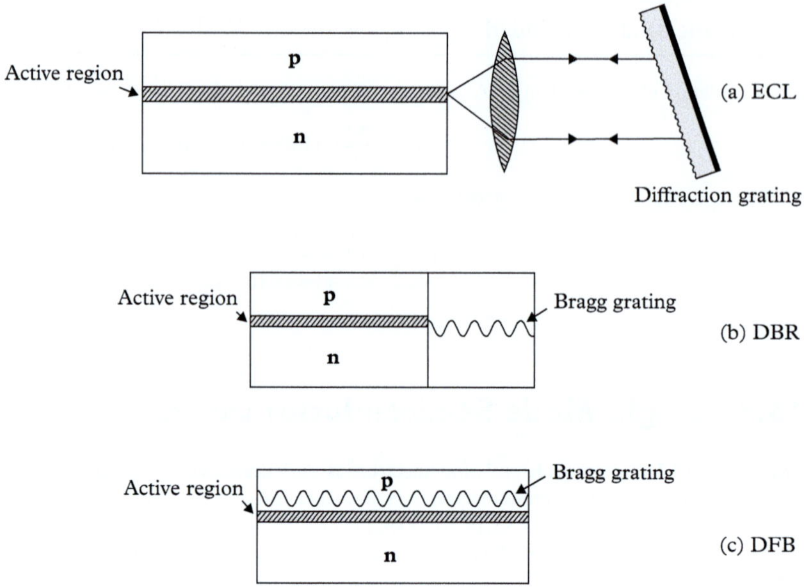

Figure 18.10 *Semiconductor laser structures able to operate in a single longitudinal mode: (a) external cavity laser (ECL); (b) DBR-type laser; (c) DFB-type laser.*

where Λ is the grating period, n_{eff} is the effective refractive index, and m is an integer representing the order of Bragg diffraction. An appropriate choice of period Λ allows one to select a single laser oscillation mode. The cavity losses are minimum for the longitudinal mode closest to λ_B and increase significantly for other longitudinal modes. Assuming $m = 1$, $n_{eff} = 3.5$, and for a wavelength of 1550 nm, the spatial period of the grating should be 221 nm. Gratings with periods of this order of magnitude are usually fabricated using holographic techniques.

As an alternative to placing it in the end regions of the laser cavity, the Bragg grating can be etched onto one of the cladding layers adjacent to the active layer, thus causing the feedback to occur through the entire cavity length. In this case, the periodic variation of the mode index along the cavity length leads to a coupling between the forward- and backward-propagating waves through Bragg diffraction, which is strongest for the first-order Bragg diffraction ($m = 1$). This laser configuration is called a *distributed feedback* (DFB) laser, as represented in Fig. 18.10(c). The facets of a DFB laser are usually coated with an antireflective film.

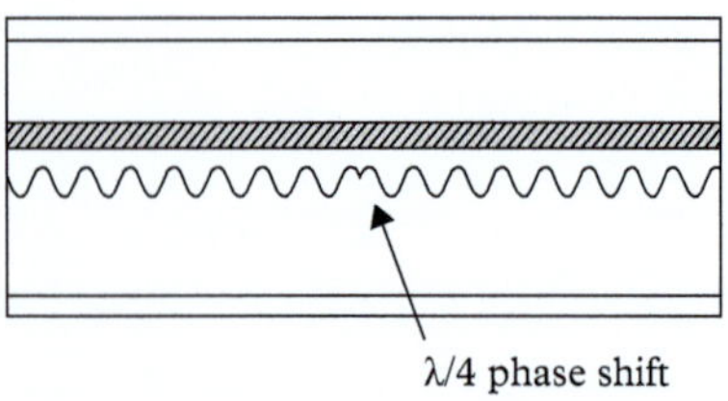

Figure 18.11 *Illustration of a λ/4 phase-shift DFB laser.*

In an idealized structure, where there are no facet reflections, longitudinal modes are spaced symmetrically around λ_B and there is a wavelength band, known as the stop band, where the reflection is the largest. Ordinarily, only the two lowest-order modes are found within the stop band, and because of the symmetry in the structure, they have the same threshold gain and would lase simultaneously. However, in practice, the randomness associated with the cleaving process creates different end phases, thus removing the degeneracy of the modal gain and providing single-mode operation only. Moreover, facet asymmetry can be increased by placing a high-reflection coating on one end facet and a low-reflection coating on the other.

Another technique to achieve stable single-mode operation consists in incorporating a phase shift in the grating of a DFB laser (Fig. 18.11). When the phase shift is $\pi/2$, corresponding to an optical path change of $\lambda/4$, the lowest threshold gain is obtained for the main mode exactly at the Bragg wavelength, and the gain difference between the central mode and the nearest side mode is maximum. Actually, the performance of the quarter-wavelength-shifted DFB laser is superior to that of the conventional DFB structure because of such a large gain difference.

18.10 Tunable Semiconductor Lasers

Single-mode and narrow-linewidth SLs with the ability of wavelength tuning over a wide range are very important in modern WDM lightwave systems. Actually, this avoids the use of a large number of SLs inside a WDM transmitter, which makes the design and maintenance of such a lightwave system expensive and impractical.

In the case of ECLs, the centre frequency for light reflected from the grating can be tuned by rotating the grating mechanically. The frequency of the resonant mode is controlled by moving the grating forwards or backwards along the optic axis. Very precise mechanical control is needed for stable single-longitudinal-mode operation, since the frequency separation between resonant modes, which is inversely proportional to the cavity length, is usually very small. On the other hand, the tuning range of this configuration is very wide. However, grating-type external cavity tunable lasers present some important limitations, concerning their mechanical instability and the necessity for very precise optical alignment.

The aforementioned disadvantages are attenuated when using monolithic semiconductor laser structures, namely DFB and DBR structures. In this case, the tuning function is achieved by carrier injection, which changes the refractive index of the laser cavity. In the case of a single-electrode DFB laser operated above threshold, most injected carriers recombine to produce photons, resulting in a very small increase in carrier density, which in turn leads to a small change in the lasing wavelength. However, the range of wavelength tuning can be improved using a two-section DFB laser (Fig. 18.12), with a large current applied to one electrode and a small current applied to the other. In the asymmetric structure of the DFB laser, the optical field is higher in the region near the output facet with lower reflectivity, and the wavelength is primarily determined by the effective index in this region. If this region is pumped with current densities at or slightly below the threshold current density, it serves mainly as a Bragg reflector. In addition, because of the low pumping level, the injected carriers do not contribute significantly to photon generation, resulting in a relatively large change of refractive index, thereby providing substantial wavelength tuning. The gain is provided by the other section pumped well above threshold.

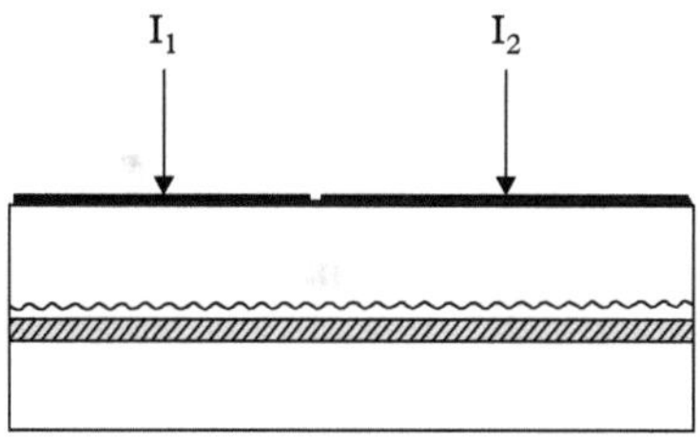

Figure 18.12 *Two-section DFB laser.*

The advantage of the tunable DFB laser is its relative ease of fabrication. The disadvantage is the limited tuning range, because the tuning section must be biased below threshold. When the tuning range is biased above threshold, both sections can oscillate independently, resulting in a two-mode operation or mode-hopping.

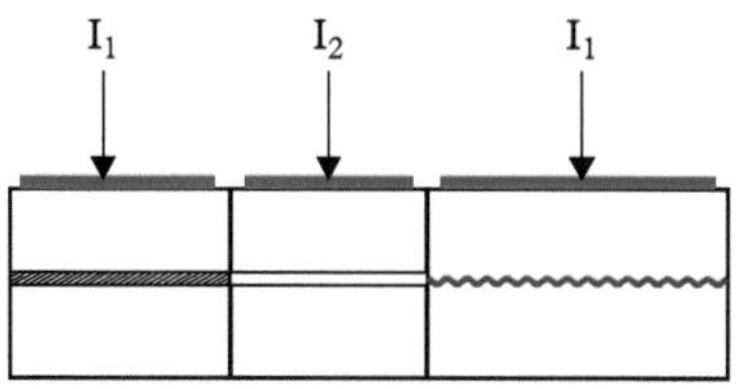

Figure 18.13 *Three-section DBR laser structure.*

A better option regarding tunability is provided by a multisection DBR laser structure, as illustrated in Fig. 18.13. It consists of three sections, referred to as the active section, the phase-control section, and the Bragg section. Each section can be biased independently by injecting different amounts of currents. The current injected into the Bragg section is used to change the Bragg wavelength through carrier-induced changes in the refractive index n. This mechanism is known as Bragg wavelength control. A continuous tuning range, however, is limited to the resonant mode spacing, which is defined by the effective cavity length of the laser. The current injected into the phase-control section is used to change the phase of the feedback from the DBR, thus allowing the lasing wavelength to be tuned around each Bragg wavelength. In this way, a significantly larger tuning range can be achieved because the deviation of the lasing wavelength from the Bragg wavelength is compensated by phase control. By appropriately controlling the currents in the phase and Bragg sections, the laser wavelength can be tuned almost continuously over a range of 10–15 nm (Ferreira et al. 1992).

18.11 Light–Current Characteristics

The number of photons, P, and the number of electrons, N, inside the active region of a single-mode laser satisfy the following rate equations:

$$\frac{dP}{dt} = GP + R_{sp} - \frac{P}{\tau_p} \tag{18.26}$$

$$\frac{dN}{dt} = \frac{I}{q} - \frac{N}{\tau_c} - GP \tag{18.27}$$

where

$$G = \Gamma v_g g = G_N(N - N_0) \tag{18.28a}$$

$$G_N = \Gamma v_g \sigma_g / V \tag{18.28b}$$

$$N_0 = N_T V \tag{18.28c}$$

In Eqs. (18.26)–(18.28), τ_p is the photon lifetime, given by Eq. (18.20); G is the net rate of stimulated emission; R_{sp} is the rate of spontaneous emission into the lasing mode; Γ is the confinement factor; and V is the active volume. One has $R_{sp} = n_{sp}G$, where n_{sp} is known as the spontaneous emission factor (~2 for SLs).

In the case of CW operation, we have

$$\frac{dP}{dt} = \frac{dN}{dt} = 0 \tag{18.29}$$

Neglecting the spontaneous emission and for currents such that

$$G\tau_p < 1, \tag{18.30}$$

one has $P = 0$ and

$$N = \tau_c \frac{I}{q} \tag{18.31}$$

Threshold is reached at a current such that

$$G\tau_p = 1 \tag{18.32}$$

The carrier population is then clamped to the threshold value

$$N_{th} = N_0 + \left(G_N\tau_p\right)^{-1} \tag{18.33}$$

The threshold current is given by Eq. (18.31) as

$$I_{th} = \frac{qN_{th}}{\tau_c} \tag{18.34}$$

Above threshold ($I > I_{th}$), the photon number increases linearly with I as

$$P = (\tau_p/q)(I - I_{th}) \tag{18.35}$$

The emitted power per facet, P_{ext}, is related to P by

$$P_{ext} = \frac{1}{2}\left(v_g\alpha_{mir}\right)h\nu P. \tag{18.36}$$

where

$$v_g\alpha_{mir} \tag{18.37}$$

is the rate at which photons of energy $h\nu$ escape from the two facets, and the factor of ½ makes P_{ext} the power emitted from each facet for a FP laser with equal facet reflectivities.

Using Eq. (18.35) in Eq. (18.36), one has

$$P_{ext} = \frac{h\nu}{2q}\frac{\eta_{\text{int}}\alpha_{mir}}{\alpha_{\text{int}} + \alpha_{mir}}(I - I_{th}) \tag{18.38}$$

Below the threshold current, the output power of an SL is very low and corresponds mainly to spontaneously emitted light, a situation similar to that of an LED. Above threshold, the output power increases linearly with the injection current. This is illustrated by the so-called P–I curves depicted in Fig. 18.14, which show the typical behaviour of the power emitted by an LED (dashed curve) and by a semiconductor laser (full curve) as a function of the injection current.

An important parameter for characterizing a semiconductor laser is its efficiency, given by the slope of the full P–I curve in Fig. 18.14 above the threshold. If dI represents the variation of current through the diode, then the variation in the number of electrons injected per unit time into the laser is dI/q, where q is the charge of the electron. If dP_{ext} represents the corresponding variation in the output power of the laser, then the variation in the number of photons is $dP_{ext}/h\nu$, where ν is the frequency of radiation.

The slope of the P–I curve for $I > I_{th}$ is given by Eq. (18.38) as

$$\frac{dP_{ext}}{dI} = \frac{h\nu}{2q}\eta_d \tag{18.39}$$

where

$$\eta_d = \frac{\eta_{\text{int}}\alpha_{mir}}{\alpha_{\text{int}} + \alpha_{mir}} \tag{18.40}$$

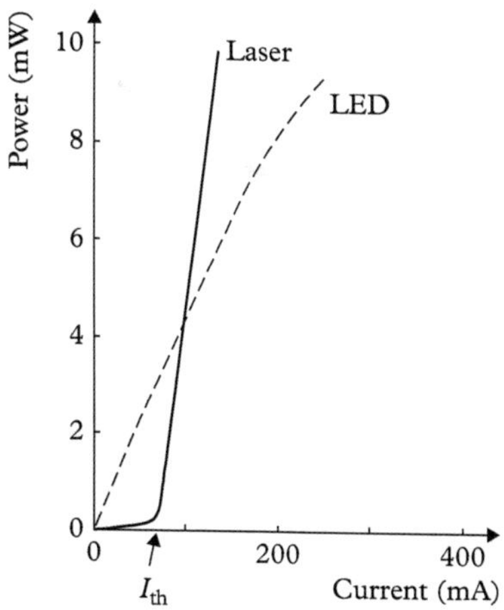

Figure 18.14 *Variation of the power emitted by a LED (dashed curve) and by a semiconductor laser (full curve) as a function of the injection current.*

is called the *differential quantum efficiency* and is a measure of the efficiency with which the light output increases with an increase in the injected current.

The external quantum efficiency is defined as

$$\eta_{ext} = \frac{2P_{ext}/hv}{I/q} = \eta_d(1 - I_{th}/I) \tag{18.41}$$

One has $\eta_{ext} \approx \eta_d$ when $I >> I_{th}$.

The P–I curve characterizes the emission properties of an SL, as it indicates not only the threshold level but also the current that needs to be applied to obtain a certain amount of power. The threshold current is found to increase exponentially with temperature:

$$I_{th} = I_0 \exp(T/T_0) \tag{18.42}$$

where I_0 is a constant and T_0 is the so-called *characteristic temperature* of the diode.

For InGaAsP lasers, T_0 is typically in the range 50–70 K, whereas it exceeds 120 K for GaAs lasers.

18.12 Modulation Response

In the context of optical communication systems, it is necessary to modulate the light emitted by the laser source. One of the most interesting properties of SLs is the possibility of performing this modulation directly through the injection current. In fact, when this current varies, the population of electrons and holes in the laser cavity also varies. As a consequence, the gain is altered, which in turn causes the power emitted by the laser to vary. The dynamics of direct modulation of a semiconductor laser are determined by several factors, namely the recombination times of the carriers and the lifetime of the photons in the laser cavity.

The modulation response of SLs can be studied by solving the rate equations (18.26) and (18.27) with a time-dependent current of the form.

$$I(t) = I_b + I_m f_p(t) \tag{18.43}$$

where I_b is the bias current, and $f_p(t)$ represents the shape of the current pulse.

Useful analytical solutions can be obtained for the case of small-signal modulation, when $I_b > I_{th}$ and $I_m << (I_b - I_{th})$. The rate equations can be linearized and solved using the Fourier-transform technique. Let us consider a sinusoidal modulation with frequency ωm such that

$$f_b(t) = \sin(\omega_m t) \tag{18.44}$$

In such a case, it can be shown that the general solution of Eqs. (18.26) and (18.27) is given by:

$$P(t) = P_b + |p_m| \sin(\omega_m t + \theta_m) \tag{18.45}$$

$$N(t) = N_b + |n_m| \sin(\omega_m t + \psi_m) \tag{18.46}$$

where P_b and N_b are the steady-state values at the bias current, I_b, $|p_m|$ and $|n_m|$ are small changes in the photon number and carrier number, respectively, occurring because of current modulation, whereas θ_m and ψ_m govern the phase lag associated with the small-signal modulation.

In particular, one has

$$p_m(\omega_m) = \frac{P_b G_N I_m / q}{(\Omega_R + \omega_m - i\Gamma_R)(\Omega_R - \omega_m + i\Gamma_R)} \tag{18.47}$$

where

$$\Omega_R = \left[G G_N P_b - (\Gamma_P - \Gamma_N)^2/4\right]^{1/2} \tag{18.48}$$

$$\Gamma_R = (\Gamma_P + \Gamma_N)/2 \tag{18.49}$$

$$\Gamma_N = \tau_c^{-1} + G_N P_b \tag{18.50}$$

$$\Gamma_p = R_{sp}/P_b \tag{18.51}$$

In Eqs. (18.48) and (18.49), Ω_R and Γ_R are the frequency and the damping rate of the relaxation oscillations, which play an important role in governing the dynamic response of SLs.

The transfer function corresponding to the photon dynamics is obtained from Eq. (18.47) by

$$H(\omega_m) = \frac{p_m(\omega_m)}{p_m(0)} = \frac{\Omega_R^2 + \Gamma_R^2}{(\Omega_R + \omega_m - i\Gamma_R)(\Omega_R - \omega_m + i\Gamma_R)} \tag{18.52}$$

The modulation response is flat ($H(\omega_m) \approx 1$) for frequencies $\omega_m << \Omega_R$, peaks at $\omega_m = \Omega_R$ and drops sharply for $\omega_m >> \Omega_R$

The 3-dB modulation bandwidth is the frequency at which $|H(\omega_m)|$ is reduced by 3 dB, and is given by

$$f_{3dB} = \frac{1}{2\pi}\left[\Omega_R^2 + \Gamma_R^2 + 2(\Omega_R^4 + \Omega_R^2\Gamma_R^2 + \Gamma_R^4)^{1/2}\right]^{1/2} \tag{18.53}$$

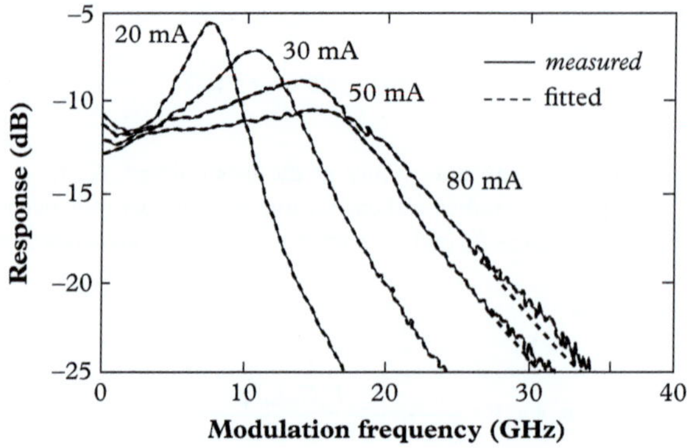

Figure 18.15 *Measured (full curves) and fitted (dashed curves) small-signal modulation response of a DFB laser operating at 1550 nm at several values of the injection current (reprinted with permission from [Lindgren et al. 1997] © IEEE).*

For most lasers $\Gamma_R << \Omega_R$ and we have

$$f_{3dB} \approx \frac{\sqrt{3}\Omega_R}{2\pi} \approx \left[\frac{3G_N}{4\pi^2 q}(I_b - I_{th})\right]^{1/2} \tag{18.54}$$

The modulation bandwidth increases with an increase in the bias level as $\sqrt{I_b - I_{th}}$. This square-root dependence has been observed experimentally. Fig. 18.15 shows how f_{3dB} can be increased to 24 GHz for a DFB laser by biasing it at 80 mA (Lindgren et al. 1997). This must be compared with a modulation bandwidth of 200 MHz or less, typical of LEDs.

18.13 Further Reading

Agrawal, G. P., and N. Dutta, *Semiconductor Lasers*, Van Nostrand Reinhold, New York, 2nd Ed., 1993.

Agrawal, G. P. *Fiber Optic Communication Systems*, John Wiley & Sons, 4th Ed., 2010.

Chen, P. K., *Fibre Optics and Optoelectronics*, Prentice Hall, 2nd Ed., 1990.

Dutta, N. K., Optical Sources for Lightwave Systems Applications, in E. E. Basch (Ed.), Optical-Fiber Transmission, H. W. Sams & Co., 1987.

Ferreira, M. F., J. F. Rocha, and J. L. Pinto, Analysis of the Frequency Noise of Tunable Multisection DBR Lasers, *IEEE J. Quantum Electron*, 28, 833–840, 1992.

Lindgren, S., H. Ahlfeldt, L. Backlin, L. Forssen, C. Vieider, H. Elderstig, M. Svensson, L. Granlund, L. Andersson, B. Kerzar, B. Broberg, O. Kjebon, R. Schatz, E. Forzelius, and S. Nilsson, High-Power 980-nm Pump Laser Module for Erbium-Doped Fiber Amplifiers, *IEEE Photon. Technol. Lett.* 9, no.3, 306–308, 1997.

Petermann, K., *Laser Diode Modulation and Noise*, KTK Scientific Publishers, 1988.

Saleh, B. A., and M. C. Teich, *Fundamentals of Photonics*, John Wiley & Sons, 3rd Ed., 2019.

Senior, J. M., *Optical Fiber Communications*, Pearson Education, 2009.

Severin, W. V., and H. J. Shaw, Gain Saturation and Noise in Traveling-Wave Optical Amplifiers, *J. Lightwave Technol.*, 3, no.5, 1041–1047, 1985.

Yariv, A., *Optical Electronics in Modern Communications*, Oxford University Press, 5th Ed., 1997.

Wyatt, R., and W. J. Devlin, *Electron. Lett.*, 19, 110, 1983.

18.14 Problems

18.1 Discuss the general requirements for a source in optical fibre communications. Describe the areas in which the LED and the semiconductor laser fulfil these requirements and comment on any drawbacks of using these devices as optical fibre communication sources.

18.2 A current of 40 mA is injected into a GaAsP LED emitting light at wavelength $\lambda = 500\,\text{nm}$. Assuming that the internal quantum efficiency of GaAsP is 0.1, determine the power generated by the LED.

18.3 A semiconductor laser has a GaAs active region with a bandgap energy of 1.43 eV. Estimate the wavelength of optical emission from the device and determine its linewidth in hertz when the measured spectral width is 0.1 nm.

18.4 The threshold current of a semiconductor laser doubles when the temperature increases by 60 °C. Determine the characteristic temperature of that laser.

18.5 Calculate the threshold gain and the photon lifetime of a semiconductor laser with length $d = 400\,\mu\text{m}$, refractive index $n = 3.5$, absorption losses $\alpha_{\text{int}} = 10\,\text{cm}^{-1}$ and facet reflectances $R_1 = 0.95$ and $R_2 = 0.30$.

18.6 An InGaAsP laser ($n = 3.5$), with length $d = 400\,\mu\text{m}$, has a spectral width $B = 1.2$ THz. Assuming negligible variation in the refractive index with frequency, determine the spacing between the longitudinal modes and the number of these modes in which the laser can oscillate. Also find the maximum length of the laser that can guarantee its operation in a single longitudinal mode.

18.7 The quaternary compound $In_{1-x}Ga_xAs_yP_{1-y}$ is lattice matched with the binary compound InP when $x = 0.45y$. The bandgap energy of the compound InGaAsP is given by

$$E_g\,(y) = 1.35 - 0.72y + 0.12y^2\ \text{eV}$$

Determine the composition of the active layer of an InGaAsP semiconductor laser designed to operate at (i) 1.3 μm and (ii) 1.55 μm.

18.8. An InGaAsP semiconductor laser operating at 1.3 μm is 250 μm long. Assuming that the internal loss is 30 cm^{-1} and both facets have a reflectance R = 0.3, find the laser threshold gain coefficient.

18.9. Determine the characteristic temperature of a semiconductor laser, assuming that its threshold current doubles when the operating temperature is increased by 50°C.

18.10 Explain the importance of single-frequency lasers in optical fibre communications and describe the main structures utilized in the fabrication of such devices.

18.11 Discuss the importance of wavelength tunability associated with single-frequency semiconductor lasers. Outline the major techniques that are being adopted to realize this feature.

18.12 A 250-μm-long InGaAsP laser operating at 1.55 μm in single mode has a refractive index $n = 3.5$, an internal loss of 40/cm, and facets r reflectances $R_1 = 0.95$ and $R_2 = 0.30$. Calculate (i) the photon lifetime and (ii) the threshold value of the electron population.

Assume that the gain varies with the electron population as given by Eq. (18.28), with $G_N = 6 \times 10^3/\text{s}$ and $N_0 = 10^8$.

18.13 Determine the threshold current for the semiconductor laser of Problem 18.11 by taking 2 ns as the carrier lifetime. Also determine the power emitted from one facet when the laser is operated twice above threshold.

18.14 Starting from the rate Eqs. (18.26) and (18.28), derive the result given by Eqs. (18.47)–(18.51) for the response of a semiconductor laser modulated by a sinusoidal injection current described by Eqs. (18.43) and (18.44).

Laser Beams and Laser Light

19

As seen in Chapter 17, the confinement of laser light in the axial direction of a laser cavity results in cavity modes that are standing waves with frequencies that depend on the separation between the two mirrors. It might be thought that side mirrors would also be necessary to confine the light in the transverse direction of the laser cavity. Actually, it will be seen in this chapter that there is a solution to the three-dimensional wave equation that provides a natural transverse confinement without the need for such mirrors. That solution is called a *Gaussian beam* and has the smallest possible angular spread for a beam with a given initial diameter.

The peculiar characteristics of laser light are determined by the process of stimulated emission, which is the basis of the operation of the laser. This process gives rise to a multiplicity of photons with identical characteristics regarding their amplitude, phase, and direction of propagation. Laser light is thus extraordinarily coherent, both in temporal and spatial terms, which contrasts clearly with the chaotic characteristics of the light emitted by spontaneous emission. The coherence of the laser light makes it possible to focus the laser beam on to an area with dimensions of the order of the wavelength. The irradiance reached in this area can be so high that it allows one to drill or cut very hard materials.

19.1 The Gaussian Beam

The three-dimensional wave equation for the electric field $E(x,y,z,t)$ in homogeneous media is given by

$$\nabla^2 E - \frac{n^2}{c^2}\frac{\partial^2 E}{\partial t^2} = 0 \tag{19.1}$$

where n is the medium refractive index and c is the velocity of light in vacuum.

The aforementioned wave equation has many possible solutions, the plane waves discussed in Section 5.5 being the simplest ones. In particular, it has a spherical wave solution given by

$$E(r) = \frac{E_0}{r} e^{i(kr - \omega t)} \tag{19.2}$$

where $r = \sqrt{x^2 + y^2 + z^2}$ is the distance to the wave point source. Assuming a transverse plane at $z = R$, one can write

$$r = \sqrt{x^2 + y^2 + R^2} \approx R + \frac{x^2 + y^2}{2R} \tag{19.3}$$

This approximation is valid if the points (x, y) are near the z-axis, such that

$$x^2 + y^2 << R^2 \tag{19.4}$$

Lasers, Optics, and Photonics. Mário F. S. Ferreira, Oxford University Press. © Mário F. S. Ferreira (2026).
DOI: 10.1093/oso/9780198952770.003.0019

Using this approximation, Eq. (19.2) can be written in the form

$$E(x,y)_{z=R} \approx Ae^{ikR}e^{i(k/2)[(x^2+y^2)/R]} \tag{19.5}$$

where A is a constant. Let us assume the following trial function to describe the electric field of a laser beam propagating along the z-axis:

$$E(r,t) = U(x,y,z)e^{i(kz - \omega t)} \tag{19.6}$$

Substituting Eq. (19.6) into Eq. (19.1) and assuming the U varies slowly with z, one obtains

$$\frac{\partial^2 U}{\partial x^2} + \frac{\partial^2 U}{\partial y^2} + 2ik\frac{\partial U}{\partial z} = 0 \tag{19.7}$$

Eq. (19.7) is known as the paraxial Helmholtz equation. We try as a solution

$$U(x,y,z) = E_0 \exp\left\{i\left(p(z) + \frac{k}{2q(z)}r^2\right)\right\} \tag{19.8}$$

where $p(z)$ and $q(z)$ are general functions to be determined and $r^2 = x^2 + y^2$ is now the distance to the z-axis. By comparing the second terms in the exponents of Eq. (19.8) and Eq. (19.5), we note that the parameter $q(z)$ is in the place of the radius of curvature R of a spherical wavefront. Thus, we will call q the *complex radius of curvature* and write it in the form

$$\frac{1}{q} = \frac{1}{R_G} + i\frac{\lambda}{\pi w^2} \tag{19.9}$$

R_G and w being parameters to be characterized. The solution given by Eq. (19.8) is called the *fundamental Gaussian beam* solution, since, upon substituting it into Eq. (19.6), the E-field becomes

$$E(x,y,z,t) = E_0 \exp\left\{i(kz + p(z) - \omega t) + \frac{k}{2R(z)}r^2\right\}\exp\left\{-\frac{r^2}{w^2(z)}\right\} \tag{19.10}$$

Thus, the field amplitude shows a Gaussian dependence on r, with $w(z)$ representing the distance to the axis where such amplitude has decreased to 1/e of its peak value on the axis.

The beam irradiance is given by

$$I = EE^* = E_0^2 \exp\{i\,[p(z) - p^*(z)]\}\exp\left\{\frac{ikr^2}{2}\left[\frac{1}{q(z)} - \frac{1}{q^*(z)}\right]\right\} \tag{19.11}$$

From Eqs. (19.9) and (19.11), one can see that the irradiance has a radial dependence described by a Gaussian function:

$$I = EE^* \propto \exp\left(\frac{2r^2}{w^2}\right) \tag{19.12}$$

Let us assume that at $z = 0$ the Gaussian beam has a planar wavefront, that is, its radius of curvature satisfies the condition $R_G(z = 0) \to \infty$. In such a case, one has

$$q(0) = \frac{\pi w_0^2}{i\lambda} \tag{19.13}$$

The plane $z = 0$ corresponds to the so-called *beam waist*, which is located inside the laser cavity, w_0 being the corresponding spot size.

Example 19.1

Show that the parameter q in Eq. (19.9) satisfies the relation

$$q(z) = q(0) + z \tag{19.14}$$

This is an important result, which corresponds to the *free propagation law for the Gaussian laser beam.*

Solution

After substituting Eq. (19.9) into Eq. (19.7), one obtains the result:

$$2k\left(\frac{i}{q} - \frac{\partial p}{\partial z}\right) + \frac{k^2}{q^2}\left(\frac{\partial q}{\partial z} - 1\right)r^2 = 0 \tag{19.15}$$

Since this equation must be satisfied for all values of r, the coefficients of different powers of r must be zero. Imposing this condition, one obtains the following results:

$$\frac{\partial q}{\partial z} = 1 \tag{19.16}$$

$$\frac{\partial p}{\partial z} = \frac{i}{q} \tag{19.17}$$

For any other transverse plane $z > 0$, Eq. (19.14) follows immediately from Eq. (19.16).

19.1.1 Characteristics of the Gaussian beam

One can rewrite Eq. (19.14) in reciprocal form to obtain:

$$\frac{1}{q(z)} = \frac{z}{\left(z^2 + \frac{\pi^2 w_0^4}{\lambda^2}\right)} + i\frac{\pi w_0^2/\lambda}{\left(z^2 + \frac{\pi^2 w_0^4}{\lambda^2}\right)} \tag{19.18}$$

Equating the real and imaginary parts of Eq. (19.18) with their counterparts in Eq. (19.9), the following expressions are obtained for the beam radius, $w(z)$, and the wavefront curvature radius, $R_G(z)$:

$$w(z) = w_0\left[1 + \left(\frac{\lambda z}{\pi w_0^2}\right)^2\right]^{1/2} \tag{19.19}$$

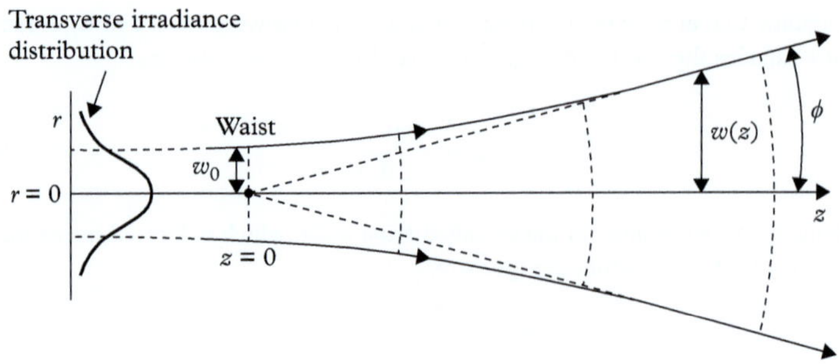

Figure 19.1 *Schematic representation of the propagation of a Gaussian beam along the z-axis.*

$$R_G(z) = z\left[1 + \left(\frac{\pi w_0^2}{\lambda z}\right)^2\right] \tag{19.20}$$

One can verify from Eq. (19.20) that the Gaussian beam has indeed a plane wavefront at its waist. At moderate values of the distance z from this waist, the wavefront cannot be considered spherical. In fact, if such wavefronts were spherical and concentric around $z = 0$, one should have $R(z) = z$, which is not the case. Actually, one has:

$$R_G(z) \approx z \tag{19.21}$$

only if

$$z >> \frac{\pi w_0^2}{\lambda} \tag{19.22}$$

a situation that corresponds to the so-called *far field*. In this region, the width of the beam increases in a practically linear way with the distance to the waist, being given by

$$w(z) \approx \frac{\lambda z}{\pi w_0} \tag{19.23}$$

Fig. 19.1 illustrates the main parameters characterizing the propagation of a Gaussian beam along the z-axis.

Beam spreading during propagation is a fundamental consequence of the wave nature of light. It occurs even for a highly collimated laser beam presenting a finite transverse extension. On the other hand, the beam propagation law given by Eq. (19.14) shows that a beam having a Gaussian profile at one location will have a Gaussian profile at all positions z along its direction of propagation.

19.2 The ABCD Propagation Law

Following the treatment in Section 2.7, let us consider an optical system described by the matrix

$$\mathbf{S} = \begin{bmatrix} A & B \\ C & D \end{bmatrix} \tag{19.24}$$

A ray incident in such a system with parameters (x_1, y_1) emerges from it with parameters (x_2, y_2), such that

$$\begin{bmatrix} x_2 \\ y_2 \end{bmatrix} = \begin{bmatrix} A & B \\ C & D \end{bmatrix} \begin{bmatrix} x_1 \\ y_1 \end{bmatrix} \tag{19.25}$$

Eq. (24) can be replaced by the following relations:

$$x_2 = Ax_1 + By_1 \tag{19.26}$$

$$y_2 = Cx_1 + Dy_1 \tag{19.27}$$

The radius of curvature R of an ordinary spherical wave can be related to the paraxial ray parameters x and y in the form

$$R = \frac{x}{y} \tag{19.28}$$

Dividing Eq. (19.26) by Eq. (19.27) and using Eq. (19.28), one can obtain the following relation:

$$R_2 = \frac{AR_1 + B}{CR_1 + D} \tag{19.29}$$

where R_1, and R_2 are the radii of curvature of the spherical wave at the input and output of the optical system, respectively. Generalizing this relation to the case of a Gaussian wave, one has

$$q_2 = \frac{Aq_1 + B}{Cq_1 + D} \tag{19.30}$$

Eq. (19.30) is a very useful and powerful result, known as the *ABCD propagation law* for Gaussian laser beams. It enables one to determine the reshaping of the laser beam after it has passes through an arbitrary optical system represented by the corresponding ABCD matrix.

19.3 Gaussian Beam in a Laser Cavity

According to the above discussion, it is natural to consider the Gaussian beam as a good option to describe the spatial distribution of light in a laser cavity, since it is self-confined in the transverse direction. Obviously, the wavefronts of such a Gaussian beam will match the mirror surfaces limiting the laser cavity. A cavity mode with a small waist size will diverge rapidly with position along the cavity axis, the beam diameter becoming much larger at the limiting mirrors. In contrast, a cavity mode with a large waist size will have a low angular divergence, and the beam diameter will remain nearly constant along the cavity.

Let us consider the particular case of a symmetric resonator, formed by two concave spherical mirrors of radius R separated by a distance d, as represented in Fig. 17.5. In this case, the radius of curvature R_G of the Gaussian beam must be equal to the mirror radius of curvature R. The beam waist is located in the centre of the cavity, at a distance $d/2$ from each mirror. Considering $z = d/2$ in Eq. (19.20), it can be shown (Problem 19.7) that the radius of such a beam at the waist is given by

$$w_0^2 = \frac{\lambda d}{2\pi}\sqrt{\frac{2R}{d} - 1} \tag{19.31}$$

Eq. (19.31) only gives a real result if $2R > d$. A cavity with $2R < d$ does not support a Gaussian beam mode and corresponds to an *unstable resonator.*

Considering $z = d/2$ in Eq. (19.19), one can obtain the following result for the beam radius w at the position of the mirrors (Problem 19.8):

$$w^2\,(d/2) = w_0^2\left[1 + \frac{1}{\frac{2R}{d} - 1}\right] \tag{19.32}$$

Once again, a real value of w is obtained only if $2R > d$.

It can be seen from Eqs. (19.31) and (19.32) that, as $R/(d/2) \rightarrow 1$ the beam waist becomes very small, the beam radius at the mirrors becomes very large, causing high losses to the cavity mode. In such a limit, the centres of curvature of both mirrors coincide at the middle of the cavity, which is said to be *concentric.*

The beam radius is also much larger than the mirror size, and the cavity becomes very lossy in the limit, $R \rightarrow \infty$ which corresponds to using flat mirrors to realize the resonator cavity. In this case, both $w^2\,(d/2)$ and w_0^2 tend to the same limiting value of $(\lambda/\pi)\sqrt{Rd/2}$ as $R \rightarrow \infty$.

A particularly interesting configuration, providing the smallest possible spot size on the mirrors, occurs for $d = R$. In this configuration, the focal points of the two mirrors coincide, and the cavity is termed *confocal.* It can be shown (Problem 19.9) that the waist diameter of the laser mode in a confocal cavity relates to the length of the cavity, d, and the wavelength, λ, in the form

$$D_0 = \left(\frac{2\lambda d}{\pi}\right)^{12} \tag{19.33}$$

If the laser cavity is realized using mirrors with different radii of curvature, the beam waist will no longer be in the middle of the cavity. Eq. (19.20) can be used again in order to achieve two wavefront-matching conditions, one for each mirror. Considering the g parameters defined by Eq. (17.34), it can be shown that the condition for stability of the laser cavity becomes as given by Eq. (17.35).

19.4 Laser Beam Directionality

The directionality of laser light will perhaps be the property that most easily holds our attention. Such directionality is due to the laser cavity mirrors, which act as collimating apertures. In fact, these mirrors generally have such a high reflectivity that the light generated in the laser cavity is reflected by them many times before being transmitted to the outside. This limitation can be overcome using mirrors, with which the effective length of the amplifier medium becomes a high multiple of the laser cavity length. The multiple reflections increase the effective length of the amplifier medium, thus providing a higher amplification of the radiation generated inside the cavity. On the other hand, due to the long distance thus travelled, the curvature of the wavefront is greatly reduced, so that the light emitted is approximately collimated.

According to Eq. (19.20), the width of the laser beam increases linearly with z when the condition (19.19) is satisfied. Given that the half angle, ϕ, of the beam divergence cone is relatively small, one can assume that $\tan\phi \approx \phi$, thus having then (see Fig. 19.1)

$$\phi = \frac{w(z)}{z} = \frac{\lambda}{\pi w_0} = \frac{0.64\lambda}{D_0} \tag{19.34}$$

where $D_0 = 2w_0$ is the diameter of the beam at the waist. It should be noted that the divergence of the beam is greater for smaller values of the diameter at the waist. The typical divergence of the beam emitted by a He–Ne laser is on the order of one milliradian, meaning that the dimension of the beam increases by about 1 mm for each metre it travels.

The result given by Eq. (19.34) must be compared with that obtained in Chapter 11 when considering the diffraction of a plane wave by a circular aperture. In this case, the light irradiance far from the aperture undergoes oscillations in the transverse direction, with the angular radius of the Airy disc given by

$$\theta = \frac{1.22\lambda}{D} \tag{19.35}$$

where D is the diameter of the aperture. Both Eq. (19.34) and Eq. (19.35) depend on the ratio of wavelength to diameter and differ only in the constant factor. In fact, it turns out that the divergence of the Gaussian beam is approximately half the divergence of a plane wave diffracted by a circular aperture with a diameter equal to that of the beam at the waist. Actually, in the case of a Gaussian beam there are no oscillations of the irradiance in the far-field transverse direction, and it has the smallest possible angular spread for a given initial diameter. The beam waist corresponds to an effective circular aperture, which is located within the laser cavity and whose diameter can be adjusted to control the divergence of the beam. The wave fronts at the beam waist are plane, but exhibit an irradiance that varies with the distance to the axis according to the Gaussian function given by Eq. (19.12).

Example 19.2

a) Obtain the half angle of divergence for the beam emitted by a He–Ne laser ($\lambda = 632.8$ nm) with a cavity length $d = 30$ cm.

b) Suppose a laser was built to emit a beam of light with a wavelength of 300 nm and whose waist has a diameter $D_0 = 1$ cm. What would be the half angle of divergence in this case?

c) In practice, since the waist diameter of the laser beam can be controlled by an appropriate design of the laser cavity, as well as selecting the wavelength by conveniently choosing the active medium, what is the lower limit that can be expected for the divergence of the beam?

Solution

a) Using Eq. (19.33), the beam waist diameter is given by

$$D_0 = \left(\frac{(2)\left(632.8 \times 10^{-9}\right)(0.3)}{3.14}\right)^{1/2} = 0.35\,\text{mm}.$$

Example 19.2 continued

Using Eq. (19.34), the half angle of divergence becomes

$$\phi = \frac{(0.64)\left(632.8 \times 10^{-9}\right)}{3.5 \times 10^{-4}} \approx 1.2\,\text{mrad}.$$

This value corresponds to the typical divergence of a laser beam.

b) Using the given parameters, the half angle of divergence becomes:

$$\phi = \frac{(0.64)\left(300 \times 10^{-9}\right)}{1 \times 10^{-2}} \approx 1.92 \times 10^{-5}\,\text{rad}$$

In this case, the beam divergence is reduced by 63 times compared with the situation in point (a).

c) Of course, since the beam waist diameter can be controlled using an appropriate design of the laser cavity and wavelengths below the ultraviolet can be chosen, there is no limit to reducing the divergence of this beam.

19.5 Laser Beam Focusing and Expansion

As seen in Chapter 2, it is common in geometrical optics to represent the focus of a collimated beam through the use of a converging lens as a single point. Such a point, however, is nothing more than an idealization, due to the aberrations that affect that lens. The objective of concentrating light on a spot with an extremely small diameter—approximately equal to the wavelength of light—can, however, be achieved by using a converging lens to focus a laser beam.

The focusing of a laser beam through a converging lens is illustrated in Fig. 19.2. If the diameter D of the lens is equal to the diameter of the laser beam at a distance z from its waist, one has from Eq. (19.16) that

$$D = 2\frac{\lambda z}{\pi w_0} \tag{19.36}$$

By the action of the lens, the beam is focused and its diameter reaches a minimum value in the focal plane, given by

$$2w_f = \frac{4f\lambda}{\pi D} = \frac{4\lambda}{\pi} f\# \tag{19.37}$$

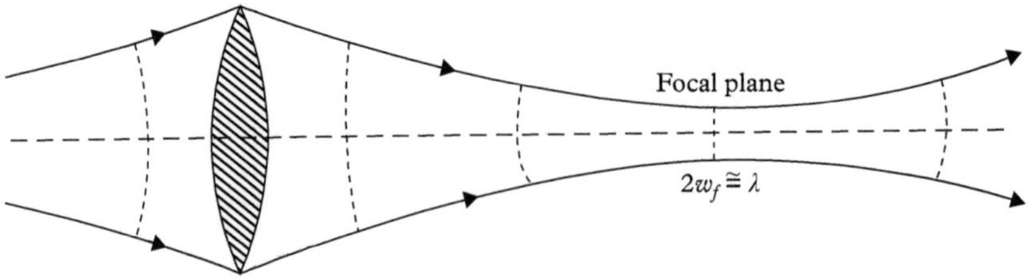

Figure 19.2 *Focusing of a laser beam through a converging lens.*

where $f/\# = f/D$ is the so-called *F-number* of the lens. In the case where $f/\# = 1$, it turns out that the diameter of the beam in the focal plane is approximately equal to the wavelength of the light.

As a consequence of laser beam focusing, the irradiance in the focal plane of the lens usually acquires very high values. This makes it possible to make very fine perforations or cuts in hard materials and, in general, to carry out industrial or medical operations on targets with dimensions similar to the wavelength of light. In the field of ophthalmology, for example, Nd:YAG lasers are commonly used in eye surgeries, and irradiances are then of the order of 10^9 to 10^{12} W/cm^2.

In order to achieve the best coupling efficiency of laser light into the core of an optical fibre, the spot size of the focused Gaussian beam should match the spot size of the fundamental mode of the fibre. This is referred to as *mode matching.* For given values of the light wavelength and fibre parameters, the fibre V parameter is found using Eq. (15.46), whereas the fibre mode field radius, w, is determined using the empirical formula given in Eq. (15.50). Such a mode field radius should match the spot size of the focused Gaussian beam to maximize the coupling efficiency. From Eq. (19.37), one can obtain the optimum beam diameter, d, and/or the focal length, f, of the lens to be used.

Beam expansion, instead of beam focusing, becomes useful in some cases. Eq. (19.31) shows that the expanded beam will have a smaller divergence angle, which makes it more collimated. A collimated beam has many practical applications, namely in free-space optical communications.

For example, let us consider the case of a He–Ne laser beam ($\lambda = 633$ nm) with a waist radius $w_0 = 1$ mm. Sensing this beam to the moon, which is at a distance $z = 384\ 000$ km from earth, the radius of the bright spot on the lunar surface is $w = 77.4$ km. However, if the beam waist at earth is $w_0 = 1$ m, one gets a bright spot of only 6 m. This allows, e.g. one to hit a retroreflector such as that placed on the lunar surface by Apollo astronauts and still get a detectable back-reflected signal at earth.

The laser beam can be expanded using two converging lenses, spaced at a distance equal to the sum of the respective focal lengths (Fig. 19.3). The first lens, with a focal length f_1, focuses the initial collimated beam, whereas the second lens, with a focal length $f_2 > f_1$, collimates the already expanded beam again. In this process, the diameter of the beam waist is increased by a factor equal to the ratio between these focal lengths:

$$\frac{D_e}{D_0} = \frac{f_2}{f_1} \tag{19.38}$$

At the same time as the beam expansion, a decrease in its angle of divergence is also obtained, given by the ratio $\frac{f_1}{f_2}$.

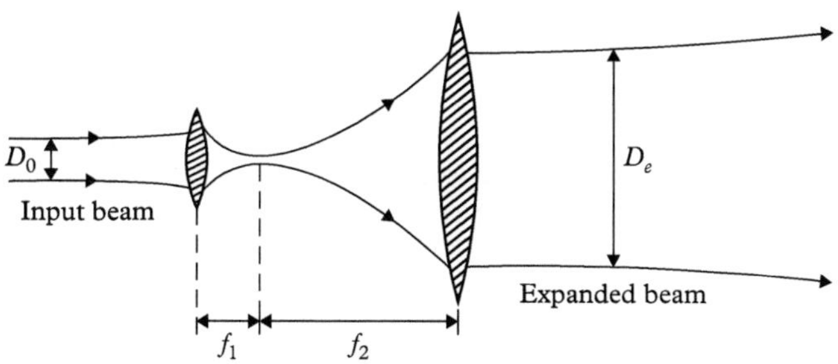

Figure 19.3 *Expansion of a laser beam using two converging lenses.*

Example 19.3

Consider the beam of a He–Ne laser with a power of 1 mW focused by a convergent lens with $f/\# = 1$. Calculate the diameter of the light spot in the focal plane of the lens and its irradiance.

Solution

The diameter of the beam in the focal plane of the lens is given by

$$2w_f = \frac{(4)\left(632.8 \times 10^{-9}\right)}{3.14}(1) = 806\,\text{nm}.$$

Assuming a uniform distribution of power, one has that the irradiance in the focal plane of the lens is given by

$$I = \frac{P}{\pi w_f^2} \approx 2 \times 10^9\,\text{W/m}^2.$$

19.6 Laser Light Characteristics

Laser light presents some unique properties, some of which are briefly described in the following.

19.6.1 Monochromaticity

From the point of view of the laboratory researcher, the most important property of laser light will perhaps be its extraordinary monochromaticity. However, even the light of a laser is not absolutely monochromatic, constituting only a good approximation to this ideal.

The degree of monochromaticity of the light emitted by a given source can be characterized by the width in wavelengths, $\Delta\lambda$, of the emission line. Depending on the source and its level of excitation, the light emitted may correspond to a more or less narrow spectrum of colours. In the case of white light, one has $\Delta\lambda \cong 300\,\text{nm}$, whereas for an ordinary discharge lamp one has $\Delta\lambda \cong 0.1\,\text{nm}$ and, for a low-pressure cadmium lamp, ($\lambda = 643.8\,\text{nm}$) one has $\Delta\lambda \cong 0.001\,\text{nm}$.

In the case of a laser, the stimulated emission process considerably restricts the band of wavelengths emitted during spontaneous emission. Thus, the spectral width $\Delta\lambda$ is greatly reduced, leading to a high degree of monochromaticity of the emitted light. A He–Ne laser ($\lambda = 632.8\,\text{nm}$) presents a width $\Delta\lambda \cong 10^{-8}\,\text{nm}$. That is, a He–Ne laser is 10 million times more monochromatic than an ordinary discharge lamp and a hundred thousand times more monochromatic than a low-pressure cadmium lamp.

The monochromatic nature of laser light is important for a number of applications. As seen in Chapter 16, group-velocity dispersion in an optical fibre is proportional to the spectral width and limits the maximum bit rate in optical communications. This makes the laser, with its narrow spectral width, an ideal light source for high-speed communications. Narrow laser linewidths also allow signals at different wavelengths to be multiplexed for transmission and later demultiplexed, a technique known as *wavelength division multiplexing* (WDM), as seen in Chapter 15.

19.6.2 Coherence

As seen in Chapter 10, the coherence property of light can be understood on the basis of the concept of correlation. If the light from a given source is completely coherent, both in spatial and temporal terms, then there will be a perfect correlation between the variations of the electric field at two arbitrary points in space. Having once measured the variations of the electric field at these two points, it will be possible, at a later time, to predict, with absolute certainty, the behaviour of the field at one of them, simply by measuring the electric field at the other point.

Only monochromatic light can be completely coherent in space and time. As a consequence, it is easy to understand why the highest degree of coherence among all types of sources is found in lasers. It is this property that allows its use in areas such as holography and interferometry.

The temporal coherence of a source is directly related to the width of its emission spectrum, $\Delta\nu$. This temporal coherence is characterized by the so-called *coherence time* τ_c, during which the components of the spectrum maintain a fixed phase relationship with each other and which is given by Eq. (10.3) as

$$\tau_c = \frac{1}{\Delta\nu} \tag{19.39}$$

The *coherence length* L_c is the distance travelled during the coherence time and is given by Eq. (10.2)

$$L_c = c\tau_c = \frac{c}{\Delta\nu} \tag{19.40}$$

The coherence length of an ordinary He–Ne laser, working simultaneously in several modes, is about 50 cm. However, in the case of a He–Ne laser operating in a single mode, its spectral width is typically 1 kHz, which corresponds to a coherence length of 300 km.

The coherence length can be measured using a Michelson interferometer. In the case of lasers with a spectral width of the order of 1 GHz, the open configuration of this interferometer, described in Chapter 9, may be used, since the differences in optical paths in its two arms are, in this case, less than 30 cm. However, for lasers with a spectral width of less than 1 MHz, the difference in optical paths is greater than 300 m. Under these circumstances, it becomes more practical to use an optical fibre in one of the arms of the Michelson interferometer.

Spatial Coherence

The degree of coherence along a wavefront (perpendicular to the direction of wave propagation) is referred to as *transverse* or *spatial coherence.* Perfect spatial coherence for an optical wave means that the wavefronts are continuous, with no interruptions in phase. The E-field at any two points along a wavefront is then correlated; knowing the field at one point allows the field at the other point to be predicted.

Spatial coherence has important consequences for the directionality of laser beams. If perfectly coherent light passes through an aperture of diameter D, it diverges with a half angle given by Eq. (19.38) for distances $z >> D$ from the aperture. Larger beam diameters correspond to smaller divergence angles, that is, more directional beams. Light that is only partially coherent can be characterized by a transverse coherence length d_c, shown in Fig. 19.4, which is the maximum separation between two points along the wavefront for

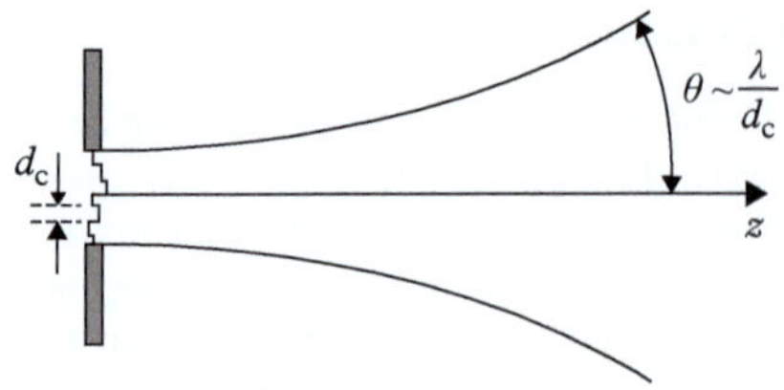

Figure 19.4 *Spatially incoherent light can be characterized by a transverse coherence length d_c.*

which the fields at the two points are correlated. The beam divergence angle for partially coherent light is similar to that of coherent light transmitted through an aperture of diameter d_c.

$$\theta \sim \frac{\lambda}{d_c} \tag{19.41}$$

19.6.3 Brightness

Another important feature of laser light is its brightness, or luminance, as defined in Chapter 4. In fact, laser light is characterized by a very high radiant power, which flows within a small solid angle in the form of a beam with a small diameter. A filament lamp, for example, generates considerably more radiant energy than a low-power continuous laser, but its emission is incoherent, the beam disperses within a large solid angle, and the bandwidth is significant.

In the case of a He–Ne laser ($\lambda = 632.8$ nm), 1 milliwatt of power corresponds to a luminous flux of 0.16 lm. Considering a beam with a diameter of 1 mm and a divergence of 1 mrad, which corresponds to a solid angle of about 10^{-6}srad, one finds that the brightness (luminance) of the light of this laser is

$$L_v \approx 2 \times 10^7 \text{ lm/sr.cm}^2 \tag{19.42}$$

Since the spectral width of the light emitted by a He–Ne laser is approximately 0.2 nm, the respective spectral luminance will be $L_v(\lambda) \approx 10^8$ lm/sr.cm^{2}nm.

On the other hand, the luminance of the sun is only $L_v \approx 1.5 \times 10^5$ lm/sr.cm^2. Considering that the visible spectrum of sunlight has a width of about 300 nm, the spectral luminance of the sun is $L_v(\lambda) \approx 5 \times 10^2$ lm/sr.cm^{2}nm. Thus, we conclude that a He–Ne laser with a power of only 1 mW has a spectral luminance five to six orders of magnitude higher than that of the sun.

19.6.4 Laser speckle

The vast majority of surfaces, synthetic or natural, are extremely rough on the scale of the wavelength. Because of the finite resolution, at any time we receive signals from a distribution of scatterers within the resolution cell. These scattered signals add coherently; that is, they add constructively and destructively depending on the relative phases of each scattered waveform. Speckle results from these patterns of constructive and destructive interference, shown as bright and dark dots in the image.

Actually, the term *speckle* refers to a random granular pattern which is generally observed when a highly coherent light beam (e.g. from a laser) is diffusely reflected at a rough surface. Even minor changes in the conditions, such as the illuminated spot or the direction of the incident laser beam, can change the detailed shape of a speckle pattern.

The speckle can be recorded on a photographic plate. A given point on this plate receives contributions from the entire illuminated surface. The relative phases of these multiple contributions vary significantly at an adjacent point on the plate, for which the difference in paths, relative to the ends of the illuminated zone, varies by one wavelength, λ.

The earlier description allows us to conclude that the dimensions of the laser speckle can be obtained based on Young's experiment described in Chapter 8. Thus, suppose that the illuminated area has a diameter h and that the distance between this area and the plane of observation is L (Fig. 19.5). Considering the two extreme points of the illuminated area, the spacing, d, between the fringes in the observation plane is given by Eq. (8.28):

$$d = \lambda \frac{L}{h} \tag{19.43}$$

Thus, for a given configuration (λ = const., L = const.), one has $d \propto 1/h$; that is, the spacing between the fringes increases when the distance between the two points decreases. The spacing given by Eq. (19.43) thus corresponds to the smallest possible dimension for the structure of the laser granite. Given that the light is diffused throughout the illuminated area and not only by the two extreme points considered, Eq. (19.43) is only an approximation.

The laser speckle considered earlier is called objective speckle, since its characteristics are not dependent on any imaging system. By contrast, subjective speckle is that which is obtained through a system of image formation, as is the case with the human eye.

When looking at a surface illuminated by laser light, each point on that surface corresponds to its image on the retina. However, the pupil of the eye functions as a circular aperture, which diffracts the light incident on it, so this image is not punctual. As a result, contributions from different points of the surface illuminated by laser light arrive at a given point on the retina, the phase of each of these contributions being random. This forms an interference pattern with a speckle aspect.

The size of the speckle observed by the eye, d_V, is given approximately by

$$d_v \approx \lambda \frac{L}{s} \tag{19.44}$$

where λ is the wavelength of light, L is the distance from the eye to the diffusing surface, and s is the diameter of the pupil. This relation can be easily verified by experience. In fact, when the eyes are slightly closed, the speckle becomes larger, as does the distance between the surface and the observer.

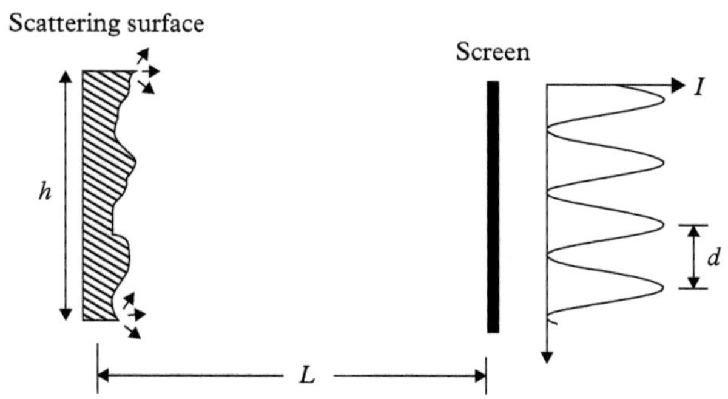

Figure 19.5 *Schematic representation for the analysis of the dimension of the objective laser speckle.*

The speckle does not disappear when the eye is blurred or when the glasses are removed. When the head moves, the optical paths from the points on the illuminated surface to the retina are altered, so that a corresponding movement of the speckle pattern is generally observed. In the case of a hypermetropic observer, this pattern moves in the same direction as the head movement. However, for a myopic observer, this movement occurs in the opposite direction. An observer with perfect vision will not observe any movement.

Speckle patterns are used in a wide range of metrology techniques, as they generally allow high sensitivity and simple setups. However, the effect of laser speckle is generally undesirable, being a limiting factor in imaging systems such as radar, synthetic aperture radar (SAR), medical ultrasound, and optical coherence tomography.

19.7 Further Reading

Miloni, P. W. and J. H. Eberly, *Lasers*, John Wiley & Sons, 1988.

O'Shea, D. C., *Introduction to Lasers and Their Applications*, Addison-Wesley, 2nd Ed., 1978.

Pedrotti, F., L. M. Pedrotti, and L. S. Pedrotti, *Introduction to Optics*, Cambridge University Press, 3rd Ed., 2018.

Saleh, B. A., and M. C. Teich, *Fundamentals of Photonics*, John Wiley & Sons, 3rd Ed., 2019.

Siegman, A. E., *Lasers*, University Science Books, 1986.

Svelto, O., *Principles of Lasers*, Plenum Press, 1998.

19.8 Problems

19.1 Verify that a paraboloidal wave with the complex envelope

$$U(r) = \frac{U_0}{z} \exp\left[i \frac{k}{2z}(x^2 + y^2)\right]$$

satisfies the paraxial wave equation given by Eq. (19.7).

19.2 The beam emitted by a laser is Gaussian, i.e. the amplitude of the field varies in a given cross-section with the radial coordinate r in the form

$$E = A \exp\left\{-\frac{r^2}{w^2}\right\}$$

where A is a constant and w is the so-called beam radius.

a) Obtain, as a function of w, the radius r_0 of the disc within which the flux density is more than half of its maximum value.

b) It is found that the beam is diffracted during its propagation. Get the distribution of the flux density corresponding to the Fraunhofer diffraction of the beam.

c) Determine the angle of diffraction θ_0 corresponding to the disc within which the flux density is more than half of its maximum value. Assume the values $\lambda = 632.8$ nm and $w = 0.5$ mm.

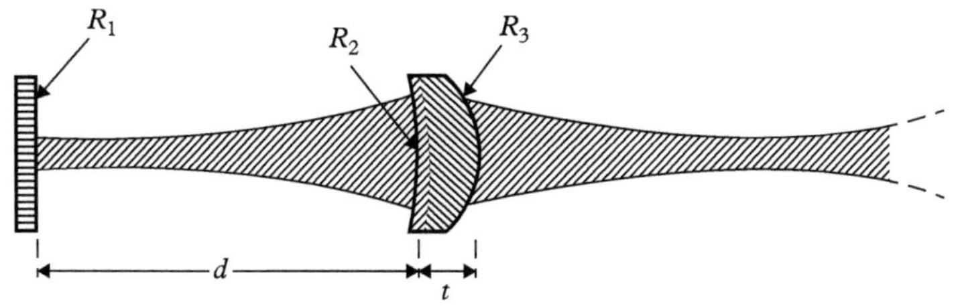

Figure 19.6 *Laser cavity and the emitted Gaussian beam.*

19.3 The beam emitted by a ruby laser ($\lambda = 694.3\,\text{nm}$) is directed at a target situated at a distance of 5000 km. Assuming that the beam has a diameter of 1 cm at the output of the laser, what value will this diameter have at the position of the target?

19.4 Compare the coherence length of the following sources: (a) a filament lamp emitting white light in a wavelength band 400–700 nm; (b) a He–Ne laser with a cavity of 40 cm oscillating with three longitudinal modes; (c) a stabilized Nd-YAG laser operating in a single longitudinal mode, with a spectral width $\Delta\nu = 30\,\text{kHz}$.

19.5 A He–Cd laser emits a beam of light with a radius of 1 mm and containing the wavelengths 325 nm and 488 nm. Assuming that the light propagates in free space, at what distance from the laser are the two components of the beam separated by 2 cm?

19.6 Consider the laser geometry in Fig. 19.6. The laser cavity has a length $d = 0.6$ m and is formed by a plane mirror ($R_1 = \infty$) and a concave mirror of radius $R_2 = 1$ m, which corresponds to the face of a lens of thickness $t = 5$ mm. The other face of the lens has a radius $R_3 = 0.5$ m. Using the ABCD propagation law, determine the radius and the location of the external beam waist.

19.7 Demonstrate the result given by Eq. (19.31) for the radius of the Gaussian beam at the waist inside the laser cavity.

19.8 Demonstrate the result given by Eq. (19.32) for the radius of the Gaussian beam at the mirror position.

19.9 Demonstrate the result given by Eq. (19.33) for the waist diameter of the laser mode in a confocal cavity of length d.

19.10 A semiconductor laser operates at a free-space wavelength of 790 nm and has a longitudinal coherence length of 1 mm. Determine the linewidth in terms of both frequency and wavelength.

19.11 Consider an elemental plane surface, with diameter h, illuminated uniformly by the light emitted by a He–Ne laser ($\lambda = 632.8\,\text{nm}$). The pattern formed by the laser speckle on a screen located at a distance of 1 m from the diffusing surface is observed at a distance of 25 cm from that screen. Determine for what value of the diameter of the diffusing surface the thinner structure of the laser speckle can no longer be observed. Consider that the minimum resolution of the view is 47 arcseconds.

20

Non-Linear Optics

In the preceding chapters, we have assumed that light interacts with matter in a linear fashion. In this context, it is assumed that the propagation of light waves in a given medium can be described by a linear wave equation. As a consequence, the propagation of two harmonic waves is described by the principle of superposition, and there is no distortion caused by this medium or by the mutual interference of the waves.

However, just as a simple mechanical system (e.g. a spring) can manifest a non-linear response when it is subjected to sufficiently intense forces, it is also reasonable to assume that a high-intensity light beam, propagating in a given material medium, shows non-linear effects. Actually, such effects have become easily observable with the advent of lasers. In these circumstances, it is found that the optical properties of the medium vary with the intensity of the light and that the interference of two or more light waves in such a medium ceases to satisfy the principle of superposition. The first experiment in the field of non-linear optics was carried out by Peter A. Franken and his collaborators at the University of Michigan in 1961, when they first observed the phenomena of second-harmonic generation and optical rectification.

This chapter will describe a few phenomena in the area of non-linear optics, some of which are mentioned in the case where the non-linear medium is an optical fibre. In fact, the field of non-linear optics is currently quite extensive, covering many other phenomena besides those mentioned in this chapter.

20.1 The Non-Linear Polarization

The most common mechanism that is at the origin of the non-linear optical behaviour of a given material is related to the distortion of the electron clouds in the atoms that constitute such material. Actually, the presence of a light wave with an electric field **E** determines a charge distribution within the atom, such that the positively charged nucleus is displaced in the direction of **E**, whereas the negatively charged electron cloud is displaced in the opposite direction. This is illustrated in Fig. 20.1. As a consequence, each atom produces an *electric dipole* **p**, which is in the same direction as **E**. At a macroscopic level, this induces a *polarization density* $\mathbf{P} = N\mathbf{p}$ in the medium, where N is the number of atoms per unit volume.

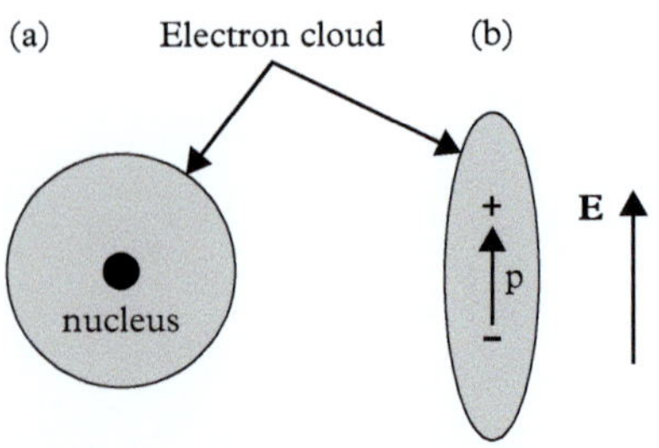

Figure 20.1 *(a) Unpolarized atom and (b) polarized atom.*

The induced dipoles are forced to oscillate at the frequency of the incident light, as in a driven harmonic oscillator. The oscillating dipoles radiate light with the same frequency and direction as the incident light, but with a shift in phase. The non-linear response of this mechanism is on the order of femtoseconds, which is the time scale at which the electron distribution in an atom can change.

The harmonic oscillator model was used in Section 6.7 to describe wave propagation in a dielectric medium. The equation of motion derived there, given by Eq. (6.64), assumes that the restoring force of this oscillator is a linear function of the electron displacement from the mean position. If the displacement is in the x-direction, the restoring force is given by

$$F_r = -kx \tag{20.1}$$

Lasers, Optics, and Photonics. Mário F. S. Ferreira, Oxford University Press. © Mário F. S. Ferreira (2026).
DOI: 10.1093/oso/9780198952770.003.0020

In physics, the linear dependence of one physical quantity on another is generally valid over only a limited range of values. In the present case, the motion of the charged particles in a dielectric medium can be considered to be linear with the applied field only if this field is relatively weak and the displacement x is small. If these conditions are not satisfied, it becomes necessary to include additional anharmonic terms in the restoring force. The first correction term depends quadratically on x, with which the restoring force becomes

$$F_r = -kx + ax^2 \tag{20.2}$$

where the coefficient a describes the strength of the non-linearity. The potential energy corresponding to this force is

$$U_p(x) = \frac{1}{2}kx^2 - \frac{1}{3}ax^3 \tag{20.3}$$

In materials with inversion symmetry, the carriers are isotropically distributed and the potential function must satisfy the symmetry condition $U_p(x) = U_p(-x)$. This is not the case for Eq. (20.3). In order to satisfy the aforementioned condition, the lowest-order correction term in the restoring force must depend on x^3, with which it becomes

$$F_r = -kx + bx^3 \tag{20.4}$$

where the coefficient b describes the strength of the third-order non-linearity.

The considerations discussed earlier can be applied to the propagation of light in a dielectric medium. In this case, the expansion of the restoring force in a power series given by Eq. (20.4) corresponds, on a macroscopic scale and in the frequency domain, to a power-series expansion of the medium polarization **P** in the electric field vector **E**:

$$\mathbf{P}(\mathbf{r}, \omega) = \varepsilon_0 \left\lfloor \chi^{(1)}\mathbf{E} + \chi^{(2)}\mathbf{EE} + \chi^{(3)}\mathbf{EEE} + \ldots \right\rfloor \tag{20.5}$$

where $\chi^{(j)}$ $(j = 1, 2, \ldots)$ is the jth-order susceptibility and ε_0 the permittivity of the vacuum. The first-order susceptibility $\chi^{(1)}$ determines the linear part of the polarization, $\mathbf{P}_L$, which is given by

$$\mathbf{P}_L(\mathbf{r}, \omega) = \varepsilon_0 \chi^{(1)}\mathbf{E} \tag{20.6}$$

On the other hand, terms of second and higher order in (20.5) determine the non-linear polarization, $\mathbf{P}_{NL}$, given by

$$\mathbf{P}_{NL}(\mathbf{r}, \omega) = \varepsilon_0 \left\lfloor \chi^{(2)}\mathbf{EE} + \chi^{(3)}\mathbf{EEE} + \ldots \right\rfloor \tag{20.7}$$

In the time domain, the form of the expansion of the polarization **P** is identical to (20.5) if the non-linear response is assumed to be instantaneous. If not, then the successive terms in Eq. (20.5) become convolutions of one, two, and three dimensions.

To account for the light polarization effects, $\chi^{(j)}$ is a tensor of rank j+1. For example, since the product **EEE** in general contains 27 terms, the susceptibility $\chi^{(3)}$ is a fourth-rank tensor that could contain 81 different terms. In practice, however, most materials of interest for technical applications exhibit symmetries that drastically reduce the number of independent tensor elements. For an isotropic material, for example, the first-order susceptibility $\chi^{(1)}$ contains only a single element. This element leads to the refractive index and attenuation coefficient, as shown in Chapter 6.

The second-order polarization term in Eq. (20.7) is proportional to the square of the electric field and is responsible for phenomena like second-harmonic generation, three-wave mixing, optical rectification and parametric amplification. Fig. 20.2 shows the polarization as a function of the electric field for the linear case and the deviation from linearity due to this second-order term. For equal positive and negative fields, the response of the optical medium is not symmetrical in the non-linear case (dashed curve). In this case, the negative field produces a greater polarization than a positive field of the same magnitude.

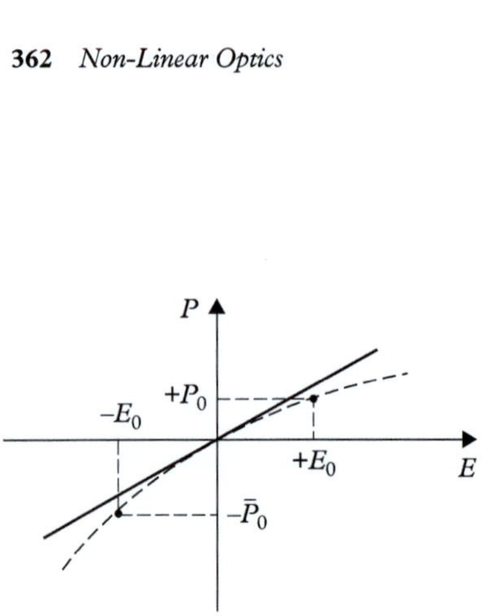

Figure 20.2 *Linear (full line) and typical non-linear (dashed curve) responses of the medium's second-order polarization to an applied electric field E.*

A crystal having inversion symmetry is characterized by an inversion centre, such that if the radial coordinate becomes symmetric, the crystal's atomic arrangement remains unchanged and so the crystal responds in the same way to a physical influence. In such a crystal, reversing the applied field direction should also reverse the direction of the second-order non-linear polarization. However, because the polarization is proportional to the square of the field, such a reversal would not be allowed mathematically. It therefore follows that in a material with inversion symmetry all 27 elements of the second-order susceptibility must be zero.

In the case of third-order non-linear polarization, the three fields can interact with each other. If the fields have different frequencies, they produce a polarization of the material at the sum and difference frequencies or at multiples of the input frequencies. The third-order non-linear polarization is given by

$$\mathbf{P}_{NL}^{(3)}\left(\omega=\omega_1+\omega_2+\omega_3\right)=\varepsilon_0\chi^{(3)}\left(\omega_1,\omega_2,\omega_3\right)\mathbf{E}\left(\omega_1\right)\mathbf{E}\left(\omega_2\right)\mathbf{E}\left(\omega_3\right) \tag{20.8}$$

which can be written in the matrix form

$$\begin{bmatrix} P_{NLx}^{(3)}(\omega) \\ P_{NLy}^{(3)}(\omega) \\ P_{NLz}^{(3)}(\omega) \end{bmatrix} = \varepsilon_0 \begin{bmatrix} \chi_{xxxx}^{(3)} & \chi_{xxxy}^{(3)} & \cdots & \chi_{xzzz}^{(3)} \\ \chi_{yxxx}^{(3)} & \chi_{yxxy}^{(3)} & \cdots & \chi_{yzzz}^{(3)} \\ \chi_{zxxx}^{(3)} & \chi_{zxxy}^{(3)} & \cdots & \chi_{zzzz}^{(3)} \end{bmatrix} \begin{bmatrix} E_x(\omega_1)E_x(\omega_2)E_x(\omega_3) \\ E_x(\omega_1)E_x(\omega_2)E_y(\omega_3) \\ \cdots \\ \cdots \\ E_z(\omega_1)E_z(\omega_2)E_y(\omega_3) \\ E_z(\omega_1)E_z(\omega_2)E_z(\omega_3) \end{bmatrix} \tag{20.9}$$

If the medium is isotropic, as happens with the silica glass, it is found that 60 of the 81 elements of $\chi^{(3)}$ are zero. Moreover, the remaining 21 elements are not independent of each other, and the tensor $\chi^{(3)}$ can be reduced to only three elements. If, in addition, all frequencies are far from any resonance frequency, the third-order susceptibility possesses only one independent element.

20.1.1 Other non-linear optical mechanisms

Although distortion of the electron cloud is the most common type of optical non-linearity, there are other non-linear mechanisms that become important in certain materials. Some of them will be briefly mentioned here. Other mechanisms are described in a simple way using the quantum concept of the photon and will be addressed in the following section.

Molecular orientation is a non-linear mechanism that occurs for asymmetrical molecules that are free to rotate in a liquid. In the presence of a strong electric field, these molecules are oriented due to the interaction of the E field with the molecular dipoles. This alignment creates a macroscopic asymmetry in the material that changes its optical properties. In particular, it induces *birefringence,* making the refractive index dependent on the direction of polarization of an incoming light wave. Changes in the refractive index due to this process occur on a picosecond time scale, a typical time scale for molecular rotations.

Electrostriction is another phenomenon that occurs quite generally in solids, liquids, and gases and involves the translation of atoms or molecules into a region of higher optical intensity. This process causes an increase in mass density and, consequently, in the refractive index in the high-intensity region of an optical beam. In the case of optical fibres, *electrostriction* is at the origin of the stimulated Brillouin scattering process, through which an optical (pump) wave creates a periodic modulation of the refractive index along the fibre. This index grating determines the scattering of part of the pump power, which goes to a newly generated optical Stokes wave with lower frequency. The response time for electrostriction is dependent on the transit time of an acoustic wave, which is on the order of 1 ns in optical fibres.

Inelastic scattering occurs when an optical (pump) wave interacts with the vibrational resonances of the material. Such interaction can be considered non-linear in the sense that there is coupling between the electronic and vibrational motions in the molecule or solid, which generates new frequencies of light. As the scattered light increases in intensity, the scattering process eventually becomes stimulated. In this regime, the scattered field interacts coherently with the pump field to further excite the resonances, which significantly enhances the transfer of power between the two optical waves. This is at the origin of Raman scattering, which can occur in all materials. In silica glass—the material from which optical fibres are made—the dominant Raman lines are due to the bending motion of the Si–O–Si bond.

20.2 Quantum Description of Non-Linear Processes

In classical terms, non-linear effects are described based on the characteristic non-linear susceptibilities of the medium. However, due to the intrinsically quantum nature of the interaction between light and matter, it turns out that these effects can be described in a simple way using the quantum concept of the photon. In this section, we will briefly address some non-linear optical phenomena from the perspective of quantum optics.

20.2.1 Interaction of two photons

By passing through a given optical medium, a photon can be absorbed and eventually cause the ionization of some atom of that medium, provided that its energy is adequate to produce such an effect. At high intensities, it may happen that not one, but two photons together carry the appropriate energy to achieve any of these results. In such cases, the involved phenomena are called *two-photon absorption* (TPA) or *two-photon ionization*, respectively.

Consider an optical medium that has an energy level $\Delta E = 2h\nu_1$ above the ground level. When a harmonic wave with frequency ν_1 passes through this medium, it can attain the excited state by the simultaneous absorption of two photons. In this process, an atom or molecule is excited from the ground state, via a virtual energy level, to a higher energy level (Fig. 20.3). The probability of this process occurring increases at higher intensity.

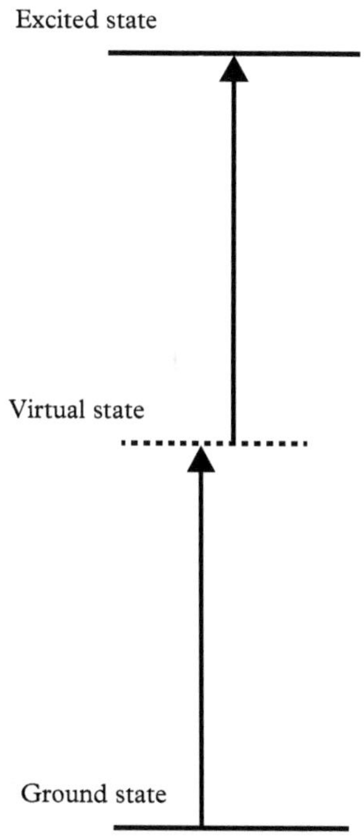

Figure 20.3 *Schematic of the energy levels involved in the two-photon absorption process.*

That is, the medium shows an absorption of photons with frequency ν_1, although it does not present any transition at this frequency.

The process of *TPA* is said to be degenerate if the two involved photons have the same frequency, However, it can be generalized when two photons with different frequencies, ν_1 and ν_2, propagate in some media. In this case, the simultaneous absorption of two photons, one with frequency ν_1 and another with frequency ν_2, can occur so that the medium achieves an excited state with energy $\Delta E = h(\nu_1 + \nu_2)$ above the ground state. Absorption of two photons with different frequencies is called *non-degenerate TPA*. The TPA phenomenon was originally predicted by Maria Goeppert-Mayer in 1931, whereas its experimental verification was made only 30 years later, after the invention of the laser.

The process of absorption of two photons has found applications in the field of spectroscopy, allowing, in particular, the exploration of the structure of the energy levels of matter. In fact, some of these levels are not accessible through the absorption of a single photon because they correspond to the so-called forbidden transitions. However, this is made possible through the contribution of two photons, since in this case the transition rules are different.

When the joint energy of two photons in the optical field is greater than the binding energy of the electron to the respective atom or molecule, it can be ionized by the simultaneous action of those photons. This process is known as *two-photon ionization*.

20.2.2 Interaction of three photons

When two light waves propagate through a medium with non-linear properties, new waves with frequencies that result from the combination of the original frequencies can be generated. In many cases, only three waves take part in this process.

The *second-harmonic generation* (SHG) is a process in which two photons with the same frequency interact with a non-linear medium, generating a new photon with twice the energy of the initial photons (and twice the frequency). Fig. 20.4 shows a schematic of the energy levels involved in this process. It turns out that this conversion is efficient only when the initial and the new waves have the same propagation velocity. The process of SHG is widely used in doubling laser frequencies. For example, it can be used to convert the emission of the Nd:YAG laser from the infrared ($\lambda = 1064\,\text{nm}$) to the visible ($\lambda = 532\,\text{nm}$). SHG was first demonstrated by Peter Franken et al. at the University of Michigan in 1961. The demonstration was made possible by the invention of the laser, which created the required high-intensity coherent light.

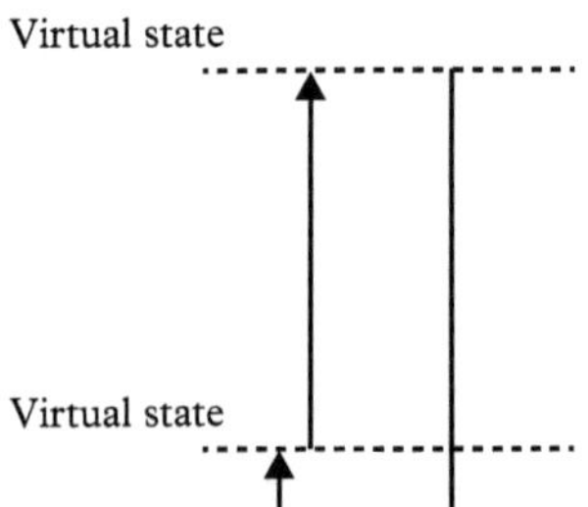

Figure 20.4 *Schematic of the energy levels involved in the second-harmonic generation process.*

In the process of *sum-frequency generation*, a photon with frequency ν_1 and a photon with frequency ν_2 are converted into a photon with frequency $\nu_3 = \nu_1 + \nu_2$. This process constitutes a generalization of second harmonic generation. In the process of *difference-frequency generation*, a photon with frequency ν_1 and a photon with frequency ν_2 are converted into a photon with frequency $\nu_3 = \nu_1 - \nu_2$. A particular case of this process is the so-called *optical rectification*, in which the incident photons have equal frequencies, so that the resulting field has a zero frequency.

Another non-linear process involving three photons corresponds to the so-called *parametric amplification*. In this case, a suitable non-linear medium is illuminated by a sufficiently intense pumping wave with frequency ν_3, which ends up decaying into a wave with frequency ν_1 (signal wave) and another wave (idler wave) with frequency equal to the difference between the previous two: $\nu_2 = \nu_3 - \nu_1$. That is, instead of having $\nu_1 + \nu_2 \rightarrow \nu_3$, as in the process of sum-frequency generation, one has, in this case, $\nu_3 \rightarrow \nu_1 + \nu_2$. In these circumstances, both the signal wave and the idler wave are amplified at the expense of the power carried by the pumping wave.

20.2.3 Interaction of multiple photons

In the processes of interaction with multiple photons, one usually has a wide variety of possible combinations. Among these processes is the *multiplication of frequencies*, in which n photons with equal or different frequencies interact with the atoms of a suitable non-linear medium, resulting in a photon with frequency $\nu_m = n\nu_1$ or, in the general case, $\nu_m = \nu_1 + \nu_2 + \cdots + \nu_n$. As particular cases, we have the process of *third-harmonic generation*, in which $\nu_m = 3\nu_1$, and the process of *fourth harmonic generation*, in which $\nu_m = 4\nu_1$. These processes can be used to produce coherent light with short wavelengths.

Other processes worth mentioning are *multiple photon absorption* and *ionization*. In the first case, n photons with generally different frequencies are simultaneously absorbed by the non-linear medium, which passes into an excited state, corresponding to an increase in energy $\Delta E = h(\nu_1 + \nu_2 + \cdots + \nu_n)$. In the ionization process, the only difference lies in the fact that the total energy of the intervening photons is higher than the binding energy of the electrons, resulting in the ionization of the atoms in the medium.

20.3 The Non-Linear Wave Equation

By taking the curl of Eq. (6.3) and using Eqs. (6.4), (6.5), and (6.6), the wave equation for the electric field can be obtained in the form:

$$\nabla^2 \mathbf{E} - \frac{1}{c^2}\frac{\partial^2 \mathbf{E}}{\partial t^2} = \mu_0 \frac{\partial^2 \mathbf{P}}{\partial t^2} \tag{20.10}$$

where c is the velocity of light in vacuum. Eq. (20.10) shows that the time-varying polarization serves as a driving term in the wave equation.

Considering Eqs. (20.5)–(20.7), the wave Eq. (20.10) can be written in the form

$$\nabla^2 \mathbf{E} - \frac{1}{c^2}\frac{\partial^2 \mathbf{E}}{\partial t^2} = \mu_0 \frac{\partial^2 \mathbf{P}_L}{\partial t^2} + \mu_0 \frac{\partial^2 \mathbf{P}_{NL}}{\partial t^2} \tag{20.11}$$

Using Eq. (20.6) and the relation $\varepsilon = \varepsilon_0 (1 + \chi_1)$, Eq. (20.11) becomes

$$\nabla^2 \mathbf{E} - \mu_0 \varepsilon \frac{\partial^2 \mathbf{E}}{\partial t^2} = \mu_0 \frac{\partial^2 \mathbf{P}_{NL}}{\partial t^2} \tag{20.12}$$

The electric field **E** in Eq. (20.12) represents the total field, that is, the sum of all incident fields and those generated by the non-linear polarization.

In contrast to the linear wave equation, it is not possible to obtain general analytical solutions for Eq. (20.12). For this reason, a specific case corresponding to the problem of the interaction of three waves in the presence of a quadratic non-linearity will be analysed in the following section.

20.3.1 Three-Wave Interaction

In general, there are several waves interacting in a non-linear medium. In some cases, two initial waves overlapping spatially may each modify the properties of the other and create additional waves with different frequencies. Fig. 20.5 illustrates this case.

Let us consider the case where the electric field is composed of three plane waves with angular frequencies ω_1, ω_2 and ω_3, such that

$$\omega_3 = \omega_1 + \omega_2 \tag{20.13}$$

Assuming that the three waves propagate along the z-axis and have the same polarization, the respective scalar fields can be written in the form:

$$E(\omega_i) = E_i = \frac{1}{2}\left[E_i(z)\exp\{i(k_i z - \omega_i t)\} + E_i^*(z)\exp\{-i(k_i z - \omega_i t)\}\right], i = 1, 2, 3 \tag{20.14}$$

The complex amplitude of the waves is assumed to vary slowly with the distance z, i.e.

$$\frac{\partial^2 E_i}{\partial z^2} << k_i \frac{\partial E_i}{\partial z} \tag{20.15}$$

This is known as the slowly varying amplitude (SVA) approximation, under which the first term on the left-hand side of Eq. (20.12) becomes

$$\nabla^2 E_i \approx -\frac{1}{2}\left(k_i^2 E_i - 2ik_i \frac{\partial E_i}{\partial z}\right)\exp\{i(k_i z - \omega_i t)\} + c.c. \tag{20.16}$$

where *c.c.* indicates the complex conjugate of the previous expression.

The second time derivative of the electric field can be written as

$$\frac{\partial^2 E_i}{\partial t^2} = -\omega_i^2 E_i, i = 1, 2, 3 \tag{20.17}$$

Moreover, considering the linear dispersion law, one can write the following relation for each wave:

$$k_i^2 = \mu_0 \varepsilon_i \omega_i^2, i = 1, 2, 3 \tag{20.18}$$

Let us consider that the non-linear part of the polarization is given only by the second-order term in Eq. (20.7), and can be written in the form:

$$\begin{aligned} P_{NL} &= P_{NL}(\omega_1) + P_{NL}(\omega_2) + P_{NL}(\omega_3) \\ &= d\left[E^*(\omega_2)E(\omega_3) + E^*(\omega_1)E(\omega_3) + E(\omega_1)E(\omega_2)\right] \end{aligned} \tag{20.19}$$

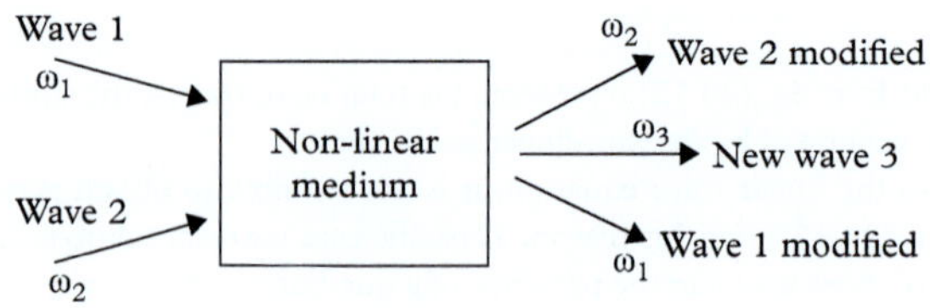

Figure 20.5 *Interaction of three waves in a non-linear medium.*

where d is the so-called *non-linear coefficient*. Considering one of the three waves with a given frequency, only that part of the non-linear polarization that has the same frequency will be a source for such a wave. For example, considering the frequency $\omega_1 = \omega_3 - \omega_2$, the source term is

$$P_{NL}(\omega_1) = \frac{1}{2} d E_2^* E_3 \exp\{i(k_3 - k_2)z - i(\omega_3 - \omega_2)t\} + c.c. \tag{20.20}$$

Considering Eqs. (20.17) and (20.20), the time derivatives of the polarization can be easily calculated. For example, the second derivative of $P_{NL}(\omega_1)$ becomes

$$\frac{\partial^2 P_{NL}(\omega_1)}{\partial t^2} = -(\omega_3 - \omega_2)^2 d E_2^* E_3 \exp\{i(k_3 - k_2)z - i(\omega_3 - \omega_2)t\} + c.c. \tag{20.21}$$

Using Eqs. (20.15), (20.16), and (20.20), the non-linear equation for the wave with frequency ω_1 becomes

$$\begin{aligned} \frac{1}{2}\left(k_1^2 E_1 - 2ik_1 \frac{\partial E_1}{\partial z}\right) \exp\{i(k_1 z - \omega_1 t)\} + c.c. - \frac{1}{2}\mu_0 \varepsilon_1 \omega_1^2 E_1 \exp\{i(k_1 z - \omega_1 t)\} + c.c. = \\ = \frac{1}{2}\mu_0 {\omega_1}^2 d E_2^* E_3 \exp\{i(k_3 - k_2)z - i(\omega_3 - \omega_2)t\} + c.c. \end{aligned} \tag{20.22}$$

Considering complex quantities and using the dispersion relation given by Eq. (20.18), Eq. (20.22) can be simplified and written in the form:

$$\frac{\partial E_1}{\partial z} = \frac{i\omega_1}{2}\sqrt{\frac{\mu_0}{\varepsilon_1}} d E_3 E_2^* \exp\{i(k_3 - k_2 - k_1)z)\} \tag{20.23}$$

There is another equation for the complex conjugate E_1^*, but it does not contain any further information.

Following a similar procedure, one can obtain the following equations for waves with frequencies ω_2 and ω_3:

$$\frac{\partial E_2}{\partial z} = \frac{i\omega_2}{2}\sqrt{\frac{\mu_0}{\varepsilon_2}} d E_3 {E_1}^* \exp\{i(k_3 - k_2 - k_1)z)\} \tag{20.24}$$

$$\frac{\partial E_3}{\partial z} = \frac{i\omega_3}{2}\sqrt{\frac{\mu_0}{\varepsilon_3}} d E_1 E_2 \exp\{i(k_1 + k_2 - k_3)z)\} \tag{20.25}$$

The quantity

$$\Delta k = k_1 + k_2 - k_3 \tag{20.26}$$

corresponds to the phase mismatch of the three waves. Assuming the phase-matching case, $\Delta k = 0$, one can multiply each of Eqs. (20.23)–(20.25) by ${E_i}^*$ and obtain the following identity:

$$\frac{1}{\omega_1}\sqrt{\frac{\varepsilon_1}{\mu}}\frac{\partial}{\partial z}\left(E_1 {E_1}^*\right) = \frac{1}{\omega_2}\sqrt{\frac{\varepsilon_2}{\mu}}\frac{\partial}{\partial z}\left(E_2 {E_2}^*\right) = -\frac{1}{\omega_3}\sqrt{\frac{\varepsilon_3}{\mu}}\frac{\partial}{\partial z}\left(E_3 {E_3}^*\right) \tag{20.27}$$

Eq. (20.27) corresponds to the law of conservation of energy and is known as *the Manley–Rowe relation*.

Example 20.1

Consider that two waves, with frequencies ω_1 and ω_2, are incident in a non-linear medium located in the region $z \geq 0$. Assuming that the amplitude of both waves remains constant during propagation, obtain the irradiance of the wave with sum frequency, $\omega_3 = \omega_1 + \omega_2$, considering that it is zero in the plane $z = 0$. Discuss the validity of the result.

Solution

Assuming that the amplitude of the incident waves remains constant, one can integrate Eq. (20.25) and get the following result for the wave with frequency $\omega_3 = \omega_1 + \omega_2$:

$$E_3 = \frac{i\omega_3}{2}\sqrt{\frac{\mu_0}{\varepsilon_3}} dE_1 E_2 \frac{\exp\{i\Delta kz\} - 1}{i\Delta k} \tag{20.28}$$

The irradiance of this wave is given by

$$I_3(z) \propto E_3(z) E_3^*(z) = \omega_3^2 \frac{\mu}{\varepsilon_3} d^2 |E_1|^2 |E_2|^2 \frac{\sin^2[\Delta kz/2]}{(\Delta k)^2}, \quad \Delta k \neq 0 \tag{20.29}$$

Eq. (20.29) shows that the evolution of the irradiance of the wave with sum frequency follows the sine-squared function, the maximum irradiance being inversely proportional to the square of Δk. For $\Delta k = 0$, the irradiance is proportional to the square of the distance, a result that is valid only for sufficiently small values of z. In fact, for large values of z, it is not physically reasonable to assume the incident wave amplitudes E_1 and E_2 to be constant.

The results given earlier can be easily adapted to the case of *difference-frequency generation*, since the formalism of this process is similar to that of the generation of the frequency-sum wave. It suffices to assume that ω_1 and ω_3 are the frequencies of the incident waves and that $\omega_2 = \omega_3 - \omega_1$ is the frequency of the generated wave with difference frequency.

20.4 Second Harmonic Generation

Let us consider the case $\omega_1 = \omega_2 = \omega$ and $\omega_3 = 2\omega$, together with the phase-matching condition $\Delta k = 0$, and assume also that $\varepsilon_1 = \varepsilon_2 = \varepsilon_3 = \varepsilon$. The fields are written as $E_1 = E_2 = E_\omega \exp\{i\phi_\omega\}$ and $E_3 = E_{2\omega} \exp\{i\phi_{2\omega}\}$, considering that $E_1 = E_{0\omega} \exp\{i\phi_\omega\}$ and $E_3 = 0$ at $z = 0$. The non-linear medium is assumed to be located in the region $z \geq 0$.

Under the aforementioned conditions, Eqs. (20.23)–(20.25) are reduced to only two equations, which can be written in the form:

$$\frac{\partial E_\omega}{\partial z} = \frac{i}{2} A E_\omega E_{2\omega} \exp\{i\phi_{2\omega}\} \tag{20.30}$$

$$\frac{\partial E_{2\omega}}{\partial z} = iAE_\omega^2 \exp\{i\Delta\psi\} \tag{20.31}$$

where

$$A = \sqrt{\frac{\mu}{\varepsilon}} d\omega \tag{20.32a}$$

and

$$\Delta\psi = 2\phi_\omega - \phi_{2\omega} \tag{20.32b}$$

In order to favour the generation of the second harmonic wave, we will assume that

$$\Delta\psi = -\frac{\pi}{2}$$

In this situation, and using the law of conservation of energy, one can rewrite Eq. (20.30) in the form:

$$\frac{\partial E_{2\omega}}{\partial z} = A\left[E_{0\omega}^2 - E_{2\omega}^2\right] \tag{20.33}$$

Integrating Eq. (20.30), the amplitude of the second harmonic wave at a distance z in the non-linear medium is found to be given by

$$E_{2\omega}(z) = E_{0\omega} \tanh\left[AE_{0\omega}z\right] \tag{20.34}$$

Using Eq. (20.33) and the principle of conservation of energy, one can also obtain the following solution for the amplitude of the incident wave:

$$E_\omega(z) = E_{0\omega} \operatorname{sech}\left[AE_{0\omega}z\right] \tag{20.35}$$

Figure 20.6 illustrates the behaviour of the irradiances of the second harmonic wave, $I_{2\omega} \propto E_{2\omega}^2$, and the incident (pumping) wave, $I_\omega(\omega) \propto E_\omega^2$, normalized by the input wave irradiance, I_0.

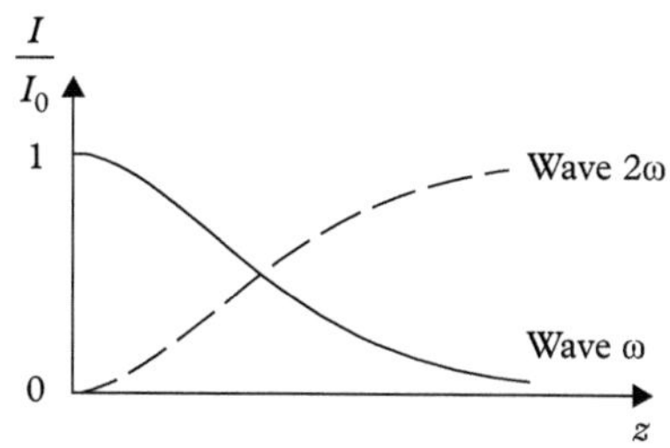

Figure 20.6 *Evolution of the normalized irradiances for the second-harmonic wave (2ω) and the pumping wave (ω).*

20.4.1 Dispersion effects on SHG

In order to study the impact of dispersion on SHG, let us consider that the propagation constants for the pumping wave (k_ω) and the second harmonic wave ($k_{2\omega}$) are given by

$$k_\omega = \frac{2\pi}{\lambda_\omega} = n_\omega \frac{\omega}{c} \tag{20.36a}$$

$$k_{2\omega} = \frac{2\pi}{\lambda_{2\omega}} = n_{2\omega} \frac{2\omega}{c} \tag{20.36b}$$

where λ_ω and $\lambda_{2\omega}$ are the wavelengths in the medium at the two frequencies ω and 2ω, and n_ω and $n_{2\omega}$ are the corresponding indices of refraction. The wave vector mismatch can be written as

$$\Delta k = (n_{2\omega} - n_\omega)\frac{2\omega}{c} \tag{20.37}$$

If $n_{2\omega} = n_\omega$, then $\Delta k = 0$, which would mean that the light radiated by atoms at different positions z would be in phase, resulting in constructive interference and an efficient SHG. Unfortunately, dispersion generally results in the condition $n_{2\omega} > n_\omega$, so that $\Delta k \neq 0$. In this case, the second harmonic waves from different atoms will be out of phase. The efficiency of SHG will start to decrease for a propagation distance L_c, the so-called *coherence length*, given by the condition

$$L_c \Delta k = \pi \tag{20.38}$$

So, in order to achieve an optimum SHG efficiency, the pump light must pass through a crystal of length L_c.

Example 20.2

Consider a situation in which a fundamental optical wave of wavelength $\lambda_0 = 1.0\ \mu\text{m}$ is incident on a potassium dihydrogen phosphate (KDP) crystal with refractive indices 1.4802 for the second harmonic and 1.5019 for the fundamental wave. Determine the maximum crystal thickness that is useful in generating second harmonic light.

Solution

From Eqs. (20.37) and (20.38), one has

$$L_c = \frac{\pi}{\Delta k} = \frac{\lambda_0}{4\Delta n} = \frac{1.0}{(4)(1.5019 - 1.4802)} = 11.5\ \mu\text{m}$$

This calculation shows that the maximum crystal thickness imposed by the coherence length, L_c is typically quite small, which leads to very inefficient conversion of the fundamental frequency into the second harmonic. To increase the efficiency, some type of *phase matching* is usually employed, as explained in the subsequent section.

20.4.2 Phase-matching techniques

As seen in Chapter 14, for crystals with certain symmetries, the refractive index varies with the direction of polarization of the light wave. Such birefringence provides a way of adjusting the indices to be equal at frequencies ω and 2ω, this technique being known as *birefringent phase matching*. The fundamental and the second harmonic waves remain in step as long as we choose a direction through the crystal such that $n_{2\omega}$ for the extraordinary ray equals n_ω for the ordinary ray. Using such a condition, the crystal can be a centimetre or so thick. Crystals commonly used for frequency doubling in the visible and near-IR regions are lithium niobate ($LiNbO_3$) and KDP.

An alternative approach to phase matching is that of *quasi-phase matching*, which is illustrated in Fig. 20.7. In this method, no attempt is made to match the refractive indices of the fundamental and second harmonic waves. However, the phase mismatch between the waves is compensated for by using a non-linear crystal with a periodic structure such that the sign of $\chi^{(2)}$ changes every coherence length L_c of the crystal. Such a periodic structure could be constructed by slicing a crystal into many thin slabs, each having a width equal to L_c, and then placing the slabs back together in such a manner that each slab is rotated 180° relative to its neighbours. Since materials with nonzero second-order susceptibilities lack inversion symmetry, the resulting crystal will have a second-order susceptibility that changes sign at each coherence length of the crystal.

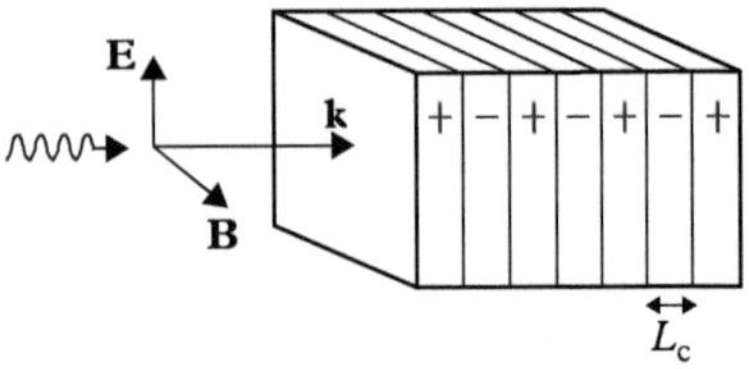

Figure 20.7 *Quasi-phase-matching crystal.*

The most common material used for quasi-phase matching is lithium niobate ($LiNbO_3$), which possesses a permanent electric dipole moment. The alternating crystalline symmetry is created by poling, a process in which a high voltage is applied for a short time to patterned electrodes deposited on the surface. The resulting material is commonly referred to as periodically poled lithium niobate, or PPLN.

20.5 Parametric Amplification

The theory presented in Section 20.4 for the three-wave interaction in a non-linear medium can be used to describe the amplification of an optical signal with a given frequency, say ω_1. For this purpose, a sufficiently intense pumping wave should be used with frequency $\omega_3 > \omega_1$. Then, a new wave—the so-called idler wave—is generated with frequency $\omega_2 = \omega_3 - \omega_1$.

Let us consider the phase-matching condition $\Delta k = k_3 - k_2 - k_1 = 0$, $\varepsilon_1 = \varepsilon_2 = \varepsilon$ and that the pumping wave has a constant amplitude along the non-linear medium:

$$E_3(z) = E_3(0) \tag{20.39}$$

As in previous cases, the non-linear medium is assumed to be located in the region $z \geq 0$. In these circumstances, one can write Eq. (20.23) and the conjugate of Eq. (20.24) in the form:

$$\frac{\partial E_1}{\partial z} = \frac{i\omega_1}{2}\sqrt{\frac{\mu}{\varepsilon}} dE_3 E_2^* \tag{20.40}$$

$$\frac{\partial E_2^*}{\partial z} = -\frac{i\omega_2}{2}\sqrt{\frac{\mu}{\varepsilon}} dE_3 E_1 \tag{20.41}$$

In order to solve Eqs. (20.40) and (20.41), the existence of a small input signal $E_1(0) \neq 0$ will be assumed, whereas $E_2(0) = 0$. The following solutions are then obtained:

$$E_1(z) = E_1(0)\cosh(Kz) \tag{20.42}$$

$$E_2^*(z) = -i\sqrt{\frac{\omega_2}{\omega_1}} E_1(0)\sinh(Kz) \tag{20.43}$$

where

$$K = \frac{1}{2}\sqrt{\frac{\omega_1\omega_2\mu_0}{\varepsilon}} dE_3 \tag{20.44}$$

Fig. 20.8 shows the evolution of the normalized irradiances

$$\bar{I}_1 \equiv \frac{I_1(z)}{I_{10}} = \cosh^2(Kz) \tag{20.45}$$

and

$$\bar{I}_2 \equiv \frac{\omega_1}{\omega_2}\frac{I_2(z)}{I_{10}} = \sinh^2(Kz) \tag{20.46}$$

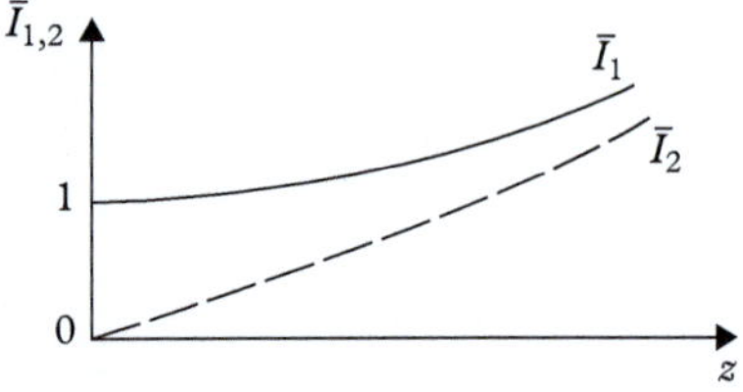

Figure 20.8 *Evolution of the normalized irradiances for the signal* $(\bar{I}_1)$ *and the idler* $(\bar{I}_2)$ *waves in a parametric optical amplifier.*

where $I_i(z) \propto E_i^2(z)$ and $I_{10} \propto E_1^2(0)$. It can be observed from Fig. 20.8 that, at an early stage, while the idler wave with frequency ω_2 is generated, the amplification of the signal with frequency ω_1 is modest. However, this amplification becomes significant at a later stage. A unique feature of the parametric optical amplifier lies in the possibility it offers to generate light with frequencies that do not correspond to any atomic transition.

20.6 Optical Phase Conjugation

Optical phase conjugation (OPC) is a non-linear phenomenon that owes its name to the fact that the spatial part of the wave resulting from the non-linear interaction is equal to the complex conjugate of the spatial part of the incident wave. That is, the new wave propagates in the opposite direction to the original wave, exactly reproducing the shape of that wave at each position. Thus, one can consider the non-linear medium producing such a conjugated wave as a mirror, usually called a *phase conjugation mirror* (PCM). The characteristics of this mirror are generally distinct from those of ordinary mirrors.

Let us consider the case of a plane wave:

$$E_i = E_0 \exp\{i(kz - \omega t)\} \tag{20.47}$$

which is incident normally on an ordinary flat mirror. In this case, the reflected wave is also a plane wave, given by

$$E_r = E_0 \exp\{i(-kz - \omega t)\} \tag{20.48}$$

That is, from the mathematical point of view, if one excepts the term ωt, the reflected wave is the complex conjugate of the incident wave and presents the characteristics of the wave with conjugate phase referred to before.

If the wave incident on the ordinary flat mirror is not a plane wave, or if the incidence is not normal, the wave reflected by that mirror no longer has the characteristics of the conjugate phase wave. In the case of a spherical wave generated at a given point, it is possible to have a wave with conjugate phase, which converges to that same point, only if one has a concave mirror with a curvature exactly equal to that of the wavefront incident on it.

In contrast to ordinary mirrors, a PCM can produce a phase-conjugated wave from an incident wavefront with arbitrary shape and is still able to respond immediately to spatial or temporal changes of that same wavefront.

OPC can be understood as real-time holography. As seen in Chapter 13, when the hologram is illuminated by a reconstruction wave, E_R, exactly equal to the reference wave used in the recording of this hologram, two images of the original object are obtained: one virtual and one real. The virtual image is located at the position where the original object was. However, if the hologram is illuminated by the conjugate reference wave, E_R^*, then Eq. (13.9) for the diffracted field is replaced by:

$$E_t(x,y) = \tau(x,y)E_R^* = \tau_0 E_R^* + \beta t_B E_R^* \left(E_R E_R^* + E_O E_O^* + E_O E_R^* + E_R E_O^*\right) \tag{20.49}$$

The last term in Eq. (20.49), which is proportional to $|E_R|^2 E_O^*$, corresponds to the object wave with conjugated phase, E_O^*, which provides a real image at the very position of the original object.

In the OPC process, the holographic recording and reconstruction processes take place in real time. Fig. 20.9 illustrates the PCM exposed to the original and conjugated reference waves, E_R and E_R^* (which function as pumping waves), as well as the object and image waves, E_O and E_O^*, respectively. One can consider that the waves E_R and E_O interfere and produce a hologram in real time. The wave E_R^* is diffracted by this hologram, producing the wave with conjugated phase E_O^*. This technique of producing OPC is called *four-wave mixing*, since it corresponds, in fact, to the interference of four waves in the non-linear medium.

When a wavefront is distorted, for example due to fluctuations in the refractive index in a given region, the conjugated phase wave, when propagating in the opposite direction, encounters the same variations, but in reverse. As a consequence, the conjugated phase wave leaves this region with the original wavefront completely restored.

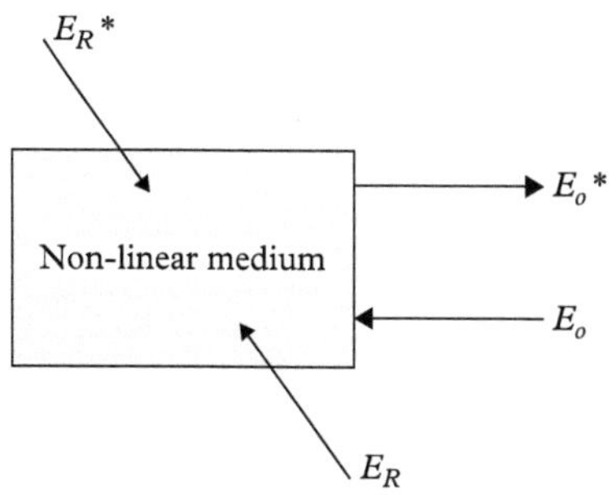

Figure 20.9 *Geometry for optical phase conjugation by mixing four waves.*

20.7 Electro-Optic Effects

In the preceding sections, we saw how two or three light waves could interact while propagating through a non-linear medium. In this section, we consider the effect of applying a static electric field on the material's refractive index. This is known as the *electro-optic effect* and can be considered as a special case of multiwave interaction in which one of the waves has zero frequency. Depending on the material, the refractive index can vary linearly with the field (*Pockels effect*) or quadratically with field (*electro-optic Kerr effect*). Both effects were already discussed in Chapter 14, dedicated to anisotropic materials. Actually, the application of an electric field can make anisotropic a material that was previously isotropic, or cause new optical axes to appear in already anisotropic materials. In this section, we look specifically at the relation of both effects with the involved non-linear susceptibilities of the medium.

In general, the refractive index of the non-linear medium varies with the applied static field E_0 as

$$n = n_0 + aE_0 + bE_0^2 \tag{20.50}$$

where the coefficients a and b are related to the non-linear susceptibilities $\chi^{(2)}$ and $\chi^{(3)}$. If a lightwave of amplitude A and frequency ω propagates in the non-linear medium, the total field in the material is the sum of the static field E_0 and the lightwave's field $A\cos\omega t$, such that

$$E(t) = E_0 + A\cos\omega t \tag{20.51}$$

Considering the medium polarization as given by Eq. (20.5), limited to the second-order term, it can be shown that the change in refractive index due to the static field is (Problem 20.8)

$$\Delta n = \frac{1}{n_0}\chi^{(2)}E_0 \tag{20.52}$$

Comparing with Eq. (20.50), one has that the coefficient a is related to the second-order susceptibility in the form:

$$a = \frac{\chi^{(2)}}{n_0} \tag{20.53}$$

As seen in Chapter 14, it is customary to write the refractive index change in the Pockels effect as

$$\Delta n = n_o^3 gE, \tag{20.54}$$

where g is the so-called *linear electro-optical constant.* Using Eqs. (20.52) and (20.54), one finds that

$$g = \frac{\chi^{(2)}}{n_0^4} \tag{20.55}$$

A device in which a material is made birefringent via the Pockels effect is called a *Pockels cell.* It is very useful in several applications, namely as an *electro-optical modulator* and as a *Q-switch* that allows the sudden dumping of the energy stored in a laser.

In materials having inversion symmetry, such as liquids and glasses, the second-order susceptibility is zero and the Pockels effect is absent. However, in such materials, there is still an electro-optic effect that depends on the third-order susceptibility, $\chi^{(3)}$. To consider this case, one can assume a lightwave of amplitude A and frequency ω propagating in the non-linear medium, such that the total field is given by Eq. (20.51). Substituting this into Eq. (20.5), limited to the third-order term, and assuming that $\chi^{(2)} = 0$, it can be shown (see Problem 20.9) that the coefficient b in Eq. (20.50) is given by

$$b = \frac{3\chi^{(3)}}{2n_0} \tag{20.56}$$

As seen in Chapter 14, it is conventional to write the refractive index change in the Kerr effect as

$$\Delta n = \lambda K E^2 \tag{20.57}$$

where λ is the wavelength in vacuum and K is the *Kerr electro-optic coefficient.* From Eqs. (20.56) and (20.57), one can find that

$$K = \frac{3\chi^{(3)}}{2\lambda n_0} \tag{20.58}$$

A device in which a material is made birefringent via the Kerr electro-optic effect is called a *Kerr cell.* As seen in Chapter 14, a Kerr cell can be used as an electrically controlled light switch. Actually, Kerr cells are often used as *Q*-factor switches in pulsed lasers. One advantage of the Kerr cell is that it can use a variety of isotropic materials such as liquids and glasses. However, compared with the Pockels cell, it has some drawbacks. In particular, its response is not linear with the applied voltage and much higher voltages are needed.

20.8 Further Reading

Boyd, R. W., *Nonlinear Optics*, Academic Press, 4th Ed., 2020.

Butcher, P. N., and D. Cotter, *The Elements of Nonlinear Optics*, Cambridge University Press, 2008.

He, G. S., *Nonlinear Optics and Photonics*, Oxford University Press, 2015.
Lugiato, L., F. Prati, and M. Brambilla, *Nonlinear Optical Systems*, Cambridge University Press, 2015.
Mandel, P., and R. W. Boyd, *Nonlinear Optics*, Academic Press/Elsevier, 3rd Ed., 2008.
Mills, P. L., *Nonlinear Optics*, Springer-Verlag, 1998.
New, G., *Introduction to Nonlinear Optics*, Cambridge University Press, 2011.
Powers, P. E., *Fundamentals of Nonlinear Optics*, CRC Press/Taylor & Francis, 2011.
Shen, Y. R., *Principles of Nonlinear Optics*, John Wiley & Sons, 2002.
Stegeman, G. I., and R. A. Stegeman, *Nonlinear Optics: Phenomena, Materials and Devices*, John Wiley & Sons, 2012.

20.9 Problems

20.1 Starting from Eqs. (20.5)–(20.7) and (20.9), derive the non-linear wave equation for the electric field given by Eq. (20.11).

20.2 Starting from Eqs. (20.23)–(20.25), obtain the Manley–Rowe relation given by Eq. (20.27).

20.3
  a) Verify, by substitution, that Eq. (20.34) is indeed a solution of Eq. (20.33).
  b) Using Eq. (20.34) and the principle of conservation of energy, show that the evolution of the amplitude of the fundamental wave is given by Eq. (20.35).

20.4 Using Eqs. (20.34) and (20.35) for the generation of the second harmonic, show that the sum of the irradiances of the pumping and the second harmonic waves is constant along the path.

20.5 Determine the optimum length of a lithium niobate ($LiNbO_3$) crystal for second harmonic generation when using light emitted by an Nd:YAG laser (wavelength 1064 nm). The refractive index of the crystal is 2.234 at 1064 nm and is 2.325 at 532 nm.

20.6 A laser beam ($\lambda = 700$ nm) has an irradiance $I_\omega = 100\,\mathrm{MWcm}^{-2}$ and enters a crystal of potassium dihydrogenated phosphate (KDP). Determine the length that this crystal should be so that the irradiance of the second harmonic reaches half that value ($I_{2\omega} = I_\omega/2$). Consider that the refractive index of the crystal is $n = 1.51$, $A = \sqrt{\mu/\varepsilon}d\omega = 1.2 \times 10^{-6}V^{-1}$ and that the phase-matching condition $\Delta k = 0$ is verified.

20.7 Show that Eqs. (20.42) and (20.43) are indeed solutions of Eqs. (20.40) and (20.41), describing the parametric optical amplification process.

20.8 Demonstrate the result given by Eq. (20.55) for the linear electro-optic constant.

20.9 Considering a non-linear medium having inversion symmetry, demonstrate the result given by Eq. (20.56) for the coefficient b of the Kerr electro-optic effect.

20.10 Describe the operation of a Kerr cell and compare its performance with that of a Pockels cell.

21 Non-Linear Effects in Optical Fibres

Modern glass fibres for optical communications are made of fused silica and show a very low attenuation over a broad optical window, which covers about 60 THz. Since SiO_2 is a symmetric molecule, the second-order susceptibility $\chi^{(2)}$ vanishes for silica glasses. However, a number of third-order non-linear processes can occur in optical fibres, which can grow to appreciable magnitudes over the long lengths available in fibres, even though the non-linearity of silica glass is very small.

Non-linear effects in optical fibres fall into two general categories. The first category arises from modulation of the refractive index of silica by intensity changes in the signal (Kerr effect). This gives rise to non-linear effects such as self-phase modulation, cross-phase modulation, and four-wave mixing. The second category of non-linearities corresponds to stimulated scattering processes, such as stimulated Brillouin scattering (SBS) and stimulated Raman scattering (SRS), which are interactions between optical signals and acoustic or molecular vibrations in the fibre.

The non-linear effects generated in the fibres severely affect the performance of optical communications systems, since they impose limits on the launched power of the signals, channel bit rate, channel spacing, transmission bandwidth, and hence the entire information capacity of such systems. However, the same non-linear effects also offer a variety of novel possibilities for ultrafast all-optical signal processing, such as switching, wavelength conversion, amplification, and regeneration.

21.1 Fibre Non-Linearity

Considering only the third-order term in Eq. (20.7), it is possible to express the dependence of the refractive index of optical fibres on the irradiance I of the optical wave in the form:

$$n = n_0 + n_2 I \tag{21.1}$$

where n_0 is the linear part of the refractive index and n_2 is the refractive index non-linear coefficient, also known as the *Kerr coefficient*. The dependence of the refractive index on the field intensity is known as the *non-linear Kerr effect*.

In the case of optical fibres, one has

$$n_0 \approx 1.46,\ n_2 \approx 3.2 \times 10^{-20}\,\mathrm{m^2/W} \tag{21.2}$$

Considering a single-mode fibre with an effective area $A_{ef} = 50\mu m^2$ and a power $P = 100\,\mathrm{mW}$, the value of the non-linear part of the refractive index is

$$n_2 \frac{P}{A_{ef}} \approx 6.4 \times 10^{-11} \tag{21.3}$$

Lasers, Optics, and Photonics. Mário F. S. Ferreira, Oxford University Press.
DOI: 10.1093/oso/9780198952770.003.0021

Although this value is quite small, its effects become significant due to the long interaction lengths provided by optical fibres.

Actually, the relevance of a non-linear process depends not only on the irradiance, I, of the propagating wave but also on the effective interaction length along which this irradiance remains sufficiently high. In the case of an optical fibre of real length L, the field is guided, and the power decreases due to attenuation, as discussed in Chapter 15, in the form

$$P(z) = P(0)e^{-\alpha z}. \tag{21.4}$$

The product of the irradiance and the effective length becomes

$$\left(IL_{eff}\right)_{fiber} = \int_0^L \frac{P(0)}{A_{eff}} e^{-\alpha z} dz = \frac{P(0)}{A_{eff}} \frac{1 - e^{-\alpha L}}{\alpha} = \frac{P(0)}{A_{eff}} L_{eff} \tag{21.5}$$

where

$$L_{eff} = \frac{1 - e^{-\alpha L}}{\alpha} \tag{21.6}$$

is the effective interaction length. If the fibre real length, L, is sufficiently small such that $\alpha L << 1$, one has $L_{eff} \approx L$. However, the difference between the real and the effective length increases with L. If $\alpha L >> 1$, the effective length approaches a limiting value, given by

$$L_{eff}^{\max} = \frac{1}{\alpha} \tag{21.7}$$

In the case of a fibre with an attenuation of 0.2 dB/km, the maximum effective length is $L_{eff}^{\max} = 21.72$ km. Fig. 21.1 shows the effective interaction length against the real fibre length for an attenuation of 0.2 dB/km.

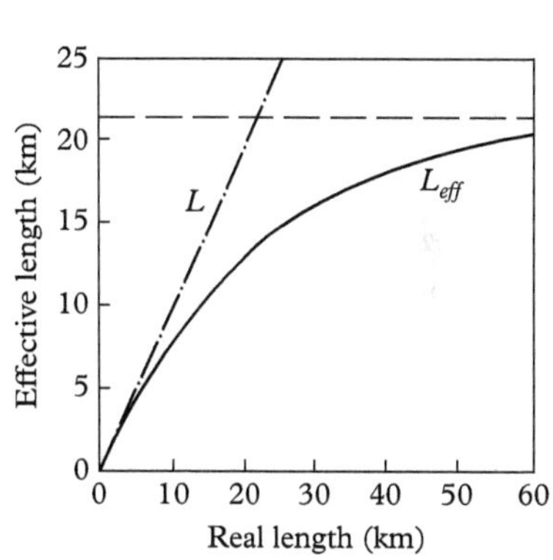

Figure 21.1 *Effective length of interaction versus the real fibre length for an attenuation of 0.2 dB/km.*

Example 21.1

Starting from Eq. (20.5), obtain an expression for the polarization of a medium with inversion symmetry, in which third-order non-linearity is dominant. Consider that the phase-matching conditions for the generation of the third harmonic are not satisfied.

Solution

In a medium with inversion symmetry, the second-order susceptibility $\chi^{(2)}$ vanishes. On the other hand, the third-order non-linear polarization is given by

$$P_{NL}^{(3)} = \varepsilon_0 \chi^{(3)} EEE \tag{21.8}$$

Let us consider the electric field of a plane wave, given by

$$E = \frac{1}{2}\left(\hat{E}(z)\, e^{-i\omega t} + c.c.\right) \tag{21.9}$$

Example 21.1 continued

where c.c. indicates the complex conjugate. Substituting Eq. (21.9) into Eq. (21.8) gives

$$P_{NL}^{(3)} = \varepsilon_0 \chi^{(3)} \left(\hat{E} e^{-i\omega t} + \hat{E}^* e^{i\omega t}\right)^3 = \frac{1}{8} \varepsilon_0 \chi^{(3)} \left(\left(\hat{E}^3 e^{-i3\omega t} + c.c.\right) + 3\left|\hat{E}\right|^2 \left(\hat{E} e^{-i\omega t} + c.c.\right) \right) \tag{21.10}$$

If the necessary phase-matching conditions for third-harmonic generation are not satisfied, the first term on the right-hand side of Eq. (21.10) can be neglected, and one has

$$P_{NL}^{(3)} = \frac{3}{4} \varepsilon_0 \chi^{(3)} \left|\hat{E}\right|^2 E \tag{21.11}$$

Taking this result into account, the polarization of the material becomes

$$P = \varepsilon_0 \chi_1 E + P_{NL}^{(3)} = \varepsilon_0 \left(\chi^{(1)} + \frac{3}{4} \chi^{(3)} \left|\hat{E}\right|^2 \right) E \tag{21.12}$$

Example 21.2

Using the result obtained in Example 21.1, show that the refractive index of the non-linear medium varies with irradiance, I, according to Eq. (21.1). Find expressions for the linear part of the refractive index, n_0, and the Kerr non-linear coefficient, n_2.

Solution

The irradiance of the plane wave is given by the modulus of the mean value of the Poynting vector as:

$$I = \frac{1}{2} c \varepsilon_0 n_0 \left|\hat{E}\right|^2 \tag{21.13}$$

where n_0 is the linear part of the refractive index. Substituting Eq. (21.13) into Eq. (21.12), the polarization of the medium can be written in the form:

$$P = \varepsilon_0 \left(\chi^{(1)} + \frac{3}{2} \frac{\chi^{(3)}}{c \varepsilon_0 n_0} I \right) E \tag{21.14}$$

From Eqs. (6.67) and (6.70), and neglecting the absorption by the medium, one has the following general relation between the polarization and the refractive index:

$$P = \varepsilon_0 \left(n^2 - 1\right) E \tag{21.15}$$

From Eqs. (21.14) and (21.15), the following result can be obtained for the refractive index:

$$n = \sqrt{1 + \chi^{(1)} + \frac{3}{2} \frac{\chi^{(3)}}{c \varepsilon_0 n_0} I} \approx n_0 + n_2 I \tag{21.16}$$

where

$$n_0 = \sqrt{1 + \chi^{(1)}} \tag{21.17}$$

is the linear refractive index, and

$$n_2 = \frac{3}{4} \frac{\chi^{(3)}}{c \varepsilon_0 n_0^2} \tag{21.18}$$

is the Kerr non-linear coefficient.

21.2 The Non-Linear Schrödinger Equation

The electric field associated with a bit stream propagating in a single-mode fibre along the z-direction can be written in the form

$$E = \frac{1}{2}\left(F(r,\phi)\,U(z,t)\,e^{i(\beta_0 z-\omega_0 t)} + c.c.\right) \tag{21.19}$$

where $U(z,t)$ describes the complex amplitude of the field envelope at a distance z inside the fibre, $F(r,\phi)$ gives the transverse distribution of the fundamental fibre mode, and β_0 is the mode propagation constant at the carrier frequency. The vector nature of the waves is neglected in this section, assuming that they are linearly polarized in the same direction.

The only quantity that changes during propagation is the amplitude $U(z,t)$. Since each frequency component of the optical field propagates with a slightly different propagation constant, it is useful to work in the spectral domain. The evolution of a specific spectral component $\widetilde{U}(z,\omega)$ is given by

$$\widetilde{U}(z,\omega) = \widetilde{U}(0,\omega)\exp\left[i\beta(\omega)z - i\beta_0 z\right] \tag{21.20}$$

where $\widetilde{U}(0,\omega)$ is the Fourier transform of the input signal $U(z=0,t)$, and $\beta(\omega)$ is the propagation constant, which can be written as the sum of a linear part, β_L, and a non-linear part, β_{NL}:

$$\beta(\omega) = \beta_L(\omega) + \beta_{NL}(\omega_0) \tag{21.21}$$

with

$$\beta_L(\omega) = \frac{\tilde{n}(\omega)\,\omega}{c}, \qquad \beta_{NL}(\omega_0) = n_2 I\frac{\omega_0}{c} \tag{21.22}$$

$\tilde{n}$ being the effective mode index. In Eq. (21.22), β_{NL} has been written as a frequency-independent quantity, considering that the spectrum of optical pulses used as bits in communication systems is usually much narrower than the carrier frequency ω_0.

We can expand $\beta_L(\omega)$ in a Taylor series around the carrier frequency ω_0, as given by Eq. (16.20). Retaining only terms up to second order in $(\omega-\omega_0)$ in such an expansion, substituting Eqs. (16.20), (21.21), and (21.22) into Eq. (21.20), calculating the derivative $\partial\widetilde{U}/\partial z$, and converting the resultant equation into the time domain by using the inverse Fourier transform and the relation

$$\Delta\omega \leftrightarrow i\frac{\partial}{\partial t} \tag{21.23}$$

we obtain the equation:

$$i\left(\frac{\partial}{\partial z} + \frac{1}{v_g}\frac{\partial}{\partial t}\right)U - \frac{1}{2}\beta_2\frac{\partial^2}{\partial t^2}U + \beta_{NL}U = 0 \tag{21.24}$$

The amplitude U in Eq. (21.24) can be normalized such that the pulse power is $P = |U|^2$. Making such normalization, we have $\beta_{NL} = \gamma|U|^2$, where

$$\gamma = \frac{\omega_0 n_2}{cA_{eff}} \tag{21.25}$$

is known as the *fibre non-linear parameter*.

Considering a moving frame propagating with the group velocity and using the new time variable

$$\tau = t - \frac{z}{v_g} \tag{21.26}$$

Eq. (21.24) can be written in the form:

$$i\frac{\partial U}{\partial z} - \frac{1}{2}\beta_2\frac{\partial^2 U}{\partial \tau^2} + \gamma|U|^2 U = 0 \tag{21.27}$$

Eq. (21.27) is usually called the *non-linear Schrödinger equation* (NLSE) due to its similarity to the Schrödinger equation of quantum mechanics. The NLSE describes the propagation of pulses in optical fibres taking into account both the group-velocity dispersion and the fibre non-linearity.

Example 21.3

Starting from Eq. (21.27) and neglecting the effects of fibre dispersion and non-linearity, show that a pulse propagates with the group velocity v_g, keeping its shape unchanged. Also show that the energy of the pulse propagates with the group velocity.

Solution

Neglecting the effects of fibre dispersion and non-linearity, Eq. (21.27) is reduced to

$$\frac{\partial U(z,\tau)}{\partial z} = 0 \tag{21.28}$$

or

$$\frac{\partial U}{\partial z} + \frac{1}{v_g}\frac{\partial U}{\partial t} = 0 \tag{21.29}$$

Eq. (21.28) has, as its general solution,

$$U = U_0(\tau) = U_0\left(t - z/v_g\right) \tag{21.30}$$

where U_0 represents the initial value ($z = 0$) of the amplitude U. It is concluded from Eq. (21.30) that, in the absence of dispersion and non-linearity, the pulse propagates without changing its profile at a group speed v_g.

Multiplying both members of Eq. (21.29) by U^* and the complex conjugate of the same equation by U, and adding the two resulting equations, one has

$$\frac{\partial |U|^2}{\partial z} + \frac{1}{v_g}\frac{\partial |U|^2}{\partial t} = 0 \tag{21.31}$$

One can conclude that the energy of the pulse, which is proportional to $|U|^2$, also propagates with the group velocity v_g.

21.3 Self-Phase Modulation

Assuming that the pulse temporal width is sufficiently large, one can neglect the dispersive term in Eq. (21.27), which reduces to

$$i\frac{\partial U}{\partial z} + \gamma|U|^2 U = 0 \tag{21.32}$$

Example 21.4

Show that, in the absence of dispersion, (i) the square of the amplitude modulus of the field envelope maintains its profile during propagation and (ii) the non-linearity of the fibre determines a phase modulation that is proportional to the pulse power.

Solution

i) Multiplying Eq. (21.32) by U^* and its complex conjugate by U and subtracting the two equations, one obtains

$$\frac{\partial|U|^2}{\partial z} = 0. \tag{21.33}$$

Eq. (21.33) has the general solution

$$|U|^2 = f(\tau) = f(t - z/v_g) \tag{21.34}$$

This result shows that, in the presence of non-linearity and neglecting the fibre dispersion, the absolute square of the wave envelope maintains its shape during propagation.

ii) Based on the previous result, the solution of Eq. (21.32) can be written in the form

$$U(z,\tau) = U_0(\tau)e^{i\phi_{NL}(z,\tau)} \tag{21.35}$$

where $U_0(\tau)$ and $\phi(z,\tau)$ are real functions. From Eqs. (21.32) and (21.35), one has

$$\phi_{NL}(z,\tau) = \gamma P(\tau)z \tag{21.36}$$

where $P(\tau) = |U(\tau)|^2$ is the pulse power. Neglecting the fibre loss, this corresponds to the input pulse power. Eq. (21.32) shows that the fibre non-linearity determines a phase modulation that is directly proportional to the pulse power and the distance of propagation.

Using Eq. (21.35), the electric field at the output of a fibre of length L can be written in the form

$$E(L,t) = \frac{1}{2}\left[\hat{E}(0,t)\exp\{i[(\beta_0 + \gamma P)L - \omega_0 t]\} + c.c\right] \tag{21.37}$$

where ω_0 is the pulse central frequency. The phase of the wave described by Eq. (21.37) is

$$\phi = (\beta_0 + \gamma P)L - \omega_0 t \tag{21.38}$$

According to Eq. (21.38), the phase alteration due to the non-linearity is proportional to the power P of the wave. If the incident wave is a pulse with a given temporal power profile, the power variation leads to the pulse's own phase modulation. Hence, this phenomenon is appropriately called *self-phase modulation* (SPM).

The instantaneous frequency within the pulse described by Eq. (21.37) is given by

$$\omega(t) = -\frac{\partial\phi}{\partial t} = \omega_0 - \gamma L\frac{\partial P}{\partial t} \tag{21.39}$$

Example 21.5

Find an expression for the instantaneous frequency in the case of a Gaussian pulse, whose electric field is given by

$$E(0,t) = \frac{1}{2}E_0 \exp\left\{-\frac{1}{2}\left(\frac{t}{t_0}\right)^2\right\}\exp\{-i\omega_0 t\} + c.c. \tag{21.40}$$

where E_0 is its peak amplitude and t_0 is the input pulse width. Represent the pulse power profile as well as its temporal derivative.

Solution

After propagating a distance z, the pulse becomes

$$E(z,t) = \frac{1}{2}E_0 \exp\left\{-\frac{1}{2}\left(\frac{\tau}{t_0}\right)^2\right\}\exp\{i[\beta_0 + \gamma P(\tau)]z - i\omega_0 t\} + c.c. \tag{21.41}$$

where

$$P(\tau) = P_0 \exp\left[-\left(\frac{\tau}{t_0}\right)^2\right] \tag{21.42}$$

and

$$\tau = t - \frac{z}{v_g} \tag{21.43}$$

The instantaneous frequency is given by

$$\omega(t) = \omega_0 + \gamma z \frac{2\tau}{t_0^2} P_0 e^{-\tau^2/t_0^2} \tag{21.44}$$

Fig. 21.2 shows the temporal variation of $P(\tau)$ and $dP/d\tau$ for a Gaussian pulse.

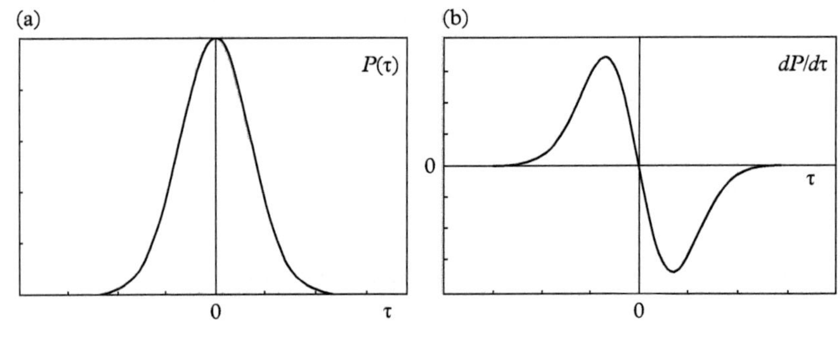

Figure 21.2 *Temporal variation of (a) $P(\tau)$ and (b) $dP/d\tau$ for a Gaussian pulse.*

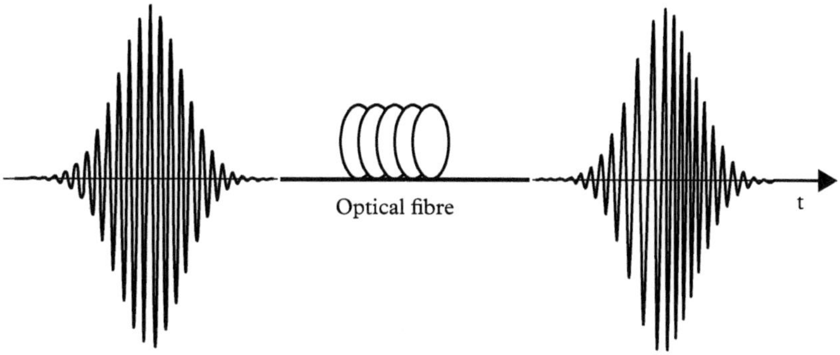

Figure 21.3 *Schematic representation of the chirp imposed on the pulse by the self-phase modulation effect.*

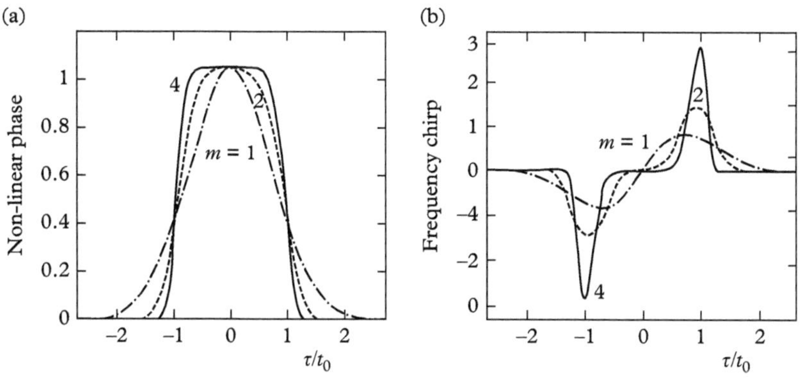

Figure 21.4 *Temporal variations of (a) the non-linear phase shift and (b) the SPM-induced frequency chirp at a distance* $L_{eff} = L_{NL}$ *for Gaussian* ($m = 1$) *and super-Gaussian pulses* (m = 2, 4).

It can be seen from Eq. (21.39) that, at the leading edge of the pulse, where $dP/d\tau > 0$, the instantaneous frequency is down-shifted from ω_0, whereas at the trailing edge, where $dP/d\tau < 0$, the instantaneous frequency is up-shifted from ω_0. The frequency at the centre of the pulse remains unchanged from ω_0. Thus, the SPM phenomenon imposes a chirp on the pulse, as illustrated in Fig. 21.3.

As a consequence of the SPM-induced frequency chirping, new frequency components are generated as the optical pulse propagates down the fibre, thereby broadening the spectrum of the bit stream. The spectral broadening depends not only on the effective length but also on the temporal variation of the input pulse power. As a consequence, short pulses are much more affected by SPM than long pulses.

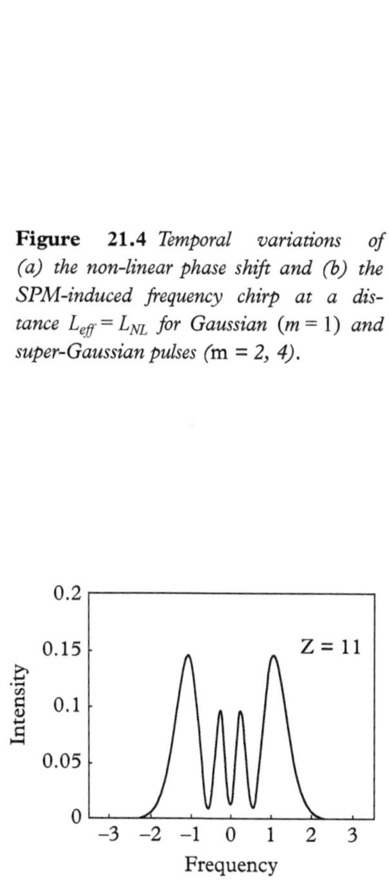

Figure 21.5 *SPM-induced spectral broadening of an initially unchirped Gaussian pulse at a distance* $L_{eff} = 11L_{NL}$.

The dependence of the SPM-induced frequency chirp on the pulse shape is well illustrated by considering the case of super-Gaussian pulses. The amplitude of a super-Gaussian pulse can be described in the form

$$U(0,\tau) = U_0 \exp\left\{-\frac{1}{2}\left(\frac{\tau}{t_0}\right)^{2m}\right\} \tag{21.45}$$

where t_0 is the input pulse width, and the parameter m controls the pulse shape. The Gaussian pulse corresponds to the case $m = 1$. For larger values of m, the pulses become nearly rectangular and develop steeper leading and trailing edges, which considerably affects the SPM-induced frequency chirp. Fig. 21.4 shows the numerical results obtained for (a) the non-linear phase shift and (b) the SPM-induced frequency chirp at a distance $L_{eff} = L_{NL}$, where $L_{NL} = (\gamma P_0)^{-1}$ is the non-linear length. The results shown in Fig. 21.4 correspond to Gaussian ($m = 1$) and super-Gaussian pulses ($m = 2, 4$). As the non-linear phase shift, given by Eq. (21.32), is proportional to the pulse intensity, Fig. 21.4(a) also provides the intensity profile for the considered Gaussian and super-Gaussian pulses. Fig. 21.4(b) shows that the frequency chirp is negative (downshift) at the leading edge of the pulse, whereas at the trailing edge it is positive (upshift). In contrast with Gaussian pulses, the frequency chirp is larger and occurs only near the leading and trailing edges in the case of super-Gaussian pulses.

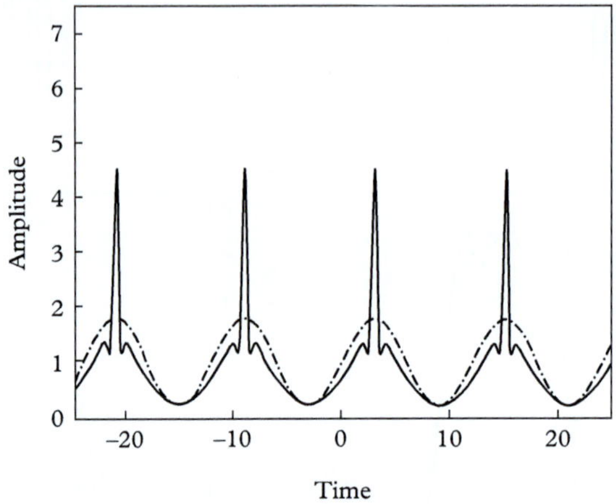

Figure 21.6 *Result of a computer simulation of induced modulational instability.*

The chirping due to non-linearity without any increase in pulse width leads to increased spectral broadening. Fig. 21.5 illustrates the SPM-induced spectral broadening experienced by a Gaussian pulse, at a distance $L_{eff} = 11L_{NL}$. One observes that the spectrum develops multiple peaks such that the two outermost peaks are the most intense.

In the presence of dispersion, the spectral broadening due to SPM determines two qualitatively different situations. In the normal dispersion region (wavelength shorter than the zero-dispersion wavelength, λ_{ZD}) the chirping due to dispersion is to downshift the leading edge and to upshift the trailing edge of the pulse, which is a similar effect to that due to SPM. Thus, in this regime, the chirping due to dispersion and SPM add. On the other hand, in the anomalous dispersion region, the chirping due to dispersion is opposite to that due to SPM. Consequently, non-linearity- and dispersion-induced chirpings can partially or even completely cancel each other. When this cancellation is total, the pulse neither broadens in time nor in its spectrum, and it is called a *soliton*.

A continuous wave propagating in the anomalous dispersion regime is known to produce *modulation instability* (MI), a phenomenon originated from the SPM effect, in which the side-band component of an amplitude-modulated light grows exponentially. Such modulation instability can be seeded by the broadband noise added by optical amplifiers (*spontaneous modulation instability*), or it can be induced by applying a small modulation to an input signal *(induced modulation instability)*. An ultrahigh-bit-rate pulse train can be generated by using such induced modulational instability. Fig. 21.6 shows an example obtained by numerically solving the NLSE, assuming an input signal modulated with a period of 12.

Besides its fundamental role in the formation of optical solitons, which will be discussed in Chapter 24 of this book, SPM has been widely used to realize various optical signal-processing functions, some of which are described in Chapter 23.

21.4 Cross-Phase Modulation

Waves with different wavelengths propagating in the same fibre can interact with each other since the refractive index that a wave experiences depends on the intensities of all the other waves. Hence, a pulse at one wavelength influences the phase of a pulse at another wavelength. This non-linear phenomenon is known as cross-phase modulation (XPM), and it can significantly limit the performance of wavelength-division multiplexing (WDM) systems.

Let us consider two waves with different wavelengths propagating simultaneously in the fibre. The two optical fields can be written in the form

$$E_j(r,t) = \frac{1}{2}F(r,\phi)\,U_j(z)\exp\left[i\left(\beta_j z - \omega_j t\right)\right] + c.c.\,(j = 1,2) \tag{21.46}$$

where $F(r,\phi)$ gives the transverse spatial distribution of the single mode supported by the fibre, $U_j(z)$ is the normalized slowly varying amplitude of the wave with frequency ω_j, and β_j is the propagation constant.

Assuming that the pulse durations are sufficiently long and that the phase-matching conditions for the occurrence of the four-wave mixing phenomenon are not satisfied, the following system of coupled equations can be obtained for the slowly varying amplitudes U_1 and U_2:

$$\frac{\partial U_1}{\partial z} = i\gamma_1\left(P_1 + 2P_2\right)U_1 \tag{21.47}$$

$$\frac{\partial U_2}{\partial z} = i\gamma_2\left(P_2 + 2P_1\right)U_2 \tag{21.48}$$

where $P_j = |U_j|^2$ is the pulse power, and

$$\gamma_j = \frac{n_2\omega_j}{cA_{eff}} \tag{21.49}$$

is the non-linear fibre parameter at frequency ω_j, $j = 1,2$.

If, for simplicity, the powers of both waves are assumed to be constant, the solution of Eq. (21.47) is given by

$$U_1(L) = U_1(0)\exp\{i\gamma_1\left(P_1 + 2P_2\right)L\} \tag{21.50}$$

A similar solution can be obtained from Eq. (21.48) for U_2. From Eq. (21.50), it is apparent that the phase of the signal at frequency ω_1 is modified not only due to its own power—which corresponds to the SPM effect—but also due to the power of the signal at frequency ω_2. This phenomenon is referred to as cross-phase modulation (XPM), and Eq. (21.50) shows that it is twice as effective as SPM.

The non-linear phase shift of the signal at frequency ω_1 resulting from the combination of SPM and XPM at the output of a fibre with length L is given by

$$\phi_{1NL} = \gamma_1\left(P_1 + 2P_2\right)L \tag{21.51}$$

The XPM-induced frequency shift, $\Delta\omega_{1XPM}$, is given by

$$\Delta\omega_{1XPM} = -2\gamma_1 L \frac{\partial P_2}{\partial t} \tag{21.52}$$

Eq. (21.52) shows that the part of the signal at ω_1 that is affected by the leading edge of the signal at ω_2 will be down-shifted in frequency, whereas the part overlapping with the trailing edge will be up-shifted in frequency. This determines a spectral broadening of the signal at ω_1, which is twice the spectral broadening caused by SPM.

The XPM effect determines a mutual influence between two pulses only if they overlap to some extent. However, in the presence of finite dispersion, the two pulses with different wavelengths will move with different velocities and thus will walk off from each other. If the pulses enter the fibre separately, walk through each other, and again become separated, it is said that they experience a complete collision. In a lossless fibre, such a collision is perfectly symmetric and no residual phase shift remains, since the pulses would have interacted equally with both the leading and the trailing edges of the other pulse. However, if the pulses enter the fibre together, the result is a partial collision, since each pulse will see only the trailing or the leading edge of the other pulse, which will lead to chirping. Moreover, in the case of a periodically amplified system, power variations also make complete collisions asymmetric, resulting in a net frequency shift that depends on the wavelength difference between the interacting pulses. Such frequency shifts lead to timing jitter in multichannel systems, since their magnitude depends on the bit pattern as well as on the channel wavelengths. The combination of amplitude and timing jitter significantly degrades the system performance.

One can define a parameter called the interaction length, L_I, which gives the distance between the beginning and the end of the overlap at half the maximum power of the two pulses, given by

$$L_I = \frac{2t_{FWHM}}{D\Delta\lambda} \tag{21.53}$$

where t_{FWHM} is the pulse full width at half maximum (FWHM), D is the fibre dispersion parameter, and $\Delta\lambda$ is the wavelength spacing between the interfering channels. Using dispersion-shifted fibres, which have a reduced dispersion, will increase the interaction length and hence will enhance the XPM effects. Eq. (21.53) also shows that closely spaced channels will interact over longer fibre lengths and hence suffer greater XPM effects. This is actually the case when $L_I << L_{eff}$, where L_{eff} is the effective fibre length. In this case, the pulse power does not change significantly over the interaction length. However, for $L_I >> L_{eff}$ the interaction length is determined by the fibre losses, and the XPM effects become almost independent of the channel spacing.

Similar to the SPM case, XPM can also induce modulation instability. A main difference concerns the fact that the XPM-induced modulation instability can occur even if one of the waves propagates in the normal–group velocity dispersion (GVD) region, whereas the other wave propagates in the anomalous-GVD region.

Besides the limitations imposed on multichannel optical communication systems, the XPM effect has been widely explored to implement several optical signal-processing functions, such as all-optical switching, wavelength conversion, and full pulse regeneration, as discussed in Chapter 23. XPM also has an important role in the supercontinuum generation phenomenon, as will be seen in Chapter 25.

21.5 Four-Wave Mixing

Four-wave mixing (FWM) is a parametric process in which four waves or photons interact with each other due to the third-order non-linearity of the material. As a result, when several channels with frequencies $\omega_1, \ldots, \omega_n$ are transmitted simultaneously over the same fibre, the intensity dependence of the refractive index leads not only to phase shifts within a channel, as discussed in previous sections, but also gives rise to signals at new frequencies.

If three waves with frequencies ω_i, ω_j, and ω_k are propagating in a fibre, they can interact non-linearly and give rise to new waves with the frequencies

$$\omega_{i,j,k} = \omega_i + \omega_j - \omega_k \tag{21.54}$$

In principle, three waves with different frequencies would produce 27 possible mixing frequencies. However, considering Eq. (6.1), we can see that changing the places of the first two frequencies ($\omega_i \leftrightarrow \omega_j$), or if the third frequency (ω_k) is equal to the first or to the second frequencies (ω_i, ω_j), no new frequencies are generated. As a consequence, only 9 new waves will be produced from the 27 possible variations. If the original frequencies are ω_1, ω_2, and ω_3, the frequencies of the 9 new waves generated due to the FWM process will be, according to Eq. (21.54): ω_{112}, ω_{113}, ω_{123}, ω_{132}, ω_{221}, ω_{223}, ω_{231}, ω_{331}, and ω_{332}. In general, considering M original channels of a WDM system, the number of possible mixing products is given by Forghieri et al. (1997):

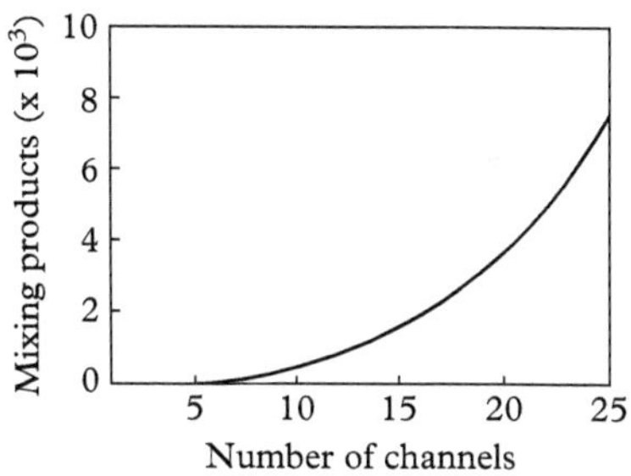

Figure 21.7 *Number of possible FWM mixing terms versus the number of channels.*

$$N = \frac{1}{2} M^2 (M-1) \tag{21.55}$$

Fig. 21.7 illustrates the variation of N with M.

If the channels of a WDM system have an equidistant frequency spacing, the newly generated waves will have the same spacing, and a large number of them will coincide exactly with the original channels or with each other. This possibility is illustrated in Fig. 21.8, where the mixing frequencies due to FWM are represented considering the existence of three original channels. In this case, from the expected nine new waves, only four new frequencies are actually generated.

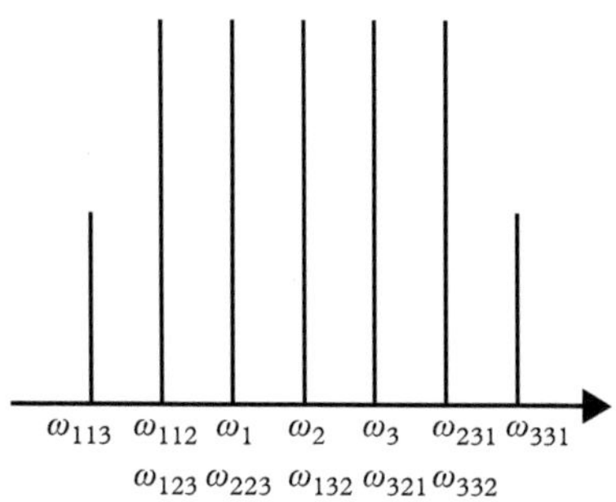

Figure 21.8 *Four-wave mixing terms for three channels equally spaced in frequency.*

The efficiency of the FMW depends on the relative phase among the interacting optical waves. In the quantum mechanical description, FWM occurs when photons from one or more waves are annihilated and new photons are created at different frequencies. In this process, the rules of conservation of energy and momentum must be satisfied. The conservation of momentum leads to the phase-matching condition.

Considering the case in which two photons at frequencies ω_1 and ω_2 are annihilated, with the simultaneous creation of two photons at frequencies ω_3 and ω_4, the conservation of energy imposes the condition

$$\omega_1 + \omega_2 = \omega_3 + \omega_4 \tag{21.56}$$

On the other hand, the phase mismatch among the four waves is given by

$$\begin{aligned} \Delta k &= \beta_3 + \beta_4 - \beta_1 - \beta_2 \\ &= (n_3\omega_3 + n_4\omega_4 - n_1\omega_1 - n_2\omega_2)/c \end{aligned} \tag{21.57}$$

The phase-matching condition is achieved when $\Delta k = 0$. The FWM efficiency becomes significantly higher near the zero-dispersion wavelength. In fact, it is higher in dispersion-shifted fibres than in standard fibres in the wavelength region around 1.5μ m due to the low value of the dispersion in the former case.

In practice, it is relatively easy to satisfy the phase-matching condition in the degenerate case $\omega_1 = \omega_2$. In this situation, a strong pump at $\omega_1 = \omega_2 \equiv \omega_p$ generates a low-frequency side band at ω_3 and a high-frequency side band at ω_4, when we assume $\omega_4 > \omega_3$. In analogy with Raman scattering, these side bands are referred to as the Stokes and anti-Stokes bands, respectively, which are also often called the *signal* and *idler* bands. The frequency shift of the two side bands is given by

$$\Omega_s = \omega_p - \omega_3 = \omega_4 - \omega_p \tag{21.58}$$

Assuming that the pulse durations are sufficiently long, we can neglect the linear term corresponding to the second-order dispersion in Eq. (21.27), which then becomes

$$\frac{\partial U}{\partial z} = i\gamma |U|^2 U \tag{21.59}$$

In order to consider the FWM mixing process in the nondegenerate case ($\omega_1 \neq \omega_2$), the amplitude U in Eq. (21.59) is assumed to be the result of the superposition of four waves:

$$U = U_1 e^{i(\beta_1 z - \omega_1 \tau)} + U_2 e^{i(\beta_2 z - \omega_2 \tau)} + U_3 e^{i(\beta_3 z - \omega_3 \tau)} + U_4 e^{i(\beta_4 z - \omega_4 \tau)} \tag{21.60}$$

Substituting Eq. (21.60) into Eq. (21.59) and considering that the four frequencies satisfy Eq. (21.56), the following set of coupled equations is obtained for the normalized amplitudes U_j:

$$\frac{\partial U_1}{\partial z} = i\gamma \left[\left(|U_1|^2 + 2\sum_{j\neq 1} |U_j|^2 \right) U_1 + 2U_3 U_4 U_2^* e^{i\Delta kz} \right] \tag{21.61}$$

$$\frac{\partial U_2}{\partial z} = i\gamma \left[\left(|U_2|^2 + 2\sum_{j\neq 2} |U_j|^2 \right) U_2 + 2U_3 U_4 U_1^* e^{i\Delta kz} \right] \tag{21.62}$$

$$\frac{\partial U_3}{\partial z} = i\gamma \left[\left(|U_3|^2 + 2\sum_{j\neq 3} |U_j|^2 \right) U_3 + 2U_1 U_2 U_4^* e^{-i\Delta kz} \right] \tag{21.63}$$

$$\frac{\partial U_4}{\partial z} = i\gamma \left[\left(|U_4|^2 + 2\sum_{j\neq 4} |U_j|^2 \right) U_4 + 2U_1 U_2 U_3^* e^{-i\Delta kz} \right] \tag{21.64}$$

where Δk is the wave-vector mismatch, given by Eq. (21.57), and γ is an averaged non-linear parameter that ignores the small variation due to the slightly different frequencies of the involved waves. In deriving Eqs. (21.61)–(21.64), only nearly phase-matched terms were kept. The first term inside the brackets in Eqs. (21.61)–(21.64) describes the effect of SPM, whereas the second term is responsible for XPM. Since these terms can lead only to a phase alteration, the generation of new frequency components is provided by the remaining FWM terms. If only the two pump waves (with frequencies ω_1 and ω_2) are present at the fibre input, the signal and idler waves are generated from the wideband noise that is always present in communication systems. However, in some cases, the signal wave can already be present at the fibre input. If the phase-matching condition is satisfied, both the signal and the idler waves grow during propagation due to the optical power transferred from the two pump waves.

Let us assume that the pump waves are much more intense than the signal and idler waves and that their phases are not matched. In such a case, the power transfer between these waves is very ineffective, and we can assume that the pump waves remain undepleted during the FWM process. The solutions of Eqs. (21.61) and (21.62) are then given by

$$U_j(z) = \sqrt{P_j} \exp\left[i\gamma \left(P_j + 2P_{3-j}\right) z\right], j = 1, 2 \tag{21.65}$$

where $P_j = |U_j(0)|^2$ are the input pump powers. In this approximation, the pump waves experience only a phase shift due to SPM and XPM. We will also consider the case in which the signal wave has a finite value at the fibre input. However, the input signal power is assumed to be much less than the input pump powers. In these circumstances, due to the phase mismatch between the involved waves, Eq. (21.63) has the following approximate solution:

$$U_3(z) = \sqrt{P_3} \exp\left[i\gamma 2 \left(P_1 + P_2\right) z\right] \tag{21.66}$$

Substituting Eqs. (21.65) and (21.66) into Eq. (21.64), the following result can be obtained for the power of the idler wave:

$$P_4(L) = 4\gamma^2 P_1 P_2 P_3 L^2 \left[\frac{\sin(\kappa L/2)}{\kappa L/2}\right]^2 \tag{21.67}$$

where the parameter

$$\kappa = \Delta k + \gamma \left(P_1 + P_2\right) \tag{21.68}$$

corresponds to the effective phase mismatch, and $P_1 + P_2 = P_p$ is the total pump power.

If the phase-matching condition $\kappa = 0$ is fulfilled, Eq. (21.67) shows that the power of the idler increases quadratically with the fibre length, as represented in Fig. 21.9 (dashed curve). However, in such a case it is not reasonable to consider that the amplitudes of the pump and signal waves do remain constant along the fibre, as assumed in deriving Eq. (21.66). If the phases are not matched, the intensity of the idler wave shows a periodic evolution along the fibre, as described by Eq. (21.67) (full curve in Fig. 21.9).

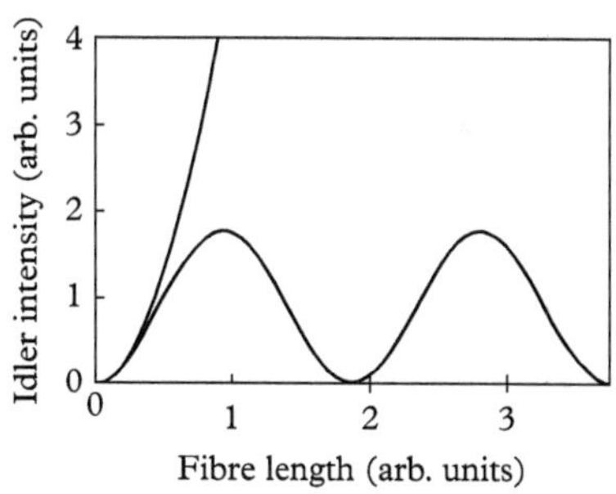

Figure 21.9 *Idler versus the fibre length for phase matching (full curve) and a phase mismatch between the waves (dashed curve).*

Let us consider the degenerate case $\omega_1 = \omega_2 \equiv \omega_p$, where ω_p denotes the pump frequency. In this case, the effective phase mismatch is given by

$$\kappa = \Delta k + 2\gamma P_p \tag{21.69}$$

where P_p is the pump power and Δk is the linear phase mismatch. The linear phase mismatch given by Eq. (21.57) can be expressed in terms of the frequency shift Ω_s given by Eq. (21.58) if we use an expansion of Δk about the pump frequency ω_p. By retaining terms up to quadratic in Ω_s in this expansion, one obtains

$$\Delta k = \beta_2 \Omega_s^2 \tag{21.70}$$

where $\beta_2 = d^2\beta/d\omega^2$ is the GVD coefficient at the pump frequency.

For a pump wavelength in the anomalous dispersion region ($\beta_2 < 0$), the negative value of Δk can be compensated by the fibre non-linearity. For phase matching one has

$$\kappa \approx \beta_2 \Omega_s^2 + 2\gamma P_p = 0 \tag{21.71}$$

From this, we obtain that the frequency shift is given by

$$\Omega_s = \sqrt{2\gamma P_p / |\beta_2|} \tag{21.72}$$

This frequency shift coincides with the frequency of the perturbations with maximum gain in the modulation instability phenomenon occurring in the anomalous dispersion regime.

FWM in optical fibres has been widely explored for various functionalities of optical signal processing, such as all-optical switching, wavelength conversion, pulse regeneration, and optical phase conjugation (OPC). In particular, OPC offers great potential to compensate Kerr non-linearity distortions in high-capacity WDM systems. As described in Chapter 22, FWM can also be used to realize fibre-optical parametric amplifiers (FOPAs), which may be tailored to operate at any wavelength.

21.6 Stimulated Raman Scattering

In experiments on light scattering in different media, one can observe the existence of a frequency-shifted field in addition to the incident field. The frequency shift of the scattered light is determined by the vibrational oscillations that occur between constituent atoms within the molecules of the material. A slight excitation of the molecular resonances due to the presence of the input field, referred to as the pump wave, results in spontaneous Raman scattering. As the scattered light increases in intensity, the scattering process eventually becomes stimulated. In this regime, the scattered field interacts coherently with the pump field to further excite the resonances, which significantly enhances the transfer of power between the two optical waves. Raman scattering can occur in all materials, but in silica glass the dominant Raman lines are due to the bending motion of the Si–O–Si bond.

Both down-shifted (Stokes) and up-shifted (anti-Stokes) light can, in principle, be generated through the Raman scattering process. However, in the case of optical fibres, the Stokes radiation is generally much stronger than the anti-Stokes radiation due to two main reasons. On the one hand, the anti-Stokes process is phase-mismatched, whereas the Stokes process is phase-matched for collinear propagation. On the other hand, the anti-Stokes process involves the interaction of the pump light with previously excited molecular resonances, which is not a condition for the Stokes process. Due to these facts, only the Stokes wave will be considered in the following.

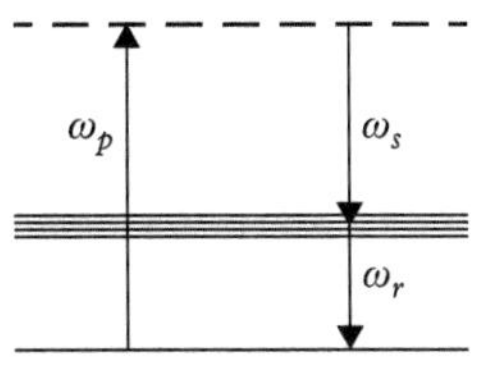

Figure 21.10 *Energy-level illustration of the stimulated Raman scattering process.*

Considering the quantum mechanical picture, in the SRS process one has simultaneously the absorption of a photon from the pump beam at frequency ω_p and the emission of a photon at the Stokes frequency, ω_S (Fig. 21.10). The difference in energy is taken up by a high-energy phonon (molecular vibration) at frequency ω_r. Thus, SRS provides energy gain at the Stokes frequency at the expense of the pump. This process is considered non-resonant because the upper state is a short-lived virtual state. If the Stokes wave becomes sufficiently strong, it can act as a pump source itself, resulting in the generation of higher-order Stokes waves.

In the case of standard optical fibres, they are basically made of fused silica, which is an amorphous material. Several vibrational modes occur in the structure of amorphous silica, with resonance frequencies which overlap with each other and form broad frequency bands. As a consequence, Raman scattering in optical fibres occurs over a relatively large frequency range.

The pump power (P_p) and the Stokes power (P_S) propagating along an optical fibre satisfy the following equations:

$$\frac{dP_s}{dz} = -\alpha_s P_s + \frac{g_R}{A_{eff}} P_p P_s \tag{21.73}$$

$$\frac{dP_p}{dz} = -\alpha_p P_p - \frac{\omega_p}{\omega_s}\frac{g_R}{A_{eff}} P_s P_p \tag{21.74}$$

where A_{eff} is the effective core area, α takes into account the fibre losses, and g_R is the Raman gain coefficient.

Fig. 21.11 shows the normalized Raman gain spectra for bulk fused silica, measured when the pump and signal light were either copolarized (full curve) or orthogonally polarized (dotted curve) (Bromage, 2004). The most significant feature of the Raman gain in silica fibres is that g_R extends over a large frequency range (up to 40 THz). Optical signals whose bandwidths are of this order or less can be amplified using the Raman effect if a pump wave with the appropriate wavelength is available. The Raman shift, corresponding to the location of the main peak in Fig. 21.11, is close to 13 THz for silica fibres. The Raman gain depends on the relative state of polarization of the pump and signal fields. The copolarized gain is almost an order of magnitude larger than the orthogonally polarized gain near the peak of the gain curve. The peak value of the Raman gain decreases with increasing the pump wavelength, and it is about 6×10^{-14}m/W in the wavelength region around 1.5 μm.

In Eqs. (21.73) and (21.74), the terms containing a product of the pump and signal powers describe their coupling via SRS. The strength of this coupling is determined by the ratio between the Raman gain coefficient and the effective core area, $C_R = g_R/A_{eff}$, which is known as the Raman gain efficiency.

If only the pump wave is launched into the fibre, the generation of the Stokes wave begins with spontaneous Raman scattering. As this scattered light increases in intensity,

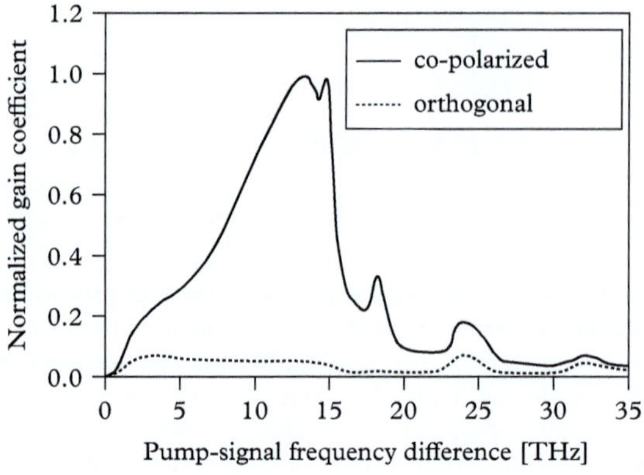

Figure 21.11 *Normalized Raman gain coefficient for copolarized (full curve) and orthogonally polarized (dotted curve) pump and signal beams (reprinted with permission from [Bromage, 2004]© IEEE).*

the process becomes stimulated, which can lead to Stokes power levels similar to those of the pump wave.

Assuming that $\alpha_S = \alpha_p = \alpha$ (which is a reasonable approximation around the 1.5 μm wavelength region in a low-loss fibre), the following analytical solutions of Eqs. (21.73) and (21.74) can be obtained:

$$P_S(z) = \frac{\omega_S}{\omega_p} \frac{F(z)}{1 + F(z)} P_0 \exp(-\alpha z) \tag{21.75}$$

$$P_p(z) = \frac{P_0 \exp(-\alpha z)}{1 + F(z)} \tag{21.76}$$

where

$$F(z) = \frac{\omega_p}{\omega_S} \frac{P_{S0}}{P_{p0}} \exp\left\{ g_R \frac{P_0}{A_{eff}} \frac{1 - \exp(-\alpha z)}{\alpha} \right\} \tag{21.77}$$

In Eqs. (21.75)–(21.77), one has $P_0 = P_{p0} + (\omega_S/\omega_p) P_{S0}$, whereas P_{S0} and P_{p0} represent the input powers of the Stokes and pump waves, respectively.

When the input power of the Stokes wave is weak, such that $P_{S0} << P_{p0}$, Eqs. (21.75) and (21.76) become:

$$P_S(z) \approx P_{S0} \exp\left\{ g_R \frac{P_{p0}}{A_{eff}} \frac{1 - \exp(-\alpha z)}{\alpha} - \alpha z \right\} \tag{21.78}$$

$$P_p(z) \approx P_{p0} \exp(-\alpha z) \tag{21.79}$$

In the absence of an input signal P_{S0}, the Stokes wave arises from spontaneous Raman scattering along the fibre.

The threshold for SRS is defined as the input pump power at which the output powers of the pump and Stokes waves become equal. It is given approximately by Smith (1972):

$$P_{p0}^{th} = 16 \frac{A_{eff}}{g_R L_{eff}} \tag{21.80}$$

where L_{eff} is the effective fibre length, given by Eq. (21.6). For example, the threshold for SRS is $P_{P0}^{th} \approx 600\,\text{mW}$ at $\lambda_P = 1.55\,\mu\text{m}$ in long polarization-maintaining fibres, such that $L_{eff} \approx 22\,\text{km}$, considering an effective core area of $A_{eff} = 50\,\mu\text{m}^2$. If the Stokes wave becomes sufficiently strong, it can act as a pump source itself, resulting in the generation of higher-order Stokes waves.

Because SRS has a relatively high threshold, it is not of concern for single-channel systems. However, in WDM systems, SRS can cause crosstalk between channel signals whose wavelength separation falls within the Raman gain curve. Specifically, the long-wavelength signals are amplified by the short-wavelength signals, leading to power penalties for the latter signals. The shortest-wavelength signal is depleted most, as it acts as a pump for all other channels. The Raman-induced power transfer between two channels depends on the bit pattern, which leads to power fluctuations and determines additional receiver noise. The magnitude of these deleterious effects depends on several parameters, such as the number of channels, their frequency spacing, and the power in each of them.

Fig. 21.12 illustrates the influence of SRS for a simple two-channel case. It is assumed that both channels are launched with the same input power and that the wavelength of channel 1 is shorter than the wavelength of channel 2. As can be observed from Fig. 21.12(b), if both channels transmit a logical 'one', it is amplified in channel 2, whereas it is depleted in channel 1. However, if one of the channels transmits a logical 'zero', no interaction occurs between the channels.

As described in Chapter 22, SRS in optical fibres can be used with advantage to implement both lumped and distributed wideband optical amplifiers. The same effect significantly impacts the propagation of ultrashort pulses and plays a significant role in the supercontinuum generation process, as described in Chapter 25.

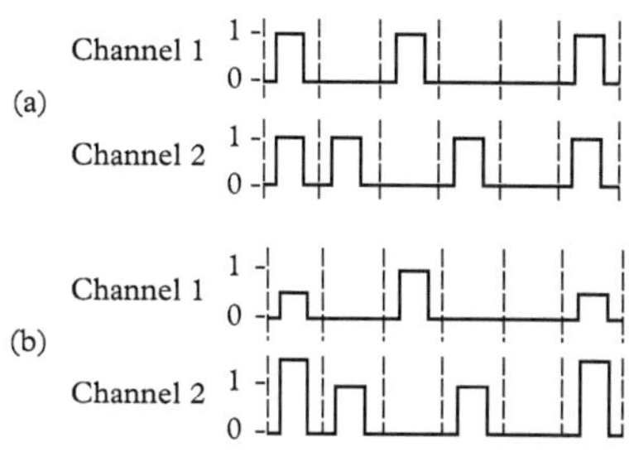

Figure 21.12 *Schematic of the influence of SRS in a two-channel WDM system at (a) fibre input and (b) fibre output for* $\lambda_1 < \lambda_2$.

21.7 Stimulated Brillouin Scattering

Brillouin scattering is a phenomenon named after the French physicist Leon Brillouin, who investigated the scattering of light by acoustic waves during the 1920s. For low incident intensities, the scattered part of the field remains very weak. However, the process becomes stimulated and strong scattered fields are generated at high input intensities, which are readily available with lasers.

The SBS process can be described as a classical three-wave interaction involving the incident (pump) wave of frequency ω_p, the Stokes wave of frequency ω_s, and an acoustic wave of frequency ω_a. The pump creates a pressure wave in the medium owing to electrostriction, which in turn causes a periodic modulation of the refractive index. This process is illustrated in Fig. 21.13. Physically, each pump photon in the SBS process gives up its energy to create simultaneously a Stokes photon and an acoustic phonon.

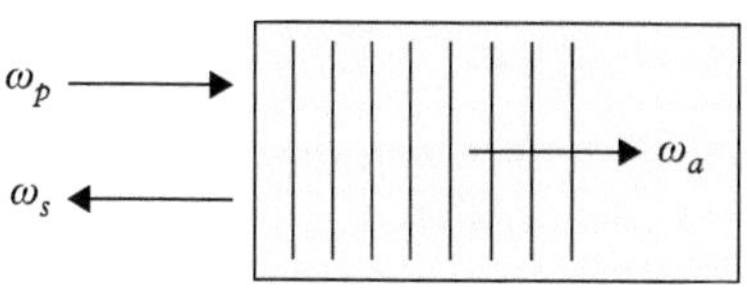

Figure 21.13 *Schematic illustration of the stimulated Brillouin scattering process.*

The three waves involved in the SBS process must conserve both energy and momentum. The conservation of energy requires that $\omega_p - \omega_s = 2\pi f_a$, where f_a is the linear frequency of the acoustic wave, which is about 11.1 GHz in standard fibres. The conservation of momentum requires that the wave vectors of the three waves satisfy the condition

$$\mathbf{k}_a = \mathbf{k}_p - \mathbf{k}_s. \tag{21.81}$$

In a single-mode fibre, optical waves can propagate only along the direction of the fibre axis. Since the acoustic wave velocity $v_a \approx 5.96\,\text{km/s}$ is by far smaller than the velocity of light, we have

$$|\mathbf{k}_a| = 2\pi f_a/v_a > |\mathbf{k}_p| \approx |\mathbf{k}_s| \tag{21.82}$$

Given this condition, the momentum conservation has the important consequence that the Brillouin effect occurs only if the Stokes and the pump waves propagate in opposite directions.

The response of the material to the interference of the pump and Stokes fields tends to increase the amplitude of the acoustic wave. Therefore, the beating of the pump wave with the acoustic wave tends to reinforce the Stokes wave, whereas the beating of the pump wave and the Stokes wave tends to reinforce the acoustic wave. This explains the appearance of the SBS process.

The pump power (P_p) and the Stokes power (P_S) propagating along an optical fibre satisfy the following equations:

$$\frac{dP_S}{dz} = -\frac{g_B(\Delta\omega_S)}{A_{eff}} P_p P_S + \alpha P_S \tag{21.83}$$

$$\frac{dP_p}{dz} = -\frac{g_B(\Delta\omega_S)}{A_{eff}} P_S P_p + \alpha P_p \tag{21.84}$$

where $\Delta\omega_s = \omega - \omega_s$ and $g_B(\Delta\omega_S)$ are the Brillouin gain coefficient, given by

$$g_B(\Delta\omega_S) = \frac{(\Gamma_B/2)^2}{(\Delta\omega_S)^2 + (\Gamma_B/2)^2} g_{B0} \tag{21.85}$$

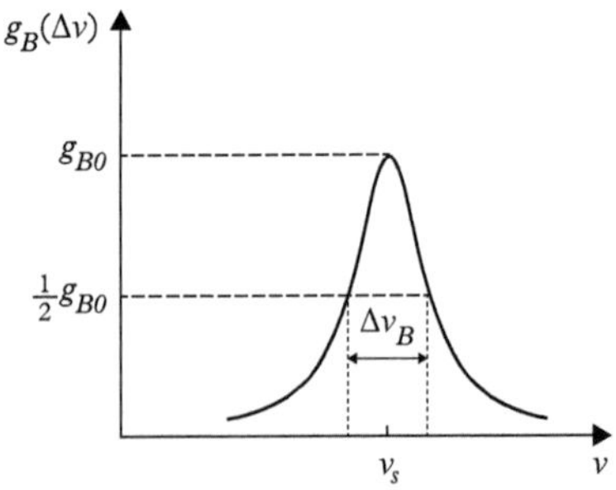

Figure 21.14 *Lorentzian profile of the Brillouin gain coefficient.*

where Γ_B is a damping coefficient. Eq. (21.85) shows that the spectrum of the Brillouin gain coefficient has a Lorentzian profile with an FWHM $\Delta\nu_B = \Gamma_B/2\pi$, as illustrated in Fig. 21.14. For bulk silica, the width is expected to be about 17 MHz at 1.5 μm. However, several experiments have shown much larger bandwidths for silica-based fibres, which in some cases can exceed 100 MHz. Such larger bandwidths are generally attributed both to the guided nature of acoustic modes and to inhomogeneities in the fibre-core cross section along the fibre length.

Considering the parameter values typical of fused silica, the peak value of the Brillouin coefficient g_{B0} is estimated to be about 2.5×10^{-11} m/W, which is between two and three orders of magnitude larger than the Raman gain coefficient in silica fibres at $\lambda_p = 1.55$ μm. Even larger values of the Brillouin gain coefficient can be obtained in some non-silica-based fibres.

The aforementioned peak value of the Brillouin coefficient, g_{B0}, is valid only if the spectral width of the pump beam $(\Delta\nu_p)$ is much narrower than the Brillouin linewidth ($\Delta\nu_B$). In general, however, to take into account the finite spectral width of the pump beam, the convolution of the power spectrum of the pump field with the intrinsic Brillouin-frequency spectrum must be calculated. Assuming that the pump spectrum has a Lorentzian profile, such a convolution is a Lorentzian of width $\Delta\nu_B + \Delta\nu_p$, and thus the Brillouin gain coefficient is reduced according to the relation.

$$\tilde{g}_{B0} = \frac{\Delta\nu_B}{\Delta\nu_B + \Delta\nu_p} g_{B0} \tag{21.86}$$

The result in Eq. (21.82) also applies to the case of a laser output comprising a very large number of longitudinal modes with completely random phase relationships and contained within a spectral envelope of width greater than the Brillouin linewidth.

In order to estimate the threshold in the Brillouin scattering process, we can consider the case of a nondepleted pump wave, where the solutions of Eqs. (21.83) and (21.40) are given by

$$P_p(z) = P_{p0} \exp(-\alpha z) \tag{21.87}$$

$$P_s(z) = P_{s0} \exp(-\alpha z) \exp\left[-\frac{g_B P_{p0} (1 - \exp(-\alpha z))}{A_{eff} \alpha} \right] \tag{21.88}$$

where $P_{p0} = P_p(z = 0)$ and $P_{s0} = P_s(z = 0)$ are the input pump power and the output Stokes power, respectively. Actually, a Stokes input signal at $z = L$ grows to produce an output signal at $z = 0$, as given by

$$P_{s0} = P_s(L) \exp(-\alpha L) \exp\left[\frac{g_B P_{p0} L_{eff}}{A_{eff}} \right] \tag{21.89}$$

where $L_{eff} = [1 - \exp(-\alpha L)]/\alpha$ is the effective fibre length.

In the absence of an input signal, the Stokes wave builds up from spontaneous scattering. The noise power provided by spontaneous Brillouin scattering is equivalent to injecting a fictitious photon per mode at the distance where the gain is equal to the fibre loss. The threshold pump power is defined as the input pump power that is equal to the backward output Stokes power, which gives the result (Smith, 1972):

$$P_{p0}^{th} = 21 \frac{K_B A_{eff}}{g_{B0} L_{eff}} \tag{21.90}$$

where K_B is a factor varying between 1 and 2 that takes into account the polarization dependence of the pump and Stokes waves. If both waves propagate in a polarization-maintaining fibre and have the same polarization, one has $K_B = 1$. In the case of a non-polarization-maintaining fibre, it is $K_B = 1.5$. Considering an effective area $A_{eff} = 50\ \mu\text{m}^2$, an attenuation constant of $\alpha = 0.2$ dB/km, a gain coefficient $g_{B0} = 2.5 \times 10^{-11}$ m/W, and $K_B = 1$, one obtains from Eq. (21.86) a threshold of $P_{p0}^{th} \approx 2$ mW. This value is around three orders of magnitude smaller than the threshold required for Raman scattering, which makes SBS the dominant non-linear effect in some circumstances.

Launch powers well above 10 mW are needed in some applications, namely in repeaterless transmission systems. In order to make possible the transmission of such power levels, the SBS threshold is increased using several schemes. In general, these schemes are based on increasing either the spectral width of the optical carrier or the Brillouin bandwidth of the fibre. Modulating the phase of the optical carrier at a frequency $\Delta\nu_m$ increases the SBS threshold by a factor $(1 + \Delta\nu_m/\Delta\nu_B)$. This factor can be greater than 10 for typical modulation frequencies of some hundreds of MHz.

Besides causing severe signal attenuation, SBS can also produce interchannel crosstalk in bidirectional transmission systems. However, Brillouin gain can be useful in some important applications, namely optical amplification, lasing, channel selection in closely spaced wavelength-multiplexed networks, OPC, temperature and strain sensing, all-optical slow-light control, optical storage, etc.

21.8 Further Reading

Agrawal, G. P., *Nonlinear Fiber Optics*, Academic Press, 5th Ed., 2012.

Bromage, J., Raman Amplification for Fiber Communications Systems, *J. Lightwave Technol.* 22, no. 1, 79–93, 2004.

Ferreira, M. F., *Nonlinear Effects in Optical Fibers*, John Wiley & Sons, 2011.

Ferreira, M. F., *Optical Signal Processing in Highly Nonlinear Fibers*, CRC Press, 2020.

Forghieri, F., R. W. Tkach, and A. R. Chraplyvy, WDM Systems and Networks, in I. P. Kaminow and T. L. Koch, Eds., Optical Fiber Telecommunications IIIA, pp. 343–371, Academic Press, 1997.

Mitschke, F., *Fiber Optics—Physics and Technology*, Springer, 2009.

Smith, R. G., Optical Power Handling Capacity of Low-Loss Optical Fibers as Determined by Stimulated Raman and Brillouin Scattering, *Appl. Opt.* **11**, no. 11, 2489–2494, 1972.

21.9 Problems

21.1 Determine the effective length of a 10-km fibre with an attenuation constant $\alpha = 0.3$ dB/km. Calculate the limiting value of the effective length for this type of fibre.

21.2 Show that Eq. (21.24) is converted into Eq. (21.27) using the transformation given by Eq. (21.26).

21.3 Calculate the maximum value of the SPM-induced non-linear phase shift experienced by (i) 10-ps pulses with 300-mW peak power and (ii) 1-ps pulses with 1-W peak power at the output of a 3-km-long fibre having a loss of 1 dB/km and an effective mode area of 40 μm^2.

21.4 Consider a super-Gaussian pulse with amplitude given by Eq. (21.45). Derive expressions for the maximum frequency shifts from the carrier induced by self-phase modulation. Determine the positions on the pulse envelope at which the maximum shifts occur.

21.5 Consider two pulses with an FWHM width of 30 ps belonging to two adjacent WDM channels with a spacing of 50 GHz. Assuming that the two pulses are introduced simultaneously at the fibre input, calculate the distance after which they are walked off for the cases of (i) a standard transmission fibre ($D = 17\,\text{ps}/(\text{nm}-\text{km})$) and (ii) a non-zero dispersion-shifted fibre with $D = 3\,\text{ps}/(\text{nm}-\text{km})$.

21.6 Derive the coupled equations for the pump, signal, and idler waves similar to those in Eqs. (21.61)–(21.64) for the case of a single pump.

21.7 Explain how FWM affects the performance of a WDM system. Discuss the merits of using unequal channel spacing and/or dispersion management to suppress the impact of FWM.

21.8 Consider pump and signal waves with a frequency spacing Δv between them. Show that the linear phase mismatch in such a case is given by $\Delta k = \beta_2(2\pi\Delta v)^2$, where β_2 is the dispersion at the pump frequency.

21.9 Consider a pump beam at a wavelength of 1550 nm carrying a power of 10 W in a single-mode fibre that has a non-linear parameter $\gamma = 4\text{W}^{-1}/\text{km}$ and a GVD $\beta_2 = -20\,\text{ps}^2/\text{km}$. Find the signal wavelength for which perfect phase matching occurs.

21.10 Two signals in the vicinity of 1.55 μm and with a frequency difference of 100 GHz are launched into a 50-km fibre with an attenuation constant $\alpha = 0.2$ dB/km, a GVD parameter $\beta_2 = -1\,\text{ps}^2/\text{km}$ and negligible dispersion slope. Estimate the value of the FWM efficiency in such a case.

21.11 Verify that Eqs. (21.75)–(21.77) are indeed solutions of Eqs. (21.73) and (21.74) if $\alpha_s = \alpha_p = \alpha$.

21.12 Starting from Eqs. (21.75)–(21.77) and assuming a sufficiently low-loss coefficient, such that $\alpha z << 1$ for all the distances considered, find an expression for the distance at which the pump and Stokes powers become equal.

21.13 Estimate the threshold power for stimulated Raman scattering in a single-mode fibre with a length of 50 km and an effective core area $A_{eff} = 50\,\mu\text{m}^2$ at (i) 1.55 μm and (ii) 1.3 μm. Assume a Raman gain coefficient $g_R = 5 \times 10^{-14}\,\text{m.W}^{-1}$ and that the fibre loss is, in case (i) $\alpha = 0.2$ dB/km and in case (ii) $\alpha = 0.6$ dB/km.

21.14 Consider two WDM channels at wavelengths $\lambda_1 = 1535$ nm $\lambda_2 = 1550$nm transmitted in a standard single-mode fibre, which has an attenuation constant $\alpha = 0.3$ dB/km, an effective area $A_{eff} = 50\,\mu\text{m}^2$ and a Raman gain coefficient $g_R = 4 \times 10^{-14}\,\text{m.W}^{-1}$. Above which power level does one of the channels experience growth during propagation?

21.15 Describe the similarities and the main differences between SBS and SRS. Explain the origin of these differences.

21.16 Obtain an expression for the Brillouin-frequency shift starting from the phase-matching condition. Explain why SBS occurs only in the backward direction.

21.17 Estimate the SBS threshold at a pump wavelength $\lambda_p = 1.55\,\mu\text{m}$ of a 50-km-long standard single-mode fibre ($K_B = 2$) with an effective core area $A_{eff} = 50\,\mu\text{m}^2$, an attenuation constant of $\alpha = 0.2$ dB/km, and a gain coefficient $g_{B0} = 5 \times 10^{-11}$ m/W. Obtain the SBS threshold of the same fibre at $\lambda_p = 1.3\,\mu\text{m}$, assuming an attenuation constant $\alpha = 0.6$ dB/km.

22

Fibre Amplifiers and Lasers

Despite being relatively small, fibre loss imposes important limitations on long-haul lightwave systems. These limitations have traditionally been overcome by periodic regeneration of the optical signals at repeaters, applying conversion to an intermediate electrical signal. However, because of the complexity and high cost of such regenerators, the need for optical amplifiers soon became obvious.

Several means of obtaining optical amplification have been suggested since the 1970s, including the direct use of the transmission fibre as a gain medium through non-linear effects, semiconductor amplifiers, or doping optical waveguides with an active material (rare-earth ions) that could provide gain. An optical amplifier can be converted into a laser simply by placing it inside a cavity designed to provide optical feedback. In this chapter, we provide a review of the main features of both optical fibre amplifiers and lasers.

22.1 General Concepts on Optical Amplifiers

An ideal lumped optical amplifier is a transparent box that provides gain and is also insensitive to the bit rate, modulation format, power, and wavelength of the signal passing through it. In practice, however, the optical gain depends not only on the wavelength (or frequency) of the incident signal, but also on the electromagnetic field irradiance at any point inside the amplifier. The details of the wavelength and irradiance dependence of the optical signal depend on the amplifying medium.

If the gain medium is modelled as a homogeneously broadened two-level system, the *gain coefficient* (i.e. the gain per unit length) can be written as

$$g(v, P) = \frac{g_0}{1 + \frac{(v-v_0)^2}{\Delta v_0^2} + \frac{P}{P_{sat}}} \tag{22.1}$$

where g_0 is the peak value of the gain coefficient determined by the pumping level of the amplifier, v is the optical frequency, v_0 is the atomic transition frequency, Δv_0 is the 3-dB local gain bandwidth, P is the signal power, and P_{sat} is the saturation power, which depends on the parameters of the gain medium. The parameters Δv_0 and P_{sat} refer to the local gain. However, from the communication system point of view, it is more useful to use the related concepts of *amplifier bandwidth* and *amplifier saturation power*.

The *amplifier gain* G is defined as

$$G = \frac{P_{out}}{P_{in}} \tag{22.2}$$

where P_{in} is the input power and P_{out} is the output power of a continuous-wave (CW) signal being amplified. The amplifier gain G may be found by using the relation

$$\frac{dP}{dz} = g(v, P)P \tag{22.3}$$

Lasers, Optics, and Photonics. Mário F. S. Ferreira, Oxford University Press. © Mário F. S. Ferreira (2026).
DOI: 10.1093/oso/9780198952770.003.0022

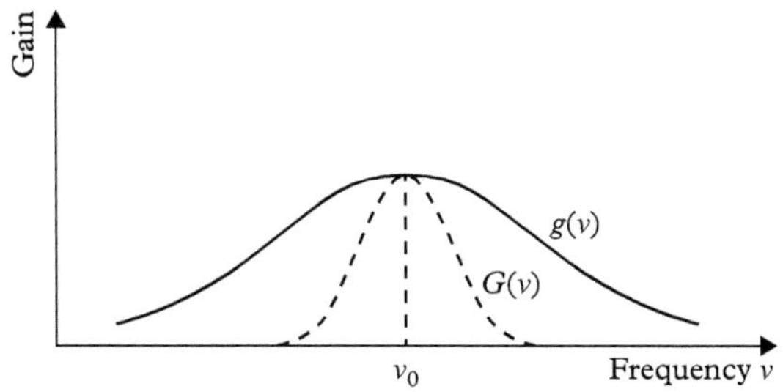

Figure 22.1 *Gain coefficient profile g(ν) and the corresponding amplifier gain spectrum G(ν).*

where $P(z)$ is the optical power at a distance z from the amplifier input end.

If the signal power obeys the condition $P \ll P_{sat}$ throughout the amplifier, the gain coefficient given by Eq. (22.1) can be considered independent of the signal power. In such a case, the amplifier is said to be operated in the unsaturated regime, and the solution of Eq. (22.3) is an exponentially growing signal power, given by $P(z) = P_{in} \exp(gz)$. For an amplifier of length L, we then find the linear amplifier gain, given by

$$G(\nu) = \exp(gL) = \exp\left[\frac{g_0 L}{1 + (\nu - \nu_0)^2 / \Delta\nu_0^2}\right] \tag{22.4}$$

Both the amplifier gain $G(\nu)$ and the gain coefficient $g(\nu)$ are maximum when $\nu = \nu_0$. However, $G(\nu)$ decreases much faster than $g(\nu)$ with the signal detuning $\nu - \nu_0$ because of the exponential dependence of G on g. As a consequence, the *amplifier bandwidth* $\Delta\nu_A$ is generally much smaller than the gain bandwidth $\Delta\nu_0$. This is illustrated in Fig. 22.1.

An important limitation of an optical amplifier is related to the power dependence of the gain coefficient given by Eq. (22.1). This property is known as *gain saturation*, and it appears when the signal power ratio P/P_{sat} is non-negligible. Since the gain coefficient is reduced when the signal power P becomes comparable to the saturation power P_{sat}, the amplifier gain G will also decrease. The *output saturation power* P_{out}^s is defined as the output power for which the amplifier gain G at $\nu = \nu_0$ is reduced by a factor of 2 from its unsaturated value, given by Eq. (22.4). Fig. 22.2 illustrates this concept. In practice, P_{out}^s is found to be about 30% smaller than P_{sat}. Gain saturation can be seen as a serious limitation, particularly for multichannel communication systems.

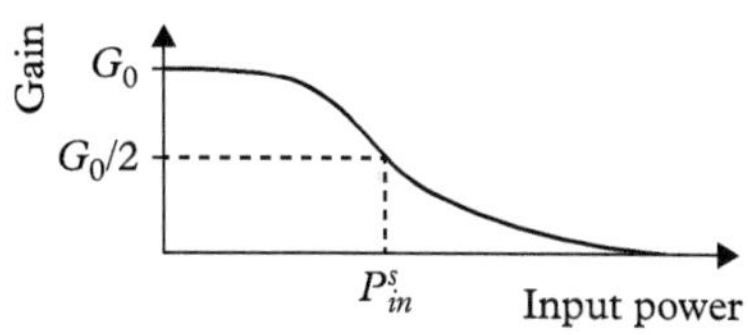

Figure 22.2 *Saturated amplifier gain as a function of the input power.*

Besides the bandwidth and gain saturation limitations, the performance of practical optical amplifiers is also limited by their noise characteristics. In fact, optical amplifiers always add spontaneously emitted photons to the signal during the amplification process. These photons are amplified along with the signal photons so that, at the amplifier output, amplified spontaneous emission (ASE) noise is present. Since spontaneous emission always takes place, ASE noise is unavoidable.

The ASE determines a degradation of the so-called *signal-to-noise ratio* (*SNR*). The *SNR* degradation is usually characterized by the *amplifier noise figure*, which is defined as the *SNR* ratio between input and output:

$$NF = \frac{SNR_{in}}{SNR_{out}} \tag{22.5}$$

The relative importance of the different limiting factors discussed earlier depends on the actual amplifier application. There are four main configurations for these applications, which are illustrated in Fig. 22.3. The first configuration is to place the amplifier

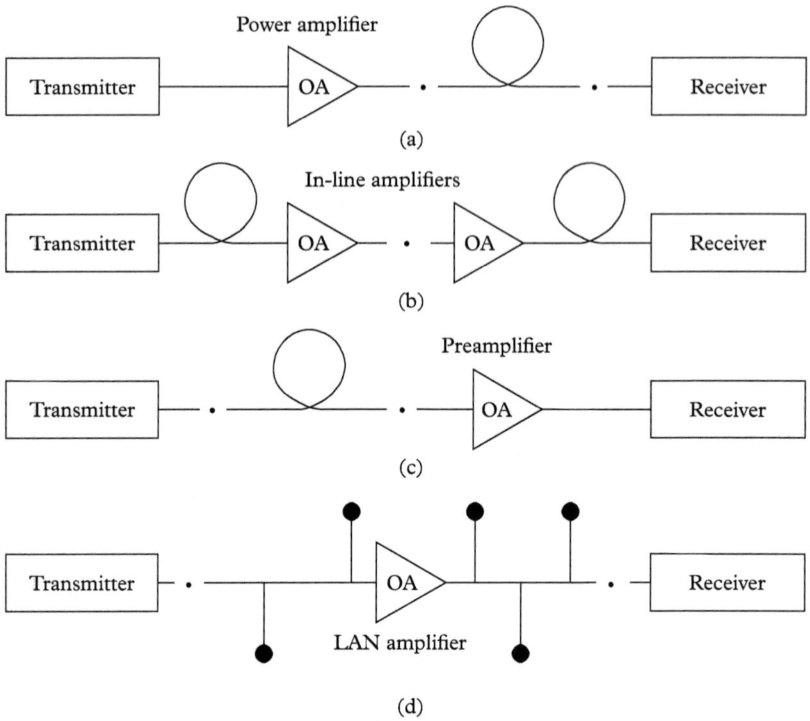

Figure 22.3 *Four generic configurations for incorporating optical amplifiers into transmission systems: (a) as a power amplifier; (b) as an in-line amplifier; (c) as a preamplifier; (d) for compensation of distribution losses in local-area networks (LANs).*

immediately following the laser transmitter to act as a *power amplifier* or *booster* (Fig. 22.3(a)). The main purpose of such amplifiers is to boost the signal power, which can increase the transmission distance by 100 km or more. The second configuration is to place the amplifier in-line (Fig. 22.3(b)), perhaps incorporating it at one or more places along the transmission path, replacing the electronic regenerators. The *in-line amplifier* compensates for periodic signal attenuation and may exist in cascade form. The third configuration consists of using the amplifier immediately before the receiver (Fig. 22.3(c)), so that it functions as a *preamplifier*. The purpose of such an amplifier is to improve the receiver sensitivity. The fourth application of optical amplifiers consists of using them to compensate for distribution losses in local area networks (LANs), as represented in Fig. 22.3(d).

22.2 Rare-Earth-Doped Fibre Amplifiers

Rare-earth-doped fibre amplifiers use ions such as Er^{3+} (erbium), Nd^{3+} (neodymium), or Pr^{3+} (praseodymium) as the active lasing elements, and their principle of operation is based on stimulated-emission transitions between energy levels. These elements have been used since 1963 in solid-state lasers, but the first rare-earth-doped fibre amplifiers were demonstrated only in 1987.

The operational wavelengths of rare-earth-doped fibre amplifiers are determined by the emission spectra of the rare-earth ions, moderately modified by the host material in which they are embedded. Gain is provided in the 1.5-μm band in erbium and in the 1.06-μm and 1.3-μm bands in neodymium, respectively, showing the promise of amplification in wavelength bands widely used in optical communication systems. The element praseodymium provides amplification at 1.3 μm.

When a rare-earth ion is excited to higher energy levels, it can decay to lower energy levels either radiatively or non-radiatively. Non-radiative decay involves the creation of phonons, whereas in radiative decay photons are emitted. The emission of these photons can assume two forms: spontaneous or stimulated. The spontaneous emission of photons is responsible for amplifier noise. However, in the stimulated-emission process, the emitted photons are in phase with an incident photon carrying an energy equal to the difference in energy between the excited and ground states. This is the process that allows signal amplification to occur and therefore is the desired property of a fibre amplifier.

Erbium-doped fibre amplifiers (EDFAs) have attracted particular attention because they operate in the wavelength region near 1.55 μm, the main window of present-day communication systems. Moreover, EDFAs present other intriguing characteristics, such as high gain, high saturation output power, polarization-independent gain, absence of crosstalk, and low noise figure, in addition to low insertion loss.

In general, erbium ions act as a three-level system in the amplification of 1550-nm light. Fig. 22.4 illustrates the relevant energy levels of erbium ions in silica fibres. The three levels involved are the ground state, ${}^4I_{15/2}$, the first excited state, ${}^4I_{13/2}$, and one of the higher excited states. Fig. 22.4 illustrates two possible pumping schemes. In the case of 1480-nm pumping, the upper portion of the ${}^4I_{13/2}$ state is used as a pumping level, and the Er ions operate as a quasi-three-level system. Gain in the Er-doped fibre occurs when an inverted population is achieved between the ${}^4I_{13/2}$ and ${}^4I_{15/2}$ states.

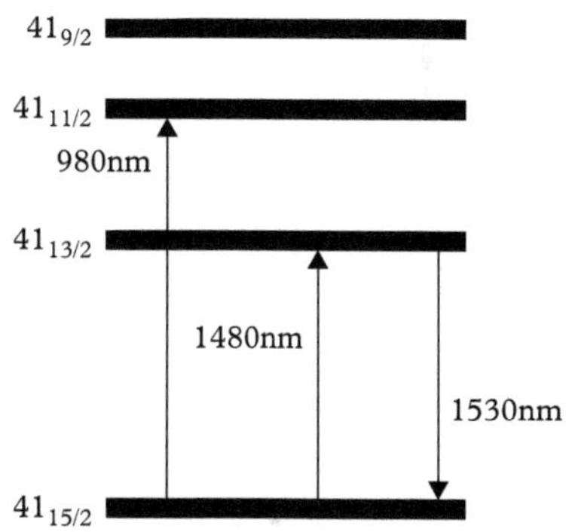

Figure 22.4 *Schematic of energy levels of erbium ions in silica fibres.*

The shape of the gain spectrum is affected considerably by the amorphous nature of silica and by the presence of other co-dopants within the fibre core such as germanium and alumina. The gain spectrum is quite broad, but it is not uniform over the entire bandwidth. For a given amplifier length, the amplifier gain begins to increase exponentially with the pump power, but such increase becomes much slower when the pump power is above a given value. On the other hand, for a fixed pump power, the amplifier gain is maximum at an optimum amplifier length L, decreasing for longer values of L.

A typical EDFA configuration consists of an Er-doped fibre (EDF), a pumping light source, and a wavelength-selective coupler (WSC). Polarization-independent optical isolators are also generally used in order to reject any unwanted scattered light. Pump light is input by way of a WSC, which can be configured for forward, backward, or bidirectional pumping. A better noise figure can be obtained with the co-propagating pump scheme as compared to that of the counter-propagating pump. However, the counter-propagating scheme is suitable as a booster application, since a higher conversion efficiency is obtained with this scheme. Fig. 22.5 illustrates the general configuration of an EDF amplifier with bidirectional pumping. A bandpass filter is usually added in the output path to eliminate the ASE light.

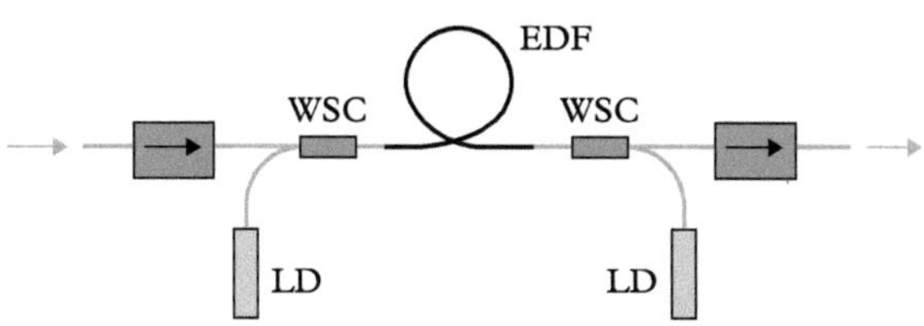

Figure 22.5 *General erbium-doped fibre amplifier configuration showing bidirectional pumping. EDF: erbium-doped fibre; WSC: wavelength-selective coupler; LD: laser diode.*

Depending on the intended use of the amplifier, the EDFA can operate in different regimes. In the small-signal or linear regime, low input signal levels (< 1μW) are amplified with negligible gain saturation. Using an optimum amplifier length, typical EDFA gains of 25–35 dB can be achieved in this regime. In the saturation regime, the input signal level is high enough to cause a measurable reduction in the net gain. This is typical of power amplifier applications, in which the input signal levels are relatively high.

Semiconductor pump sources are generally used to pump EDFAs in the 800-, 980-, and 1480-nm absorption bands. It has been noted, however, that in the 800-nm band the occurrence of the excited-state absorption (ESA) phenomenon becomes significant. In this case, once an ion is excited through the absorption of a pump photon, it relaxes to a metastable state, where, instead of causing stimulated emission by decaying towards the ground state, another pump photon is absorbed so as to make the ion jump to a different excited state. This phenomenon strongly reduces the pumping efficiency at 800 nm.

The candidates for pump wavelength are therefore currently restricted to either 980nm or 1480 nm, and gains in excess of 30 dB can be obtained in both cases. The gain efficiency of a rare-earth-doped fibre is defined as the ratio of the maximum small-signal gain to the input pump power, using the optimized fibre length. EDFA efficiencies are typically on the order of 10 dB/mW for pumping at 980 nm. Moreover, since a noise figure very close to 3 dB can be achieved in this case, the 980 nm pumping solution is also the preferred choice in cases where very low noise figures are required. For pumping at 1480 nm, efficiencies are about half the values obtainable at 980 nm and require about twice the fibre length.

Depending on the fibre parameters, the 3-dB bandwidth of a standard EDFA in a small-signal amplification regime ranges from 25 to 44 nm, which enables its use to amplify several channels simultaneously, as required for dense wavelength-division multiplexing (WDM). The same feature could also allow its use in the L-band of the fibre transmission window. However, the non-uniform gain spectrum of EDFA causes problems when multiple EDFAs are cascaded in the system. Actually, if several channels are located in the relatively flat shoulder region of the gain spectrum, then the gain differential after a single amplifier will be within a few decibels. However, when a cascade of EDFAs is used to periodically compensate for losses, the differential gain and the resultant optical signal-to-noise ratio (OSNR) can become quite severe. This feature could limit the usable bandwidth of EDFAs and, hence, the amount of data transmission by the system. Several approaches have been used to mitigate this limitation, namely by using gain equalizing filters, long-period fibre gratings, and chirped fibre Bragg gratings.

22.3 Fibre Raman Amplifiers

Due principally to its poor pumping efficiency and the scarcity of high-power pumps at appropriate wavelengths, much of the early work on Raman amplification, carried out during the 1970s and mid-1980s, was overtaken by erbium-doped fibre amplifiers (EDFAs) by the late 1980s. Then, in the mid-1990s, the development of suitable high-power pumps and the availability of higher Raman-gain fibres motivated a renewed interest in Raman amplification. Actually, high Raman gain has been observed in several materials and structures.

Fig. 22.6 shows schematically a fibre Raman amplifier. The pump and the signal beams are injected into the fibre through a WDM fibre coupler. The case illustrated in Fig. 21.6 shows the two beams co-propagating inside the fibre, but the counter-propagating

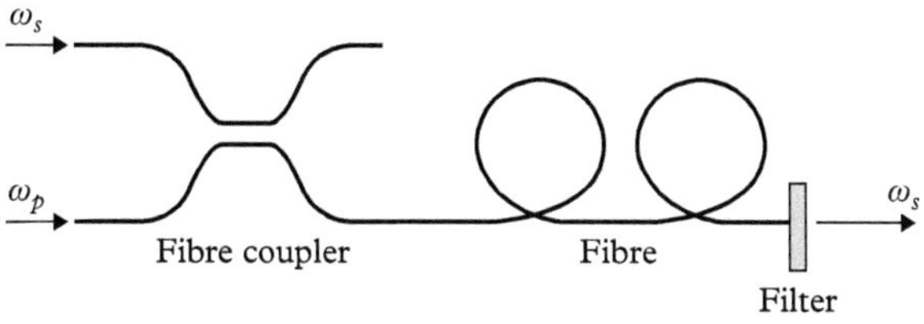

Figure 22.6 *Schematic of a fibre Raman amplifier.*

configuration is also possible. In fact, SRS can occur in both directions, forwards and backwards, the Raman gain being almost the same in the two cases.

As seen in Chapter 21, the Raman gain in silica fibres extends over a frequency range of about 40 THz. Optical signals whose bandwidths are of this order or less can be amplified using the Raman effect if a pump wave with the right wavelength is available. The Raman gain depends on the relative state of polarization of the pump and signal fields. The peak value of the Raman gain decreases with increasing pump wavelength and is about 6×10^{-14} m/W in the wavelength region around 1.5 μm.

In the SRS process, the evolution of the pump and a co-propagating signal power is governed by the coupled Eqs. (21.73) and (21.74), whose solutions are given by Eqs. (21.75)–(21.73). Such results can be used to study the main characteristics of a fibre Ram amplifier.

Fig. 22.7 shows the output signal power $P_s(L)$ versus the input signal power $P_s(0)$ for a Raman amplifier with length L = 8 km and several values of the input pump power. Typical values $g_R = 6.7 \times 10^{-14}\,\mathrm{m.W^{-1}}$, $\alpha = 0.2\,\mathrm{dB.km^{-1}}$, $A_{eff} = 30 \times 10^{-12}\,\mathrm{m^2}$, $\lambda_p = 1.46\,\mu\mathrm{m}$ and $\lambda_s = 1.55\,\mu\mathrm{m}$ were assumed. The transfer characteristics are linear for $P_p(0) = 0.1, 0.2$, and 0.4 W, but the effects of pump depletion become discernible for input signal levels $P_s(0) > -15\,\mathrm{dBm}$ when $P_p(0) = 0.6\,\mathrm{W}$. For a pump power $P_p(0) = 0.8\,\mathrm{W}$, the linear behaviour of the Raman amplifier cannot be observed even for signal power levels as low as −25 dBm.

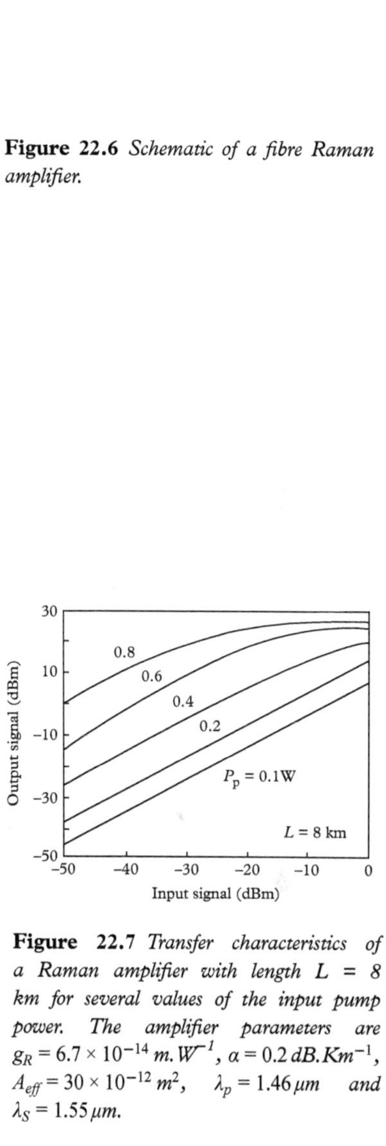

Figure 22.7 *Transfer characteristics of a Raman amplifier with length L = 8 km for several values of the input pump power. The amplifier parameters are* $g_R = 6.7 \times 10^{-14}\,m.W^{-1}$, $\alpha = 0.2\,dB.Km^{-1}$, $A_{eff} = 30 \times 10^{-12}\,m^2$, $\lambda_p = 1.46\,\mu m$ *and* $\lambda_S = 1.55\,\mu m$.

Since in the absence of Raman amplification the signal power at the amplifier output would be $P_s(L) = P_s(0)\exp(-\alpha L)$, the amplifier gain is given by

$$G_R = \frac{P_s(L)}{P_s(0)\exp(-\alpha L)} = \frac{\exp\left(g_R P_p(0) L_{eff}/A_{eff}\right)}{1 + F(L)} \tag{22.6}$$

where L_{eff} is the effective interaction length, given by Eq. (21.6). The amplifier gain is seen to be a function of the input signal power (i.e. a saturation non-linearity) through the term $F(L)$.

Fig. 22.8 illustrates the dependence of the amplifier gain on the pump power for three values of the fibre length. The amplifier gain (in dB) increases linearly with the pump power in the case of a short fibre length (L = 2 km). However, for a fibre length L = 8 km, the saturation effects are observable at values of the pump power as low as 400 mW. The saturated amplifier gain is seen to be about 30 dB. Dashed curves in Fig. 22.8 represent the factor $F(L)$, given by Eq. (21.77), and the dots indicate the pump power at which $F(L) = 1$.

Assuming $F(L) \ll 1$ corresponds to neglecting the gain saturation, in which case the amplifier gain becomes

$$G_R = \frac{P_s(L)}{P_s(0)\exp(-\alpha L)} = \exp\left(g_R L_{eff} P_0/A_{eff}\right) \tag{22.7}$$

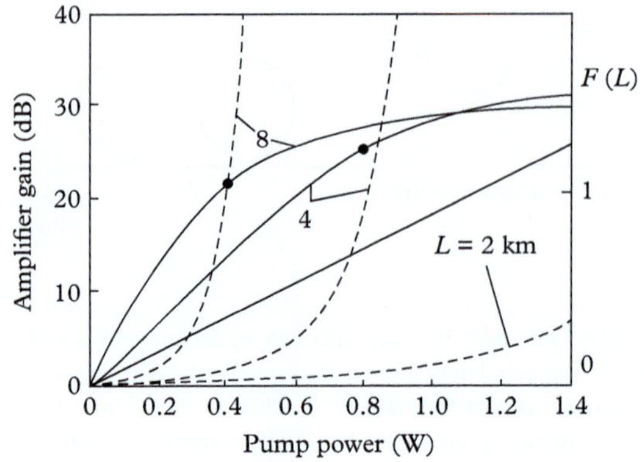

Figure 22.8 *Gain of a Raman amplifier as a function of pump power for three values of the fibre length L = 2, 4, and 8 km. Dashed curves represent the factor F(L), given by Eq. (21.77). The amplifier parameters are the same as for Fig. 22.7.*

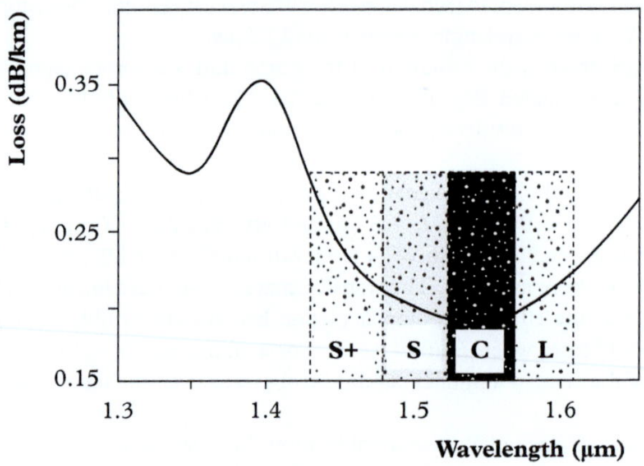

Figure 22.9 *Representation of wavelength bands S+, S, C, and L used in fibre communication systems.*

From Eq. (22.7), the Raman gain in decibels is expected to increase linearly with the pump input power.

Considering its broad bandwidth (>5 THz), fibre Raman amplifiers can be used to amplify several channels simultaneously in a multichannel communication system. The same characteristic also makes such amplifiers suitable for the amplification of short optical pulses. Moreover, Raman amplifiers can in principle have a fully configurable gain profile determined by the Raman-gain spectra produced by multiple pump lasers. In practice, CW power at several wavelengths is provided by a set of high-power semiconductor lasers located at the pump stations. The wavelengths of the pump lasers should be in the vicinity of 1450 nm for amplifying optical signals in the 1550 nm spectral region.

Raman amplifiers can be realized considering two main schemes. One is the lumped Raman amplifier scheme, in which all the pump power is confined to a relatively short fibre element that is inserted into the transmission line to provide gain. The primary use of the lumped amplifier is to open new wavelength bands beyond the commonly used band C (see Fig. 22.9), which are inaccessible by EDFAs.

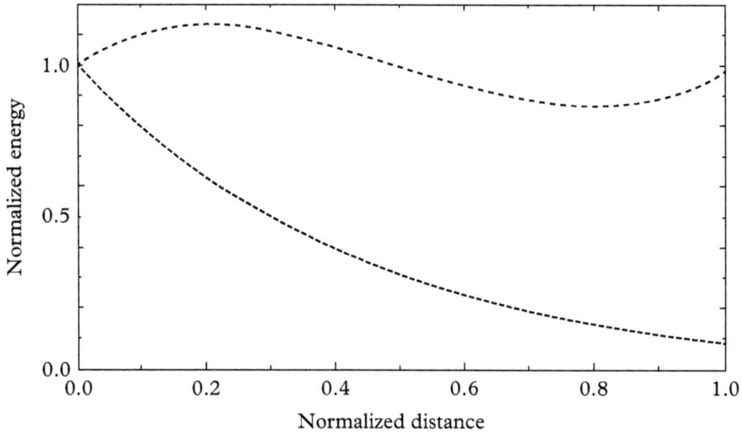

Figure 22.10 *Normalized signal energy against the normalized distance, z/L_A. A comparison is made between a purely lumped amplified system (dotted curve) and a system using bidirectional distributed Raman amplification (dashed curve).*

The other option is the distributed Raman amplifier, which utilizes the transmission fibre itself as the Raman-gain medium to obtain amplification. In this case, pump lasers are included periodically along the transmission link in order to inject optical power at suitable wavelengths.

When using a lumped amplification scheme, the length of the fibre amplifier is usually much shorter than the amplifier spacing, L_A. After propagating this distance, the average signal power is reduced from its initial value P_0 to $P(L_A) = P_0 \exp(-\alpha L_A)$. For example, this means a reduction of 20 dB for a standard 100-km-long fibre, presenting a minimum attenuation of 0.2 dB/km. In order to compensate for this loss, the lumped amplifier gain must be $G = \exp(\alpha L_A)$. In this case, there is a significant variation in the signal power, which shows a saw-tooth-like profile. This behaviour is illustrated by the dotted curve in Fig. 22.10.

Backward distributed Raman amplification can be used to assist the lumped amplifier in order to reduce the overall excursion of the signal power. This scheme provides two main advantages. On the one hand, the top signal level is reduced, thus minimizing the impact of non-linear effects. On the other hand, as the signal does not dip down so profoundly, the SNR ratio remains higher than when using only lumped amplifiers. Actually, a further reduction of the signal power excursion can be achieved using distributed Raman amplification with a bidirectional pumping scheme, as illustrated by the dashed curve in Fig. 22.10.

22.4 Fibre Brillouin Amplifiers

Stimulated Brillouin scattering (SBS) in optical fibres is a highly efficient non-linear amplification mechanism with which large gains can be achieved using pump powers of only a few milliwatts. Such a non-linear process can be used to construct a fibre Brillouin amplifier (FBA), which is basically an optical fibre in which the pump and signal waves propagate in opposite directions. The only required condition is that the spacing between the pump and signal frequencies must correspond to the Brillouin shift in that fibre.

In the SBS process, the power evolution of the pump and signal waves is governed by Eqs. (21.83)–(21.84). Assuming that the attenuation coefficient satisfies the condition

$\alpha << 1$, the following approximate analytical solutions can be found:

$$P_S(z) = P_S(0)\,D(z)\,e^{\alpha z} \tag{22.8}$$

$$P_p(z) = P_p(0)\,D(z)\,H(z)\,e^{-\alpha z} \tag{22.9}$$

where

$$D(z) = \frac{\lfloor P_p(0) - P_S(0) \rfloor}{\left[P_p(0)\,H(z) - P_S(0)\right]} \tag{22.10}$$

and

$$H(z) = \exp\{g_B \lfloor P_p(0) - P_S(0) \rfloor [1 - \exp(-\alpha z)]/(A_{eff}\alpha)\} \tag{22.11}$$

The Brillouin amplifier gain is given by Eq. (22.8) as

$$G_B = \frac{P_S(0)}{P_S(L)\,e^{-\alpha L}} = \frac{1}{D(L)} \tag{22.12}$$

Fig. 22.11 shows the transfer characteristics for a Brillouin amplifier with length L = 8 km, considering different input pump powers and typical parameter values: $g_B(0) = 2.5 \times 10^{-11}$ m.W^{-1}, $\alpha = 0.2$ dB.km^{-1}, $A_{eff} = 30 \times 10^{-12}$ m^2. The transfer characteristics are linear for $P_p(0) = 0.2$ mW, but the effects of pump depletion become clearly observable for higher pump levels. For a pump power $P_p(0) = 2$mW, a linear behaviour of the Brillouin amplifier cannot be observed even for signal powers as low as ~ 0.1 μW.

Fig. 22.12 illustrates the dependence of the Brillouin amplifier gain on the input pump power for several amounts of the signal detuning from the gain peak and for two values of the signal power. A fibre length L = 5 km, a gain bandwidth $\Delta\nu_B = 30$ MHz, and typical parameter values $g_B(0) = 2.5 \times 10^{-11}$ m.W^{-1}, $\alpha = 0.2$ dB.km^{-1}, and $A_{eff} = 30 \times 10^{-12}$ m^2 were assumed. We observe that the saturated gain shows a significant dependence on the

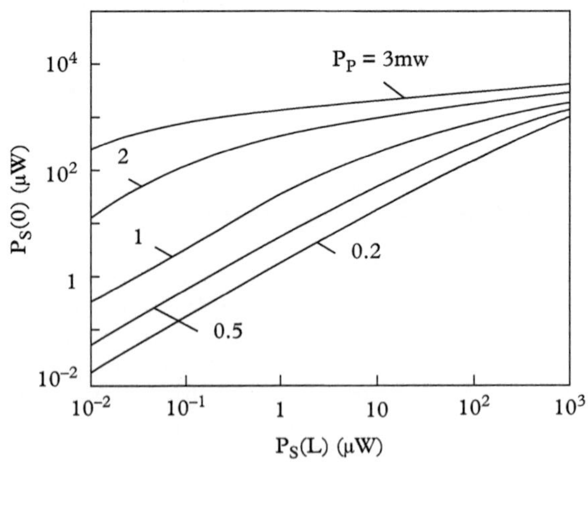

Figure 22.11 *Transfer characteristics for a Brillouin amplifier with a length L = 8 km and several values of the input pump power.*

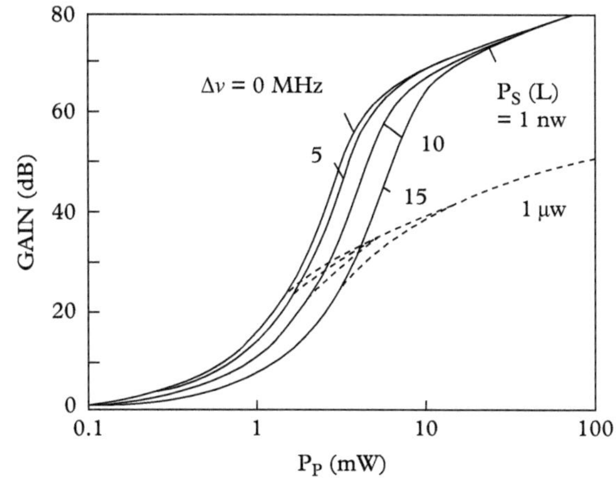

Figure 22.12 *Brillouin amplifier gain against the input pump power for an amplifier length L = 5 km and different amounts of detuning from the gain centre. The full curves and the dashed curves correspond to input signal powers $P_s(L) = 1\,nW$ and $P_s(L) = 1\,\mu W$, respectively.*

input signal power. An input signal of 1 nW, which is comparable to the original value of spontaneous emission, can be amplified by about 60 dB for a pump power $P_p(0) \approx 5\,\text{mW}$. However, the amplifier gain can be considerably reduced when the signal is detuned from the gain peak. For example, for an input signal power $P_s(L) = 1\,\text{nW}$ and a pump power $P_p(0) \approx 3\,\text{mW}$, a detuning of 15 MHz (which corresponds to a reduction of the gain coefficient to half of its maximum value) determines a reduction of about 25 dB in the amplifier gain.

Fibre Brillouin amplifiers can provide high gains at low pump powers. However, the gain bandwidth is small, which strictly limits the bandwidth of data signals that can be amplified in a FBA. However, as mentioned in Chapter 21, the intrinsic Brillouin linewidth is generally enhanced by compositional inhomogeneities in the fibre and can be intentionally extended by more than one order of magnitude by applying frequency modulation to the pump laser. Of course, the bandwidth enhancement is accompanied by a reduction in the peak gain. Consequently, a higher pump power will be necessary to achieve the same gain.

22.5 Fibre-Optic Parametric Amplifiers

Four-wave mixing can be used with advantage to realize fibre-optical parametric amplifiers (FOPAs). A FOPA offers a wide gain bandwidth and may be tailored to operate at any wavelength.

In the case of a single pump, the unsaturated single-pass gain may be written as (Ferreira, 2011)

$$G_p = 1 + \left[\frac{\gamma P_p}{g}\sinh(gL)\right]^2 \tag{22.13}$$

where P_p is the pump power, g is the parametric gain coefficient given by:

$$g^2 = \left[(\gamma P_p)^2 - (\kappa/2)^2\right] = -\Delta k\left(\frac{\Delta k}{4} + \gamma P_p\right) \tag{22.14}$$

and Δk is the effective phase mismatch, given by Eq. (21.56). The maximum value of the parametric gain occurs when $\Delta k = -2\gamma P_p$ and is given by

$$g_{\max} = \gamma P_p \tag{22.15}$$

For signal wavelengths close to the pump wavelength, one has $\Delta k \approx 0$ and

$$G_p \approx 1 + (\gamma P_p L)^2 \tag{22.16}$$

In the case of perfect phase matching ($\kappa = 0$) and assuming that $\gamma P_p L << 1$, the unsaturated single-pass gain becomes

$$G_p \approx \sinh^2(gL) \approx \frac{1}{4}\exp(2\gamma P_p L) \tag{22.17}$$

A very simple expression for the FOPA peak gain may be obtained if Eq. (22.17) is rewritten in decibel units

$$G_p\,(dB) = 10\log_{10}\left[\frac{1}{4}\exp(2\gamma P_p L)\right] = P_p L S_p - 6 \tag{22.18}$$

where

$$S_p = 10\log_{10}(e^2)\,\gamma \approx 8.7\gamma \tag{22.19}$$

is the parametric gain slope in (dB/W/km).

Fig. 22.13 shows the small-signal gain against the pump power provided by a parametric amplifier constituted by a 500-m long highly non-linear fibre with $\gamma = 11\text{W}^{-1}/\text{km}$ (Hansryd et al. 2002). The gain slope in the undepleted pump region is 100 dB/W/km. The input signal power was −19.5 dBm, and the maximum achieved output signal power was 21 dBm, limited by signal-induced SBS. Fig. 22.14 shows the amplifier gain versus the

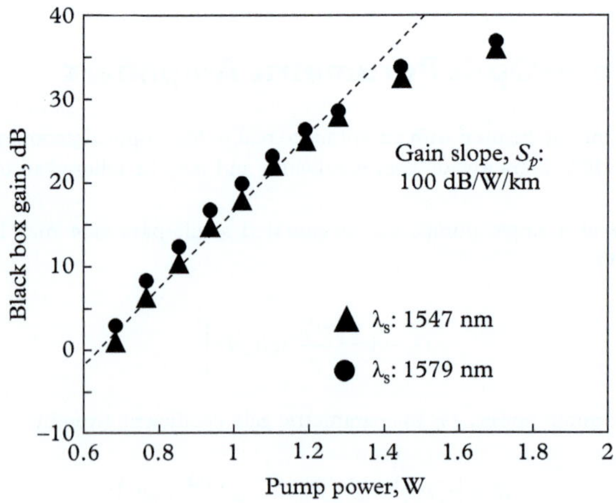

Figure 22.13 *Small-signal parametric gain using a 500 m long fibre with $\gamma = 11W^{-1}$ / km (reprinted with permission from [Hansryd et al. 2002]© IEEE).*

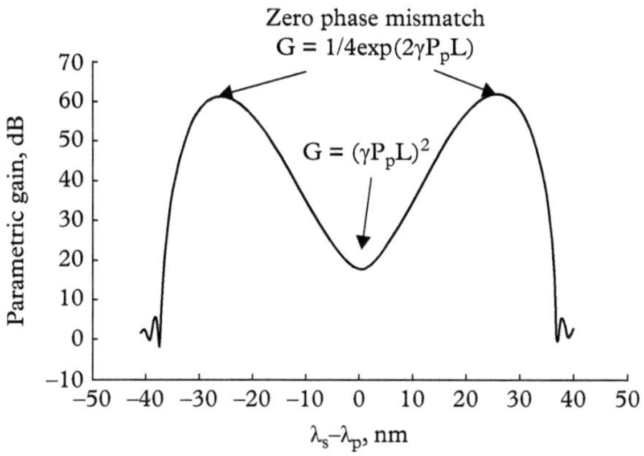

Figure 22.14 *Calculated amplifier gain versus the wavelength difference between pump and signal for the same fibre as in Fig. 22.13, considering a pump power of 1.4 W (reprinted with permission from [Hansryd et al. 2002]© IEEE).*

wavelength difference between pump and signal for the same fibre of Fig. 22.13, considering a pump power of 1.4 W. The region with exponential gain (corresponding to perfect phase matching) and the region with quadratic gain (corresponding to $\lambda_3 \approx \lambda_1$) are marked in the figure.

Considering the approximation given by Eq. (21.70) for the linear phase mismatch, the FOPA bandwidth $\Delta\Omega$ can be obtained from the maximum effective phase mismatch,

$$\kappa_m = \beta_2(\Omega_s + \Delta\Omega)^2 + 2\gamma P_p, \tag{22.20}$$

with $\Delta\Omega \ll \Omega_s$ and Ω_s given by Eq. (21.58). This maximum value of the effective phase mismatch occurs when the parametric gain in Eq. (22.14) vanishes, which gives $\kappa_m = 2\gamma P_p$. In these circumstances, the FOPA bandwidth is approximately given by

$$\Delta\Omega \approx \sqrt{\frac{\gamma P_p}{2\,|\beta_2|}} \tag{22.21}$$

Eq. (22.21) shows that the FOPA bandwidth can be increased by increasing the non-linear parameter γ and reducing the $|\beta_2|$. This is the reason why modern FOPAs use highly non-linear fibres (HNLFs), and the pump wavelengths are chosen near the ZDW of the fibre. Moreover, Eq. (21.16) shows that, for a fixed value of the FOPA gain, the value of $\gamma P_p L$ must be kept constant. In such a case, the amplifier bandwidth will increase with decreasing L. Using a HNLF provides the possibility of simultaneously decreasing L and increasing the non-linear parameter γ. For example, a value $\gamma P_p L = 10$ can be achieved with a pump power of 1 W and a HNLF with L = 1 km and $\gamma = 10\,\mathrm{W}^{-1}$/km. The bandwidth will be increased 20 times using for instance, a fibre length of 50 m, a pump power of 5 W, and a non-linear parameter $\gamma = 40\,\mathrm{W}^{-1}$/km. Such a high value of γ can be achieved using some types of HNLFs.

As the FOPA bandwidth is proportional to the square root of the pump power, it could, in principle, be arbitrarily increased if enough optical power is available. However, such high values of the pump power are limited in practice by SBS. SBS shows a reduced threshold in the case of HNLFs due to the small effective core area of this type of fibre.

On the other hand, it can be seen from Fig. 22.14 that the gain of a single-pump FOPA is far from being uniform over the entire bandwidth. Actually, only a small portion of the gain spectrum near the two gain peaks can be used in practice.

The limitations imposed by the SBS process and the nonuniformity of the gain spectrum can be circumvented if the pump power is distributed between two pumps instead of being concentrated in one pump. Moreover, a two-pump FOPA offers additional degrees of design freedom while also allowing polarization-independent operation of the device.

22.6 Optical Fibre Lasers

The advent and rapid progress of optical fibre amplifiers have enabled one to use them as a gain medium to fabricate fibre lasers. In fact, a fibre laser can be realized simply by placing a fibre amplifier inside a cavity in order to provide optical feedback. Actually, a fibre laser has a number of advantages over a conventional laser. Light in the fibre core is trapped by total internal reflection, which provides a natural way of confining the lasing mode laterally. Moreover, a fibre laser is compact, lightweight, and provides a very stable output beam.

22.6.1 Doped fibre lasers

A doped fibre laser can be considered as an optical wavelength converter in which the photons at the pumping wavelength are absorbed to produce the required population inversion and stimulated emission; this provides a lasing output at a wavelength which is characterized by the dopant in the fibre. Because of the small core area of a typical optical fibre, the intensity of the guided field can be very high even for modest optical powers. These high intensities are sufficient to produce steady-state population inversion and lasing. As a consequence, the threshold pumping powers are much smaller in doped fibre lasers compared with other conventional lasers. The two major dopants currently employed for fibre lasers are neodymium and erbium. The former element provides a four-level laser scheme, whereas the latter element gives a three-level scheme. The first fibre laser, developed by Elias Snitzer in 1961 (one year after Maiman demonstrated the ruby laser), made use of a core of neodymium-doped glass.

Nd-doped fibre lasers can provide significant lasing outputs at wavelengths of 900, 1060 and 1320 nm and are particularly interesting as they can be pumped using GaAs semiconductor lasers operating near 800 nm. A Nd-doped fibre laser was first demonstrated in 1973 using a graded-index silica fibre whose core was co-doped with alumina. The laser threshold was reached at an absorbed pump power of 0.6 mW. A single-mode silica fibre laser was first realized in 1985 using a 2-m-long Nd-doped fibre. The laser cavity was made by butt-coupling the cleaved fibre ends against dielectric mirrors, and the threshold was reached at only 0.1 mW of pump power.

EDFLs present major useful lasing transitions at 800, 980 and 1550 nm. The 1550 nm region has attracted the most attention because it coincides with the low-loss region of silica fibres used for optical communications. Since a three-level laser requires that at least half of the ion population be raised to the excited state, it is expected that EDFLs present a high threshold. Indeed, early attempts to make EDFLs used high-power argon-ion lasers as a pump source and presented threshold pump powers of ~100 mW. As noted in Section 22.2, EDFLs pumped near 800 nm using GaAs semiconductor lasers suffer from the problem of ESA. This limitation is absent, and the performance of EDFLs improves considerably when they are pumped at the 980 or 1480 nm wavelength.

Both neodymium-doped fibre and EDF lasers usually employ co-dopants such as phosphorus pentoxide (P2O5), germania (e.g. GeO2, GeCl4) or alumina (Al2O5). Silica-based glasses are the host material most employed, because they are endowed with optical and mechanical properties superior to those of most other glass fibres. However, the transparency of silica glass diminishes appreciably for wavelengths longer than about 2.2 μm so that fibres fabricated from fluorides and other glasses are also used. Actually, light emission in the mid-infrared wavelength range has been achieved using fluoride glasses with other dopants. Fluorozirconate fibres doped with both neodymium and erbium ions have also been used to produce lasers emitting at wavelengths of 1050 and 1350 nm, and 1550 nm, respectively.

Besides those mentioned earlier, many other dopants, such as holmium, samarium, thulium, and ytterbium, have also been used to make fibre lasers emitting at wavelengths ranging from visible to infrared. For example, Pr-doped fibre lasers were developed in an attempt to achieve emission at 1300 nm. Holmium-doped fibre lasers have also attracted considerable attention because they operate near 2000 nm, a wavelength useful for medical and other eye-safe applications. Thulium co-oping permits these lasers to be pumped with GaAs lasers.

22.6.2 Cavity configurations

The optical feedback can be realized considering different configurations for the laser cavity. In the linear configuration, optical feedback can be achieved simply by butt-coupling high-reflecting mirrors to the ends of a doped fibre, thus providing a *Fabry–Perot cavity*, as illustrated in Fig. 22.15. Light from a pumping laser source is launched into the fibre core through one of the mirrors (mirror 1), which is highly transmitting at the pump wavelength but is highly reflecting at the lasing wavelength. The optical output from the device is coupled through the mirror on the other fibre end face (mirror 2 in Fig. 22.15). However, the alignment of such a cavity generally is not easy, which causes an increase in the cavity losses.

An alternative to using the aforementioned butt-coupling mirrors consists in depositing adequate dielectric coatings directly onto the polished ends of the fibre. A drawback of this solution is the possibility of damage to the dielectric coatings when high-power pump light is coupled into the fibre. Furthermore, in spite of the gain spectrum of most rare-earth ions extending over a wavelength range of some 50 nm, such dielectric coatings restrict that range to 5–10 nm. Such a linewidth is too narrow for the provision of a broadband optical source but too wide to be used in some single-frequency laser applications.

A number of alternative fibre laser structures have therefore been fabricated which do not require dielectric or metallic mirrors. For example, one can use fibre Bragg gratings (FBGs) located at the ends of the doped fibre, thus forming an all-fibre Fabry–Perot cavity.

Figure 22.15 *Schematic of a fibre laser with a Fabry–Perot cavity.*

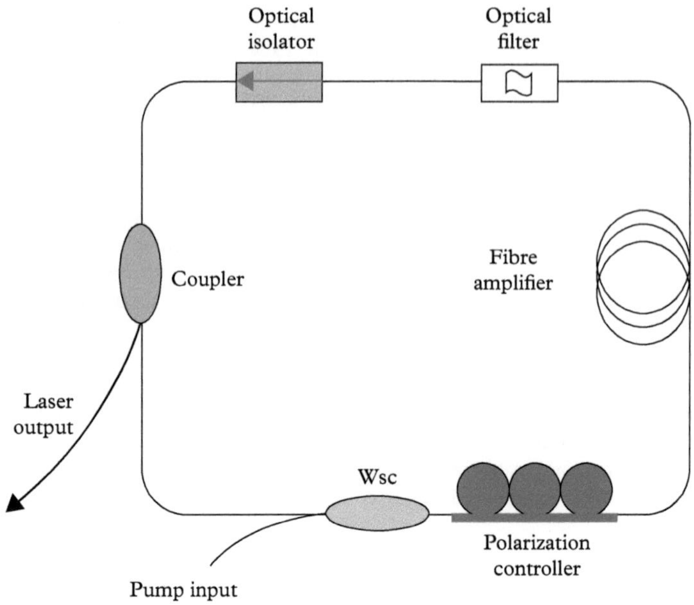

Figure 22.16 *Schematic of a ring fibre laser. WSC: wavelength-selective coupler.*

Some main characteristics of FBGs have been discussed in Chapter 16. Actually, besides acting as a high-reflectivity mirror for the laser light, the FBG can be designed in order to be transparent to the pump light. FBGs are also commonly used for dispersion and spectral control in fibre lasers, providing the possibility of laser operation in a single longitudinal mode. Another solution to provide optical feedback makes use of *fibre-loop mirrors*. A fibre-loop mirror can be designed to reflect the laser light but transmit pump radiation, as described in the next section.

The laser cavity can also assume a ring configuration when the fibre is sliced into a loop, as represented in Fig. 20.16. This is an all-fibre laser cavity which does not use any mirrors and can be implemented simply by connecting two ports of a fibre coupler. The inclusion of an isolator within the loop is necessary to force unidirectional operation. Coupling the pump light into the fibre amplifier without coupling the signal out is realized using a WSC. An output coupler is also used to out-couple a portion of the signal in the laser cavity. An optical filter is used for wavelength selection and tuning, to reject noise, or to modify the cavity dispersion, whereas a polarization controller is also incorporated in the case of a conventional doped fibre that does not preserve polarization.

22.6.3 Non-linear optical loop mirrors

The non-linear optical loop mirror (NOLM) is basically an anti-resonant non-linear Sagnac interferometer with an intensity-dependent transmission. The Sagnac interferometer is represented in Fig. 22.17 and is made by connecting the two outputs of a fibre coupler through a piece of long fibre. The input beam is divided in the fibre coupler into two beams counter-propagating in the fibre-loop. After one roundtrip, both beams arrive at the coupler at the same time, where they interfere coherently, according to their relative phase difference.

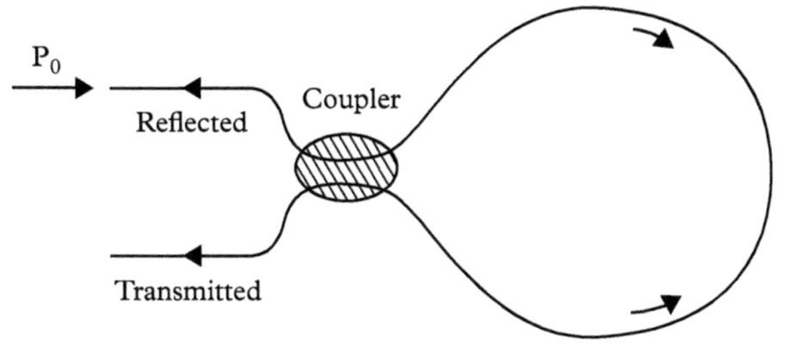

Figure 22.17 *Schematic illustration of an all-fibre Sagnac interferometer.*

Considering an input beam with power $P_0 = |A_0|^2$ incident at one input port, the amplitudes of the optical fields at the two output ports of the coupler are given by:

$$A_1 = \sqrt{f}A_0, \qquad A_2 = i\sqrt{1-f}A_0 \tag{22.22}$$

where f is the coupler power-splitting fraction. After one round trip, both the forward- and backward-propagating fields acquire linear and non-linear phase shifts, taking the form:

$$A'_f = A_f \exp\left[i\beta L + i\gamma\left(|A_f|^2 + 2|A_b|^2\right)L\right] \tag{22.23}$$

$$A'_b = A_b \exp\left[i\beta L + i\gamma\left(|A_b|^2 + 2|A_f|^2\right)L\right] \tag{22.24}$$

where L is the loop length, γ is the fibre non-linear parameter, and β is the propagation constant.

The reflected and transmitted fields can be obtained by using the transfer matrix of the fibre coupler and are given by

$$\begin{pmatrix} A_t \\ A_r \end{pmatrix} = \begin{pmatrix} \sqrt{f} & i\sqrt{1-f} \\ i\sqrt{1-f} & \sqrt{f} \end{pmatrix} \begin{pmatrix} A'_f \\ A'_b \end{pmatrix} \tag{22.25}$$

From Eqs. (22.22)–(22.25) we obtain the following result for the transmittivity $T = |A_t|^2 / |A_0|^2$ of the Sagnac loop:

$$T = 1 - 4f(1-f)\cos^2\left[\left(f - \frac{1}{2}\right)\gamma P_0 L\right], \tag{22.26}$$

From Eq. (22.26) one can see that $T = 0$ for a 3-dB coupler. In this case, the field is totally reflected and the Sagnac interferometer serves as an ideal mirror. For this reason, it is also known as a *fibre-loop mirror.* In such a case, the power is equally divided between the counter-propagating waves and the non-linear phase shift is equal for both waves, resulting in no relative phase difference between the counter-propagating waves.

An asymmetry between the clockwise and anticlockwise waves propagating around the loop in Fig. 22.17 can be created by using a coupler with an uneven coupling ratio ($f \neq 0.5$). Such asymmetry can also be provided by a fibre amplifier which is placed closer to one of the coupler arms than to the other. In this case, the device is known as a *non-linear amplifying loop mirror* (NALM). In either case, light travelling clockwise around the loop has a different intensity from light travelling anticlockwise, and hence experiences different amounts of self-phase modulation in the loop due to fibre non-linearity. The interference at

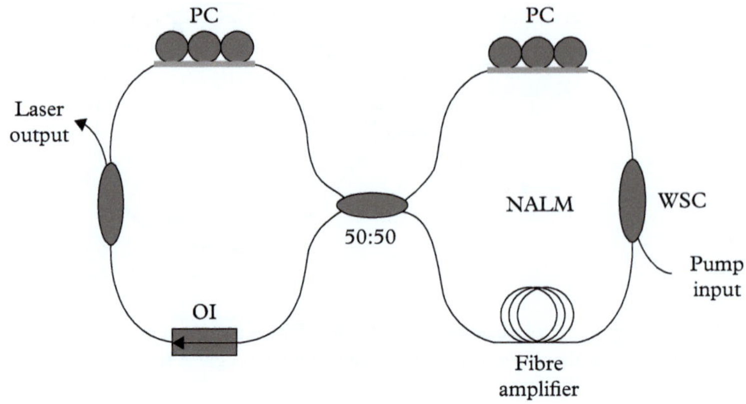

Figure 22.18 *Schematic of a figure-eight laser. PC: polarization controller; WSC: wavelength-selective coupler; OI: optical isolator.*

the coupler, constructive or destructive, depends on the phase difference between the two waves, which is proportional to the power of the incoming signal. If the phase shift is equal to π for the central part of the pulse, this part is transmitted, but its wings are reflected because of their lower phase shifts. As a consequence, the exiting pulse is narrower than the input pulse.

The use of a NALM provides passively mode-locked fibre lasers with a cavity geometry which makes them known as figure-eight lasers. A schematic of this laser is illustrated in Fig. 22.18. It can be regarded as a ring laser (to the left of the 50:50 coupler), with a NALM on the right, which provides fast saturable gain. The NALM causes additive-pulse mode locking of the laser and consists of a 3-dB fibre coupler, a doped fibre, and a wavelength-division multiplexer for pumping. Mode locking is favoured since the loop becomes fully transmissive for pulses whose power attains a critical value. An isolator in the left cavity ensures unidirectional operation. The laser output is provided by a fibre coupler with low transmission to minimize the cavity losses. A further reduction of the cavity losses can be achieved if the central coupler in Fig. 22.18 is unbalanced.

22.6.4 Short pulse fibre lasers

Fibre lasers can be used to generate short pulses based on two techniques: Q-switching and mode locking. As described in Section 17.7, Q-switching makes use of a device that increases the cavity losses but opens periodically for a short time, during which such losses are reduced. Nanosecond pulses with high peak powers (>1 kW) can be generated in this way.

Q-switching produces relatively broad optical pulses (~ 100 ns). In contrast, mode locking generates ultrashort pulses, with durations from hundreds of picoseconds down to a few femtoseconds. However, the finite gain bandwidth of optical fibre amplifiers usually limits the minimum possible pulse duration that can be generated to tens of femtoseconds.

Mode Locking

The longitudinal modes of a cavity laser occur at frequencies $\nu_j = jc/2nl$, where j is an integer, c is the speed of light in vacuum, n is the refractive index in the cavity, and l is the cavity length. A train of short pulses, spaced by the cavity round trip time, can be obtained by phase-locking several of these modes. This can be achieved using an adequate active or passive element in the laser cavity. Since optical fibre lasers are relatively long, the spacing between longitudinal modes is relatively small (typically between 1 GHz and 1 MHz). To generate ultrashort optical pulses in these lasers at high repetition rates therefore requires mode locking of many harmonically related modes.

Active mode-locking is typically achieved by inserting a modulator into the cavity to modulate either the amplitude or the phase of the intracavity optical field at a frequency equal to the mode spacing. In both cases, modulation sidebands are generated, spaced apart by the modulation frequency. These sidebands overlap with the neighbouring modes, leading to a phase synchronization.

Harmonic mode locking occurs when the modulation frequency is an integer multiple of the mode spacing. Both acousto-optic and electro-optic modulators have been used for this purpose. Active mode locking is an attractive technique for producing ultra-low-jitter pulses in the picosecond range at repetition rates as high as tens of GHz. Actually, the generation of picosecond pulses at repetition rates above 100 GHz has already been demonstrated using this technique.

In practice, $LiNbO_3$ modulators are commonly used for active mode-locking fibre lasers, not only because they can be integrated within the laser cavity with relatively low insertion losses, but also because they can be modulated at speeds of some tens of GHz.

Passive mode locking is a widely used technique which does not require any externally modulated media or devices but only makes use of a non-linear device whose response depends on the intensity of the input pulse. To promote pulse generation and shortening, it uses a fast saturable absorber, the function of which is to reject low-intensity light from the cavity, but to pass high intensities. This idea can be implemented using different approaches, namely using a saturable absorber medium, a non-linear fibre-loop mirror, or a non-linear polarization-rotation scheme. Non-linear effects such as self-phase modulation and cross-phase modulation, as well as fibre dispersion, play an important role in the operation of most passively mode-locked fibre lasers.

In the case of a fast saturable absorber, the wings of an optical pulse experience more loss than its central part, which is intense enough to saturate the absorber. As a consequence, the exiting pulse is narrower than the input pulse. A semiconductor absorbing medium is often used to realize such a fast saturable absorber.

With the inclusion of some types of saturable absorbers, fibre lasers lose their all-fibre nature. However, this undesirable feature can be avoided if the fast saturable absorber function is implemented in optical fibre using a non-linear fibre-loop mirror. Actually, the power-dependent transmission achieved with this technique can shorten an optical pulse just as a saturable absorber material does. The corresponding passive mode-locking mechanisms are called interferometric or additive-pulse mode locking.

22.6.5 Non-linear fibre lasers

Any of the non-linear fibre amplifiers discussed in Sections 22.3–22.5 can be used to realize fibre lasers by placing them inside a cavity designed to provide optical feedback. For example, one of the first works on stimulated Raman scattering also demonstrated a Raman oscillator using mirrors to provide feedback. After this pioneer work, a great interest was shown in the development of other configurations of Raman fibre oscillators. As a result of such efforts, low-threshold tunable Raman oscillators for both visible and near-IR regions were created in the second half of the 1970s.

At high pump powers, higher-order Stokes wavelengths are generated inside the fibre cavity. The use of an intracavity prism allows the spatial dispersion of these wavelengths. The *Raman fibre laser (*RFL) can operate at several wavelengths simultaneously if different mirrors are used for each Stokes wave. However, the use of bulk optics (prisms and mirrors) inside the laser cavity is not desirable, since it introduces additional losses as well as mechanical instabilities.

The integration of cavity mirrors within the fibre in order to achieve a compact structure became one main goal in the research about RFLs. In an early approach, an all-single-mode RFL using a ring optical cavity and a fibre coupler was proposed. Afterwards, the creation of FBGs represented an important step in the development of Raman lasers. High-power CW radiation can be efficiently converted to several higher-order Stokes beams by utilizing highly reflective FBGs as mirrors in RFLs. Using this approach, a cascaded Raman laser can be developed such that each intermediate Stokes radiation is efficiently converted to the next higher-order Stokes until the desired order is achieved. Thus, the use of FBGs allowed the transformation of a bulky set-up of the multi-resonant RFL into an elegant all-fibre structure.

A *Brillouin fibre laser* (BFL) can also be built by placing the fibre into the cavity of a resonator. A BFL with very narrow linewidth can be designed for virtually any wavelength, provided that the required pump laser is available, thus showing a high potential for applications in metrology and spectroscopy. In the Fabry–Perot geometry, both forward- and backward-propagating components are present, associated with the pump and Stokes waves. With sufficient pump intensities, higher-order Stokes wave generation is possible through a cascaded SBS process, whereas four-wave mixing between co-propagating pump and Stokes waves provides the generation of anti-Stokes waves.

BFLs using a ring cavity are often preferred in order to avoid the generation of higher-order Stokes waves through cascaded SBS. Using such a configuration, submilliwatt thresholds are easily achievable if the pump signal coincides with one of the resonator modes. Ring-cavity BFLs can be used as high-performance laser gyroscopes, the so-called *Brillouin fibre-optic gyroscopes* (FOGs). The advantage of the Brillouin FOG over the interferometric and resonant FOGs is its direct readout of rotation rate, measured as a beat frequency between counter-propagating lasing waves inside the fibre resonator.

Hybrid *Brillouin–erbium fibre lasers* (BEFLs) have been the object of much attention since the 1990s, considering their operation either at a single wavelength or at several wavelengths simultaneously, which can be tuned over a wide range. The BEFL presents two gain media, which contribute to different aspects of the overall operation. The linear gain of the EDF allows for large output power generation and contributes the majority of the output power. The SBS process provides additional gain at the characteristic Stokes-shifted frequency from the injected Brillouin pump, to determine the wavelength of operation.

A *fibre-optic parametric oscillator* (FOPO) can be realized by placing a fibre inside an optical cavity and pumping it with a single-pump beam. The presence of the cavity allows for the coherent buildup of the ASE, from which the signal and idler waves are created at frequencies determined by the phase-matching condition. Both the signal and idler waves are amplified through the FWM process and are subsequently emitted by the FOPO. Several FOPOs have been constructed using standard and highly non-linear dispersion-shifted optical fibres. In particular, the use of microstructured optical fibres (MOFs) allows the construction of very short devices and the possibility of extending the wavelengths of operation and efficiency of FOPOs

22.7 Further Reading

Agrawal, G. P., *Nonlinear Fiber Optics*, Academic Press, 5th Ed., 2012.
Agrawal, G. P., *Applications of Nonlinear Fiber Optics*, Academic Press, 3rd Ed., 2020.
Bjarklev, A., *Optical Fiber Amplifiers: Design and System Applications*, Artech House, 1993.
Digonnet, M. F. Ed., *Rare-Earth Doped Fiber Lasers and Amplifiers*, Marcel Dekker, 1993.
Desurvire, E. *Erbium-Doped Fiber Amplifiers*, Wiley, 1994.
Ferreira, M. F., *Nonlinear Effects in Optical Fibers*, John Wiley & Sons, 2011.
Ferreira, M. F., *Optical Signal Processing in Highly Nonlinear Fibers*, CRC Press, 2020.
Ferreira, M. F., and M. Paul (Eds), *Optical Fiber Sources and Amplifiers*, Special Issue of 'Fiber and Integrated Optics', Taylor & Francis, 2021.
Ferreira, M. F., and M. C, Paul, *Optical Fiber Lasers—Recent Advances and Applications*, Elsevier, 2026.
Hansryd, J., P. A. Andrekson, M. Westlund, J. Li, and P. O. Hedekvist, Fiber-Based Optical Parametric Amplifiers and Their Applications, *IEEE J. Sel. Top. Quantum Electron.* 8, no.3, 506–520, 2002.
Marhic, M. E., *Fiber Optical Parametric Amplifiers, Oscillators and Related Devices*, Cambridge University Press, 2008.
Suhara, T., and M. Fujimura, *Waveguide Nonlinear-Optic Devices*, Springer-Verlag, 2003.

22.8 Problems

22.1 The gain coefficient $g(v)$ of an amplifying medium has a 3-dB bandwidth of 1 THz. Neglecting the gain saturation effect, calculate the 3-dB bandwidth when the optical amplifier is operated to provide 20 dB signal gain.

22.2 A fibre amplifier can amplify a 1 μW signal to the 1 mW level. Assuming a saturation power of 20 mW, calculate the output power for an input signal of 2 mW.

22.3 Discuss the different pumping options for erbium-doping fibre lasers, considering their relative advantages.

22.4 Describe the main differences between Raman fibre amplifiers and erbium-doped fibre amplifiers.

22.5 Consider a Raman amplifier constituted by a 4-km-long fibre with an effective core area $A_{eff} = 30 \times 10^{-12}\,\text{m}^2$, a loss coefficient $\alpha = 0.2\,\text{dB.km}^{-1}$, and a Raman coefficient $g_R = 6.7 \times 10^{-14}\,\text{m.W}^{-1}$. The pump and signal wavelengths are $\lambda_p = 1.46\,\mu\text{m}$ and $\lambda_S = 1.55\,\mu\text{m}$, respectively. Assuming a pump power $P_p(0) = 800\,\text{mW}$, calculate the signal power at which the depletion parameter F given by Eq. (21.77) becomes unity. Calculate the amplifier gain in such a case.

22.6 Discuss the merits of using a bidirectional pumping scheme compared with unidirectional pumping ones in fibre Raman amplifiers.

22.7 Prove that Eqs. (22.8)–(22.11) indeed constitute an approximate solution of Eqs. (21.83) and (21.84). Verify the validity of the result given by Eq. (22.12) for the fibre Brillouin amplifier gain.

22.8 Describe the main differences between fibre Brillouin amplifiers and fibre Raman amplifiers.

22.9 Demonstrate the result given by Eq. (22.13) for the gain of a fibre-optic parametric amplifier.

22.10 Derive the result given by Eq. (22.26) for the transmittivity of a NOLM and explain how it can operate as a mirror.

22.11 Discuss two different design possibilities for the Fabry–Perot cavity of a fibre laser without using actual mirrors. Explain the operation of such fibre lasers.

All-Optical Signal Processing

23

One of the key issues in ultrafast optical networks is how we can achieve various kinds of signal processing functions such as amplification, reshaping and regeneration, clock extraction and retiming, wavelength conversion, switching, etc. One approach to realize these functions is electronic signal processing relying on optoelectronic/electro-optic conversion. The most flexible functions are available in the electronic domain; however, this approach presents bandwidth limitations, prohibitive cost, and manufacturing complexity of high-speed optoelectronics, which become more important as data modulation rates increase.

The all-optical non-linear approach conveniently avoids the need for optical-to-electronic conversions and, therefore, circumvents the above limitations. In this case, ultrafast non-linear optical effects are applied to signal processing, and the response speed can go beyond 1 THz. Moreover, since optical techniques do not need to operate on every individual bit, as electronic transistors do, a single photonic element has the ability to transparently process a data channel regardless of its data rate and modulation format, also allowing for simultaneous operation on multiple data channels.

Various optical materials and devices have been utilized for optical signal processing, including optical fibres, semiconductor optical amplifiers (SOAs), silicon waveguides, highly non-linear waveguides, periodically poled lithium niobate (PPLN) waveguides, and photonic crystals. Regarding the first two options, while the speed of SOAs is ultimately limited by carrier recovery effects, optical fibre exploiting the third-order $\chi^{(3)}$ optical non-linearity is nearly instantaneous. Moreover, fibre-based solutions have the advantage of being directly integrated with existing fibre networks and utilizing cheap fibre components for their implementation.

Considering the device dimensions, while SOA and PPLN have typical lengths on the order of centimetres, fibre devices based on common dispersion-shifted fibre (DSF) usually require a length of several kilometres due to their relatively small non-linear coefficient γ of $\sim 1-2\,\mathrm{W}^{-1}$/km at around 1550 nm. However, using highly non-linear silica fibres (HNLFs), to be discussed in Chapter 25, the required fibre length for several non-linear processing applications can be impressively reduced to the order of centimetres.

This Chapter is intended to provide an overview of some all-optical signal processing functions that can be performed using non-linear effects in optical fibres, namely optical switching, wavelength conversion, and optical regeneration. Pulse amplification has already been discussed in Chapter 22, whereas pulse compression, supercontinuum generation, and spectrum slicing will be discussed in Chapters 24 and 25.

23.1 Optical Switching

An optical switch with fully transparent features in both the time and wavelength domains is a key device for providing several functions required in optical signal processing. A practical optical switch should have a broad bandwidth over the entire transmission band and should be capable of ultrahigh-speed operation at a data rate of 100 Gb/s or higher. Some of the

Lasers, Optics, and Photonics. Mário F. S. Ferreira, Oxford University Press. © Mário F. S. Ferreira (2026).
DOI: 10.1093/oso/9780198952770.003.0023

first demonstrations of all-optical switching made use of in-line SOAs. However, since the performance of an SOA depends on the carrier dynamics, its operation is limited to some tens of gigahertz. Ultrafast switching operations with speeds beyond the limits of electrical devices can only be achieved through an all-optical approach. The third-order non-linearity in optical fibres can provide such features. Optical switching can be achieved based on self-phase modulation (SPM), cross-phase modulation (XPM), and four-wave mixing (FWM) effects described in Chapter 21.

23.1.1 SPM-based optical switching

SPM-induced optical switching can be realized using a Mach–Zehnder interferometer (MZI), as represented in Fig. 23.1. The first coupler splits the input signal into two parts, which are recombined in the second coupler. In general, the two arms of the interferometer can have different lengths and propagation constants. The two couplers can also have different power-splitting fractions, f_1 and f_2. Considering a continuous-wave (CW) input beam with power $P_0 = |A_0|^2$ incident at one input port, the amplitudes of the optical fields at the two output ports of the first coupler are given by

$$A_1 = \sqrt{f_1}A_0,\ A_2 = i\sqrt{1-f_1}A_0 \tag{23.1}$$

During propagation along the interferometer arms, both fields acquire both linear and SPM-induced phase shifts. As a result, the two fields reaching the second coupler take the following form:

$$A_1' = A_1 \exp\left(i\beta_1 L_1 + if_1\gamma P_0 L_1\right) \tag{23.2}$$

$$A_2' = A_2 \exp\left(i\beta_2 L_2 + i(1-f_1)\gamma P_0 L_2\right) \tag{23.3}$$

where L_1 and L_2 are the lengths, and β_1 and β_2 are the propagation constants for the two arms of the interferometer.

The optical fields at the output ports of the MZI can be obtained using the transfer matrix of the second fibre coupler and are given by

$$\begin{pmatrix} A_3 \\ A_4 \end{pmatrix} = \begin{pmatrix} \sqrt{f_2} & i\sqrt{1-f_2} \\ i\sqrt{1-f_2} & \sqrt{f_2} \end{pmatrix} \begin{pmatrix} A_1' \\ A_2' \end{pmatrix} \tag{23.4}$$

Figure 23.1 *Schematic illustration of a Mach–Zehnder interferometer.*

Using Eqs. (23.1)–(23.4), the transmittivity $T_3 = |A_3|^2 / |A_0|^2$ through the bar port of the MZI is given by

$$T_3 = f_1 f_2 + (1 - f_1)(1 - f_2) - 2[f_1 f_2 (1 - f_1)(1 - f_2)]^{1/2} \cos(\Delta\phi_L + \Delta\phi_{NL}) \tag{23.5}$$

where

$$\Delta\phi_L = \beta_1 L_1 - \beta_2 L_2 \tag{23.6}$$

and

$$\Delta\phi_{NL} = \gamma P_0 [f_1 L_1 - (1 - f_1) L_2] \tag{23.7}$$

are the linear and non-linear parts of the relative phase shift, respectively.

The result in Eq. (23.5) becomes particularly simple if the MZI is made using two 3-dB couplers, so that $f_1 = f_2 = 1/2$. In this case, the transmittivity is given by

$$T = \sin^2 [(\Delta\phi_L + \Delta\phi_{NL})/2] \tag{23.8}$$

As long as the two arms of the interferometer have different lengths, the SPM-induced phase shift allows switching the output from one of the ports from low to high (or vice versa) by changing the input peak power of the incident signal.

Another scheme to realize SPM-based optical switching makes use of the Sagnac interferometer, as represented in Fig. 22.17. It is made by connecting the two outputs of a fibre coupler through a piece of long fibre. The input beam is divided in the fibre coupler into two beams counter-propagating in the fibre loop. After one round trip, both beams arrive at the same time at the coupler, where they interfere coherently, according to their relative phase difference.

The transmittivity T of the Sagnac loop is given by Eq. (22.26). From such a result, it can be seen that $T = 0$ for a 3-dB coupler (the loop acts as a perfect mirror). However, if the coupler power-splitting fraction $f \neq 0.5$ the fibre loop exhibits different behaviour at low and high powers and can act as an optical switch. At low input powers, one has $T \approx 1 - 4f(1 - f)$ and little light is transmitted if $f \approx 0.5$. At high input powers, one can have $T = 1$ if the condition

$$\left| f - \frac{1}{2} \right| \gamma P_0 L = (2N - 1)\frac{\pi}{2}, \ (N \text{integer}) \tag{23.9}$$

is satisfied. For $N = 1$, a fibre loop 50 m long will need a switching power of 63 W when $f = 0.45$ and $\gamma = 10\,\text{W}^{-1}/\text{km}$.

The switching power of a Sagnac interferometer can be reduced to levels below 1 mW if a fibre amplifier is included within the loop. This device is referred to as the *non-linear amplifying loop mirror* (NALM), and it can operate as a switch even if a 3-dB coupler ($f = 0.5$) is used. In the NALM, the asymmetry is provided by the fibre amplifier, which is placed closer to one of the coupler arms than the other. In this case, light travelling clockwise around the loop has a different intensity from light travelling anticlockwise, and hence experiences different amounts of SPM in the loop due to fibre non-linearity. If the clockwise wave is amplified by a factor G, the transmittivity becomes

$$T = 1 - 4f(1 - f)\cos^2\left[\left(\frac{Gf + f - 1}{2}\right)\gamma P_0 L\right] \tag{23.10}$$

In this case, one can have $T = 1$ if the condition

$$\left|\frac{Gf+f-1}{2}\right|\gamma P_0 L = (2N-1)\frac{\pi}{2} \tag{23.11}$$

is satisfied. Assuming $f = 0.5$ and $N = 1$, Eq. (23.11) provides the following result for the switching power:

$$P_0 = \frac{2\pi}{(G-1)\gamma L} \tag{23.12}$$

It can be seen that the switching power is reduced by a factor up to 1000 for an amplification factor $G \approx 30$ dB.

Using short pulses in Sagnac interferometers, the power dependence of loop transmittivity, as given by Eq. (22.26), can lead to significant distortion and pulse narrowing. This happens because only the central part of the pulse is sufficiently intense to undergo switching. The same feature can be used with advantage for *pulse shaping* and *pulse cleanup.* For example, a low-intensity pedestal accompanying a short optical pulse can be removed by passing it through a Sagnac interferometer.

Non-linear switching without the deformation effect can be achieved by using optical solitons, since they have a uniform non-linear phase across the entire pulse. Any low-energy dispersive radiation accompanying a train of optical solitons is reflected back, while the solitons are transmitted by the loop. Optical solitons will be discussed in Chapter 24.

23.1.2 XPM-based optical switching

As discussed in Section 21.4, the effect of cross-phase modulation (XPM) produces an alteration in the phase of one pulse due to the intensity of another pulse propagating at the same time in the fibre. This effect can be used in a Sagnac interferometer, as shown in Fig. 22.17, to realize an optical switch. Let us consider a control pulse that is injected into the Sagnac loop such that it propagates in only one direction. Due to the XPM effect in the fibre, the signal pulse that propagates in that direction experiences a phase shift, while the counter-propagating pulse remains unaffected. From Eq. (21.51), the XPM-induced phase shift is given by $\Delta\phi_{NL} = 2\gamma P L_{eff}$, where P is the power of the control pulse. If this power is such that $\Delta\phi_{NL} = \pi$, the signal pulse will be transmitted instead of being reflected.

Sagnac interferometers can be used for demultiplexing optical time-domain multiplexed (OTDM) signals. In this case, the control signal consists of a train of optical pulses that is injected into the loop such that it propagates in a given direction. If the control (clock) signal is timed such that it overlaps with pulses belonging to a specific OTDM channel, the XPM-induced phase shift allows the transmission of this channel, whereas all the remaining channels are reflected. Different channels can be demultiplexed simultaneously by using several Sagnac loops.

The main limitation of a Sagnac interferometer used to provide XPM-induced switching and demultiplexing stems from the weak fibre non-linearity. Fibres with a length of several kilometres are necessary for a control pulse power in the mW range. In these circumstances, if the signal and control pulses have different wavelengths, the walk-off effects due to the group-velocity mismatch must be taken into consideration. This problem can be avoided by reducing the fibre length, which is possible if highly non-linear optical fibres are employed.

XPM-induced optical switching can also be realized using a Mach–Zehnder interferometer. If both branches have the same properties and, in the absence of any control pulse, the two partial pulses created in the first fibre coupler experience the same phase shift, they will interfere constructively in the second fibre coupler of the interferometer. However, if a control pulse at a different wavelength is injected into one branch of the interferometer, it will change the signal phase through XPM in that arm and switch it.

If an OTDM signal is introduced at the interferometer input and a control (clock) signal is injected into one of its branches, a specific OTDM channel can be demultiplexed. Demultiplexers based on XPM-induced phase shift have attracted special attention, since they present some advantages relative to SPM-based devices. In particular, multiple MZIs can be cascaded, as the remaining channels appear at the output end of the MZI, rather than being reflected.

Considering a typical single-mode fibre with a non-linear parameter $\gamma = 1.3\,\mathrm{W}^{-1}/\mathrm{km}$ and an effective length $L_{eff} = 1$ km, a phase shift $\Delta\phi_{NL} = \pi$ requires an input power of 1.2 W for the control pulse. Reducing the effective core area A_{eff} increases the non-linear parameter γ and, consequently, reduces the power required for XPM-induced switching. For example, using fibres with $A_{eff} = 2\ \mu\mathrm{m}^2$ in each arm of a MZI, an XPM-induced phase shift of 10° is achieved with a pump power of only 15 mW.

XPM-induced optical switching can also be realized with a Kerr shutter, consisting of an optical fibre and a polarizer. Both the signal and the control pulses are linearly polarized, making an angle of 45° between their polarization directions. The polarizer at the fibre output is adjusted perpendicular to the polarization of the input signal. As a consequence, no signal light can pass through the polarizer in the absence of a control pulse.

The Kerr shutter makes use of the XPM-induced non-linear birefringence. In the presence of the control pulse, a slower axis is induced parallel to its polarization direction, which determines a rotation of the signal state of polarization. A fraction of the signal pulse can now pass through the polarizer, depending on the intensity of the control pulse.

Since the signal and control pulses have different wavelengths, the performance of a Kerr shutter used for optical switching depends on the group-velocity mismatch between them. This mismatch can be reduced by using highly non-linear fibres, since much shorter lengths are then needed.

23.1.3 FWM-based optical switching

FWM can also be used to switch a given sequence of signal pulses propagating along an optical fibre. To realize such an operation, a control signal consisting of a train of optical pulses with a frequency (ω_c) different from that of the signal pulses (ω_s) is also injected into the fibre. When a signal and control pulse are present together in the fibre, a switched signal appears at the idler frequency, ω_i, given by

$$\omega_i = 2\omega_c - \omega_i \tag{23.13}$$

Using a filter at the fibre output, the switched signal can be isolated from the control and the original signal sequence.

Optical bandwidths of the order of 200 nm can be achieved with a FWM-based fibre switch, in spite of the limitations imposed by the phase-matching condition. Such a device can also be used as a demultiplexer. In this case, the pump for the FWM process is provided by the clock signal. A pulse at the idler frequency appears only when a clock pulse overlaps with a signal pulse of the channel to be demultiplexed. In this way, a replica of the demultiplexed channel is obtained at the idler frequency.

A main problem associated with conventional FWM-based switching is related to the different wavelengths of the original and of the switched signal sequences. A technique to solve this problem consists in launching the optical signal wave E_S and the control (pump) wave E_P into a non-linear fibre, whose output is connected to a polarizer. Using a polarization controller at the fibre input, the state of polarization (SOP) of E_S is adjusted such that no light passes through the polarizer in the absence of pump light. The SOP of E_P is also adjusted to linear polarization and aligned at ~45° with respect to the polarizer's main axis.

The configuration of this FWM-based fibre switch is similar to that of the conventional optical Kerr shutter, which is based on the XPM effect. However, in contrast to the Kerr shutter, the actual switch has the advantage of using optical parametric amplification provided by FWM. To achieve the necessary phase matching, the pump wave is tuned to the zero-dispersion wavelength, λ_{ZD}, of the fibre. The efficiency of the FWM process is maximum when the two waves have the same SOP, whereas it is negligible when the SOPs of the two waves are orthogonal to each other. Therefore, decomposing E_S into two linearly polarized components, parallel $(E_{S\|})$ and orthogonal $(E_{S\perp})$ with respect to E_P (see Fig. 23.2), only $E_{S\|}$ will be amplified by FWM. The output power of the OPA fibre switch is

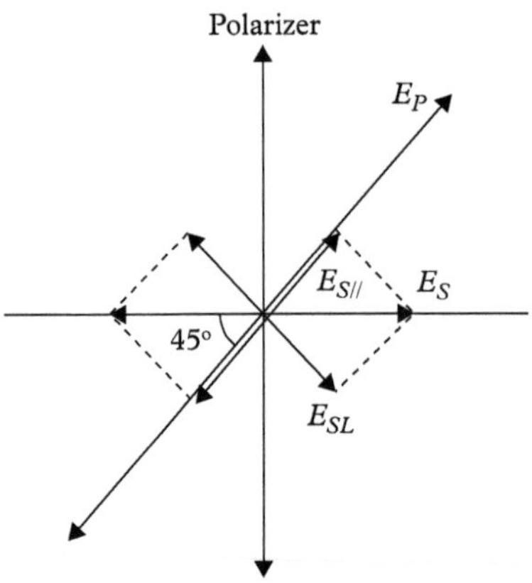

Figure 23.2 *Polarization setup of a FOPA fibre switch.*

$$P_{S-out} = \frac{1}{2}\Delta P_{S\|} \tag{23.14}$$

where the factor 1/2 is due to the polarizer and $\Delta P_{S\|}$ is the power increase of $E_{S\|}$ due to parametric amplification. For high pump powers and linear phase matching ($\Delta k \approx 0$), the switching gain is approximately given by

$$G_S \equiv \frac{P_{S-out}}{P_{S-in}} = \frac{1}{4}\exp(-\alpha L)\,\phi_{NL}^2 \tag{23.15}$$

where P_{S-in} is the signal input power, L is the fibre length, α is the fibre loss coefficient, and $\phi_{NL} = \gamma P_P L_{eff}$ is the non-linear phase shift, L_{eff} being the effective interaction length, given by Eq. (21.6).

Besides the fact that there is no frequency shift of the signal, the fibre-optic parametric amplifier (FOPA) switch described earlier also provides an extinction ratio of more than 30 dB because the unswitched component of E_S is eliminated by the polarizer. Such an extinction ratio could be improved by increasing the switching gain. Considering the ultrafast response of FWM in the fibre, such a FOPA switch has the potential for application in signal processing at data rates of 1 Tb/s and higher.

23.2 Wavelength Conversion

The possibility of generating signals at a desired frequency by all-optical wavelength conversion is extremely attractive for many different applications, including telecommunications, sensing, medicine, and defence. In particular, wavelength-conversion technologies are essential in future optical communication networks for improving the effective use of fibre transmission bandwidths and for solving the problem of network congestion caused by wavelength contention. The capability of converting cheap and reliable optical sources into very different frequency bands compared with the original is also very important.

Wavelength conversion has been performed using optical/electrical/optical (O/E/O) converters, with which the input optical signal is converted to an electrical signal by using photo-detectors. Electrical signal processing is performed and, finally, the electrical signal is used to modulate an optical source delivering a different wavelength. However, besides increasing the cost and complexity of the network, these devices achieve limited transparency and typically introduce latency into a system. To improve performance, all-optical wavelength converters (WCs) have been extensively investigated using non-linear optical devices.

All-optical wavelength conversion can be realized using both SOA-based converters and fibre-based converters. WCs based on SOAs exhibit many attractive features but are generally limited to signal rates of ~40 Gb/s owing to carrier recombination lifetimes or intraband dynamics. By contrast, converters based on non-linear fibre effects can support almost unlimited bit rates and are transparent to the signal modulation format due to the femtosecond response time of the fibre non-linearity.

23.2.1 FWM-based wavelength converters

Fibre-based FWM has been shown to be very efficient in performing wavelength conversion. Besides being transparent to both data rate and modulation format, FWM can work in any spectral region as long as phase matching is satisfied. In a single-mode fibre, phase matching can be achieved either over a relatively broad spectral region around the pump wavelengths or at narrow bands that can be far away from the pump wavelengths.

As discussed in Section 21.5, the FWM process generates an idler wave that is a replica of the signal wave, but at a different frequency. This effect can be used to realize wavelength conversion. Considering the case of degenerate FWM, if the old frequency corresponds to the signal, the new frequency provided by the idler wave is given by:

$$f_i = 2f_p - f_s \tag{23.16}$$

In order to obtain the desired wavelength, we need to launch both the signal and a CW pump with the appropriate wavelength inside the fibre. An optical filter placed at the output of the fibre blocks the pump and signal waves while allowing the passage of the idler wave. Fig. 23.3 illustrates this device.

Assuming that the intensity of the pump wave is much higher than that of the signal and idler waves, Eqs. (21.61)–(21.64), with $U_1 = U_2 = U_p$, $U_3 = U_s$, and $U_4 = U_i$, give

$$\frac{\partial U_p}{\partial z} = i\gamma |U_p|^2 U_p \tag{23.17}$$

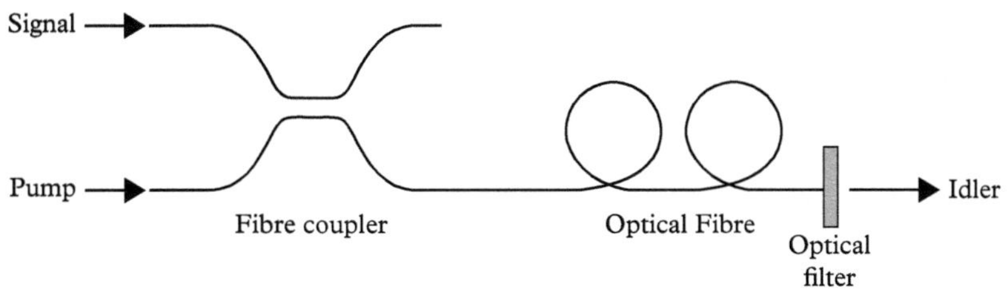

Figure 23.3 *Schematic of a FWM-based optical fibre wavelength converter.*

$$\frac{\partial U_s}{\partial z} = i2\gamma\left[\left|U_p\right|^2 U_s + U_p^2 U_i^* e^{-i\Delta kz}\right] \tag{23.18}$$

$$\frac{\partial U_i}{\partial z} = i2\gamma\left[\left|U_p\right|^2 U_i + U_p^2 U_s^* e^{-i\Delta kz}\right] \tag{23.19}$$

where γ is the non-linear parameter and $\Delta k = \beta_s + \beta_i - 2\beta_p$ is the phase mismatch of the wave vectors. As seen in Chapter 21, the effective phase mismatch with the induced non-linearity is given by

$$\kappa = \Delta k + 2\gamma P_p \tag{23.20}$$

where P_p is the pump power. Using the results obtained in Section 22.5, the efficiency of the FWM converter can be written as

$$G_i = \frac{P_i(L)}{P_s(0)} = \left(\frac{\gamma P_p}{g}\right)^2 \sinh^2(gL) \tag{23.21}$$

where L is the fibre interaction length, and the parametric gain coefficient g is given by

$$g^2 = \left[(\gamma P_p)^2 - (\kappa/2)^2\right] = -\Delta k\left(\frac{\Delta k}{4} + \gamma P_p\right) \tag{23.22}$$

The parameter g represents real gain over a conversion bandwidth corresponding to $-4\gamma P_p < \Delta k < 0$. The maximum gain for the parametric process is $g_{\max} = \gamma P_p$ and occurs if $\kappa = 0$ is satisfied. The conversion efficiency G_i can be much greater than 1 in such a case. Thus, the FWM converter provides the amplification of a bit stream while switching its wavelength, which is a very useful feature.

As seen in Section 21.5, the linear phase mismatch can be written as $\Delta k = \beta_2 \Omega_s^2$, where $\Omega_s = \omega_p - \omega_s$. Eq. (23.20) shows that, for a given value of Ω_s, the phase mismatch is zero and the conversion efficiency becomes maximum when the pump wavelength is in the anomalous dispersion regime such that $\beta_2 = -2\gamma P_p / \Omega_s^2$. However, the conversion efficiency decreases if the signal wavelength deviates from the specific value of Ω_s. This determines the bandwidth over which wavelength conversion can be realized, which decreases for longer fibre lengths.

Fibres with a high non-linear parameter γ will be advantageous for efficient wavelength conversion. On the other hand, in order to fulfil the phase-matching condition, fibres must be operated at their dispersion minimum. Fibres with a high value of γ and exhibiting the ZDW around 1.5 μm are called highly non-linear dispersion-shifted fibres (HNL-DSFs).

The derivation of Eq. (23.21) ignores pump depletion, fibre loss, competing non-linear processes, and walk-off between the pump and the signal pulses. Since these effects would lower the parametric gain by lowering the pump power or reducing the interaction length, the conversion efficiency given by Eq. (23.21) effectively corresponds to an optimum value.

When the GVD $|\beta_2|$ is small, we must take into account the fourth-order term in the expansion of Δk, which becomes

$$\Delta k = \beta_2 \Omega_s^2 + \frac{\beta_4}{12}\Omega_s^4 \tag{23.23}$$

Eq. (23.23) can be modified and written in terms of the wavelengths involved as

$$\Delta k = \left[2\pi c\left(\frac{1}{\lambda_p} - \frac{1}{\lambda_s}\right)\right]^2 \left[\beta_2 + \frac{1}{3}\beta_4 \pi^2 c^2 \left(\frac{1}{\lambda_p} - \frac{1}{\lambda_s}\right)^2\right] \tag{23.24}$$

For low pump powers, one can neglect the non-linear term in Eq. (23.20), and the phase-matching condition is given simply by $\Delta k = 0$. In order to realize this condition, Eq. (23.24) shows that the relation

$$\frac{1}{\lambda_p} - \frac{1}{\lambda_s} = \pm \frac{1}{\pi c}\sqrt{\frac{-3\beta_2}{\beta_4}} \tag{23.25}$$

should be satisfied. As seen from Eq. (23.23), when $\beta_2 = 0$, the conversion bandwidth is proportional to $|\beta_4|^{-4}$. Hence, depending on the value of β_4, FWM-based wavelength conversion can be classified into three types. The first one corresponds to conventional conversion and uses fibres with a moderate value $|\beta_4| \approx 1 \times 10^{-55} \mathrm{s}^4/\mathrm{m}$. The second type corresponds to large values β_4 and provides narrowband conversion. The third one is realized using fibres with a small absolute value of β_4, which provides broadband conversion.

In addition to broadband wavelength conversion, there is also a growing interest in developing wavelength-selective devices for multiple signals, such as reconfigurable optical add-drop multiplexers. In order to realize selective wavelength conversion, Δk should be small in a narrow bandwidth around λ_s, as determined by Eq. (23.25). Therefore, besides β_2/β_4 being negative, one finds from Eq. (23.23) that both $|\beta_2|$ and $|\beta_4|$ should be large. The desired value of β_2/β_4 and tunable wavelength conversion can be realized by tuning the pump wavelength near the ZDW.

Broadband FWM-based WCs, using phase matching provided by positive fourth-order dispersion, have enabled distant translation of CW and/or amplitude-modulated signals from the near-infrared (NIR) to the short-wave infrared (SWIR) regions. Actually, the SWIR and the mid-infrared (MIR) wavelength ranges are deserving of great interest due to the large number of potential applications that can be developed in these spectral regions, such as free-space communications, biological and chemical sensing, and spectroscopy. Moreover, 2 μm laser sources can be used for pumping non-linear media to achieve all-optical ultrafast processing and generate light at targeted wavelengths deeper into the MIR through FWM.

Since silica glass presents too much losses beyond 2 μm, novel highly non-linear materials with high transparency over a broad wavelength range have to be considered for operation in the SWIR and MIR wavelength ranges. An approach relies on using microstructured optical fibres (MOFs) made of chalcogenide glasses. These glasses have a third-order non-linearity up to 1000 times that of silica, as well as a broad transparency window up to 10 μm or 15 μm in the MIR, depending on the glass composition. In addition, as seen in Chapter 16, MOFs offer great design flexibility. By controlling the structural parameters of a MOF, one can adjust its dispersion characteristics, which become especially important in the case of the FWM process.

SBS has a threshold of only some mW for long fibres, and this constitutes one of the main effects limiting the efficiency of FWM-based WCs. Since FOPAs need pump powers around 1 W, some suitable technique to increase the SBS threshold becomes necessary. One commonly used technique for this purpose consists in modulating the phase of the pump wave in order to increase its spectrum. However, this technique also leads to the broadening of the idler spectrum, which is clearly undesirable for WCs.

The use of two pumps phase-modulated 180° out of phase can cancel out the broadening of the converted signal, while providing nearly uniform gain over a wide bandwidth and polarization-independent operation of the converter. This can be understood by noting that the complex amplitude of the idler wave has the form $A_i \propto A_{p1}A_{p2}A_s^*$, where A_{p1} and A_{p2} are the pump amplitudes. By modulating the pump waves in such a way that their phases remain equal but opposite in sign, the product $A_{p1}A_{p2}$ will not experience any modulation and the idler spectrum will not be affected.

23.2.2 XPM-based wavelength conversion

One main limitation of a FWM-based converter comes from the fact that it is not possible to convert an unknown wavelength to a predetermined wavelength. This limitation is avoided by using a XPM-based converter. A NOLM acting as a Sagnac interferometer can be used for this purpose. The data channel whose wavelength λ_2 needs to be converted is launched together with a CW whose wavelength λ_1 is chosen to coincide with the desired new wavelength. It affects the CW seed in only one direction, imposing an XPM-induced phase shift only in the time slots associated with the 1 bits. Such a phase shift is converted into amplitude modulation by the interferometer, resulting in a bit pattern at the new wavelength λ_1 which is a replica of the original bit stream.

Using an optical interferometer is not essential to realize a XPM-based wavelength converter. A simpler approach consists in using a DSF, in which a CW seed is launched along with the control signal. The XPM will act to broaden the spectrum of the CW lightwave when a mark is copropagated with it. In this way, red- and blue-shifted sidebands are generated on the CW lightwave. By filtering out one of these sidebands, the wavelength-converted signals can be obtained. The pulse width of the converted pulse based on XPM in the DSF is determined by the power of the control pulse, dispersion effects, and walk-off time between the control pulse and the CW lightwave. Broadband pulsewidth-maintained wavelength conversion can be realized by the use of a short HNL-DSF because dispersion and walk-off effects can be reduced.

The use of Raman gain can significantly enhance the performance of an XPM-based fibre wavelength converter (Wang et al., 2005). Fig. 23.4 shows a schematic of such a Raman-gain-enhanced wavelength converter (RE-WC). It is basically constituted by a non-linear fibre, a Raman pump laser, and a band-pass filter (BPF). The data pump light at wavelength λ_2 is combined with the CW probe light at wavelength λ_1 and injected into the fibre. The Raman pump simultaneously amplifies the pump light and the probe light and significantly enhances the amount of XPM. At the output of the fibre, a BPF is used to extract one of the sidebands of the CW probe, generating an amplitude-modulated signal from the phase modulation of the CW probe. This amplitude-modulated signal thus forms a wavelength-converted replica of the original data signal.

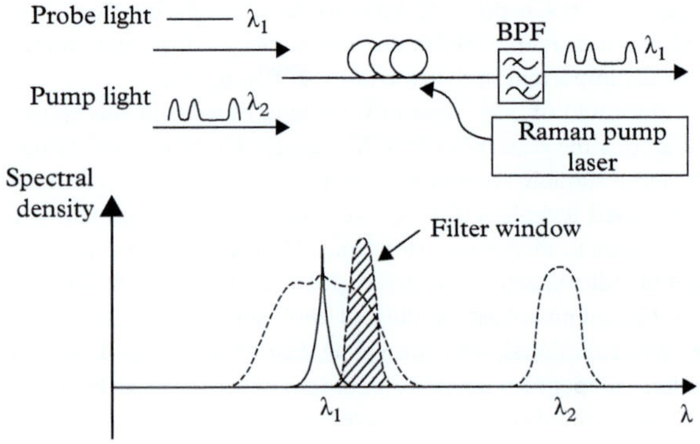

Figure 23.4 *Schematic of an XPM-based Raman-gain-enhanced wavelength converter (after Wang (2005)).*

In the absence of Raman gain, and considering the pulsed pump light and the CW probe light propagating in the fibre with the same polarization, the following equations governing their behaviours can be obtained from Eq. (21.27) by neglecting both the GVD and the FWM effects:

$$\frac{\partial U_s}{\partial z} + \frac{1}{v_{gs}} \frac{\partial U_s}{\partial t} = -\frac{\alpha}{2} U_s + i\gamma \left(|U_s|^2 + 2|U_p|^2 \right) U_s \tag{23.26}$$

$$\frac{\partial U_p}{\partial z} + \frac{1}{v_{gp}} \frac{\partial U_p}{\partial t} = -\frac{\alpha}{2} U_p + i\gamma \left(|U_p|^2 + 2|U_s|^2 \right) U_p \tag{23.27}$$

where $U_j, j = S, P$ is the slowly varying complex amplitude, v_{gj} is the group velocity, α is the attenuation coefficient, and γ is the fibre non-linear parameter. The solution of Eq. (23.26) at the output of the fibre with length L is given by

$$U_s(L, t) = U_s\left(0, t - L/v_{gs}\right) \exp(-\alpha L/2) \exp\{i\Delta\phi(L, t)\} \tag{23.28}$$

where

$$\Delta\phi = \gamma L_{eff}\left(P_s + 2P_p\right) \tag{23.29}$$

is the non-linear phase shift, L_{eff} is the effective interaction length, and $P_j = |U_j|^2$ is the power of the signal ($j = s$) or pump ($j = p$) waves. The first term in Eq. (2.29) is caused by SPM, whereas the second term is due to the XPM effect. This term determines the conversion efficiency of the XPM-based wavelength conversion. When Raman gain is present, the accumulated phase shift for a given signal power level is increased. This effect is due to the increased effective interaction length, L_{eff}, which becomes

$$L_{eff} = \frac{1 - \exp\{-(\alpha - g)L\}}{\alpha - g} \tag{23.30}$$

where g is the gain provided by the Raman amplification.

From Eqs. (23.29) and (23.30) it can be seen that, in order to achieve a given phase shift, the signal launch power can be reduced in the presence of Raman gain. On the other hand, for a given CW launch power, more power will be available in the wavelength-converted output signal if it is amplified while passing through the fibre. As a consequence, less amplified spontaneous emission noise will be generated when the converted signal is amplified at the output of the converter.

A NOLM acting as a Sagnac interferometer can also be used as an XPM wavelength converter, as illustrated schematically in Fig. 23.5. A CW beam with wavelength λ_1 which coincides with the desired wavelength of the converted signal is launched inside the fibre loop together with the data channel whose wavelength λ_2 needs to be changed. This data signal acts as a pump and imposes an XPM-induced phase shift on the CW beam only in the time slots associated with the 1 bits. The interferometer effect converts such a phase shift into amplitude modulation. The data channel is launched such that it affects the CW seed in only one direction, resulting in a differential phase shift that can be used to copy the bit pattern of the data channel onto the transmitted λ_1 signal.

A different scheme for wavelength conversion of complex (amplitude and phase) optical signals can be demonstrated using an all-optical implementation of the concept of time-domain holography. In such a scheme, wavelength conversion is considered as a

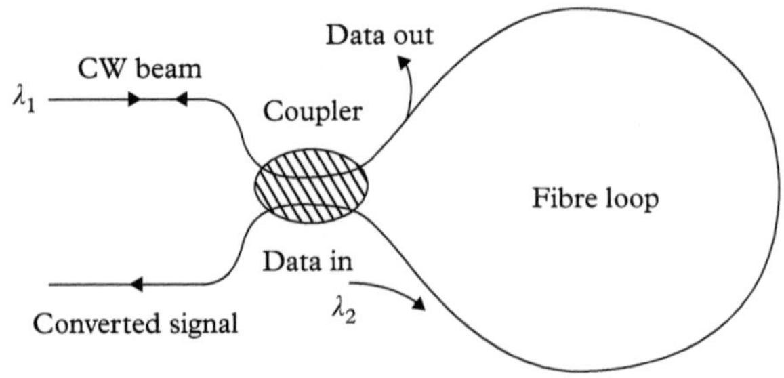

Figure 23.5 *A NOLM acting as a Sagnac interferometer for XPM-based wavelength conversion.*

modulation process where the complex envelope of the information signal modulates a different carrier signal. The temporal holograms are generated by phase-only modulation, namely XPM, in a HNLF. The scheme inherently provides a wavelength-converted temporally conjugated copy of the original signal in addition to the non-conjugated copy. Although such a configuration is similar to that of a FWM-based dual pump configuration, it avoids the need to satisfy the phase-matching condition and also relaxes significantly the power requirements.

A different type of XPM-based wavelength converter consists in the use of a CW pump that is sinusoidally modulated at a frequency f_m higher than the signal bit rate. The pump is amplified and interacts with the signal channel through XPM. As a consequence, the spectrum of the signal channel develops AM sidebands separated in frequency by f_m. The use of a suitable optical filter allows the selection of a specific sideband, which represents the wavelength-converted signal.

In a variant of this technique, the sinusoidally modulated pump can be obtained from beating two CW lasers with a small wavelength spacing. Then the data channel will be phase-modulated with the beat-note frequency, through which it will develop the sidebands. The optical frequency separation of the two pump waves determines the amount of optical frequency shift. Different polarization schemes can be implemented for XPM generation. XPM is driven most strongly when the two pump waves are in the parallel polarization state. On the other hand, the phase modulation index is maximal when the signal is polarized parallel to the pump waves and is minimal when the signal is polarized orthogonal to the pump waves. In the last case, the power of the generated signal is lower than that generated by the parallel polarization inputs by a factor of 4 (6 dB). Despite the decrease in conversion efficiency, an advantage of this alignment is to minimize additional FWM generation, which sometimes results in harmful noise to the converted signal.

23.3 Optical Regeneration

Optical signals propagating in fibre-optic transmission systems are affected by several effects, namely amplified spontaneous emission (ASE) from optical amplifiers, chromatic dispersion, polarization-mode dispersion, and non-linear phenomena. Considering the impairments imposed by these effects, the use of signal regenerators becomes necessary in order to extend the maximum transmission distance of the systems.

Signal regeneration is conventionally performed by optical-electrical-optical (OEO) regenerators that involve signal conversion between the optical and electrical domains and

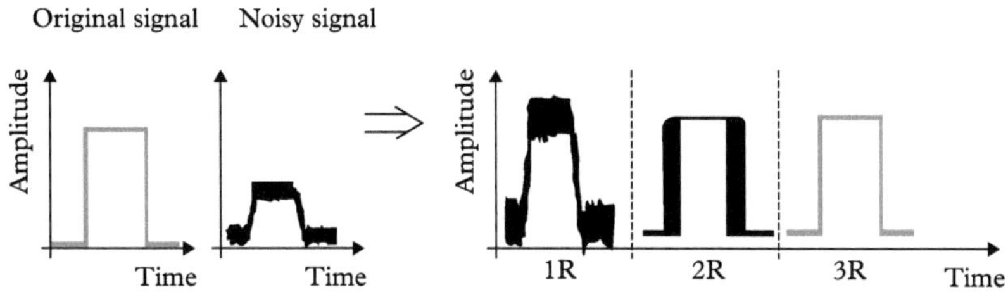

Figure 23.6 *Different levels of optical regeneration.*

noise removal from the electrical signal. Another type of regeneration that does not rely on OEO conversion is provided by all-optical regenerators. Compared with conventional techniques, all-optical signal regeneration offers much higher response speeds, lower power consumption, lower costs, and flexibility in handling signals in different modulation formats.

Conventionally, one uses the letter R together with a number to classify the type of regeneration according to its features and also its increasing complexity. Devices that perform only the reamplification process are called 1R regenerators, whereas those able to perform the reamplification and reshaping processes are called 2R regenerators. Finally, those devices that perform the reamplification, reshaping and retiming of the degraded signal are called 3R regenerators.

The right-hand side of Fig. 23.6 illustrates the different stages of the regeneration process operated on the degraded (noisy) signal pulse, which is represented on the left-hand side. Optical amplification can increase the amplitude of the signal and noise, as identified by 1R, or reamplification only. A further stage of improvement to restore the rectangular shape of the original signal can be provided by 2R devices. The dark areas in the pulse, shown by vertical columns, represent the timing jitter of the received signal, which can be corrected by using 3R devices.

The main characteristics of different types of fibre 1R regenerators (optical amplifiers) were already discussed in Chapter 22. In the following sections, we discuss the possibility of performing 2R and/or 3R regeneration using some of the non-linear effects in fibres described in Chapter 21.

23.3.1 2R Regenerators

All three major non-linear effects, SPM, XPM, and FWM, can be used to realize 2R regeneration. Among them, SPM is most often used. An advantage of the SPM-based 2R regenerators is that neither pump nor probe optical sources are needed inside the regenerator, which makes it particularly simple.

SPM-Based 2R Regenerators

2R optical regeneration using a HNLF with subsequent spectral filtering was first proposed by Mamyshev (Mamyshev, 1998). This scheme relies on SPM to broaden the spectrum of a return-to-zero (RZ) input signal and has been the subject of a lot of attention, since it presents a number of key features. Among them are the simplicity, a bandwidth which is limited only by the intrinsic material non-linear response, and the possibility to achieve a direct bit error rate (BER) improvement. Actually, an SPM-based regenerator has the important characteristic of discriminating pulses of different widths. This capacity

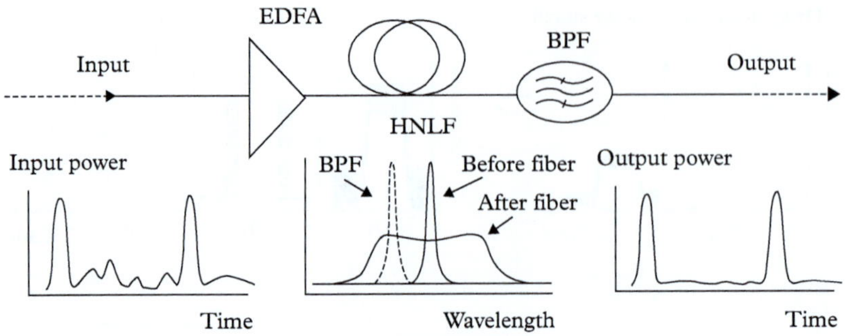

Figure 23.7 *An SPM-based 2R regenerator (top) and its action on a bit stream (bottom). HNLF stands for a highly non-linear fibre (after Mamyshev (1998)).*

is enabled by the slope dependence of the spectral broadening in SPM, which allows one to discriminate logical ones from logical zeros and thereby provides a BER improvement.

Fig. 23.7 illustrates the operation of a Mamyshev regenerator. A noisy input RZ signal is phase modulated by its own waveform (SPM) while it propagates through a non-linear fibre. The temporal phase variation induces an instantaneous frequency shift, and the signal spectrum is broadened. The intensity slope point of the original signal is spectrally shifted and can be selected using a BPF offset from the input centre wavelength. Furthermore, above a certain level, any variation in the signal intensity will translate into a change in the width of the SPM-induced broadened spectrum. Since the bandwidth of the optical filter is fixed, the intensity of the filtered output is insensitive to those variations in the signal power, which results in an effective suppression of the amplitude noise at the bit 'one' level. On the other hand, optical noise at the bit 'zero' level is low and does not induce any spectral shift, being filtered out by the BPF. This results in an improvement in the optical signal-to-noise ratio (SNR).

The SPM-based signal processing can be characterized by three parameters: the instantaneous frequency shift induced by phase modulation, f_{PM}, the BPF offset frequency relative to the original signal centre frequency, f_{BPF}, and the BPF bandwidth, Δf_{BPF}. The requirement for f_{PM} in order to guarantee proper operation is given by (Igarashi and Kikuchi, 2008):

$$f_{PM} > 2\Delta f_{in} \tag{23.31}$$

where Δf_{in} is the bandwidth of the incoming optical signal. The maximum value of the SPM-induced frequency shift depends on the pulse shape. Assuming that Gaussian pulses with a peak power P_0 are launched into an optical fibre of length L, Eq. (23.31) provides the following condition (Igarashi, 2008):

$$\gamma P_0 L > \frac{4\pi}{3.45} \tag{23.32}$$

where γ is the non-linear parameter of the fibre. Considering a 500-m-long silica-based HNLF, such that $\gamma = 20\,\mathrm{W}^{-1}/\mathrm{km}$, the required peak power is $P_0 \approx 0.4\,\mathrm{W}$.

2R regenerators based on SPM with subsequent spectral filtering have been demonstrated using different types of HNLFs. Since logical 'ones' (e.g. pulses) and logical 'zeros' (e.g. ASE noise) are processed with distinct transfer functions, such regenerators are able to directly improve the BER of a noisy signal.

SPM-based 2R regenerators can also be used when soliton-like pulses propagate in the anomalous dispersion regime of an optical fibre. In such a case, the input pulse is compressed or decompressed, depending on whether its amplitude is larger or smaller than that of the fundamental soliton in the fibre. As a consequence, the spectral width of the pulse is broadened or narrowed, respectively. By using a band-pass optical filter at the fibre output, larger (smaller) energy loss is induced in the pulse having larger (smaller) initial amplitude, which provides a mechanism for self-amplitude stabilization.

It can be seen from Eq. (22.26) that the transmittivity of a NOLM has a minimum value at vanishing input power, given by

$$T_{\min} = 1 - 4f(1 - f) \tag{23.33}$$

where f is the coupler power-splitting fraction. This shows that the NOLM can be used to suppress the zero-level noise, which becomes more effective for a coupling ratio near 0.5. On the other hand, the NOLM can also suppress the amplitude fluctuations of the input signal if its power is set to one of the values at which the output power is locally constant.

Besides the regeneration function described earlier, the SPM-based pulse reshaping in HNLFs can also be used with advantage in the generation of optical pulse trains. This application is particularly important for 160-Gb/s OTDM transmitters, where the generation of 10-GHz picosecond optical pulse trains is one of the main issues. High-quality pulse trains are required for such OTDM transmitters, namely a reduced pulse width, a low timing jitter, a long-term stability, as well as wavelength and repetition-rate tunability.

Wavelength-widely tunable (~400 nm) femtosecond sources have also been achieved using the Mamyshev regenerator idea. In this case, the SPM-dominated non-linearity is employed to dramatically broaden an input narrowband optical spectrum to a spectral width of 100s of nm. By using optical filters to select the leftmost/rightmost spectral lobes of the broadened spectrum, nearly transform-limited ~100-fs pulses with a widely tunable centre wavelength can be obtained.

FWM-Based 2R Regenerators

As seen in Section 21.5, when a signal at frequency f_s is launched into a FOPA together with a pump at frequency f_p, the FWM process amplifies the signal and generates an idler wave with frequency $f_i = 2f_p - f_s$. As the input signal power is increased, the direction of power flow from the pump to the signal can be reversed and higher-order FWM products are generated. These lead to gain saturation of the parametric amplifier, which can be used to reduce the fluctuations in the signal pulse peak power. An important feature of this type of FWM-based regenerator is that the optical carrier phase of the input signal is basically preserved, which indicates that it can be used for the regeneration of phase-shift keying (PSK) signals. A basic single-pump FOPA consists of a pump source, a highly non-linear fibre, and an optical BPF for extracting the output signal wave.

The idler wave and the higher-order FWM products can also be used to realize simultaneously the 2R regeneration and the wavelength conversion of an input signal. The output powers of the higher-order FWM components at $f_2 = 2f_s - f_p$ and $f_3 = 2f_s - f_i$ grow proportionally to the square and cube of the input signal power, respectively, in the small-signal regime. This means that the low-power noise is more effectively suppressed using these components. On the other hand, the saturation that these waves also experience at higher input powers provides the mechanism for pulse amplitude equalization and the realization of the 2R function. An issue in using higher-order FWM products is that the input signal phase is not preserved, which precludes the use of this regeneration scheme in the case of PSK signals.

23.3.2 3R regenerators

In spite of all the advantages offered by 2R regeneration, it is not able to retime the signal and therefore cannot reduce the timing jitter. Full regeneration (3R) is accomplished when the retiming function is added to the reamplification and reshaping of the optical signals. An SPM-based 3R regenerator can be realized by adding a synchronous amplitude modulator to the scheme shown in Fig. 23.7. An electrical clock signal recovered from the incoming bit stream itself is used to drive the modulator.

Full regeneration can also be achieved using XPM. The configuration of an XPM-based 3R regenerator has been proposed, which consists basically of an optical clock source, an erbium-doped fibre amplifier (EDFA), a HNLF, and a BPF, as illustrated in Fig. 23.8. An impaired signal with a centre wavelength λ_1 is strongly pumped by an EDFA. Such a signal is launched into a highly non-linear dispersion-shifted fibre (HNL-DSF) together with a clean clock pulse train at λ_2. A BPF with a centre wavelength λ_2 is placed behind the HNL-DSF. The clock is generated from a clock-recovery circuit and produces pulses of shorter duration than the pulses to be regenerated. The repetition rate of the clock can be chosen equal to the signal data rate when regeneration is desired. However, the repetition rate of the clock can be smaller than the signal data rate when regeneration and demultiplexing are desired. As the signal bits and clock pulses temporally overlap in the HNLF, the clock spectrum experiences a cross-phase modulation (XPM)-induced frequency shift proportional to the slope of the superimposed signal, which depends on the peak power, as represented in Fig. 23.8(b)–(d). The output power of the clock pulse from the BPF is then a non-linear function of the signal peak power, as shown in Fig. 23.8(e). The output filter

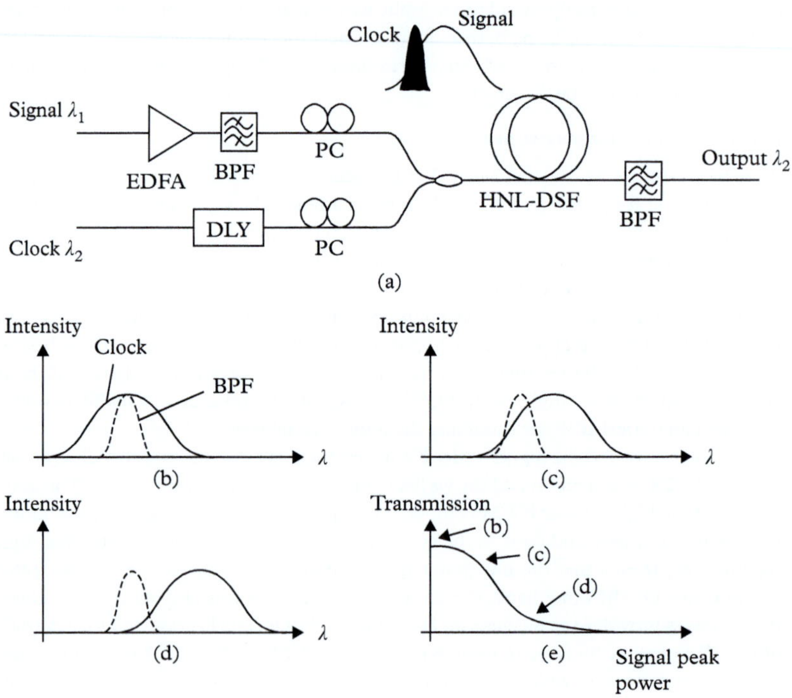

Figure 23.8 *Schematic and principle of the 3R regenerator based on XPM-induced spectral shift followed by filtering. The peak power of the signal pulse is increased from (b) to (d) (after Suzuki et al. (2005)).*

BPF filters out the spectral components that have been shifted by XPM and keeps the original frequency components at wavelength λ_2. Note that this regeneration scheme inverts the bits, i.e. logical ones are transformed into logical zeros and vice versa. A non-inverting mode can also be achieved by offsetting the centre frequency of the output BPF away from the original clock wavelength, in order to transmit only XPM-generated frequencies.

A common challenge of non-linear effects such as FWM and cross-phase modulation is their inherent sensitivity to the SOP of the input signal. Since the SOP of a data signal varies randomly with time in practical systems, it is important that the all-optical retiming function be polarization independent. Several techniques have been proposed to address this problem, namely depolarization of the clock signal, polarization diversity, fibre twisting for circularly polarizing fibre, and the use of fibre birefringence. However, these techniques often require additional complexity and/or careful control and stability of the clock signal SOP.

Polarization-independent all-optical retiming without additional complexity has been demonstrated using XPM in an HNLF and offset spectral filtering (Ito and Cartledge, 2008). This technique is based on the fact that there exist wavelength regions for which the XPM broadened spectrum is polarization independent. The regenerator used a self-pulsating DFB laser for clock recovery, XPM in an HNLF and offset spectral slicing for retiming, and SPM in an HNLF and offset spectral slicing for reshaping. In the retiming stage, an optical clock signal was recovered and used to induce XPM on the input data signal in an HNLF, followed by offset spectral filtering. A tunable optical delay was implemented to control the temporal alignment of the clock and data signals at the input of the HNLF. The retimed signal was obtained by filtering the broadened spectrum of the data signal with a BPF offset from the carrier wavelength of the data signal by $\Delta\lambda$. The subsequent reshaping stage used SPM in an HNLF and spectral filtering with an offset of $-\Delta\lambda$, thus preserving the signal wavelength.

23.3.3 All-optical regeneration of phase-encoded signals

Increasing needs for bandwidth and power-efficient data transmission in long-distance lightwave systems have recently motivated a change in modulation formats from on-off keying (OOK) to PSK and more advanced formats. As a consequence, several schemes for all-optical regeneration of PSK signals have been proposed and demonstrated.

In comparison with the OOK format, the differential binary PSK (DPSK) format offers better sensitivity and better tolerance to fibre non-linearity. In one class of DPSK regenerators, the phase information of the signal is converted into amplitude information through the use of a delay interferometer (DI). The phase noise of the signal is thus transferred to the amplitude of the demodulated OOK signal, whose noise is afterwards removed by an amplitude regenerator. The regenerated OOK signal is used as a control signal to modulate the phase of probe pulses, thus providing a regenerated PSK signal.

Phase-preserving amplitude-only regeneration has also been shown to be effective in suppressing the non-linear phase noise. This is induced by the in-phase component of the ASE relative to the signal (amplitude noise), which is converted to phase noise via the effect of SPM in the transmission fibre. The non-linear phase noise dominates over the linear phase noise in long-distance systems. It can be controlled by suppression of the amplitude noise in the system, which can be realized by inserting an amplitude limiter after the transmitter and at every amplifier span.

23.4 Further Reading

Agrawal, G. P., *Fiber Optic Communication Systems*, John Wiley & Sons, 4th Ed., 2010.

Ferreira, M. F., *Nonlinear Effects in Optical Fibers*, John Wiley & Sons, 2011.

Ferreira, M. F., *Optical Signal Processing in Highly Nonlinear Fibers*, CRC Press, 2020.

Igarashi, K., and K. Kikuchi, Coherent Optical Transmission Systems, *IEEE J. Sel. Topics Quantum Electron*, 14, no. 3, 551, 2008.

Ito, C., and J. C. Cartledge, Performance of DPSK and DQPSK Signals in the Presence of Nonlinear Phase Noise, *IEEE J. Sel. Top. Quantum Electron.* 14, no. 3, 616, 2008.

Mamyshev, P. V., Eur. Conf. Opt. Commun. (ECOC98), Madrid, Spain, 475, 1998.

Senior, J., *Optical Fiber Communications – Principles and Practice*, Pearson Education, 2009.

Suzuki, J., T. Tanemura, K. Taira et al. Polarization-Insensitive Fiber Optical Parametric Amplifier Using Orthogonally Pumped Configuration, *IEEE Photon. Technol. Lett.* **17**, no. 2, 423, 2005.

W. Wang, W., H. N. Poulsen, L. Rau, H.-F. Chou, J. E. Bowers, and D. J. Blumenthal, Experimental Demonstration of Optical Parametric Amplification for High-Speed Optical Transmission, *J. Lightwave Technol.* 23, no. 3, 1105, 2005.

23.5 Problems

23.1 Describe the operation of a SPM-based switch using the Mach–Zehnder interferometer.

23.2 Show that the transmittivity through the bar port of the Mach–Zehnder interferometer is given by Eq. (23.5).

23.3 Describe the operation of a SPM-based optical switch using the Sagnac interferometer.

23.4 Demonstrate the result given by Eq. (23.10) for the transmittivity of a NALM, which includes an amplifier providing a gain G.

23.5 Demonstrate Eq. (23.15) for the FWM-based switching gain.

23.6 Describe the operation of the XPM-based wavelength-conversion scheme illustrated in Fig. 23.4.

23.7 Discuss the role of fourth-order dispersion in FWM-based wavelength conversion.

23.8 Describe the three optical regeneration stages for a noisy optical signal and outline their applications in optical fibre transmission systems.

23.9 Describe the principle of operation of the SPM-based 2R Mamyshev regenerator and explain how it can provide a system BER improvement.

23.10 Explain the operation of the 3R regenerator illustrated in Fig. 23.8.

Optical Solitons

24

As seen in Chapter 21, if the operating wavelength is in the anomalous-dispersion regime, the group-velocity dispersion (GVD) and the self-phase modulation (SPM) effects can compensate each other. Under certain circumstances, such effects can exactly cancel each other, so that the temporal and spectral shape of the pulses are preserved even over long propagation distances. This kind of phenomenon was first observed in the context of water waves, but later also for light propagating in optical fibres.

Soliton pulses are technically very important, in particular for long-distance optical fibre communications and in mode-locked lasers. Solitons are also applied in various techniques for all-optical signal processing, namely for pulse compression using optical fibres.

24.1 Waves Called Solitons

The first report on the observation of a solitary wave was made by the Scottish engineer John Scott Russell (1808–1882) in August 1834, when he saw a water wave about 3 kilometres long propagating without change in shape on the Edinburgh–Glasgow Canal. He called it the 'great wave of translation' and then reported his observations at the British Association in his 1844 paper 'Report on Waves' (Russell, 1844).

These observations were followed by several laboratory experiments on the generation and propagation of such waves. Russell found that they are bell-shaped and have a velocity that depends on the finite depth of the channel and on the wave amplitude above the free surface of the water. Russell also observed that a given initial water elevation can give rise to one or more solitary waves.

Russell's predictions of solitary waves were later confirmed by other authors. On the other hand, Korteweg and de Vries derived in 1895 an equation that aimed to describe the propagation of surface waves in water of finite depth. Such an equation can be obtained considering the combined effect of both dispersion and non-linearity of the propagating medium and is given by

$$\frac{\partial A}{\partial \tau} + A\frac{\partial A}{\partial \xi} + \frac{\partial^3 A}{\partial \xi^3} = 0 \tag{24.1}$$

where A is the amplitude and (ξ, τ) is a coordinate system moving with the wave. Eq. (24.1) is known as the Korteweg–de Vries (KdV) equation, and it has a solitary wave solution given by

$$A(\xi, \tau) = 3w\mathrm{sech}^2\left[\frac{\sqrt{w}}{2}(\xi - w\tau)\right] \tag{24.2}$$

Actually, this solution represents the long-time evolution of a wave in which the non-linear effect is counterbalanced by linear dispersion and corresponds to the solitary wave observed experimentally by Russell. The wave amplitude is proportional to the velocity, w and its width is inversely proportional to the square root of the amplitude. Consequently, a taller

Lasers, Optics, and Photonics. Mário F. S. Ferreira, Oxford University Press. © Mário F. S. Ferreira (2026).
DOI: 10.1093/oso/9780198952770.003.0024

(and thinner) soliton, initially behind a lower and wider one, will travel faster and eventually overtake and pass through the lower one.

After the work of Korteweg–de Vries, the solitary wave remained for a long time mainly as a simple curiosity in the area of non-linear wave theory. The subject revived in 1955, when Fermi, Pasta, and Ulam (FPU) published a Laboratory Report on a numerical model of a discrete non-linear mass–spring system. The FPU model was described by a system of coupled non-linear ordinary differential equations that, under certain approximations, can be transformed into the KdV equation. Such work inspired Martin Kruskal and Norman Zabusky, who solved the KdV equation numerically. They observed the formation of solitary waves that undergo non-linear interaction, after which they emerge without any change of shape and amplitude, but with only a small change in their phases. Due to the particle-like behaviour of such solitary waves, Kruskal and Zabusky called them for the first time 'solitons' in a famous 1965 paper (Zabusky and Kruskal, 1965). Among other features, it was numerically observed that this kind of pulses remains intact even after mutual collisions.

24.2 The NLSE Fundamental Soliton

The non-linear Schrödinger equation (NLSE), given by Eq. (21.27), describes the propagation of pulses in optical fibres, taking into account both the GVD, represented by the parameter β_2, and the fibre non-linearity, represented by the parameter γ. Introducing a normalized amplitude Q given by

$$U(z,\tau) = \sqrt{P_0}\,Q(z,\tau) \tag{24.3}$$

where P_0 is the peak power of the incident pulse, Eq. (21.27) converts to

$$i\frac{\partial Q}{\partial z} - \frac{1}{2}\beta_2\frac{\partial^2 Q}{\partial \tau^2} + \gamma P_0 |Q|^2 Q = 0 \tag{24.4}$$

It can be shown that Eq. (24.4) has a pulse solution given by

$$Q(z,\tau) = \text{sech}\,[\alpha\tau]\exp\left\{i\frac{\gamma P_0}{2}z\right\} \tag{24.5}$$

where

$$\alpha = \sqrt{\gamma P_0 / |\beta_2|} \tag{24.6}$$

Eq. (24.5) corresponds to the so-called NLSE fundamental soliton solution, which propagates along the optical fibre with a constant profile. It arises as the result of a balance between the fibre anomalous GVD and the SPM effect, presenting some remarkable properties that make it a unique candidate for optical communications and signal processing functions.

The existence of solitons in optical fibres was suggested for the first time in 1973 by Hasegawa and Tappert (Hasegawa and Tappert, 1973), whereas their first experimental observation was reported only in 1980 by Mollenauer et al. (Mollenauer et al., 1980). The experimental results confirmed that the properties of a pulse propagating along an optical fibre are described, to a remarkable degree, by the non-linear Schrödinger equation.

Example 24.1

Assume a pulse solution of Eq. (24.4) in the form

$$Q(z,\tau) = S(\tau)\, e^{i\phi(z)} \tag{24.7}$$

Confirm the result given by Eq. (24.5) by obtaining expressions for the phase, $\phi(z)$, and for the amplitude, $S(\tau)$, assuming that it is a real function of τ.

Solution

Substituting Eq. (24.7) in Eq. (24.4) and rearranging, one obtains

$$\frac{d\phi}{dz} = \gamma P_0 S^2(\tau) - \frac{1}{2}\beta_2 \frac{d^2S/d\tau^2}{S(\tau)} \tag{24.8}$$

Since the left-hand side of Eq. (24.8) depends only on z and the right-hand side depends only on τ, one can set each side equal to a constant, C_1. The equation resulting from the left-hand side has the solution

$$\phi(z) = C_1 z + \phi_0 \tag{24.9}$$

where ϕ_0 is a constant of integration, which is assumed to be equal to zero. Concerning the right-hand side of Eq. (24.8), it gives after integration:

$$\left(\frac{dS}{d\tau}\right)^2 = \frac{\gamma}{\beta_2} P_0 S^4(\tau) - \frac{2C_1}{\beta_2} S^2(\tau) + K \tag{24.10}$$

where K is a constant of integration. This constant is zero, as for a localized solution one must have

$$\lim_{\tau\to\pm\infty} S(\tau) = \lim_{\tau\to\pm\infty} \frac{dS}{d\tau} = 0 \tag{24.11}$$

Moreover, considering that the envelope function presents a single maximum such that $S(\tau) = 1$ and $\frac{dS}{d\tau} = 0$, one obtains

$$C_1 = \frac{\gamma P_0}{2} \tag{24.12}$$

Using this result, one can rewrite Eq. (24.10) in the form

$$\frac{dS}{d\tau} = \alpha S\sqrt{1 - S^2} \tag{24.13}$$

where α is given by Eq. (24.6). Eq. (24.13) can be integrated, giving the result

$$S(\tau) = \operatorname{sech}(\alpha\tau) \tag{24.14}$$

The soliton solution given by Eq. (24.5) is obtained by substituting Eqs. (24.9), (24.12), and (24.14) in Eq. (24.7).

24.2.1 The normalized NLSE

Let us define a *non-linear distance*, L_{NL}, as

$$L_{NL} = \frac{1}{\gamma P_0} \tag{24.15}$$

L_{NL} is the propagation distance required to produce a non-linear phase rotation of one radian at a power P_0. A *dispersion distance*, L_D, can also be defined as

$$L_D = \frac{t_0^2}{|\beta_2|} \tag{24.16}$$

where t_0 is the pulse duration. These two characteristic distances provide the length scales over which non-linear and dispersive effects, respectively, become important for pulse evolution.

Using a distance Z normalized by the dispersion distance L_D and a time scale T normalized by t_0, Eq. (24.4) becomes

$$i\frac{\partial Q}{\partial Z} \pm \frac{1}{2}\frac{\partial^2 Q}{\partial T^2} + N^2|Q|^2 Q = 0 \tag{24.17}$$

where

$$N^2 = \frac{L_D}{L_{NL}} = \frac{\gamma P_0 t_0^2}{|\beta_2|} \tag{24.18}$$

In the second term of Eq. (24.17), the plus sign corresponds to the case of anomalous GVD ($\text{sgn}(\beta_2) = -1$), whereas the minus sign corresponds to normal GVD ($\text{sgn}(\beta_2) = +1$). The parameter N can be eliminated from Eq. (24.17) by introducing a new normalized amplitude $q = NQ$, with which the NLSE takes the following standard normalized form:

$$i\frac{\partial q}{\partial Z} + \frac{1}{2}\frac{\partial^2 q}{\partial T^2} + |q|^2 q = 0 \tag{24.19}$$

The case of anomalous GVD ($\beta_2 < 0$) was assumed in writing Eq. (24.19).

Considering the above normalizations, the result given by Eq. (24.5) can be written in the form

$$q(Z, T) = \eta \text{sech}[\eta T] \exp\left\{i\frac{1}{2}\eta^2 Z\right\} \tag{24.20}$$

where

$$\eta = \left(\frac{\gamma P_0 t_0^2}{|\beta_2|}\right)^{1/2} \tag{24.21}$$

is simultaneously the soliton amplitude and the inverse of the pulse width. In real units, the soliton width changes with η as t_0/η, i.e. it scales inversely with the soliton amplitude. This inverse relationship between the amplitude and the width of a soliton is the most crucial property of NLSE solitons.

The canonical form of the fundamental soliton can be obtained from Eq. (24.20) considering $\eta = 1$ and is given by

$$q(Z, T) = \text{sech}(T) \exp(iZ/2) \tag{24.22}$$

It can be shown that, given a solution $q(Z, T)$, the non-linear Schrödinger equation is satisfied by another function

$$q'(Z,T) = q(T + \kappa Z, Z)\exp\left[-i\left(\kappa T + \frac{1}{2}\kappa^2 Z\right)\right] \quad (24.23)$$

Introducing a phase constant σ and a time position T_0, the fundamental soliton solution can be written in the more general form:

$$q(Z,T) = \eta \operatorname{sech}\left[\eta\left(T + \kappa Z - T_0\right)\right]\exp\left(-i\kappa T + \frac{i}{2}\left(\eta^2 - \kappa^2\right)Z + i\sigma\right) \quad (24.24)$$

The soliton solution given by Eq. (24.24) is characterized by four parameters: the amplitude η (also the inverse of the pulse width), the frequency κ (also the pulse speed), the time position T_0 and the phase σ.

In general, it is difficult to obtain an analytical solution of the non-linear Schrödinger equation for certain initial conditions. In such cases, the NLSE must be solved numerically, which is efficiently performed using the so-called split-step Fourier method. In such an approach, the waveguide is partitioned into n equally long segments, whose lengths are adjusted according to the desired accuracy. For each segment, the non-linear and dispersive effects are considered independently.

Fig. 24.1 (a) shows the numerical solution of the NLSE obtained for an initial condition $q(0, T) = \operatorname{sech}(T)$, which corresponds to the fundamental soliton. This figure illustrates the most important property of the fundamental soliton: as long as the fibre loss is negligible, it propagates undistorted, without any change in shape, for arbitrarily long distances. It is this feature of fundamental solitons that makes them attractive for optical communication systems.

An interesting feature of the fundamental soliton solution, which demonstrates its notable robustness, is that it can be obtained starting from an arbitrary input pulse, since it has an area that lies in the range between $\pi/2$ and $3\pi/2$. Fig. 24.1 (b) shows, for example, the evolution of an initially rectangular pulse towards the fundamental soliton. In the first stage of propagation, some energy of the initial pulse is radiated until the pulse achieves the fundamental soliton profile.

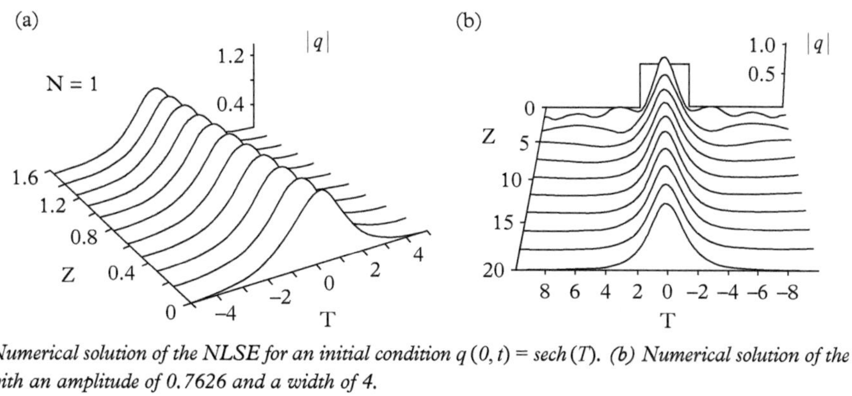

Figure 24.1 *(a) Numerical solution of the NLSE for an initial condition $q(0, t) = \operatorname{sech}(T)$. (b) Numerical solution of the NLSE for an initially rectangular pulse, with an amplitude of 0.7626 and a width of 4.*

According to Eq. (24.18), the power of a fundamental soliton ($N = 1$) in an optical fibre is given by

$$P_0 = \frac{|\beta_2|}{\gamma t_0^2} \approx \frac{3.11\,|\beta_2|}{\gamma t_{FWHM}^2} \tag{24.25}$$

where $t_{FWHM} = 1.763 t_0$ is the full width at half maximum (FWHM) of the soliton. Using typical values $\beta_2 = -20\,\text{ps}/\text{km}$ and $\gamma = 1.3\,\text{W}^{-1}/\text{km}$ for standard fibres near the 1.55-μm wavelength and considering pulses with a temporal width $t_0 = 25$ ps, we obtain a soliton power of 25 mW. This value is reduced significantly using dispersion-shifted fibres in the same wavelength region.

24.2.2 Soliton interaction

The soliton solution (24.24) exists in the range $-\infty < T < \infty$ and it remains valid for a train of solitons only if the individual solitons are well separated. This means that the soliton width must be a small fraction of the bit slot. The bit rate R is related to the soliton width t_0 as

$$R = \frac{1}{t_B} = \frac{1}{s_0 t_0} \tag{24.26}$$

where t_B is the duration of the bit slot and $s_0 = t_B/t_0$ is the separation between neighbouring solitons in normalized units.

Since a soliton solution ideally allows only one soliton in $-\infty < T < \infty$, the existence of an adjacent soliton must modify the ideal soliton solution. This modification appears as an interaction force between two neighbouring solitons. One can study numerically the interaction between two solitons by solving the NLSE with an input field of the form:

$$q(0,T) = \text{sech}(T - s_0/2) + r\,\text{sech}(T + s_0/2)\exp(i\Delta\phi) \tag{24.27}$$

where s_0 is the normalized separation between the two solitons, r is the ratio of their amplitudes, and $\Delta\phi$ is their relative phase difference. Numerical simulations show that the soliton separation evolves from its initial value in a way that depends on r and $\Delta\phi$. Fig. 24.2 shows the contour plots corresponding to a pair of solitons with equal amplitudes ($r = 1$), an initial separation $s_0 = 7$, and a phase difference $\Delta\phi = 0$ (a) or $\Delta\phi = \pi/2$ (b). Fig. 24.2 (a) shows that two in-phase solitons with equal amplitudes collide periodically. However, such periodic evolution does not happen for other values of the phase difference between the two interacting solitons. For example, Fig. 24.2 (b) shows that, by introducing a phase difference $\Delta\phi = \pi/2$, the two solitons separate monotonously during propagation.

In order to avoid soliton interaction, two adjacent solitons should be separated sufficiently (typically more than six times the pulse width in time), which represents a severe drawback for soliton-based communications. However, it can be shown that soliton interaction can be controlled and the tolerable spacing can be reduced by using adjacent solitons with slightly different amplitudes ($r \neq 1$). Since small changes in the peak power are not detrimental to the soliton nature of pulse propagation, this scheme is feasible in practice and can be useful for increasing the system capacity.

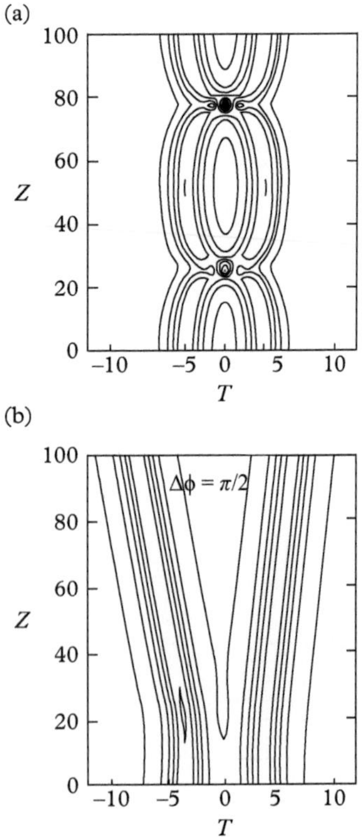

Figure 24.2 *Interaction between two solitons with equal amplitudes ($r = 1$), an initial separation $\Delta t_0 = 7$, and a phase difference $\Delta\phi = 0$ (a) or $\Delta\phi = \pi/2$ (b).*

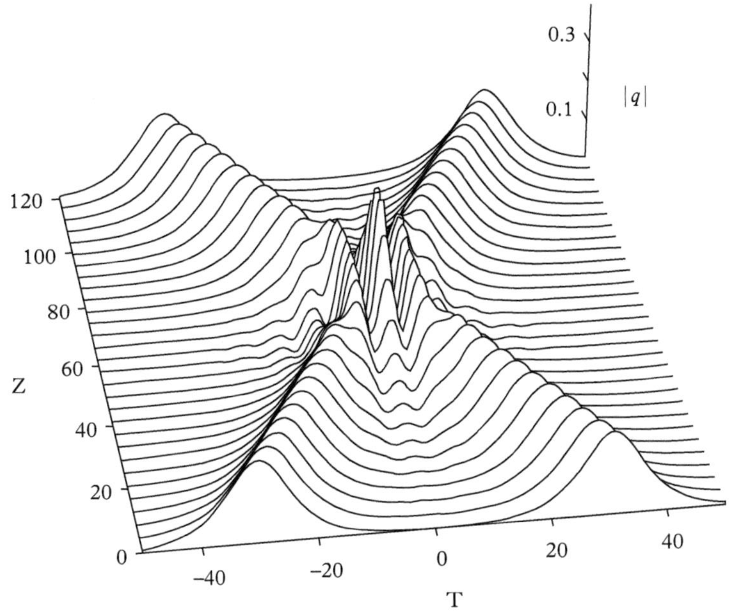

Figure 24.3 *Collision of two solitons.*

A very interesting feature of optical solitons is the fact that, in the collision between two solitons propagating at different velocities, they experience small temporal and phase shifts, but otherwise they recover fully. A collision between two solitons is illustrated in Fig. 24.3. When the two pulse envelopes overlap, beating between the two carrier frequencies can be observed.

24.2.3 Effect of fibre loss

High-speed fibre-optic communication systems are generally limited both by the fibre non-linearity and by the group dispersion that causes pulse broadening. However, since fundamental solitons are obtained by the balance between the group dispersion and Kerr non-linearity, their width can be maintained over long distances. In other words, fundamental solitons are free from either dispersive distortion or SPM. Thus, it is quite natural to use solitons as information carriers in fibres since all other formats will face distortion either from dispersion or non-linearity.

In a realistic communication system, soliton propagation is affected by many perturbations related to the input pulse shape, chirp and power, amplifier noise, optical filters, modulators, etc. However, because of its particle-like nature, it turns out that the soliton remains stable under most of these perturbations. When such perturbations are relatively small, the evolution of the four-soliton parameters can be described using soliton perturbation theory (Ferreira, 2011).

Fibre loss is the first perturbation affecting soliton propagation and one of the main causes of signal degradation in long-distance fibre-optic communication systems. This limitation can be minimized by operating near $\lambda = 1.55\,\mu\text{m}$. However, even with fibre losses as low as 0.2 dB/km, the signal power is reduced by 20 dB after transmission over 100 km of fibre.

Fibre loss can be taken into account theoretically by adding a loss term to the NLSE, which becomes

$$i\frac{\partial q}{\partial Z} + \frac{1}{2}\frac{\partial^2 q}{\partial T^2} + |q|^2 q = -\frac{i}{2}\Gamma q \tag{24.28}$$

where $\Gamma = \alpha_0 L_D$ is the loss rate per dispersion distance and α_0 is the fibre loss coefficient. If $\Gamma << 1$, the soliton perturbation theory shows that the soliton amplitude is reduced during propagation according to the relation

$$\eta(Z) = \eta(0)\exp(-\Gamma Z) \equiv \eta_0 \exp(-\Gamma Z) \tag{24.29}$$

Eq. (24.29) indicates that the soliton amplitude decreases along the fibre length at the same rate as the power amplitude. The decrease in the amplitude at a rate twice as fast as that of a linear pulse is a consequence of the non-linear property of a soliton. Since the amplitude and width are inversely related, the soliton width will increase exponentially according to

$$t_p(Z) = t_0 \exp(\Gamma Z) \tag{24.30}$$

In fact, the exponential increase of the soliton width predicted by the perturbation theory occurs only for relatively short propagation distances. Fig. 24.3 shows the variation of the broadening factor $t_p(Z)/t_0$ with distance for the fundamental soliton when $\Gamma = 0.07$. Clearly, the result predicted by the perturbation theory is valid only for distances up to $Z \approx \Gamma^{-1}$. The numerical result displayed in Fig. 24.4 shows that for large distances the soliton width increases linearly with distance but at a rate slower than that of a linear pulse.

In the context of communication systems, the fibre loss problem was solved prior to 1991 using repeaters periodically installed in the transmission line. A repeater consists of a light detector and light pulse generators. In fact, it was the most expensive unit in the transmission system, and its use also represented the bottleneck to increasing the transmission speed.

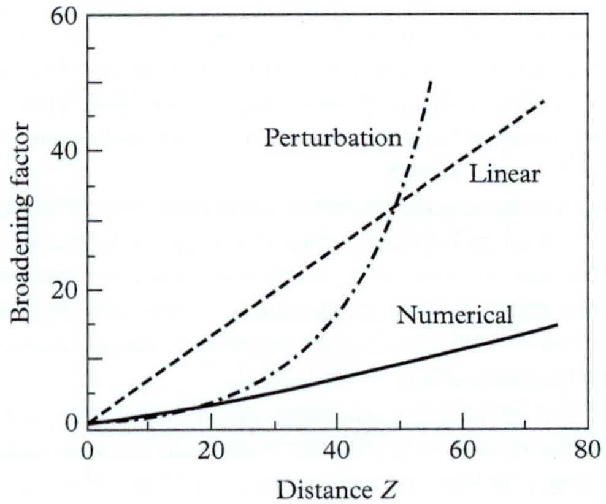

Figure 24.4 *Broadening factor $t_p(Z)/t_0$ against the propagation distance obtained numerically (full curve) and predicted by the perturbation theory (dash-dotted curve) when $\Gamma = 0.07$. The dashed curve corresponds to the case of a linear pulse.*

Distributed Raman gain provided by the fibre itself to compensate for the fibre loss was early suggested to enable repeaterless soliton transmission systems. Its use requires periodic injection of the pump power into the transmission fibre. This concept of distributed all-optical transmission gave its position to lumped amplification during the 1990s, through the use of erbium-doped fibre amplifiers (EDFAs). The main properties of both Raman and EDFAs were reviewed in Chapter 22.

Gordon and Haus anticipated in 1986 that the transmission of a signal made of optical solitons could not be extended to an unlimited distance when optical amplification is used (Gordon and Haus, 1986). In fact, the amplifiers needed to compensate for the fibre loss generate amplified spontaneous emission (ASE), and this noise is, in part, incorporated into the soliton, whose mean frequency is then shifted. Due to GVD, the arrival time of the soliton then becomes a random variable, whose variance is proportional to the cube of the propagation distance. This is the so-called *Gordon–Haus effect*. Actually, the timing jitter resulting from the ASE noise produced by optical amplifiers is the major cause of bit errors in soliton communication systems.

24.3 Higher-Order Solitons

The non-linear Schrödinger equation given by Eq. (24.19) is a completely integrable equation that can be solved exactly using the so-called inverse scattering method (Zakharov and Shabat, 1972). Using such a method, it can be shown that, in general, the N-soliton solutions that arise from an arbitrary initial waveform are asymptotically given in the form of N separated solitons:

$$q(Z,T)=\sum_{j=1}^{N}\eta_j \operatorname{sech}\left[\eta_j\left(T+\kappa_j Z-T_{0j}\right)\right]\exp\left[-i\kappa_j T+i\frac{1}{2}\left(\eta_j^2-\kappa_j^2\right)Z-i\sigma_j\right] \tag{24.31}$$

Considering an input pulse shape given by

$$q(0,T)=N\operatorname{sech}(T) \tag{24.32}$$

all the output solitons propagate at exactly the same speed. This type of solution is called the *bound-soliton solution*, and its shape evolves periodically during the propagation due to the phase interference among the constituting solitons. This is in contrast to the fundamental soliton, whose shape remains constant during propagation.

For the second-order soliton ($N=2$) the field distribution is given by

$$q(Z,T)=\frac{4\left[\cosh(3T)+3\exp(4iZ)\cosh(T)\right]\exp(iZ/2)}{\left[\cosh(4T)+4\cosh(2T)+3\cos(4Z)\right]} \tag{24.33}$$

Fig. 24.5 provides a perspective on the second-order soliton evolution. We observe that such evolution is indeed periodic and that, in the initial part, the pulse contracts in the time domain and increases its peak power.

Fig. 24.6 shows the numerical solution of the NLSE assuming an initial condition $q(0,T)=N\operatorname{sech}(T)$, with (a) $N=2$ and (b) $N=3$. As pointed out before, the pulse evolves during propagation, returning to its initial profile after one period. Meanwhile, it develops a multi-peak structure with a maximum number of peaks equal to N-1.

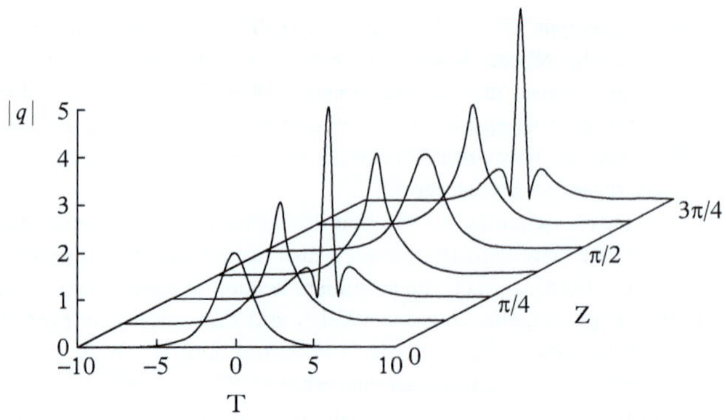

Figure 24.5 *Evolution of the second-order soliton.*

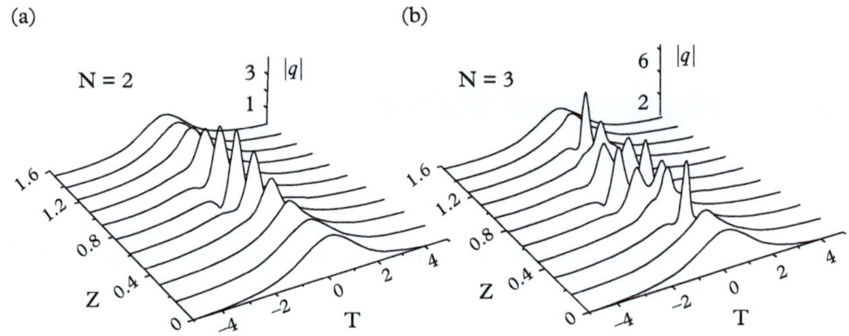

Figure 24.6 *Numerical solution of the NLSE assuming an initial condition* $q(0,T) = N\mathrm{sech}(T)$, *with (a)* $N=2$ *and (b)* $N=3$.

The period of oscillation of the higher-order solitons resulting from the initial pulse given by Eq. (24.32) is

$$Z_0 = \frac{\pi}{2} \tag{24.34}$$

which can be written in terms of real units as

$$z_0 = \frac{\pi}{2}L_D = \frac{\pi}{2}\frac{t_0^2}{|\beta_2|} \tag{24.35}$$

The length z_0 is usually called the *soliton period*. Considering a pulse with a temporal width $t_0 = 10$ ps propagating in a dispersion-shifted fibre with $\beta_2 = -1\,\mathrm{ps/km}$ in the C-band (centred at 1550 nm), one obtains a propagation period $z_0 \approx 150\,\mathrm{km}$.

The periodic evolution of higher-order solitons can be understood as follows. In the initial part of its evolution, the pulse contracts in the time domain and increases its peak power. As seen in Chapter 21, SPM generates a frequency chirp such that the leading edge of the pulse is red-shifted, while its trailing edge is blue-shifted from the central frequency, resulting in a broad spectrum in the frequency domain. In the absence of GVD, the pulse shape would have remained unchanged. However, in the presence of anomalous GVD, the red-shifted components in the leading edge of the pulse move slower, whereas the blue-shifted components in the trailing edge are faster. As a consequence, the pulse spectrum is compressed and the pulse again reaches its original width in the time domain. After that, everything starts again. It is this mutual interaction between the SPM and GVD effects that explains the periodic evolution of higher-order solitons.

Example 24.2

Calculate the power that should be used to generate (i) a fundamental soliton and (ii) a soliton of order 3, considering typical parameter values for a dispersion-shifted fibre, namely $\beta_2 = -1\,\text{ps}^2/\text{km}$ and $\gamma = 3\,\text{W}^{-1}/\text{km}$, and a temporal pulse with $t_0 = 6\,\text{ps}$. Comment on the relative advantages of using fundamental solitons for optical transmission systems.

Solution

According to Eq. (24.18), the power that should be used to generate a soliton of order N is given by

$$P_0 = \frac{N^2 |\beta_2|}{\gamma t_0^2} \tag{24.36}$$

Considering the proposed parameter values, the power necessary to generate a fundamental soliton in the dispersion-shifted fibre is $P_0 \sim 10\,\text{mW}$. In the case of a third-order soliton, one has $P_0 = 90\,\text{mW}$.

In general, the required power to launch the Nth-order soliton is N^2 times that for the fundamental one. Moreover, higher-order solitons compress periodically, resulting in soliton chirping and spectral broadening. In contrast, fundamental solitons preserve their shape during propagation. This fact, together with the lower power required for their generation, makes fundamental solitons the preferred option in soliton communication systems.

24.3.1 Soliton-effect compression

After having understood the evolution of optical pulses in silica fibres, the combination of GVD and SPM effects has been used to achieve pulse compression. This pulse compression technique is known as a *soliton-effect compressor* and uses the fibre itself as a compressor without the need for an external device. This compression mechanism is related to the behaviour of higher-order solitons. As described above, these solitons follow a periodic evolution pattern such that they go through an initial narrowing phase at the beginning of each period. If the fibre length is suitably chosen, the input pulses can be compressed by a factor that depends on the soliton order. Obviously, this compression technique is applicable only in the case of optical pulses with wavelengths exceeding 1.3 μm when propagating in standard single-mode fibres.

In practice, soliton-effect compression is carried out by initially amplifying optical pulses up to the power level required for the formation of higher-order solitons. The peak optical power of the initial pulse required for the formation of the Nth-order soliton is given by Eq. (24.31). These Nth-order solitons are then passed through the appropriate length of optical fibre to achieve a highly compressed pulse. The optimum fibre length, z_{opt}, and the optimum pulse compression factor, F_{opt}, of a soliton-effect compressor can be estimated from the following empirical relations:

$$z_{opt} \approx z_0 \left(\frac{0.32}{N} + \frac{1.1}{N^2} \right) \tag{24.37}$$

$$F_{opt} \approx 4.1N \tag{24.38}$$

where z_0 is the *soliton period*, given by Eq. (24.30).

Compression factors as large as 500 can be obtained using soliton-effect compressors. However, the pulse quality is relatively poor, since the compressed pulse carries only a fraction of the input energy, while the remaining energy appears in the form of a broad pedestal. From a practical point of view, the pedestal is deleterious, since it causes the compressed pulse to be unstable, making it unsuitable for some applications. Despite this, soliton-effect compressed pulses can still be useful because there are some techniques that can eliminate the pedestal. One of them consists of the use of a non-linear optical loop mirror (NOLM), which was described in Chapter 22. Since the transmission of a NOLM is intensity dependent, the loop length can be chosen such that the low-intensity pedestal is reflected while the higher-intensity pulse peak is transmitted, resulting in a pedestal-free transmitted pulse.

One difficulty faced when using the soliton-effect compressor is that pulses with high peak power are required for the formation of high-order solitons in conventional fibres. The use of dispersion-shifted fibres (DSFs) with small values of β_2 at the operating wavelength can reduce the peak power required for soliton generation by an order of magnitude. However, because the soliton period z_0 is inversely proportional to $|\beta_2|$, longer lengths of DSFs will be required for solitons to achieve optimum compression.

24.4 Higher-Order Effects

The propagation of short solitons ($t_0 < 5$ ps) in distributed fibre amplifiers can be described by a generalized NLSE which, besides the addition of a gain term, must also include the effects of finite gain bandwidth, third-order dispersion (TOD), and intrapulse Raman scattering (IRS). The inclusion of these effects provides a generalization of Eq. (24.19), in the form:

$$i\frac{\partial q}{\partial Z} + \frac{1}{2}\frac{\partial^2 q}{\partial T^2} + |q|^2 q = -\frac{i}{2}\Gamma q + \frac{i}{2}g(Z)L_D\left(q + \tau_2^2\frac{\partial^2 q}{\partial T^2}\right) + i\delta_3\frac{\partial^3 q}{\partial T^3} + \tau_R q\frac{\partial |q|^2}{T} \tag{24.39}$$

where τ_2 is inversely related to the gain bandwidth. The TOD and Raman effects are represented, respectively, by the parameters δ_3 and τ_R, given by

$$\delta_3 = \frac{\beta_3}{6|\beta_2|t_0}, \quad \tau_R = \frac{t_R}{t_0} \tag{24.40}$$

where t_R is a constant dependent on the Raman gain slope. The TOD term becomes important only when $|\beta_2|$ is sufficiently small. On the other hand, since $t_R \approx 5$ fs, the Raman term in Eq. (24.34) can generally be treated as a small perturbation. Considering only this term and applying perturbation theory, it is found that the soliton amplitude remains constant, whereas its frequency varies with distance as

$$\kappa(Z) = -\frac{8}{15}\tau_R\eta^4 Z. \tag{24.41}$$

The IRS effect leads to a continuous downshift of the carrier frequency, an effect known as the *soliton self-frequency shift* (SSFS). The origin of SSFS can be understood by noting that, for ultrashort solitons, the pulse spectrum becomes so broad that the high-frequency

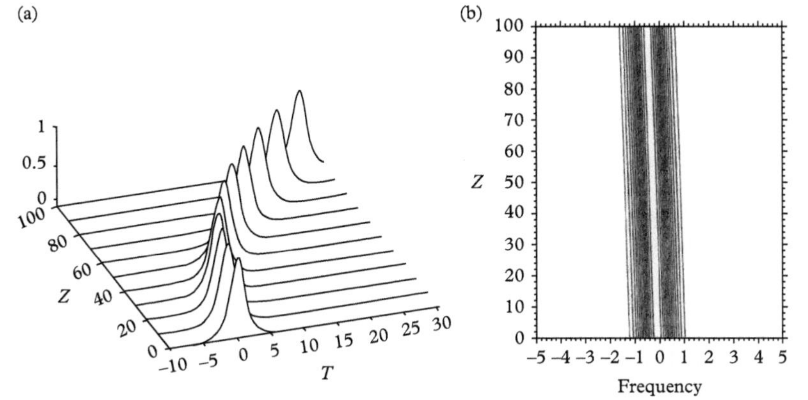

Figure 24.7 *IRS effect on the propagation of the fundamental soliton for* $\tau_R = 0.01$*: (a) evolution of the amplitude and (b) a contour plot of the soliton spectrum.*

components of the pulse can transfer energy through Raman amplification to the low-frequency components of the same pulse. Such an energy transfer appears as a red shift of the soliton spectrum, with the shift increasing with distance. Since the amplitude and width are inversely related, the frequency shift scales as t_0^4, indicating that it can become quite large for ultrashort pulses. In practice, the limited bandwidth of the gain spectrum ($\tau_2 \neq 0$) can, in some circumstances, provide SSFS compensation and adiabatic soliton trapping.

The IRS effect on the propagation of the fundamental soliton is illustrated in Fig. 24.7 for the case $\delta_3 = \tau_2 = 0$, $g(Z)L_D = \Gamma$, and $\tau_R = 0.01$. The linear variation of the frequency shift with distance is clearly illustrated by the contour plot of the soliton spectrum in Fig. 24.7(b). As a consequence, there is a continuous variation in the soliton velocity and position, as observed in Fig. 24.7(a).

24.5 Dark Solitons

In the normal-dispersion regime ($\beta_2 > 0$) the normalized non-linear Schrödinger equation becomes

$$i\frac{\partial q}{\partial Z} - \frac{1}{2}\frac{\partial^2 q}{\partial T^2} + |q|^2 q = 0 \tag{24.42}$$

Eq. (24.42) differs from Eq. (24.19) only in the sign of its dispersion term. The soliton solution of Eq. (24.42) can be written in the form:

$$q(Z,T) = \sqrt{\rho}e^{i\sigma} \tag{24.43}$$

where

$$\rho = \rho_0\left[1 - a^2\text{sech}^2\left(\sqrt{\rho_0}aT\right)\right],\ a^2 \leq 1 \tag{24.44}$$

$$\sigma = \left[\rho_0\left(1-a^2\right)\right]^{1/2} T + \tan^{-1}\left[\frac{1}{\sqrt{1-a^2}}\tanh\left(\sqrt{\rho_0}aT\right)\right] - \frac{\rho_0\left(3-a^2\right)}{2}Z \tag{24.45}$$

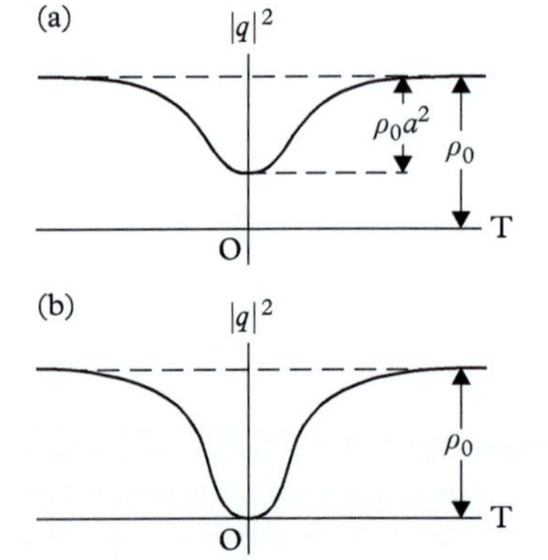

Figure 24.8 *Schematic representation of dark solitons with (a) $a < 1$ and (b) $a = 1$.*

The parameter a is known as the depth of modulation. In fact, this kind of soliton appears in the section where light waves are absent, and thus they are called *dark solitons*. Pulse-like solitons discussed in previous sections are called *bright solitons* to make the distinction clear. When $a = 1$ we have a *black soliton*, given by

$$q = \sqrt{\rho_0} \tanh\left(\sqrt{\rho_0} T\right) \tag{24.46}$$

Thus, an input pulse with a 'tanh' profile, exhibiting an intensity 'hole' at the centre, would propagate unchanged in the normal-dispersion region of optical fibres. The two types of dark solitons are illustrated in Fig. 24.8.

24.6 Dispersion-Managed Solitons

In order to minimize the impact of some non-linear effects, namely the four-wave mixing effect, different dispersion maps have been tried in linear optical fibre transmission systems. Basically, the idea is to introduce locally relatively large dispersion and to compensate for the accumulated dispersion at the end of the transmission line so that the integrated dispersion becomes zero. A method of programming dispersion is called *dispersion management*. Dispersion management has also been found to be effective in soliton transmission systems since it can significantly reduce the timing jitter, which is the major cause of bit error in such systems.

A typical periodic dispersion map uses both anomalous and normal-dispersion fibres alternately. The non-linear stationary pulse that propagates in such a fibre has a peak power larger than that of the soliton that propagates in a fibre with a constant dispersion equal to the average dispersion of the periodic map. Such a non-linear stationary pulse is commonly called a *dispersion-managed* (DM) *soliton.*

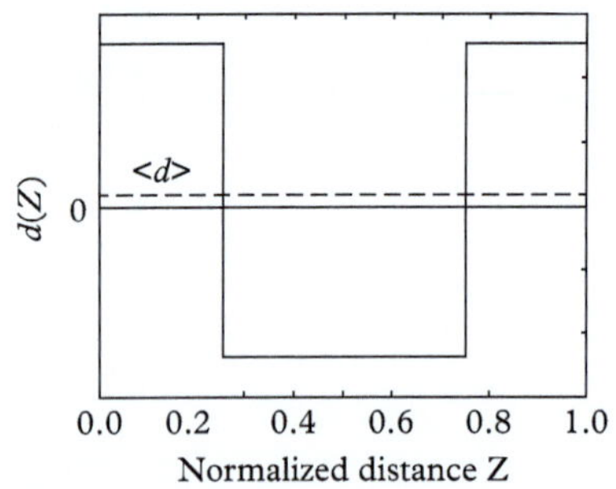

Figure 24.9 *Canonical dispersion map with anomalous- and normal-dispersion fibres. The distance Z is normalized by the map length L.*

Fig. 24.9 shows the so-called canonical dispersion map for DM soliton systems. This is a two-step symmetric dispersion map, constituted by an anomalous-dispersion fibre ($D_1 > 0$) of length L_1 and a normal-dispersion fibre ($D_2 < 0$) of length L_2, while $L_1 + L_2 = L$ is the map length.

The model equation describing the pulse propagation in a DM transmission line is the normalized non-linear Schrödinger equation, modified to include a spatially varying dispersion $d(z)$, fibre loss, and amplifier gain:

$$i\frac{\partial q}{\partial Z} + \frac{1}{2}d(Z)\frac{\partial^2 q}{\partial T^2} + |q|^2 q = -i\frac{\Gamma}{2}q + iG(Z)q \tag{24.47}$$

where Γ describes fibre losses, $G(Z)$ is the amplifier gain compensating for the fibre loss, and $d(Z)$ is the variable normalized dispersion, assuming the value D_1 in the anomalous-dispersion fibre and the value D_2 in the normal-dispersion fibre.

DM solitons are entities whose shape varies substantially over one period of the dispersion map but which return to the same shape at the end of each period. Fig. 23.10 shows the evolution of the DM soliton during one period of the dispersion map with anomalous- and normal-dispersion fibres of equal length. The DM pulse alternately spreads and compresses as the sign of the dispersion is switched. The pulse peak power varies rapidly, and the pulse width becomes minimum at the centre of each fibre where the frequency chirp vanishes.

The variational approach is one of the most powerful techniques capable of providing insight into the main characteristics of the DM soliton. The general idea underlying this

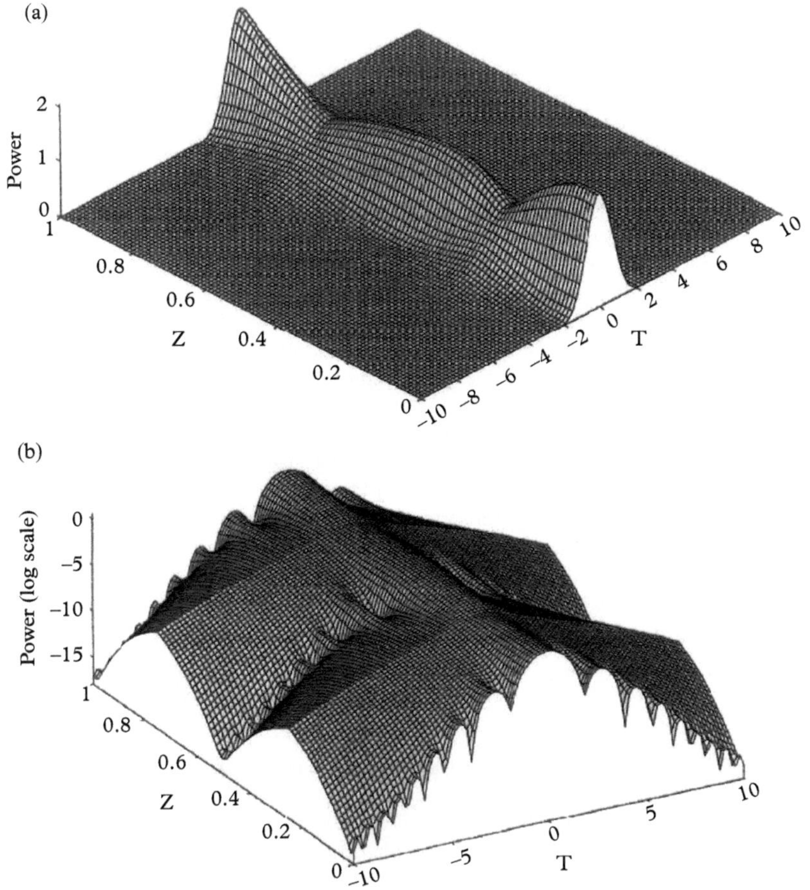

Figure 24.10 *Evolution of the DM soliton along one dispersion map shown in (a) linear scale and (b) logarithmic scale.*

approximation involves choosing an initial ansatz for the shape of the solution sought, but leaving in the ansatz a set of free parameters, which may evolve with Z. An important feature of DM solitons is that their profile in general is not given by a sech function, as it is for the NLSE fundamental soliton, but varies from sech to a Gaussian function, and an even more flat distribution.

Since a DM soliton can exist even when the average dispersion is zero, it is free from the Gordon–Haus jitter. Furthermore, the amplitude of the DM soliton can be made sufficiently large by the so-called soliton energy enhancement factor, with a proper design of the dispersion map. That is, a DMS can be free from both amplitude and timing jitter and can be very robust in periodically amplified all-optical transmission systems.

The DM soliton is quite different from a standard fundamental soliton, since its shape, width, and peak power are not constant. However, since these parameters repeat from period to period at any location within the map, DM solitons can be used in optical communications. Actually, the remarkable characteristics of such solitons make them the preferred option for use in high-capacity transmission systems.

24.7 Dissipative Optical Solitons

Although solitons are clearly a nice option as information carriers in optical fibres because of their robust nature, their interaction and the timing jitter produced by the amplifier-added noise can lead to the loss of information in transmission systems. These problems can be suppressed using some techniques to control the soliton parameters, namely its amplitude (or width), time position, velocity (or frequency), and phase.

The use of narrow-band filters was early suggested to control the Gordon–Haus jitter and other noise effects. The basic idea is that any soliton whose central frequency has strayed from the filter peak will be returned to the peak by virtue of the differential loss the filters induce across its spectrum. Suppression of the timing jitter is a consequence of the frequency-jitter damping.

In order to compensate for the filter-induced loss, some excess gain must be provided to the soliton. An effective technique to guarantee the control and stable propagation of the soliton in this case consists in the use of an amplifier having a non-linear property of gain, or gain and saturable absorption in combination, such as the non-linear loop mirror or non-linear polarization rotation in combination with a polarization-dependent loss element (Ferreira, 2022). The key property of the non-linearity in gain is to give an effective gain to the soliton and a suppression (or very small gain) to the noise.

The pulse propagation in optical fibre transmission systems where linear and non-linear amplifiers and narrow-band filters are periodically inserted may be described by the following generalized complex Ginzburg–Landau equation (CGLE):

$$i\frac{\partial q}{\partial Z}+\frac{D}{2}\frac{\partial^2 q}{\partial T^2}+|q|^2q=i\delta q+i\beta\frac{\partial^2 q}{\partial T^2}+i\varepsilon|q|^2q+i\mu|q|^4q-\nu|q|^4q, \tag{24.48}$$

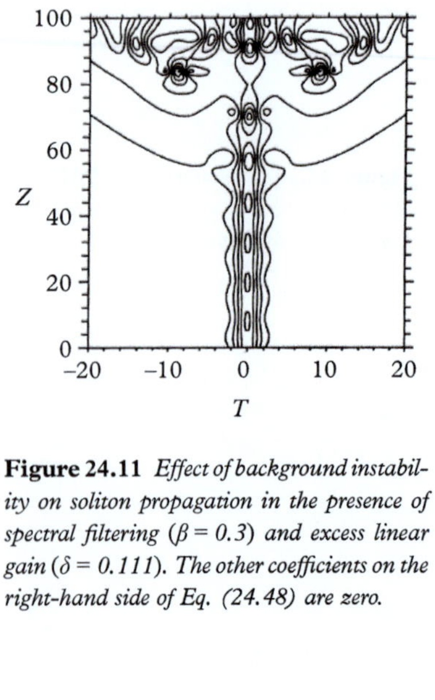

Figure 24.11 *Effect of background instability on soliton propagation in the presence of spectral filtering ($\beta = 0.3$) and excess linear gain ($\delta = 0.111$). The other coefficients on the right-hand side of Eq. (24.48) are zero.*

where D is the dispersion parameter, with $D>0$ in the anomalous regime and $D<0$ in the normal regime; β stands for the normalized filtering strength, δ is the linear excess gain, ε accounts for non-linear gain-absorption processes, μ represents a higher-order correction to the non-linear gain-absorption, and ν is a higher-order correction term to the non-linear refractive index.

Eq. (24.48) becomes the standard non-linear Schrödinger equation (NLSE) when the right-hand side is set to zero. When this is not the case, Eq. (24.48) is non-integrable, and only particular solutions can be obtained. The soliton perturbation theory can be used in the anomalous-dispersion regime ($D>0$) for small values of the parameters. Analytical solutions can be presented explicitly only for certain relations between the parameters of the equation.

It is important to mention that the use of non-linear gain is necessary to guarantee the effective control and stable propagation of the soliton. Actually, using linear excess gain only, it also amplifies the small-amplitude waves coexistent with the soliton. Such amplification results in a background instability which can significantly affect and even destroy the soliton itself. This process is illustrated in Fig. 24.11, from which we observe that the soliton is severely disturbed for transmission distances $Z>80$.

The soliton solutions of Eq. (24.48) are termed *dissipative solitons*, and they differ significantly from conventional solitons. Actually, the conventional soliton concept implies a single balance between non-linearity and dispersion, whereas a dissipative soliton arises as a result of a double balance: between non-linearity and dispersion, and also between gain and loss. The two balances result in solutions which are fixed. The shape, amplitude and

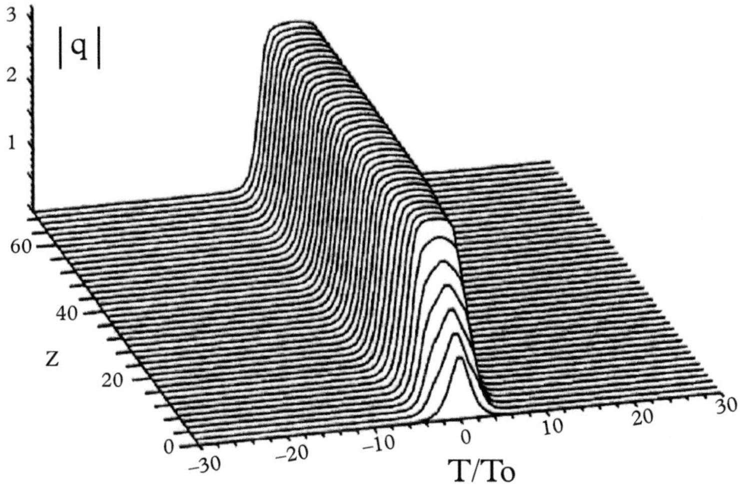

Figure 24.12 *Formation and evolution of a flat-top soliton, considering an input pulse* $q(0,T) = sech(T)$, *for* $\delta = -0.1$, $\beta = 0.5$, $\varepsilon = 0.66$, $\mu = \nu = -0.01$.

width are all fixed and depend on the parameters of the equation. In particular, the balance between gain and loss, which should be exact in order to produce stationary localized solutions, plays a crucial role in the soliton dynamics. As a consequence, compared with the case of conventional (conservative) solitons, the impact of dissipative effects reveals unusual situations in which solitons can be found.

For small values of the parameters on the right-hand side of Eq. (24.48), the stable soliton solutions of the CGLE have a sech profile, similar to the soliton solutions of the NLSE, and correspond to the so-called plain pulses (PPs). However, rather different pulse profiles can be obtained for non-small values of those parameters. As an example, Fig. 24.12 illustrates the formation and stable propagation of a flat-top soliton, starting from an initial pulse with a sech profile. A flat-top soliton can be considered as two mirror-imaged interacting fronts.

Besides the stationary solutions, the CGLE also presents different types of localized solutions with periodically or quasi-periodically varying amplitude, width, and energy. Actually, in order to observe these pulsating solitons, the parameters of the CGLE must be far enough from those of the non-linear Schrödinger equation.

Among the CGLE pulsating solitons, we may refer to the plain pulsating soliton, whose propagation is illustrated in Figure 24.13(a), assuming the following parameter values: $\delta = -0.1, \beta = 0.08, \epsilon = 0.66, \mu = -0.1$, and $v = -0.1$. The pulse shows a periodic behaviour, recovering exactly its initial shape after a given period. Figure 24.13(b) shows the soliton peak power, $|q|^2$, against the soliton energy, given by

$$E = \int_{-\infty}^{\infty} |q|^2 dT \tag{24.49}$$

The trajectory corresponds to a cycle repeating itself indefinitely, which means that the pulse energy and peak power also exhibit a periodic behaviour.

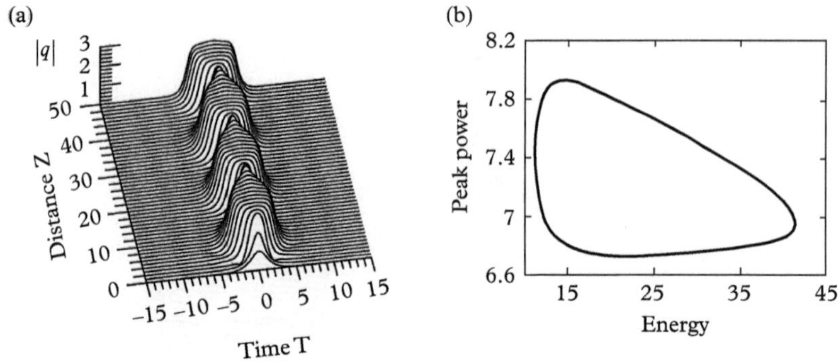

Figure 24.13 *(a) Amplitude evolution and (b) peak power versus energy, for a plain pulsating soliton, assuming the following parameter values:* $\delta = -0.1$, $\beta = 0.08$, $\varepsilon = 0.66$, $\mu = -0.1$, *and* $v = -0.1$.

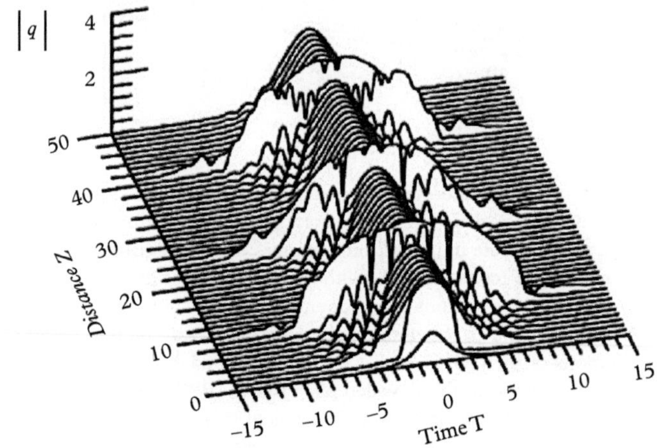

Figure 24.14 *Evolution of an erupting pulse when* $\delta = -0.1$, $\beta = 0.125$, $\varepsilon = 1.0$, $\mu = -0.1$, *and* $v = -0.6$.

One of the most striking species of pulsating solitons is that of 'erupting solitons'—soliton solutions that intermittently suffer explosive instabilities but return to their original shapes after each explosion. Such a soliton has extended intervals of almost stationary propagation, during which an instability develops and an explosion occurs. Following the eruption, there is a 'cooling' period, after which the solution becomes 'stationary' again. All successive explosions have similar features, but, being chaotic, they are not identical. Fig. 24.14 shows an example of the evolution of an erupting soliton.

The use of a novel, powerful real-time spectra measurement technique, called dispersive Fourier transform (DFT), has enabled the experimental observation in real time of the transient dynamics of diverse pulsating solitons and other complex ultrafast non-linear phenomena in fibre lasers.

24.8 The Variational Approach

The variational technique is well known in several areas, and it is particularly useful in non-linear optics to describe the propagation of chirped pulses in non-linear media. The accuracy of this approach depends essentially on the ansatz chosen to describe the evolution of the field experiencing non-linearity.

The 'action' functional is given by

$$S = \int L dz \tag{24.50}$$

where L is the Lagrangian, which is obtained from the Lagrangian density L_d in the form

$$L = \int_{-\infty}^{\infty} L_d(v, v^*) d\tau \tag{24.51}$$

In Eq. (24.51), $v(z)$ and $v^*(z)$ represent the generalized coordinates. Minimization of the action S requires that L_d satisfy the Euler–Lagrange equation

$$\frac{\partial L_d}{\partial v} - \frac{\partial}{\partial \tau}\left(\frac{\partial L_d}{\partial v_\tau}\right) - \frac{\partial}{\partial z}\left(\frac{\partial L_d}{\partial v_z}\right) = 0 \tag{24.52}$$

where v_τ and v_z represent the derivatives of v relative to τ and z, respectively.

It can be shown that the Lagrangian density corresponding to Eq. (21.27) is

$$L_d = \frac{i}{2}\left(U\frac{\partial U^*}{\partial z} - U^*\frac{\partial U}{\partial z}\right) - \frac{\beta_2}{2}\left|\frac{\partial U}{\partial \tau}\right|^2 - \frac{\gamma}{2}|U|^4 \tag{24.53}$$

In fact, Eq. (21.27) can be obtained using Eq. (24.53) in Eq. (24.52) and considering U^* as the generalized coordinate v. Minimization of the action S with respect to the pulse parameters provides the reduced Euler–Lagrange equations

$$\frac{\partial L}{\partial v} - \frac{d}{dz}\left(\frac{\partial L}{\partial v_z}\right) = 0 \tag{24.54}$$

Let us consider the case of a chirped Gaussian pulse, given by

$$U(z,\tau) = U_p(z)\exp\left[-\frac{1}{2}\left(1 + iC_p(z)\right)\left(\frac{\tau}{t_p(z)}\right)^2 - i\theta_p(z)\right] \tag{24.55}$$

where the parameters $U_p(z)$, $C_p(z)$, $t_p(z)$, and $\theta_p(z)$ represent the amplitude, the chirp parameter, the width, and the phase of the pulse, respectively. Substituting Eq. (24.55) into Eq. (24.53) and performing the time integration in Eq. (24.51) gives the following result for the Lagrangian:

$$L = -\frac{\beta_2 E_p}{4t_p^2}\left(1 + C^2\right) - \frac{\gamma E_p^2}{\sqrt{8\pi}t_p} - \frac{E_p}{4}\left(\frac{dC_p}{dz} - \frac{2C_p}{t_p}\frac{dt_p}{dz}\right) - E_p\frac{d\theta_p}{dz} \tag{24.56}$$

where $E_p = \sqrt{\pi}U_p^2 t_p$ is the pulse energy. Using $v = \theta_p$ in Eq. (24.54) gives $E_p = E_0$ = constant. On the other hand, using $v = C_p$, t_p, and E_p provides the following equations for t_p, C_p, and θ_p, respectively:

$$\frac{dt_p}{dz} = \frac{\beta_2 C_p}{t_p} \tag{24.57}$$

$$\frac{dC_p}{dz} = \frac{\gamma E_0}{\sqrt{2\pi}t_p} + \left(1 + C_p^2\right)\frac{\beta_2}{t_p^2} \tag{24.58}$$

$$\frac{d\theta_p}{dz} = -\frac{5\gamma E_0}{4\sqrt{2\pi}t_p} - \frac{\beta_2}{2t_p^2} \tag{24.59}$$

Eqs. (24.57) and (24.58) show that the SPM affects the pulse width via the chirp parameter C_p. The two terms in Eq. (24.58) are due to the SPM and GVD effects, and they have the same or the opposite sign, depending on whether the dispersion is normal or anomalous, respectively. As a consequence, the combination of the SPM and GVD effects results in reduced pulse broadening in the case of anomalous dispersion compared with the case of normal dispersion, because of smaller values of the chirp parameter in the first case.

Example 24.3

Starting from Eqs. (24.57) and (24.58), derive expressions for the chirp parameter $C_p(z)$ and the pulse width $t_p(z)$ (i) by neglecting the fibre non-linearity and (ii) by neglecting the fibre dispersion.

Solution

(i) As seen in Chapter 16, the quantity $(1 + C_p^2)/t_p^2$, corresponds to the spectral width of the pulse, and it remains constant if one neglects the fibre non-linearity. Replacing this quantity in Eq. (24.58) by its initial value $(1 + C_0^2)/t_0^2$ and performing the integration, one obtains the solution

$$C_p = C_0 + (1 + C_0^2)\frac{\beta_2 z}{t_0^2} \tag{24.60}$$

which is similar to Eq. (16.45). Using this solution in Eq. (24.57), one finds the following result for the pulse width:

$$t_p^2 = t_0^2\left[(1 + C_0\sigma)^2 + \sigma^2\right] \tag{24.61}$$

where

$$\sigma = \frac{\beta_2 z}{t_0^2} \tag{24.62}$$

(Eqs. 24.61) and (24.62) are similar to Eqs. (16.47) and (16.36), respectively. They show that the input pulse broadens monotonously during propagation if C_0 and β_2 are of the same sign. However, if these parameters are of opposite sign, the pulse compresses until a given location, after which it starts to broaden, as illustrated in Fig. 16.9.

(ii) Keeping the fibre non-linearity but neglecting the GVD, Eq. (24.57) shows that the pulse width t_p remains constant, equal to its initial value t_0. Concerning the chirp parameter, it varies with distance as

$$C_p = C_0 + \frac{1}{\sqrt{2}}\gamma P_0 z \tag{24.63}$$

Eq. (24.63) shows that the SPM-induced chirp is always positive. It adds to the initial pulse chirp when $C_0 > 0$, resulting in an enhanced oscillatory structure of the pulse spectrum. However, if $C_0 < 0$, the two contributions have opposite signs, which produces spectral narrowing.

24.9 Soliton Fibre Lasers

A soliton fibre laser is defined as any non-linear optical oscillator containing optical fibres and which generates optical pulses due to a balance between various sources of linear and non-linear gain, loss, SPM, and dispersion. Actually, owing to the simplicity, ease, and precise control of the experimental parameters, optical fibre lasers are ideal systems in which to generate and observe solitons. Ultrashort pulses can be generated directly by optical fibre soliton lasers using different mode-locking techniques.

Eq. (24.48) can be used to model any soliton fibre laser, since it includes all the main involved physical effects. In particular, linear loss and non-linear gain account for the required saturable absorber effect. On the other hand, even if the correction to the non-linear refractive index is not imposed by most laser materials under common operating conditions, it can be justified by the discrete nature of the laser cavity. An explicit relation of the coefficients in the cubic-quintic CGLE with regard to the physical parameters of a fibre laser can be obtained in some cases.

Actually, Eq. (24.48) models the fibre laser as a distributed system, which is reasonable if the pulse shape changes only slightly during each round-trip. There are many advantages in using such a distributed model, governed by a continuous equation, since it allows, to some extent, an analytic study. However, when the discrete nature of the laser cavity cannot be ignored, the cubic-quintic CGLE given by Eq. (24.48) can be used such that the various parameters vary periodically with Z, the period corresponding to a cavity round-trip.

As mentioned in Section 24.7, many different types of soliton solutions of the cubic-quintic CGLE, both stationary and nonstationary, can be found. However, stable pulses can be generated in a very narrow range of the laser parameters, which must be carefully adjusted for such a purpose. In general, the pulses may change their shape from one round-trip to another and have complicated dynamics in time. In some circumstances, the pulse shape evolution in time may even become chaotic. Soliton explosions are among the most striking phenomena that can be observed in soliton lasers.

24.9.1 High-energy dissipative solitons

Ultrashort optical pulses with extremely high energy are of great importance for a variety of applications. They can be used in fields like medicine, biology, ultrasensitive laser spectroscopy, micromachining, ultrahigh bit-rate communication systems, supercontinuum generation, metrology, and others. Thus, producing shorter and more powerful pulses is vital for further progress in science and technology.

It has been found numerically that the energy of a dissipative soliton solution of the CGLE increases indefinitely when the equation parameters converge to a given region of the parameter space, which is known as dissipative soliton resonance (DSR). The energy of a DSR pulse increases mainly due to the increase in the pulse width, while keeping the amplitude at a constant level. In order to obtain high-energy ultrashort pulses, a linear pulse compression technique has to be used outside the laser cavity.

A different kind of high-energy ultrashort pulses corresponds to the so-called very high amplitude (VHA) soliton solutions of the CGLE. Such VHA solutions occur due to a singularity of the CGLE, namely, as the non-linear gain saturation effect tends to zero. The increase in energy of these pulses is mainly due to the increase in the pulse amplitude, whereas the pulse width becomes narrower. Clearly, high-energy ultrashort pulses can be obtained in this case without using any additional linear pulse compression technique.

Fig. 24.15(a) illustrates the pulse profiles in the normal-dispersion regime ($D<0$), assuming the values $D=-0.5$, -1, and -1.5. Fig. 24.15(c) shows the pulse profiles in the anomalous-dispersion regime ($D>0$), considering that the dispersion parameter, D, assumes symmetric values of the previous ones: 0.5, 1, and 1.5. In both regimes, the peak pulse amplitudes increase with an increase in the dispersion magnitude. The pulse width increases significantly with the magnitude of D in the normal-dispersion regime, whereas it remains almost unchanged in the anomalous-dispersion regime. From Figs. 24.15(b) and (d), one can observe that the spectral range of the pulse power spectral density is almost the same for all cases in both regimes. On the other hand, the spectrum peak power becomes

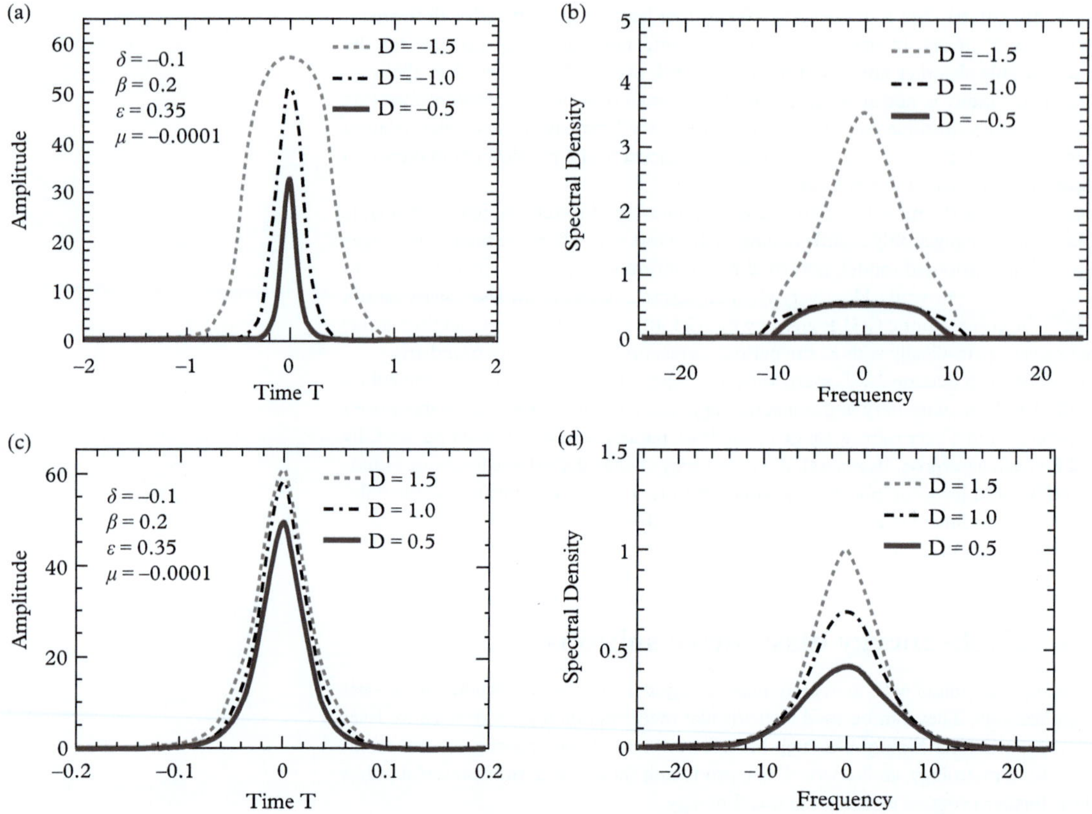

Figure 24.15 *(a, c) Amplitude profiles and (b, d) power spectral density in the normal ($D < 0$) and anomalous ($D > 0$) dispersion regimes. The dispersion parameter assumes the following values: $D = \pm 1.5$ (dashed curves), $D = \pm 1$ (dashed-dot curves), and $D = \pm 0.5$ (solid curves). The other parameter values are: $\delta = -0.1$, $\beta = 0.2$, $\varepsilon = 0.35$, $\mu = -0.0001$, and $v = 0$.*

higher when increasing the magnitude of D in the normal-dispersion regime, as observed particularly for $D = -1.5$. Actually, this corresponds to the case with the highest pulse energy.

Fig. 24.16 shows (a) the amplitude profiles and (b) the power spectral density for $D = -1.5$ and several values of the non-linear gain saturation parameter: $\mu = -0.0001$ (solid curves), $\mu = -0.00005$ (dash–dot curves), and $\mu = -0.00001$ (dashed curves). As the magnitude of μ decreases, the pulses become narrower, whereas their amplitudes increase. On the other hand, Fig. 24.16(b) shows that both the power spectrum peak values and the spectral range increase as $\mu \to 0-$.

24.9.2 Dissipative soliton molecules

Dissipative soliton molecules correspond to bound states arising from the interaction of initially separate single solitons. They are truly stationary and robust structures, whose energy is proportional to the number of involved solitons. Such structures differ significantly from

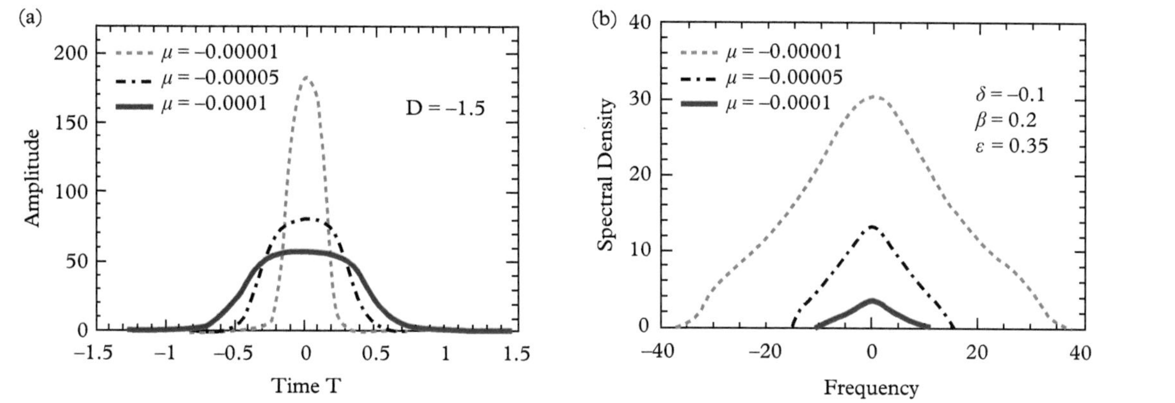

Figure 24.16 *(a) Amplitude profiles and (b) power spectral density for* $D = -1.5$ *and several values of the non-linear gain saturation parameter:* $\mu = -0.00001$ *(dashed curves),* $\mu = -0.00005$ *(dashed-dot curves), and* $\mu = -0.0001$ *(solid curves). The other parameter values are* $\delta = -0.1$, $\beta = 0.2$, $\varepsilon = 0.35$, *and* $v = 0$.

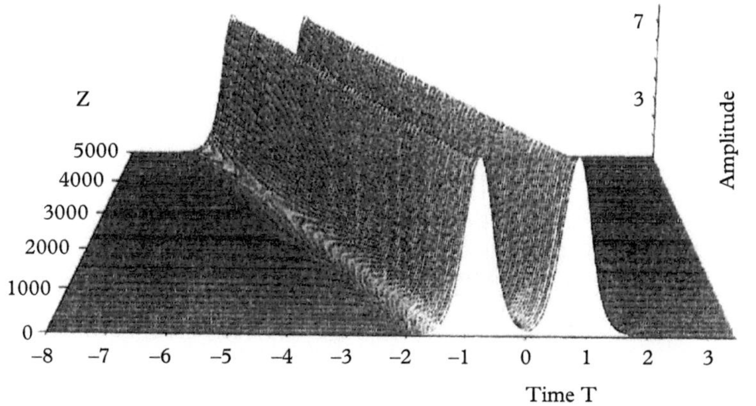

Figure 24.17 *Propagation of a soliton molecule, considering the following parameter values:* $\delta = -0.01$, $\beta = 0.5$, $\varepsilon = 1.5$, $v = 0$, $\mu = -0.03$.

higher-order soliton solutions discussed in Section 24.3, which are easily destroyed by external perturbations or higher-order effects.

The stable propagation of a soliton molecule constituted by two solitons is illustrated in Fig. 24.17. On the other hand, Fig. 24.18 shows (a) the initial (dashed curve) and final (full curve) profiles, (b) the phase, and (c) the spectrum of the soliton molecule.

The two-soliton molecule can be assumed as the building block to construct various multi-soliton molecules. An example is given in Fig. 23.19 of the four-soliton molecule with a phase difference of $\pi/2$ between adjacent pulses. As observed in the case of the two-soliton solution, multi-soliton molecules formed by PPs also move with the same constant velocity along the T axis. Soliton molecules like that illustrated in Fig. 23.19 may increase the capacity of telecommunications and can also be attractive in all-optical information storage.

Phase-locked multi-soliton molecules have been numerically and experimentally observed considering different laser configurations and dispersion regimes. Actually,

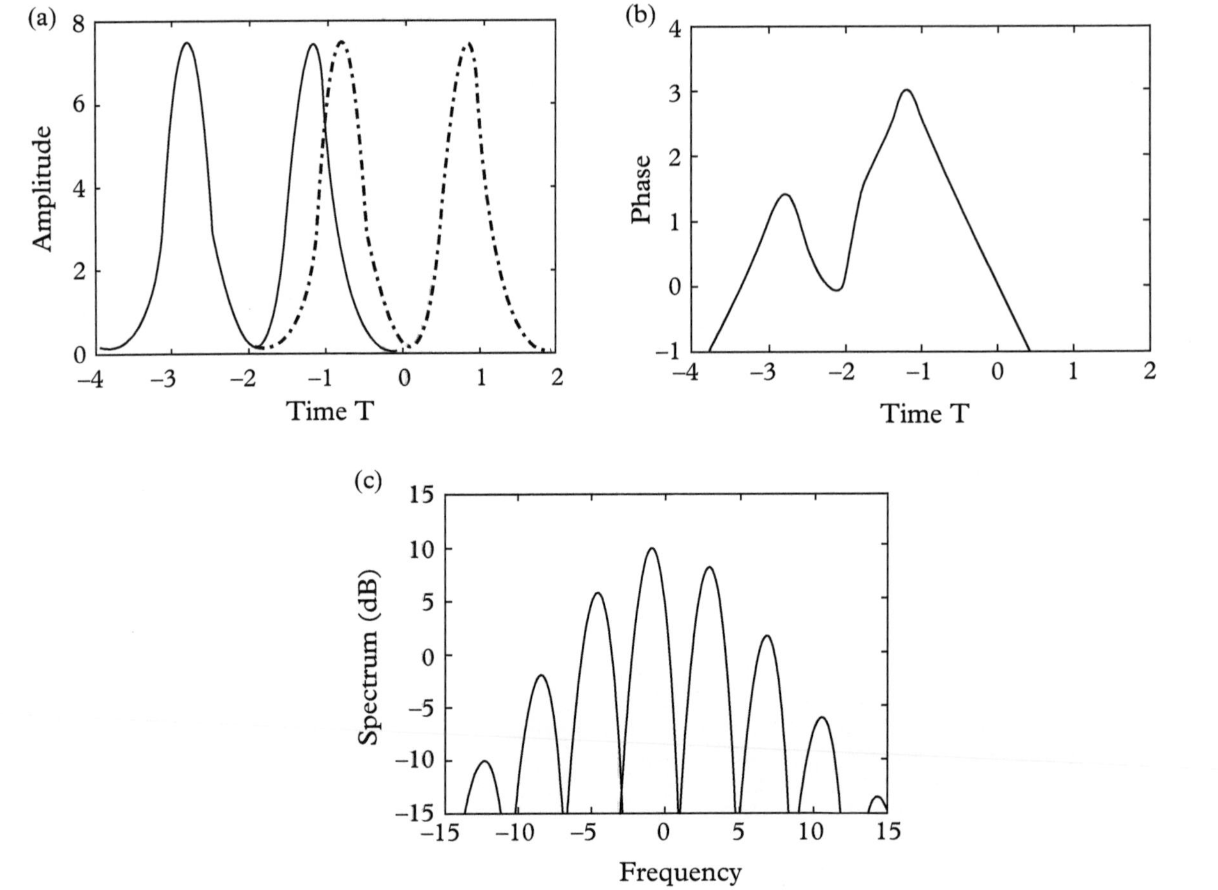

Figure 24.18 *(a) Initial (dashed curve) and final (full curve) profiles, (b) phase, and (c) spectrum of a molecule constituted by two pulses with a phase difference of $\pi/2$ between them.*

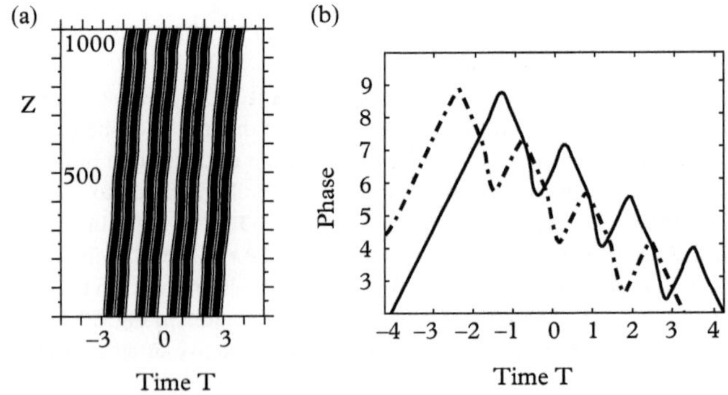

Figure 24.19 *Four-plain pulse molecule with a phase difference of $\pi 2$ between adjacent pulses. The dash-dotted (full) lines in (b) correspond to the initial (final) phase profiles.*

mode-locked soliton fibre lasers constitute an ideal testbed to study the formation and non-linear dynamics of this kind of structures.

24.10 Further Reading

Ablowitz, M. J., and P. A. Clarkson, *Solitons, Nonlinear Evolution Equations, and Inverse Scattering*, Cambridge University Press, 1991.

Agrawal, G. P., *Nonlinear Fiber Optics*, Academic Press, 5th Ed., 2012.

Akhmediev, N. and A. Ankiewicz, *Solitons: Nonlinear Pulses and Beams*, Chapman and Hall, 1997.

Drazin, P. G., *Solitons: An Introduction*, Cambridge University Press, 1993.

Ferreira, M. F., *Nonlinear Effects in Optical Fibers*, John Wiley & Sons, 2011.

Ferreira, M. F., *Solitons in Optical Fiber Systems*, John Wiley & Sons, 2022.

Ferreira, M. F. (Ed.), *Dissipative Optical Solitons*, Springer, 2022.

Gordon, G. P., and H. A. Haus, Random Walk of Coherently Amplified Solitons in Optical Fiber Transmission, *Opt. Lett.*, 11, no. 10, 665, 1986.

Hasegawa, A., and F. D. Tappert, Transmission of Stationary Nonlinear Optical Pulses in Dispersive Dielectric Fibers I. Anomalous Dispersion, *Appl. Phys. Lett.* 23, no. 3, 142, 1973.

Hasegawa, A., and Y. Kodama, *Solitons in Optical Communications*, Oxford University Press, 1995.

Mollenauer, L. F., R. H. Stolen, and J. P. Gordon, Experimental Observation of Picosecond Pulse Narrowing and Solitons in Optical Fibers, *Phys. Rev. Lett.* 45, no. 13, 1095, 1980.

Russell, J. S., *Report of 14th Meeting of the British Association for Advancement of Science*, York, 311, 1844.

Taylor, J. T. (Ed.), *Optical Solitons—Theory and Experiment*, Cambridge University Press, 1992.

Zabusky, N., and M. Kruskal, Interaction of 'Solitons' in a Collisionless Plasma and the Recurrence of Initial States, *Phys. Rev. Lett.* 15, 240, 1965.

Zakharov, V. E., and A. Shabat, Exact Theory of Two-Dimensional Self-Focusing and One-Dimensional Self-Modulation of Waves in Nonlinear Media, *Sov. Phys. JETP* 34, 62, 1972.

24.11 Problems

24.1 Verify by direct substitution that $q(Z,T)$, given by Eq. (24.20) or Eq. (24.22), is indeed a solution of the non-linear Schrödinger Eq. (24.19).

24.2 Determine the required peak power of a 10 ps (FWHM) pulse of centre wavelength 1550 nm to propagate as a fundamental soliton in a fibre having a dispersion of $D = 2\,\text{ps}/(\text{nm}-\text{km})$ and a non-linear parameter $\gamma = 5\,\text{W}^{-1}/\text{km}$. What peak power is required to propagate a $N = 5$ soliton in such a fibre?

24.3 Verify by direct substitution that the dark soliton described by Eqs. (24.43)–(24.45) is indeed a solution of the normalized NLSE given by Eq. (24.42).

24.4 Show that the NLSE (24.27) is obtained from the Euler–Lagrange Eq. (24.52), considering the Lagrangian density given by Eq. (24.53).

24.5 Obtain the Lagrangian given by Eq. (24.56) from Eq. (24.51), considering the Lagrangian density in Eq. (24.53) and $U(z,\tau)$ in Eq. (24.55).

24.6 Considering the Lagrangian given by Eq. (24.56), reproduce Eqs. (24.57)–(24.59) from the reduced Euler–Lagrangian Eq. (24.54).

24.7 Explain the intrapulse Raman scattering effect and the reason for the self-frequency shift of solitons during propagation along an optical fibre.

24.8 Explain the main advantages of using DM solitons instead of conventional solitons in optical fibre transmission systems.

24.9 Assume a sech pulse, given by

$$U(z,\tau) = U_p(z)\,\text{sech}\left(\frac{\tau}{t_p(z)}\right)\exp\left[-iC_p(z)\left(\frac{\tau}{t_p(z)}\right)^2 + i\theta_p(z)\right]$$

where the parameters $U_p(z)$, $C_p(z)$, $t_p(z)$, and $\theta_p(z)$ represent the amplitude, the chirp parameter, the width, and the phase of the pulse, respectively. Obtain the Lagrangian and derive the evolution equation for the above pulse parameters.

24.10 Using the evolution equations obtained in Problem 24.9, show that the pulse width remains constant in the case of an unchirped pulse propagating in the anomalous-dispersion region if the initial pulse parameters satisfy Eq. (24.18) for $N = 1$.

Supercontinuum and UV Light Generation

25

Besides imposing several limitations on the performance of optical communication systems, non-linear effects in optical fibres offer a variety of novel possibilities for all-optical processing, which are very promising for near-future high-capacity networks. In order to improve some of these applications, several types of highly non-linear fibres (HNLFs) have been developed during recent years.

One of the most impressive non-linear phenomena that can be observed in HNLFs is supercontinuum (SC) generation. It corresponds to an extremely wide spectrum achieved by an optical pulse while propagating in a non-linear medium, and generally results from the synergy between several fundamental non-linear processes. The spectral location and power of the pump, as well as the non-linear and dispersive characteristics of the medium, determine the relative importance and the interaction between these non-linear processes. A SC source can find applications in various areas such as spectroscopy, optical coherence tomography, frequency metrology, optical communication sources, etc.

25.1 Highly Non-Linear Optical Fibres

As defined in Chapter 21, the fibre non-linear parameter γ is given by

$$\gamma = \frac{2\pi}{\lambda}\frac{n_2}{A_{eff}} \tag{25.1}$$

where λ is the light wavelength, n_2 is the non-linear-index coefficient of the fibre core, and A_{eff} is the effective mode area. In standard single-mode silica fibres, the non-linear parameter has a typical value of only $\gamma \approx 1.3\,\mathrm{W}^{-1}/\mathrm{km}$. Such a value is too small for some applications requiring highly efficient non-linear processes, namely optical signal processing and supercontinuum generation. Actually, a fibre length of several kilometres would generally be necessary in such applications. To shorten the interaction length, HNLFs made with standard silica core and cladding compositions but with a smaller effective mode area, and hence a larger non-linear coefficient, $\left(\gamma \approx 10-20\,\mathrm{W}^{-1}/\mathrm{km}\right)$ have been widely used. About 1 km of this type of fibre is usually used for the demonstration of different non-linear effects. The fibre non-linearity can be further enhanced by appropriately tailoring its structure. Different types of silica-based microstructured optical fibres (MOFs)—already discussed in Chapters 15 and 16—have been designed to address this purpose, providing a non-linear coefficient $\gamma \approx 70\,\mathrm{W}^{-1}/\mathrm{km}$. Significantly higher values of γ can be achieved by combining tight mode confinement with the use of glasses with a greater intrinsic material non-linearity coefficient than that of silica, such as bismuth oxide, lead silicate, and chalcogenide glasses.

Lasers, Optics, and Photonics. Mário F. S. Ferreira, Oxford University Press. © Mário F. S. Ferreira (2026).
DOI: 10.1093/oso/9780198952770.003.0025

25.1.1 Highly non-linear silica fibres

Eq. (25.1) shows that, for a fixed wavelength, since n_2 is determined by the material from which the fibre is made, the most practical way to increase the non-linear parameter γ is to reduce the effective mode area A_{eff}. However, the non-linear parameter γ can also be enhanced by using the dopant dependence of the Kerr coefficient n_2. In fact, n_2 increases linearly with the relative index difference Δ_d for a GeO_2-doped bulk glass. The relative index difference is defined as $\Delta_d = (n_d^2 - n_0^2)/2n_d^2$, where n_0 and n_d denote the refractive indices of pure silica and doped glass, respectively. On the other hand, n_2 decreases linearly with Δ_d for a fluorine-doped bulk glass.

In order to enhance the non-linear parameter γ, the best option is provided by a fibre heavily doped with GeO_2 in the core and having a large refractive index difference between the core and the cladding, because such fibres exhibit a large n_2 as well as a small effective area A_{eff}. Fig. 25.1 shows the schematic refractive index profile of a typical HNLF, in which the relative refractive index difference of the core to the outer cladding is around 3%, while the core diameter is around 4 µm. A W-cladding profile with a fluorine-doped depressed cladding having a refractive index lower than that of the outer cladding allows single-mode operation in the communication bands as well as flexibility in the design of chromatic dispersion. Using such approaches, new kinds of HNLFs with specific dispersive properties, namely dispersion-shifted fibres, dispersion-compensating fibres, dispersion-decreasing fibres, and dispersion-flattened fibres, have been developed. The values of the non-linear parameter γ for most of these fibres are in the range 10–20 W^{-1}/km. Increasing γ much above these values is not possible in this kind of fibre, since the optical mode confinement is lost when the core diameter is further reduced.

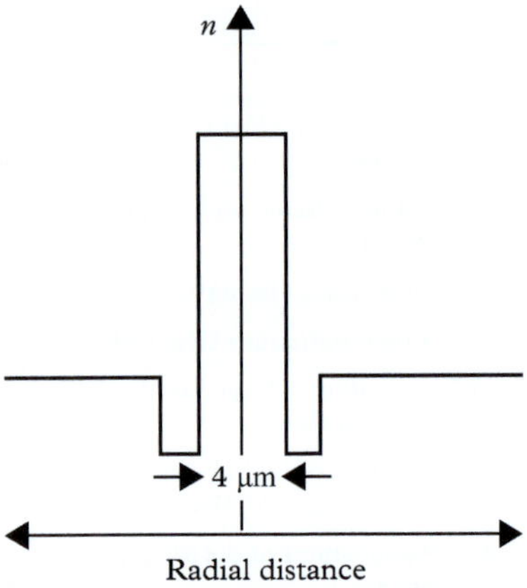

Figure 25.1 *Schematic refractive index profile of a highly non-linear optical fibre.*

25.1.2 Tapered fibres

As seen in Chapter 15, the fibre normalized frequency V can be given in the form:

$$V = k_0 a n_1 \sqrt{2\Delta} \tag{25.2}$$

where k_0 is the vacuum wavenumber, a is the core radius, n_1 is the core refractive index, and

$$\Delta = (n_1 - n_c)/n_1 \tag{25.3}$$

is the relative index difference between the core and the cladding of the fibre, n_c being the cladding refractive index. A single-mode fibre is achieved when $V < 2.405$. If Δ is kept constant and the core radius a is reduced, the normalized frequency V decreases and the mode confinement is lost. Eq. (25.2) shows that, in order to maintain a given value of V when decreasing a, Δ must increase such that $a^2\Delta$ remains constant.

The maximum value of Δ is achieved when the cladding material is replaced by air ($n_c \approx 1$). In such a case, the mode remains confined to the core even if the core diameter is close to 1 µm. However, such confinement is reduced for smaller values of the core diameter.

Narrow-core fibres with air cladding have been produced by tapering standard optical fibres, with an original cladding diameter of 125 µm. Tapering a fibre involves heating and stretching it to form a narrow waist connected to the untapered fibre by taper transitions (see Fig. 25.2). If the transitions are gradual, light propagating along the fibre suffers very little loss.

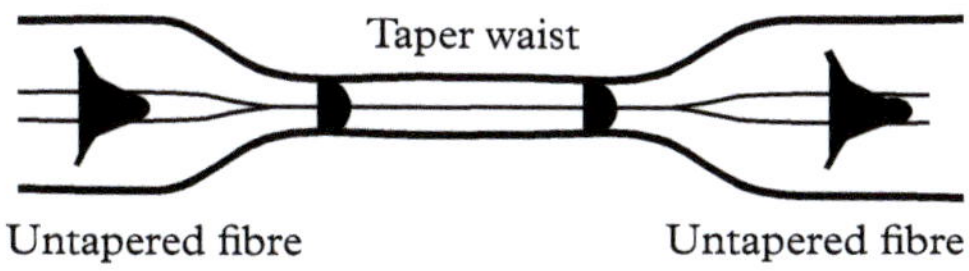

Figure 25.2 *Schematic of a tapered fibre.*

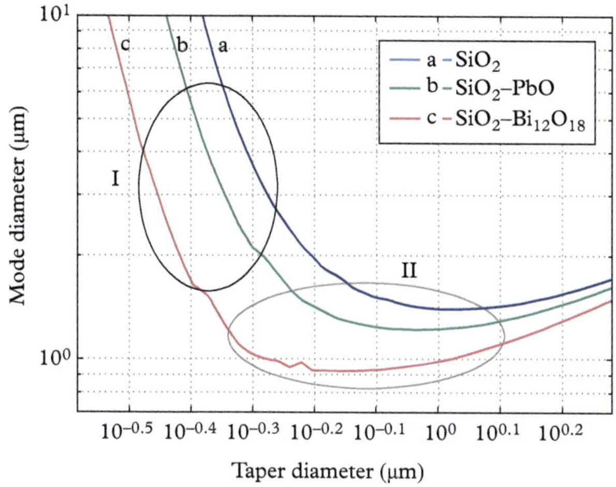

Figure 25.3 *Variation of the spot size of a tapered fibre with the fibre diameter.*

The non-linear parameter γ is significantly enhanced when the core diameter is reduced to values of the order of 2 μm. Considering $\lambda = 1$ μm for the wavelength of light transmitted through the fibre, $n_1 = 1.45$ for the silica core, and $n_c = 1$ for the air cladding, a maximum value $\gamma \approx 370\,\text{W}^{-1}/\text{km}$ is found for $V \approx 1.85$.

Fig. 25.3 shows the mode spot size as a function of the fibre diameter, considering different fibre materials: silica (SiO_2), a lead silicate glass ($SiO_2 - PbO$), and a bismuth silicate glass ($SiO_2 - Bi_{12}O_{18}$). One can observe the existence of a minimum beam waist for a given value of the fibre diameter, which depends on the fibre material. Two regions are identified in Fig. 25.3. Region I corresponds to the large evanescent-field region, which may be useful for sensing applications. Region II corresponds to the high-confinement region, which is more appropriate for non-linear applications.

The dispersive properties of a tapered fibre are very sensitive to the taper. The zero-dispersion wavelength (ZDW) can be shifted to the visible range and, in some cases, a second ZDW appears at longer wavelengths, defining a wavelength window between the two ZDWs that shows anomalous GVD. This is illustrated in Fig. 25.4, which shows the calculated dispersion spectra of taper waists for several values of the taper diameter. One can observe that, as the diameter of the taper waist decreases, the peak in dispersion shifts to shorter wavelengths and a second ZDW appears at longer wavelengths.

25.1.3 Microstructured optical fibres

As described in Chapter 15, microstructured optical fibres (MOFs) can be divided into two main types: solid-core and hollow-core MOFs. In a solid-core MOF, increasing the

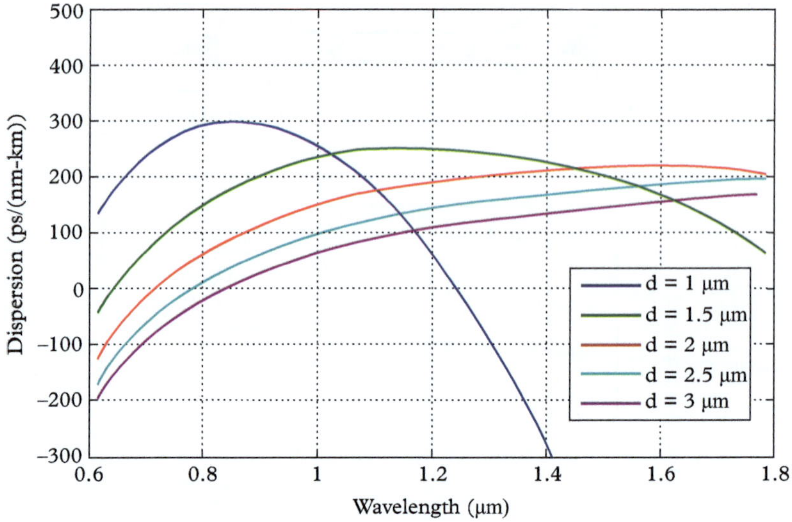

Figure 25.4 *Dispersion properties of a tapered fibre surrounded by air, for several values of the diameter.*

hole–pitch ratio reduces the effective mode area, due to the increased contrast between the refractive index of the core and the average refractive index of the cladding. This is illustrated in Fig. 25.5, which shows the spatial mode distribution for a MOF with a hexagonal pattern of holes of diameter $d = 2\,\mu m$ (a) and $d = 4\,\mu m$ (b), spaced by $\Lambda = 10\,\mu m$. Consequently, the non-linear parameter increases when the hole–pitch ratio is increased.

As the holes get larger, the core becomes more and more isolated, until it resembles an isolated strand of silica glass. The MOFs with larger cores exhibit semi-infinite anomalous dispersion above the ZDW. By decreasing the core size, the ZDW tends to shift to a shorter wavelength, leading to anomalous dispersion at near-infrared and visible wavelengths. When the core size is decreased further, a second ZDW arises on the longer wavelength side, such that the GVD is anomalous in the spectral window between the two ZDWs and normal outside it, as illustrated in Fig. 25.4. Submicron-diameter MOF cores with low dispersion have been fabricated using a conventional tapering process.

Fig. 25.6 shows the effective mode area of a MOF against the hole-to-hole spacing Λ, for different hole–pitch ratios d/Λ (full curves). The dashed curve corresponds to the case of a so-called jacketed air-suspended rod (JASR) and shows a minimum mode area for a rod diameter around $1.1 - 1.2\,\mu m$. Fig. 25.6 shows that solid-core MOFs also exhibit a minimum effective mode area for a given value of the hole-to-hole spacing. The smallest effective mode area corresponds to the largest air-filling fraction, which is achieved in Fig. 25.6 for the case $d/\Lambda = 0.9$. In this case, a minimum effective mode area $A_{eff} = 1.7\,\mu\,m^2$ is obtained, which occurs for a hole-to-hole spacing $\Lambda \approx 1\,\mu m$.

In the case of MOFs, and to take into account the different proportions of light in glass and air, the non-linear coefficient γ must be redefined as follows:

$$\gamma = k_0 \sum_i \frac{n_2^i}{A_{eff}^i} = k_0 \frac{n_2^{eff}}{A_{core}} \tag{25.4}$$

In Eq. 25.4, n_2^i is the non-linear refractive index of material i ($2.9 \times 10^{-23}\ m^2W^{-1}$ for air, $2.6 \times 10^{-20}\,m^2W^{-1}$ for silica), A_{eff}^i is the effective area of light in material i, and n_2^{eff} is the

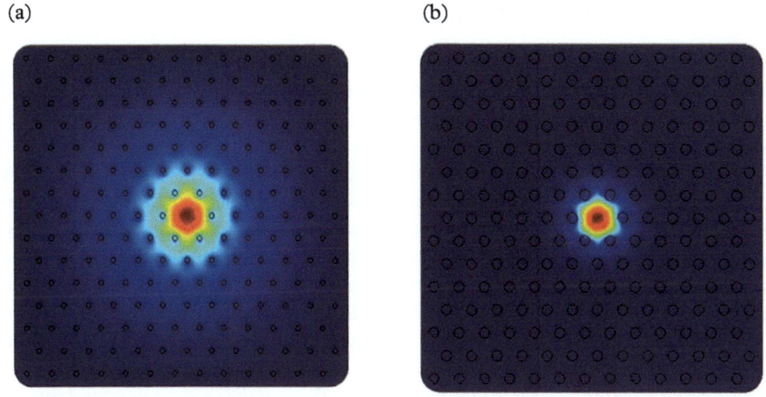

Figure 25.5 *Spatial mode distribution for a MOF with a hexagonal pattern of holes of diameter* $d = 2\,\mu m$ *(a) and* $d = 4\,\mu m$ *(b), spaced by* $\Lambda = 10\,\mu m$*, considering a wavelength of* $1.55\,\mu m$.

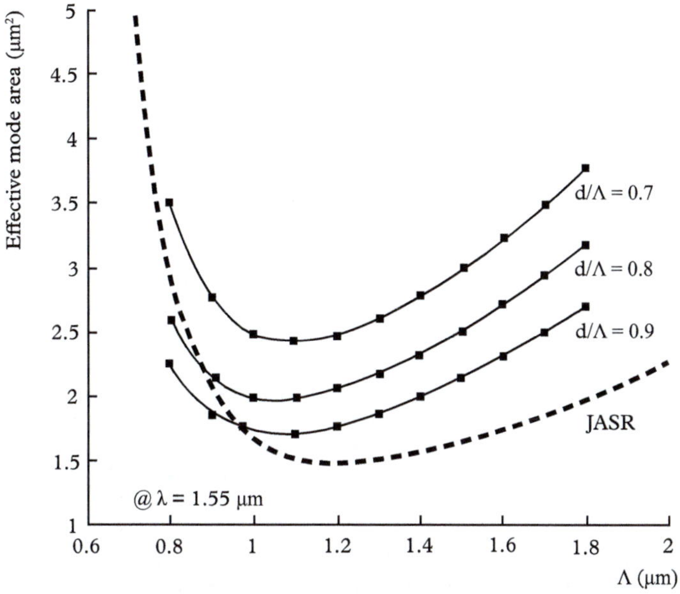

Figure 25.6 *Effective mode area of a jacketed air-suspended rod (dashed curve) and of a microstructured optical fibre (full curves), as a function of the rod diameter or the hole-to-hole spacing, respectively.*

effective non-linear index for the fibre, with a core area A_{core}. A non-linear coefficient $\gamma = 240\,W^{-1}/km$ at 850 nm has been measured for a solid-core MOF with a core diameter of 1 µm, which must be compared with the highest values $\sim 20\,W^{-1}/km$ obtainable in conventional highly non-linear silica fibres. Considering a peak power of $P_0 = 5\,kW$, the non-linear length for the same solid-core MOF will be $L_{NL} < 1\,mm$. For dispersion values in the range $-400 < \beta_2 < 400\,ps^2/km$ and pulse durations $t_0 = 300\,fs$, one has a dispersion length $L_D = t_0^2/|\beta_2| > 0.2\,m$. Considering also typical values of loss (1–100 dB/km), we conclude that both the effective fibre length $L_{eff} = (1 - \exp(-\alpha L))/\alpha$ and the dispersion length L_D are much longer than the non-linear length L_{NL}, which means that strong non-linear effects will be readily observable in solid-core MOFs.

25.1.4 Non-silica fibres

The value of the non-linear parameter γ can be enhanced if the optical fibre is made using glasses with higher non-linearities than silica, such as lead silicate, tellurite, bismuth oxide, and chalcogenide glasses. Considering several optical glasses, it has been observed that the non-linear-index coefficient n_2 increases with the linear refractive index, n. Introducing heavy atoms or ions with a large ionic radius (i.e., using chalcogen elements S, Se, or Te to replace oxygen) acts to increase the polarizability of the components in the glass matrix and also increase the non-linear index n_2. Actually, the non-linear index can be increased by a factor of almost 1000 compared with its value for silica fibres using some types of highly non-linear glasses. The ZDW of a glass also shifts to longer wavelengths with increasing linear refractive index. Heavy metal oxide glasses (lead silicate, bismuth oxide, tellurite) have linear indices in the range 1.8–2.0, non-linear indices ~10 times higher than silica, and material ZDWs of 2–3 µm. Chalcogenide glasses (GLS, As_2S_3) have linear indices of 2.2–2.4, non-linear indices significantly higher than those of the oxide glasses, and material ZDWs larger than 4 µm.

Lead silicate and tellurite glasses can offer values of n_2 which are about 10–20 times larger than those of silica. Both types of materials have been used for making microstructured optical fibres. The extrusion technique for fibre preform manufacture is particularly suitable when using compound glasses, since they exhibit low softening temperatures of ~500 °C, as opposed to ~200 °C for silica. Besides being simpler than the more common stacking technique developed for silica MOFs, extrusion allows the realization of cladding structures consisting mostly of air, as required to increase the fibre non-linearity.

Much attention has been paid in recent years to fibres based on bismuth oxide (Bi_2O_3). Among the different types of non-linear fibres, Bi-NLFs have a relatively mature technology, and the fibres can be spliced to standard single-mode fibres. They also possess good chemical, mechanical and thermal stability and have low photosensitivity as compared to chalcogenide fibres. Depending on the concentration of bismuth in the fibre, the non-linear refractive index n_2 can be varied from 30×10^{-20} m^2/W to 1.1×10^{-18} m^2/W, representing a 10–40-fold enhancement as compared with that of silica.

A non-linear parameter $\gamma = 2450\,\text{W}^{-1}$/km was measured in a 2004 experiment for an 85-cm-long chalcogenide fibre with a core diameter of 7 µm (Slusher et al., 2004). This corresponds to a non-linear refractive index $n_2 = 2.4 \times 10^{-17}$ m^2/W, which is larger by a factor of about 1000 relative to the silica case. The same fibre also exhibited a Raman gain parameter which was nearly 800 times larger than that of silica fibres. This allows the use of either a smaller fibre length or a lower pump power to attain amplifier gain characteristics similar to those of silica-based devices. Due to their high non-linearity, chalcogenide fibres have been considered a good option to enhance several non-linear effects.

In recent years, MOFs have been fabricated with various types of chalcogenide glasses for several applications. The employment of photonic crystal geometry in the cladding of As_2Se_3 fibres can give rise to a strong overlap between pump and signal, providing higher Raman gain efficiency than in conventional fibres.

25.2 Soliton Fission and Dispersive Waves

As mentioned in Chapter 24, the propagation of ultrashort pulses in optical fibres can be perturbed by several higher-order dispersive and non-linear effects, namely third-order dispersion, self-steepening, and intrapulse Raman scattering. In the case of a higher-order soliton, these perturbations can determine the fission of the pulse into its constituent fundamental solitons, a phenomenon called *soliton fission*.

In the presence of higher-order dispersion, an Nth-order soliton gives origin to N fundamental solitons. Soliton fission generally occurs after a propagation distance $L_{fiss} \sim L_D/N$, where $L_D = t_0^2/|\beta_2|$ is the dispersion distance at which the injected higher-order soliton attains its maximum compression and larger bandwidth. The fission distance L_{fiss} is a particularly significant parameter in the context of higher-order soliton effect compression, as discussed in Section 24.3.

Besides higher-order dispersion, another main effect affecting the dynamics of higher-order soliton fission is intrapulse Raman scattering (IRS). As discussed in Chapter 24, this phenomenon leads to a continuous downshift of the soliton carrier frequency, an effect known as the *soliton self-frequency shift* (SSFS). The origin of SSFS can be understood by noting that, for ultrashort solitons, the pulse spectrum becomes so broad that the high-frequency components of the pulse can transfer energy through Raman amplification to the low-frequency components of the same pulse. Such an energy transfer appears as a redshift of the soliton spectrum, with the shift increasing with distance.

The rate of frequency shift per propagation length is given in real units by

$$\frac{df}{dz} = -\frac{4t_R\,|\beta_2|}{15\pi t_0^4} = -\frac{4t_R(\gamma P_0)^2}{15\pi\,|\beta_2|} \tag{25.5}$$

where $t_R \approx 5$ fs is the Raman parameter and P_0 is the soliton peak power. Since the SSFS effect is proportional to $(\gamma P_0)^2$, it will be enhanced if short pulses with high peak power are propagated in highly non-linear fibres.

As a consequence of the SSFS induced by IRS, the solitons arising from the fission process separate from each other. Since the SSFS is the largest for the shortest soliton, its spectrum shifts the most towards the red side. In this process, new spectral peaks appear at $z = L_{fiss}$, corresponding to the so-called *non-soliton radiation* (NSR), which is emitted by the solitons resulting from the fission process in the presence of higher-order dispersion.

The NSR is emitted at a frequency such that its phase velocity matches that of the soliton. The two phases are equal when the following phase-matching condition is satisfied:

$$\beta(\omega) = \beta(\omega_s) + \beta_1(\omega - \omega_s) + \frac{1}{2}\gamma P_s \tag{25.6}$$

In Eq. (25.6), P_s and ω_s are the peak power and the frequency, respectively, of the fundamental soliton formed after the fission process. Expanding $\beta(\omega)$ around ω_s, this phase-matching condition reduces to

$$\sum_{n\geq 2} \frac{\Omega^n}{n!}\beta_n(\omega_s) = \frac{\gamma P_s}{2} \tag{25.7}$$

where $\Omega = \omega - \omega_s$ is the frequency shift between the dispersive wave and the soliton, and $\beta_n\,(n = 2, 3)$ are the dispersion parameters, calculated at the soliton central frequency ω_s.

Neglecting the fourth- and higher-order dispersion terms, the frequency shift between the frequency of the dispersive waves and that of the soliton is given approximately by (Akhmediev and Karlsson, 1995):

$$\Omega_d \approx -\frac{3\beta_2}{\beta_3} + \frac{\gamma P_0\beta_3}{3\beta_2^2} \tag{25.8}$$

For solitons propagating in the anomalous-GVD regime, one has $\beta_2 < 0$. In these circumstances, Eq. (25.8) shows that $\Omega > 0$ when $\beta_3 > 0$, in which case the NSR is emitted at wavelengths shorter than that of the soliton. On the contrary, if $\beta_3 < 0$, the NSR is emitted at a longer wavelength than that of the soliton, and it can result in the suppression of the SSFS effect.

25.3 Four-Wave Mixing

As discussed in Chapter 21, new frequencies can be generated in conventional optical fibres by parametric four-wave mixing (FWM) when they are pumped close to the ZDW. In this case, the phase-matching condition is sensitive to the exact shape of the dispersion curve. Thus, we must expect that the peculiar dispersive properties of some highly non-linear fibres have a profound impact on the FWM process through the phase-matching condition.

The phase-matching condition for the FWM process requires that dispersive effects compensate non-linear ones through the following equation:

$$\beta_2 \Omega_s^2 + \frac{\beta_4}{12} \Omega_s^4 + 2\gamma P_p = 0 \tag{25.9}$$

where β_2 and β_4 are, respectively, the second- and fourth-order dispersion terms at the pump frequency; Ω_s is the frequency shift between the signal or idler frequencies and the pump frequency; γ is the non-linear parameter; and P_P is the pump peak power. The third and all odd dispersion orders play no role in Eq. (25.9) since they cancel out from the degenerate phase-matching condition $2\beta(\omega_p) - \beta(+\Omega_s) - \beta(-\Omega_s) = 0$, where $\beta(\omega)$ is the exact wave vector.

The impact of the fourth-order dispersion term (β_4) on parametric processes has been experimentally studied both in standard telecommunications fibres and in MOFs. In the anomalous dispersion regime ($\beta_2 < 0$), the β_2 term dominates in Eq. (25.9), and the frequency shift is given by Eq. (21.68). However, modulation instability can also occur in the normal-GVD regime ($\beta_2 > 0$). In this case, the positive non-linear phase mismatch $2\gamma P$ is compensated by the negative value of the linear phase mismatch due to β_4.

25.4 Supercontinuum Generation

Supercontinuum (SC) generation in optical fibre involves the interplay between non-linear and linear effects that occur during the propagation of an optical field (pulsed or continuous wave (CW)) in a fibre waveguide. Although the complex dynamics of this interplay can sometimes make it difficult to identify each of the contributing mechanisms, an understanding of how the different effects act individually is important for a description of the spectral broadening process.

SC generation can be broadly divided into two categories, depending on the duration of the pump pulses. The first category corresponds to long (picosecond, nanosecond, and CW) pump pulses, whereas the second category is obtained with short (femtosecond) pump pulses. The dispersion regime (normal or anomalous) at the pump wavelength is also an important factor, at least in the initial propagation phase. Soliton-related effects play a significant role whenever light with sufficient power propagates in the anomalous dispersion regime.

25.4.1 Pumping with picosecond pulses

When pumping with long-duration and high peak-power pulses in the anomalous dispersion regime near the ZDW, spontaneous modulation instability (MI) becomes the dominant process in the initial propagation phase. In the frequency domain, MI causes the spread of energy from the central wave frequency to sidebands with a frequency separation given by Eq. (21.72). The MI period can be obtained from the above frequency separation as

$$T_s = \frac{2\pi}{\Omega_s} = \sqrt{\frac{2\pi^2 |\beta_2|}{\gamma P_p}} \tag{25.10}$$

The modulated picosecond pulse can then break up into a train of solitons with durations approximately given by T_s. The subsequent evolution of these solitons then leads to additional spectral broadening and supercontinuum formation through a variety of mechanisms. If such solitons are sufficiently short, they are red-shifted due to the IRS effect. As seen from Eq. (25.5), the rate of frequency shift is inversely proportional to the fourth power of the soliton width. If the solitons resulting from the initial pulse break-up are not sufficiently short, they can undergo soliton collisions, which determine the transfer of energy among them due to Raman scattering. In this process, some solitons achieve a higher peak power and then undergo a larger redshift.

If the pump wavelength is near the ZDW, the redshift of the solitons to longer wavelengths is accompanied by the generation of dispersive waves on the short-wavelength side of the ZDW in fibres with positive dispersion slopes ($\beta_3 > 0$). Since the dispersive waves have a lower group velocity, they will lag behind the solitons. However, the group velocity of the red-shifting solitons decreases during the propagation, and eventually they meet the dispersive waves. The temporal overlap between a soliton and a dispersive wave allows them to interact via the cross-phase modulation effect, which determines a blue-shift and a decrease in the group velocity of the dispersive wave. In this way, as long as this cycle continues, the blue-shifted radiation cannot escape from the soliton, which is usually referred to as a 'trapping effect'. This process is responsible for the maximum blue-shift of the supercontinuum, which is only limited by the ability of the solitons to redshift. This redshift can be limited by the existence of a second ZDW, which determines the spectral recoil of the soliton.

The solitonic effects are absent, at least in the initial propagation phase, if pumping is realized far into the normal-dispersion regime. In this case, the spectral broadening occurs initially due to stimulated Raman scattering (SRS). The contribution of the SRS process to the supercontinuum generation is especially effective on the long-wavelength side of the spectrum, which tends to make the overall spectrum asymmetric.

The FWM becomes progressively more important if the pumping is closer to the ZDW, since the parametric gain is higher than the Raman gain. Using pump pulses of sub-kilowatt peak power and a duration of some tens of ps, a spatially single-mode supercontinuum with a width of more than 600 nm can be generated by the interplay of SRS and FWM.

The dispersive properties of the fibre become especially important in the case of the FWM process. If the required phase-matching conditions are satisfied, FWM generates sidebands on the short- and long-wavelength sides of the pulse spectrum. This process can produce a wideband supercontinuum even in the normal-GVD region of an optical fibre. Actually, the phase-matching condition for the FWM process requires that dispersive effects compensate non-linear ones through Eq. (25.9). This phase-matching condition can be satisfied even for a pump wavelength in the normal-GVD regime ($\beta_2 > 0$) if the fourth-order dispersion coefficient (β_4) is negative. This can be easily realized in tapered and microstructured fibres.

A flat octave-spanning supercontinuum has been generated from the visible to the near-infrared by pumping a MOF at 1064.5 nm, corresponding to the normal-dispersion regime, with 0.6-ns pulses with a 7 kHz repetition rate and ~60 mW average power (Kudlinski et al. 2008a). Different fibre lengths were considered in this experiment. For a length $L = 1$ m, the signal and idler waves generated by degenerate FWM were clearly visible. They were located, respectively, at 810 and 1548 nm, which is in excellent agreement with Eq. (25.9). By increasing the fibre length, the spectral widths of these two waves and that of the pump increased. The broadening of the pump spectrum was expected since the injected peak power is much higher than the Raman threshold. As a consequence, the phase-matching relation given by Eq. (25.9) could be satisfied between additional wavelengths located close to the pump and other wavelengths located around the signal and idler waves. For a fibre length of 30 m, the signal and idler waves started generating their own Raman cascade. For a fibre length of 100 m, the power of the signal wave became strong enough to generate up to five Raman Stokes orders located in the normal-GVD region. In contrast, the Raman orders generated by the pump and idler waves were located in the anomalous regime and broadened rapidly, evolving into a continuum.

The general features observed in the case of long-duration pulses can also be extended to continuous-wave (CW) pumping in the anomalous dispersion regime. In this case, supercontinuum generation also starts with the MI of the continuous pump wave in the anomalous dispersion region. This instability evolves from background noise and is often stimulated by the intensity fluctuations which are typical of a CW pump. The MI leads to the splitting of the CW into fundamental solitons.

In some circumstances, the fundamental solitons arising from MI can be short enough in order to enable the occurrence of IRS, leading to the SSFS. As the MI evolves from noise, there will be some jitter in the period and hence also in the duration of the generated solitons. In this case, a range of soliton self-frequency shifts occurs, creating a smooth red-shifted continuum evolution.

Soliton collisions also constitute an important mechanism for spectral expansion in CW-pumped supercontinua. In this case, the number of solitons can be particularly large, and as red-shifted solitons travel more slowly than those at higher frequencies, they can eventually collide while propagating through a fibre. During such collisions, some energy will be transferred to the most red-shifted soliton if it is within the spectral range of Raman scattering. The energy increase leads to a corresponding compression, which results in a further redshift for that soliton, accordance with Eq. (25.5). The increased redshift also slows down the soliton, leading it to pass by more solitons and hence experience more collision events. Soliton collisions can considerably increase the extent of a supercontinuum compared to Raman self-scattering alone.

The expansion of the supercontinuum to longer wavelengths can be limited by either the dispersion magnitude increasing or the non-linearity decreasing. Depending on the dispersion profile, the existence of a second ZDW can lead to the generation of dispersive waves and the spectral recoil of the solitons. Concerning the extension to shorter wavelengths, it is determined by the excitation of phase-matched dispersive waves and soliton-trapping mechanisms. Experimentally, this phenomenon has been combined with dispersion-engineered MOFs to further extend SCs to the UV in nanosecond and picosecond pumping schemes.

A SC ranging from 670 nm to 1350 nm with 9.55 W output power has been achieved in the CW pumping regime, using a 200-m long ZDW-decreasing MOF pumped by a

20 W ytterbium fibre laser at 1.06 μm (Kudlinski and Mussot, 2008b). The fibre was composed of a 100 m long section with a constant dispersion followed by a 100 m long section with linearly decreasing ZDW. To decrease the ZDW along propagation, the outer diameter of the second-section fibre was linearly reduced from an initial value of 125 μm to a final value of 80 μm. The initial ZDW was located at 1053 nm, just below the pump wavelength, and decreased to 950 nm in a quasi-linear way. The broadening of the output spectrum was shown to increase on both sides with increasing launch power. The spectrum was limited at long wavelengths by solitons red-shifted by the SSFS effect. On the other hand, the extension of the SC towards short wavelengths was mainly due to the trapping of dispersive waves by red-shifted solitons.

Compared with long-pulse pumped systems, CW pumped systems allow a simple experimental realization. Typically, the experimental arrangement consists simply of a CW fibre laser spliced directly to a suitable fibre. This brings advantages in terms of robustness and stability. An Er-doped fibre laser is generally used for pumping around 1400–1600 nm, whereas a Yb-doped fibre laser is commonly used around 1060 nm. Typically, a fibre-coupled power of 5–50 W is used for CW continuum generation. These relatively high values for the coupled power lead to very high spectral powers in the resulting continuum, typically greater than 10 mW/nm.

25.4.2 Pumping with femtosecond pulses

Pumping in the normal-dispersion regime using femtosecond pulses, where soliton formation is not allowed, leads only to spectral broadening through self-phase modulation (SPM). As seen in Chapter 21, SPM-induced spectral broadening depends on the propagation distance, peak power, and shape of the input pulse. Its role becomes particularly significant at high input powers and for ultrashort pulses, so that the rate of power variation with time, dP/dt, is large. If the pump wavelength is located near the ZDW, the spectrum can extend into the anomalous dispersion regime. In such a case, the soliton dynamics described above can also have a significant contribution to the spectral broadening.

Soliton dynamical effects play a prominent role when pumping in the anomalous dispersion regime using femtosecond pulses. If the power of the pump pulses is high enough, they can evolve as higher-order solitons. As seen in Chapter 24, these pulses begin to experience spectral broadening and temporal compression, which are typical of higher-order solitons. However, because of perturbations such as higher-order dispersion or IRS, the dynamics of such pulses depart from the behaviour expected of ideal high-order solitons, and the pulses break up. This process is known as soliton fission and determines the separation of each higher-order soliton pulse into N fundamental solitons.

The soliton fission process is affected if the input pulse wavelength is near a ZDW. As seen in Section 25.2, in such a case the ejected solitons shed some of their energy to dispersive waves generated in the normal-dispersion regime, the so-called NSR. In spite of the fact that Raman solitons and dispersive waves have widely separated spectra, they can overlap in the time domain. In such a case, they interact through cross-phase modulation (XPM), causing the formation of new spectral components. This results in the spectral broadening of the dispersive waves and the merging of different peaks, helping in the formation of a broadband supercontinuum. This shows that the XPM effect indeed plays an important role during supercontinuum formation in highly non-linear fibres.

The group-velocity mismatch between the soliton and the dispersive wave can be relatively small if this wave propagates in the normal-dispersion regime. In such a case, if they overlap in the time domain, the Raman soliton can trap the dispersive wave due to the XPM interaction. The condition for this to occur can be obtained from the relation

$$\beta_1(\omega_d) \approx \beta_1(\omega_s) + \beta_2(\omega_s)\Omega_d + \frac{1}{2}\beta_3(\omega_s)\Omega_d^2 \tag{25.11}$$

where $\Omega_d = \omega_d - \omega_s$ is the difference between the frequency of the dispersive waves (ω_d) and that of the soliton (ω_s). Considering that $v_g = 1/\beta_1$, we have $\beta_1(\omega_d) \approx \beta_1(\omega_s)$ and the dispersive wave can be trapped by the soliton if

$$\Omega_d = -\frac{2\beta_2}{\beta_3} \tag{25.12}$$

The FWM effect also plays an important role in the process of supercontinuum formation by femtosecond pulse pumping in the anomalous dispersion regime. Provided an appropriate phase-matching condition is satisfied, a Raman soliton can act as a pump and interact through FWM with a dispersive wave, giving origin to an idler wave, as described in Chapter 21. In one possible FWM process, corresponding to conventional FWM, the phase-matching condition is given by

$$\beta_d(\omega_3) + \beta_d(\omega_4) = \beta_s(\omega_3) + \beta_s(\omega_4) \tag{25.13}$$

where $\beta_d(\omega)$ and $\beta_s(\omega)$ are the propagation constants of the dispersive wave and of the soliton, respectively. In this case, the idler frequency ω_4 lies far from that of the original dispersive wave. Considering a positive third-order dispersion, this FWM process contributes to the extension of the supercontinuum towards the infrared side of the input wavelength.

25.5 Modelling of Supercontinuum Generation

Modelling the ultrashort pulse propagation and supercontinuum generation in optical fibres can be realized by considering a generalized non-linear Schrödinger equation (NLSE) that includes higher-order dispersive and non-linear effects. Such an equation can be written as

$$\frac{\partial U}{\partial z} - i\sum_{k\geq 2}\frac{i^k\beta_k}{k!}\frac{\partial^k U}{\partial \tau^k} + \frac{\alpha(\omega)}{2}U = i\gamma\left(1 + \frac{i}{\omega_0}\frac{\partial}{\partial \tau}\right)\left(U(z,\tau)\int_{-\infty}^{+\tau} R(t')|U(z,\tau-t')|^2 dt'\right) \tag{25.14}$$

where $U(z,t)$ is the electric field envelope, ω_0 is the centre frequency, β_k are the dispersion coefficients at the centre frequency, $\alpha(\omega)$ is the frequency-dependent fibre loss, and γ is the non-linear parameter of the fibre.

The non-linear response function $R(t)$ in Eq. (25.14) can be written as

$$R(t) = (1 - f_R)\delta(t) + f_R h(t), \tag{25.15}$$

where the δ-function represents the instantaneous electron response (responsible for the Kerr effect), $h(t)$ represents the delayed ionic response (responsible for Raman scattering), and f_R is the fractional contribution of the delayed Raman response to the non-linear polarization, for which a value $f_R = 0.18$ is often assumed. It is common to approximate $h(t)$ in the form (Blow and Wood, 1989):

$$h(t) = \frac{\tau_1^2 + \tau_2^2}{\tau_1 \tau_2^2} \exp(-t/\tau_2) \sin(t/\tau_1) \tag{25.16}$$

where $\tau_1 = 12.2$ fs and $\tau_2 = 32$ fs. The Raman parameter, t_R, which characterizes the IRS effect, is related to $h(t)$ in the form $t_R \equiv f_R \int_{-\infty}^{+\infty} th(t)\,dt$. Typically, it has a value $t_R \approx 5$ fs.

Eq. (25.14) can be used to describe the propagation of femtosecond pulses in optical fibres, in both the normal and anomalous dispersion regimes. When such pulses have enough power, their spectra undergo extreme broadening. In the anomalous dispersion regime, this process is mainly influenced by the phenomenon of soliton fission, which occurs whenever a higher-order soliton is affected by third- or higher-order dispersion. The soliton order is given from Eq. (24.18) by $N = \sqrt{L_D/L_{NL}}$, where $L_D = t_0^2/|\beta_2|$ is the dispersion distance and $L_{NL} = 1/\gamma P_0$ is the non-linear length.

The simulation of the supercontinuum has been realized in (Hult, 2007) using fibre and laser parameters taken from (Dudley and Coen, 2002). Such parameters represent typical experimental conditions for supercontinuum generation in MOFs employing femtosecond lasers and pumping in the anomalous dispersion regime. An input pulse $U(0.\tau) = \sqrt{P_0}\text{sech}(\tau/t_0)$, was assumed, where $P_0 = 10$ kW, and the width $t_0 = 28.4$ fs, which corresponds to an intensity full width at half maximum (FWHM) of 50 fs. The fibre has a length L = 10 cm and a non-linear parameter length $\gamma = 45\,\text{W}^{-1}$/km. The pulse was initially centred at 850 nm and the following fibre parameters were also assumed: $\beta_2 = -12.76\,\text{ps}^2/\text{km}$, $\beta_3 = 8.119 \times 10^{-2}\,\text{ps}^3/\text{km}$, $\beta_4 = -1.321 \times 10^{-4}\,\text{ps}^4/\text{km}$, $\beta_5 = 3.032 \times 10^{-7}\,\text{ps}^5/\text{km}$, $\beta_6 = -4.196 \times 10^{-10}\,\text{ps}^6/\text{km}$, $\beta_7 = 2.570 \times 10^{-13}\,\text{ps}^7/\text{km}$. For these conditions, the soliton order is N ~5.

The spectral evolution of the supercontinuum at different positions along the MOF is shown in Fig. 25.7, which reproduces very well the experimental observations reported in (Dudley and Coen, 2002). Dramatic spectral broadening is observed in the first centimetre of the fibre. The initial broadening, which is mainly due to the interaction between the SPM and GVD, is symmetric. Afterwards, the spectrum becomes asymmetric and distinct spectral peaks develop on both sides of the pump wavelength as higher-order dispersive and non-linear perturbations cause the fission of the higher-order soliton. The resulting fundamental solitons then undergo a continuous self-frequency shifting to longer wavelengths because of IRS. The emergence and self-shifting of one of these fundamental solitons on the long-wavelength side is clearly seen in Fig. 25.7(b). In this process, each Raman soliton sheds some of its energy in the form of NRS on the short-wavelength side of the pump, which also leads to the appearance of discrete spectral components in this region of the spectrum.

Fig. 25.8 shows the temporal and spectral evolution of an optical pulse propagating along a silica MOF with a hexagonal pattern of holes of diameter $d = 1.4\,\mu$m spaced by $\Lambda = 1.6\,\mu$m. This result was obtained taking into account the full dispersion curve of such fibre, which presents a single ZDW at 735 nm. The pump is in the anomalous dispersion regime at 790 nm, where the non-linear parameter is $\gamma = 117 W^{-1} km^{-1}$. An input pulse $U(0, \tau) = \sqrt{P_0}\text{sech}(\tau/t_0)$, is assumed, where $P_0 = 5$ kW, and the width $t_0 = 14.2$ fs, which corresponds to an intensity FWHM of 25 fs. For the assumed pulse and fibre parameters, the soliton order is $N = 3.01$.

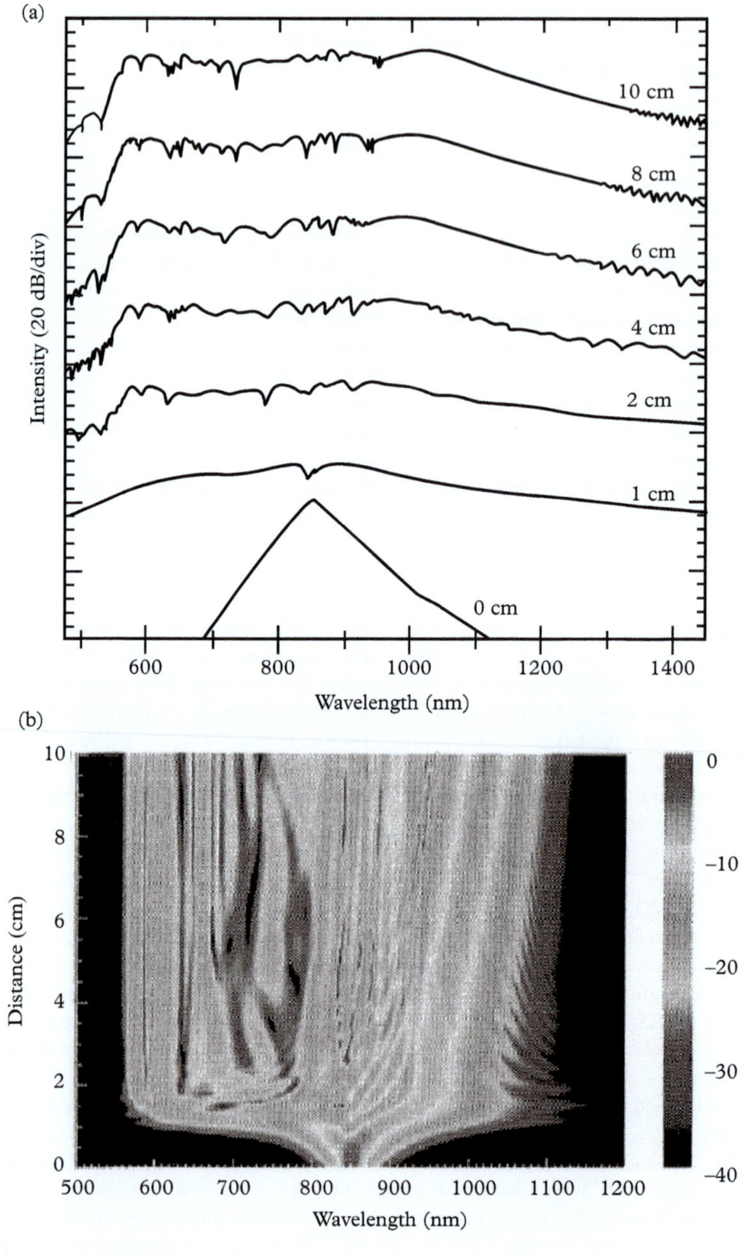

Figure 25.7 *Results from a numerical simulation of supercontinuum generation in a MOF. (a) Spectra at discrete locations along the fibre and (b) density plot as a function of the propagation distance (reprinted with permission from Hult (2007) © IEEE).*

For the pulse and fibre parameters assumed, one has a fission distance $L_{fiss} \sim 0.51$ cm, which can be confirmed in Fig. 25.8. A clear signature of soliton fission is the appearance of a new spectral peak in Fig. 25.8(a) at $z = L_{fiss}$, which corresponds to the NRS.

As a consequence of the SSFS induced by IRS, the solitons arising from the fission process separate from each other. Since the SSFS is the largest for the shortest soliton, its spectrum shifts the most towards the red side in Fig. 25.8(a). The change in the soliton's frequency determines a reduction in the soliton's speed because of dispersion. This deceleration appears as a bending of the soliton trajectory in the time domain, as observed in Fig. 25.8(b).

Fig. 25.9 shows the temporal and spectral evolution of an optical pulse for the same conditions as in Fig. 25.8 except that the pulse width is now $T_0 = 28.4$ fs, which corresponds to an intensity FWHM of 50 fs. Comparing with Fig. 25.8, one now has a wider and more uniform supercontinuum. Fig. 25.10 shows the initial and final pulse profiles, both in the frequency and time domains. Since the power of the first ejected soliton is greater than in the case of Fig. 25.8, it experiences a larger SSFS, according to Eq. (24.41). At the same time, the corresponding nonsolitonic radiation is generated at a lower wavelength in the normal region.

In most SCG studies, silica has been used as the MOFs' background material. However, the material loss in silica increases drastically at wavelengths beyond ~ 2 μm, effectively preventing the spectral evolution of SCG into the mid-infrared (MIR) region.

Research into other highly non-linear materials that can exploit the importance of the mid-IR (MIR) region has increased in recent years. Tellurite, fluoride, sapphire, As_2Se_3 and $As_{40}Se_{60}$ chalcogenide MOFs have been used to achieve such MIR SCG.

Fig. 25.11 shows a MIR SC spectrum spanning $\sim$ 2.0to 15.1 μm, obtained in a 3 cm-long chalcogenide step-index fibre pumped at 9.8 μm with 170 fs pulses (Cheng et al., 2016). Fig. 25.11(a) shows the experimental result, whereas Fig. 25.11(b) displays the simulation

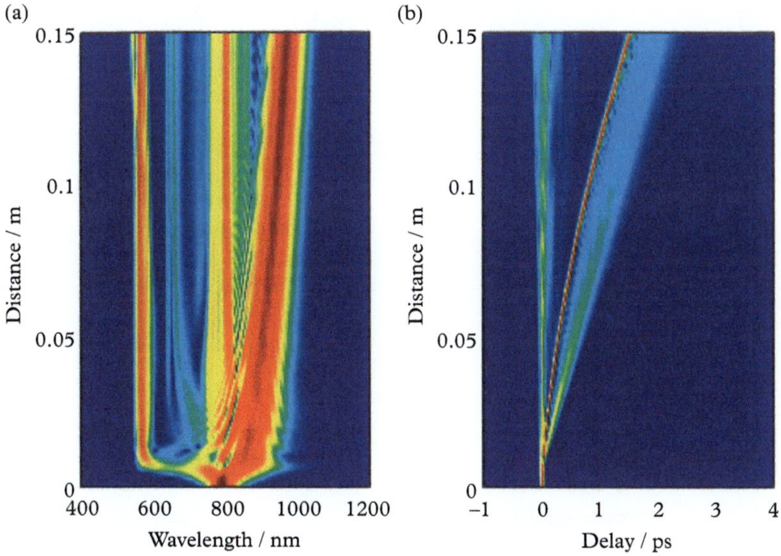

Figure 25.8 *(a) Spectral and (b) temporal evolution along a MOF with $d = 1.4$ μm and $\Lambda = 1.6$ μm. The input pulse has a peak power of 5 kW and a FWHM of 25 fs.*

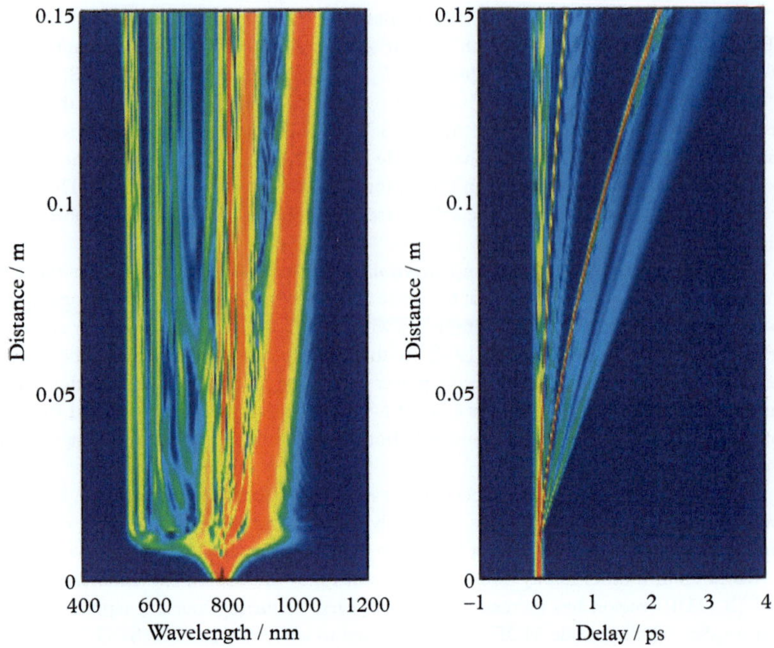

Figure 25.9 *(a) Spectral and (b) temporal evolution of an optical pulse along a MOF with $d = 1.4\ \mu m$ and $\Lambda = 1.6\ \mu m$. The input pulse has a peak power of 5 kW and a FWHM of 50 fs.*

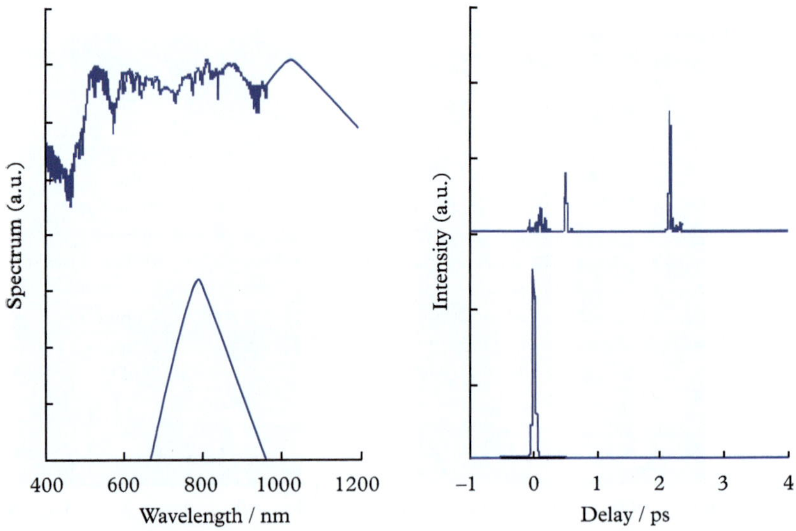

Figure 25.10 *Initial and final pulse profiles in the (a) frequency and (b) time domains for the case of Fig. 25.9.*

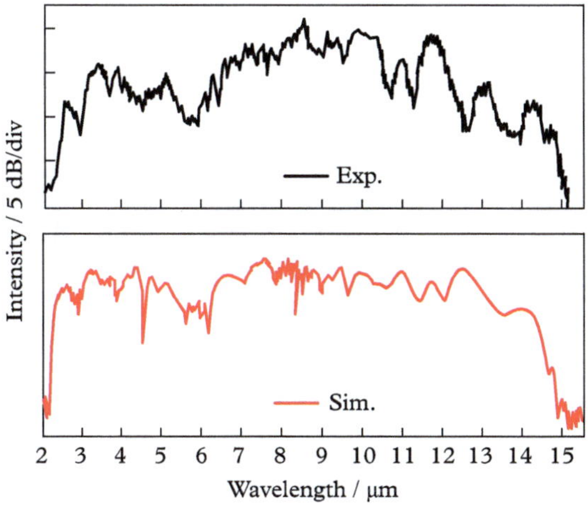

Figure 25.11 *Measured (upper curve) and simulated (lower curve) MIR SC generation in a 3 cm-long chalcogenide fibre pumped at 9.8 µm with 170 fs pulses with a peak power of ~ 2.89 MW (reprinted with permission from Cheng et al. (2016) © Optical Society of America).*

result, obtained by solving the generalized non-linear Schrödinger equation (GNLSE) given by Eq. (25.14). The non-linear response function of the glass, $R(t)$, was assumed as in Eq. (25.15), whereas the delayed Raman response was approximated by Eq. (25.16). Good agreement is observed between the two results.

25.6 SC and UV Light Generation in Hollow-Core MOFs

Supercontinuum sources based on solid-core conventional and MOFs can provide broadband light from the mid-IR to the near-UV. However, they have not been demonstrated for wavelengths shorter than 280 nm and are found to suffer from limited optical transparency and cumulative optical damage when delivering UV light. Actually, no solid-state or fibre-based supercontinuum source exists that can span the vacuum ultraviolet (VUV) (~100–200 nm) and deep-ultraviolet (DUV) (~200–300 nm) spectral ranges.

Hollow-core MOFs provide a means of eliminating the diffraction of laser light in gases while offering precise control of group-velocity dispersion and long low-loss interaction paths. In particular, kagome MOFs offer broadband transmission and weak anomalous dispersion from the UV to the near-IR. These properties not only help support ultrafast soliton dynamics but also allow the guidance of any UV light that is subsequently generated at relatively low losses. Actually, noble-gas-filled kagome MOFs have been used to demonstrate widely tunable and efficient generation of ~5 nm wide bands of UV light down to 176 nm.

Pulse propagation in a gas-filled kagome MOF can be described by Eq. (25.14), where the Kerr coefficient depends linearly on the gas pressure. In the case of a noble gas, the Raman effect is ignored and the ZDW varies with pressure, as illustrated in Fig. 16.13.

Figs. 25.12 and 25.13 show the spectral evolution of 40 fs pulses launched into a kagome MOF with a core diameter of 40µm filled with argon and helium, respectively, at different pressures. The pulses are launched at $\lambda_{\text{pump}} = 790\,\text{nm}$ and have an energy of 10 µJ. These results can be understood by considering the dependence of the ZDW, λ_{ZD}, on the pressure

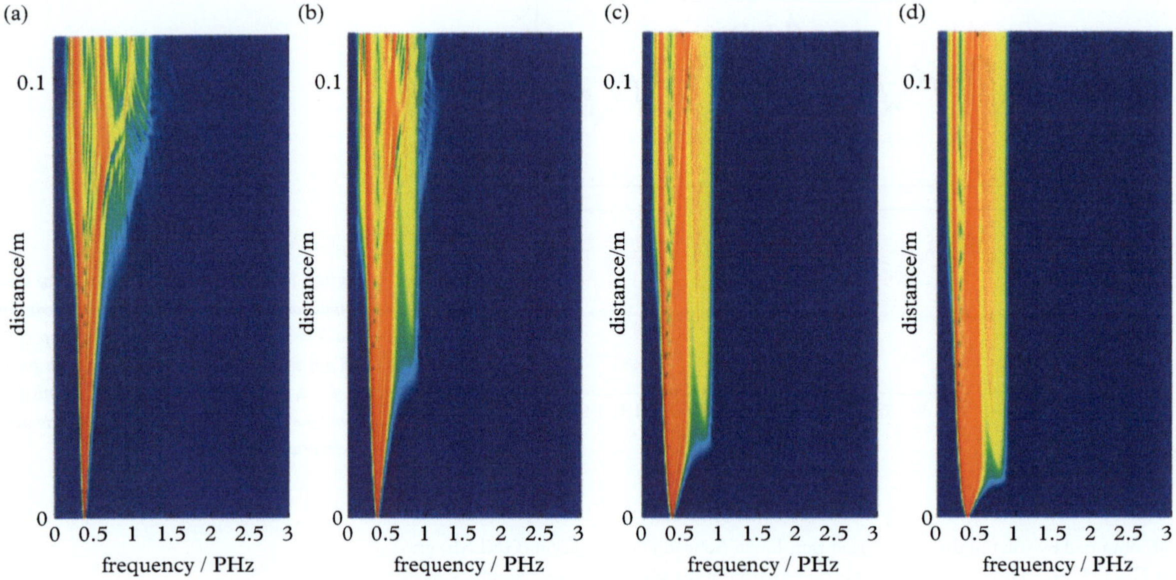

Figure 25.12 *Spectral evolution in a kagome MOF filled with argon of a 40 fs input pulse, with an energy of 10 μJ and $\lambda = 790$ nm, at pressures of (a) 10 bar, (b) 20 bar, (c) 40 bar, and (d) 80 bar.*

for each gas, as illustrated by Fig. 16.13. In the case of Fig. 25.12(a), the pumping is made in the anomalous dispersion regime and the input soliton order is $N \sim 38$. However, in Fig. 25.12(b–d), the pumping is realized in the normal-dispersion regime and the spectral expansion becomes less significant. One can observe that by increasing the pressure, the spectral broadening occurs within shorter distances, which is due to the increased non-linearity for higher pressure values.

In the case of helium-filled kagome fibre (Fig. 25.13), the pumping is always in the anomalous dispersion regime. By increasing the gas pressure, the ZDW becomes closer to the pumping wavelength and the fibre non-linear parameter also increases, so that the soliton order in Fig. 25.13(a–d) is $N = 4.9$, 7.0, 10.3 and 15.8, respectively. Consequently, the soliton fission length decreases with pressure.

As a consequence of the higher-order soliton compression, its spectrum expands and overlaps with resonant dispersive-wave frequencies, which are consequently excited in the UV region. The resonance condition is given by Eq. (25.6), where the soliton peak power P_s must be substituted by the peak power of the compressed pulse. This condition provides the resonance frequencies $\omega \sim 2.4, 2.0, 1.7$, and 1.6 PHz for the cases of Figs. 25.13(a–d), respectively, which agrees reasonably with the central frequencies of the generated UV bands. Concerning the generation of UV light, the conversion efficiency can be defined as the fraction of the total output power at wavelengths shorter than 350 nm.

The UV-generated light source can be tuned by varying the input pulse parameters or the gas pressure. Fig. 25.14 illustrates the possibility of tuning the frequency of the final UV pulse by changing the pressure of the gas in the fibre's core. 40 fs pulses with an energy of 20 μJ at $\lambda_{\text{pump}} = 800$ nm are launched into a kagome HC-MOF with a core diameter $d = 40.0$ μm. The variation of the gas pressure between 7 and 16 bar provides a tuning of the UV light main peak frequency by ~ 0.4 PHz.

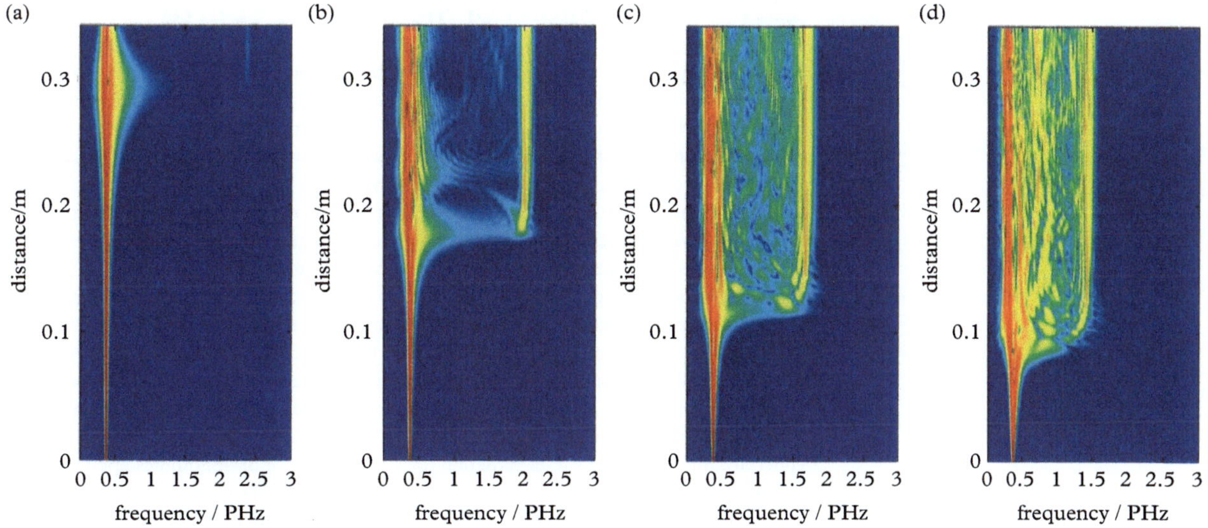

Figure 25.13 *Spectral evolution in a kagome MOF filled with helium of a 40 fs input pulse, with an energy of 10 μJ and λ = 790 nm, at pressures of (a) 10 bar, (b) 20 bar, (c) 40 bar, and (d) 80 bar.*

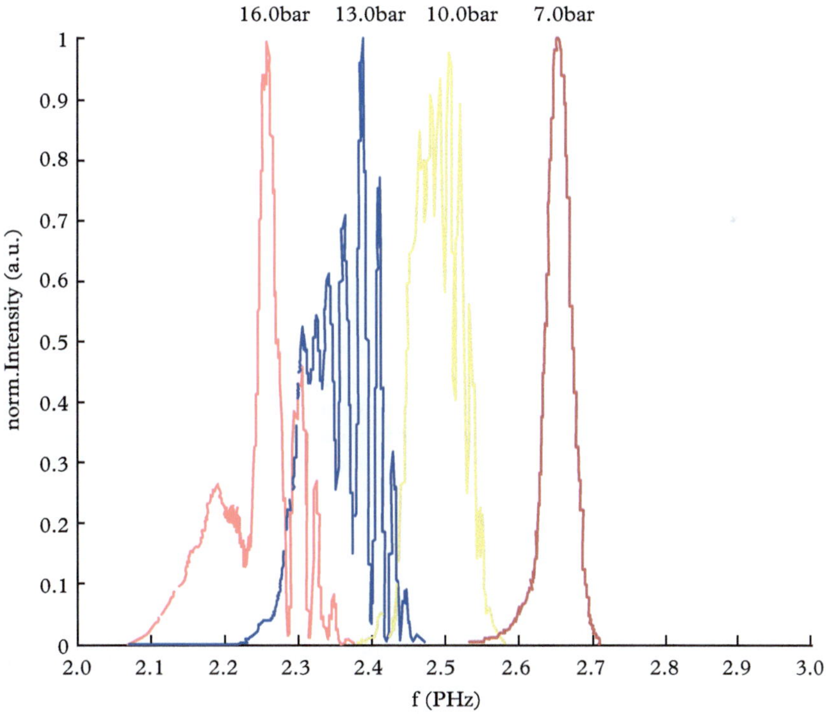

Figure 25.14 *Output spectral profile of UV light generated in a kagome MOF with a core of diameter* $d = 40.0\,\mu m$*, filled with helium; the 40 fs pump pulses have a wavelength* $\lambda_{pump} = 800$ *nm and an energy of 20 μJ. The gas pressure is indicated in the inset.*

25.7 Applications of the SC Light Source

Supercontinuum light sources are generally used for purposes where one requires light with a broad spectrum but, at the same time, a high degree of spatial coherence, so that the light can be well collimated and focused. For example, one often uses such a source in conjunction with a monochromator as a tunable light source in spectroscopy. In comparison with a tunable laser, a supercontinuum source can usually cover a much wider wavelength range. Other applications, like fluorescence microscopy, coherent anti-Stokes scattering (CARS) microscopy, optical coherence tomography, and the characterization of optical devices, can similarly profit from supercontinuum sources.

Supercontinuum light sources have been widely used in the field of bio-photonics, namely to perform spectroscopic measurements and to implement imaging instrumentation. The interaction of optical radiation with biological samples can provide molecular contrast, while structural information can be obtained from a range of microscopy techniques, many of which exploit the characteristic properties of laser radiation. Supercontinuum sources therefore have the potential to provide spectral versatility and spatially coherent, high-brightness laser radiation for a wide range of bio-photonics applications.

Supercontinuum generation also finds applications in ultrafast laser physics, for example for the detection and stabilization of the carrier–envelope phase of a mode-locked laser in both the time and frequency domains. Stabilization of the carrier–envelope phase evolution means that the frequency spectrum of the laser corresponds to a comb of well-defined frequencies, which can be used to perform optical frequency metrology or build optical atomic clocks. It has also enabled the production of single attosecond pulses.

25.7.1 The SC as a source for WDM systems

Actual wavelength division multiplexing (WDM) transmission systems use a large number of distributed feedback (DFB) lasers as transmitters. Each DFB laser corresponds to a wavelength channel and is generally used together with a modulator and a wavelength-stabilizing system. This approach is costly and becomes impractical when the number of channels becomes large.

One approach to realize a practical WDM source uses the technique of spectrum slicing. In this approach, the light of a broadband source is sliced into several narrow bands centred at the wavelengths of the WDM channels. A multipeak optical filter can be used to realize the spectrum slicing. Since some of the components are shared among all the wavelengths, the number of active elements in such multiwavelength transmitters is greatly reduced.

The spectral width of femtosecond pulses emitted by a mode-locked laser is already sufficiently large to begin with and can be used for frequency slicing. In fact, according to the Fourier theorem for bandwidth-limited pulses, the temporal duration and the bandwidth are in inverse proportion to each other. For example, a laser producing a train of 50-fs transform-limited Gaussian pulses has an optical bandwidth of 70 nm at 1550 nm. The bandwidth of such pulses can be further broadened by chirping them using 10–15 km of a standard telecommunication fibre.

Spectrum slicing for WDM applications has been reported employing amplified spontaneous emission (ASE) from light-emitting diodes, superluminescent diodes, or erbium-doped fibre amplifiers. However, the transmission capacity of such incoherent spectrum-sliced systems is generally limited by spontaneous emission beat noise. This limitation can be alleviated using coherent light sources, such as mode-locked lasers and supercontinuum generators. The coherent nature of the broadband radiation originating from these sources is crucial to achieving high transmission capacities both in multi-wavelength optical time division multiplexing (OTDM) and in dense wavelength division multiplexing (DWDM) systems.

Three characteristics of the supercontinuum are particularly important in the context of its application as a WDM system transmitter. First, its spectrum should ideally coincide with the transmission window. Second, it should be smooth enough in order to provide similar powers to each channel, resulting in a uniform impact of the non-linear effects. Third, the amplitude noise of the filtered SC should be sufficiently reduced. This last characteristic becomes especially important for high-capacity WDM systems, since the intensity variations among pulses at a given wavelength have a negative impact on the performance of that channel.

25.8 Further Reading

Agrawal, G. P., *Nonlinear Fiber Optics*, Academic Press, 5th Ed., 2012.

Akhmediev, N., and M. Karlsson, *Phys. Rev.* A 51, 2602, 1995.

Blow, K., and D. Wood, Theoretical Description of Transient Stimulated Raman Scattering in Optical Fibers, *IEEE J. Quantum Electron.* 25, no. 12, 2665, 1989.

Cheng, T., K. Nagasaka, T. Tuan, X. Xue, M. Matsumoto, H. Tezuka, T. Suzuki, and Y. Ohishi, Efficient Supercontinuum Generation in Silica Fiber Pumped at 2 μm, *Opt. Lett.* **41**, no. 9, 2117, 2016.

Dudley, J., and Taylor, J. (Eds), *Supercontinuum Generation in Optical Fibers*. Cambridge University Press, 2010.

Dudley, J., and S. Coen, Supercontinuum Generation in Photonic Crystal Fiber, *J. Sel. Top. Quantum Electron.* **8**, no. 3, 651, 2002.

Ferreira, M. F., *Nonlinear Effects in Optical Fibers*, John Wiley & Sons, 2011.

Ferreira, M. F., *Optical Signal Processing in Highly Nonlinear Fibers*, CRC Press, 2020.

Ferreira, M. F., *Solitons in Optical Fiber Systems*, John Wiley & Sons, 2022.

Hult, J., A Fourth-Order Runge–Kutta Method for Simulating Supercontinuum Generation in Optical Fibers, *J. Lightwave Technol.* **25**, no. 12, 3770, 2007.

Kudlinski, A., V. Pureur, G. Bouwrnans, and A. Mussot, Polarization-Dependent Supercontinuum Generation in Photonic Crystal Fibers, *Opt. Lett.* **33**, no. 21, 2488, 2008a.

Kudlinski, A., and A. Mussot, Polarization Modulation Instability in Birefringent Photonic Crystal Fibers, *Opt. Lett.* **33**, no. 20, 2407, 2008b.

Slusher, R., G. Lenz, J. Hodelin, J. Sanghera, L. Shaw, and I. Aggarwal, Large Raman Gain and Nonlinear Phase Shifts in High-Purity As2Se3 Chalcogenide Fibers, *J. Opt. Soc. Am. B.* **21**, no. 6, 1146, 2004.

25.9 Problems

25.1 Discuss different strategies to achieve a highly non-linear optical fibre.

25.2 Determine the non-linear parameter of a silica tapered fibre $\left(n_2 \approx 3.2 \times 10^{-20}\ \mathrm{m^2/W}\right)$, which has an effective core area $A_{eff} = 3\mu\,\mathrm{m}^2$ at 1550 nm.

25.3 Explain how the non-linear phenomenon of SPM can be used to determine the non-linear parameter γ for an optical fibre. Why is it important to know the exact shape of the input pulses used in such measurements?

25.4 Describe the phenomenon of nonsolitonic radiation (NSR) and explain the conditions under which it can occur. Comment on the new features of NSR that can be observed in fibres exhibiting two zero-dispersion wavelengths.

25.5 Give a physical explanation for the self-frequency shift effect occurring during the propagation of an ultrashort soliton and explain why this effect is enhanced in highly non-linear fibres.

25.6 Explain how the self-frequency shift effect can be suppressed in fibres which exhibit two ZDWs.

25.7 Determine the Raman-induced frequency shift experienced by a 100-fs (FWHM) input pulse, which propagates as a fundamental soliton, at the output of a 20-m-long highly non-linear fibre with $\beta_2 = -30\text{ps}^2/\text{km}$. Assume the typical value $t_R = 3$ fs. Considering that such a fibre has a non-linear parameter $\gamma = 50\,\text{W}^{-1}/\text{km}$, determine the pulse peak power.

25.8 Derive the phase-matching condition given by Eq. (25.9) and obtain the frequency shift for a pulse with a peak power $P_p = 100\text{W}$ propagating in a highly non-linear fibre with a non-linear parameter $\gamma = 200\,\text{W}^{-1}/\text{km}$ and dispersion coefficients $\beta_2 = 6\text{ps}^2/\text{km}$ and $\beta_4 = -2 \times 10^{-4}\text{ps}^4/\text{km}$.

25.9 Describe the physical processes involved in the supercontinuum generation when a CW beam is launched into a highly non-linear fibre.

25.10 Discuss the main advantages of using a hollow-core fibre filled with a noble gas, instead of a solid-core fibre, to generate the supercontinuum.

Appendix A
Optics in Nature

Throughout the chapters of this book, various topics in the fields of optics, photonics, and lasers have been addressed, which have developed thanks to the capacity of the human intellect. However, many phenomena and optical systems can be found in nature which are not due to the activity of man, but are manifestations of nature itself and the result of an admirable evolutionary process. Among such natural optical systems, one must consider the human eye, which is surely one of the wonders of nature, as described in Chapter 3.

The rainbow, the wings of butterflies, the mirages, the pictures of the sunset, and the constitution and functioning of the eyes of various animals are just some of the optical wonders that abound in nature and that provide many opportunities to contemplate and better understand our world. In this chapter, some of these natural optical phenomena and systems are briefly described, and it is hoped that the most interested reader will be able to find in the specialized literature the possibility of deepening these subjects.

A.1 The Eyes on Land Animals

There is a great variety of eyes among animals, which is undoubtedly the result of different paths of evolution over time. As for the human eye in particular, its description has already been given in Chapter 3, where some of its admirable characteristics have been mentioned.

The primary element from which the evolution of the eyes of animals is considered to have been proceeded is the simple photodetector. The photodetector is made up of molecules of a protein capable of absorbing photons in the visible band and found in the eyes of both animals and plants. These molecules are basically made up of a carbon chain with alternately single and double bonds—the so-called conjugate chain.

The next step in evolution was the formation of the eye socket. When the photodetectors mentioned earlier are lodged in a cavity, a certain directionality in vision is already achieved, and the animals can then reorient themselves or change their position in relation to the light stimuli. The performance of this elementary optical system can be further improved if the eye socket is filled with some transparent substance. As far as directionality is concerned, the improvement becomes more significant if the surface of this transparent substance is curved. A simple and primitive version of the ocular structure is then obtained.

The vision of terrestrial vertebrate animals depends heavily on the convergent power of the air–corneal interface. In the case of human eyesight, for example, the power of the lens is only about one-third of the characteristic power of the cornea. On the other hand, the sharpness of this vision depends on the chromatic and spherical aberrations of the optical system. As regards chromatic aberration, there is no known mechanism in nature that can correct it. However, this problem is mitigated due to the rather limited spectral band that characterizes the response of the photodetectors mentioned earlier.

As far as spherical aberration is concerned, however, nature offers an interesting solution that is not easy for man to reproduce artificially: the lens has a refractive index that varies gradually, while the shape of the frontal surface of the cornea is such as to minimize this aberration. Actually, in many types of eyes, including the human eye, the surface of the

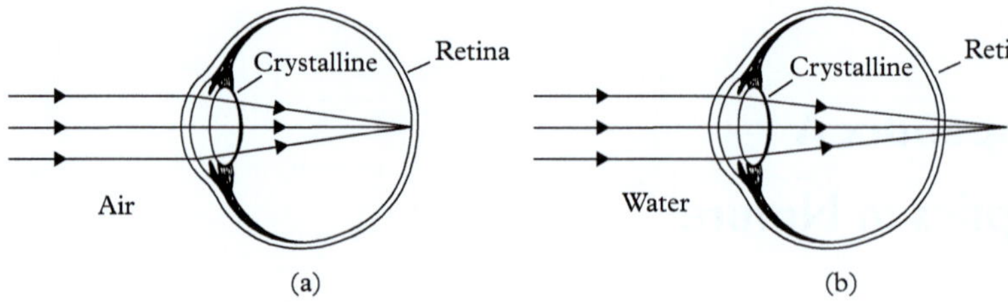

Figure A.1 *The human eye (a) in air and (b) in water.*

cornea is not strictly spherical, exhibiting a lower curvature at the periphery. The radius of curvature in this region is typically double compared to its value in the centre of the cornea, which allows to significantly reduce the effect of spherical aberration.

A.2 The Eyes on the Fish

The process of focusing light in fish and other aquatic animals cannot be the same as that in terrestrial animals because of the relatively small difference between the refractive indices of water and the aqueous humour behind the cornea. Actually, it is this fact that explains the noticeable degradation of vision that we experience when opening our eyes underwater without wearing any type of glasses. In these circumstances, due to the poor focus provided by the cornea, the light is focused somewhere behind the retina, where an ill-defined image appears (Fig. A.1).

In the case of aquatic animals, the focusing of light is carried out quite efficiently by means of a spherical lens, the so-called *fisheye lens*. As is well known, a spherical lens with a uniform refractive index has a fairly significant spherical aberration. However, the fisheye lens has a refractive index that varies gradually, eventually having a very short focal length and near-perfect focusing. Its spherical geometry also provides a very wide field of vision. Given these exceptional characteristics, it is not surprising that the fisheye lens is common to the vast majority of aquatic animals, even if they are very different from one another.

A.3 Eyes on Marine Mammals

While most terrestrial animals live exclusively in the air and aquatic animals live exclusively in water, marine mammals (dolphins, whales, seals, sea lions, walruses, etc.) can be found both above and below the water level. In this way, they must adapt to different levels of luminosity, ranging from full sunlight to almost total darkness. On the other hand, it is recognized that the structure of the eye must be able to withstand large variations in pressure, while in optical terms it has to adapt to different refractions at its interface with water or air.

Like other aquatic animals, most marine mammals have a roughly spherical lens that is the main refractive element of their eyes and provides them with excellent underwater vision (Fig. A.2 (a)). When they are out of the water, however, the cornea also exerts its ability to focus light, so it is expected that the image will be focused in front of the retina (Fig. A.2 (b)) and that the quality of vision will be greatly degraded.

However, anyone who has ever seen the dolphin or sea lion exhibits in an amusement park will have been able to see the excellent vision they also have in the air. Therefore, it must be admitted that there are some mechanisms of adaptation of the eyes of these animals to the variation of refractive effects at the interface between the cornea and water

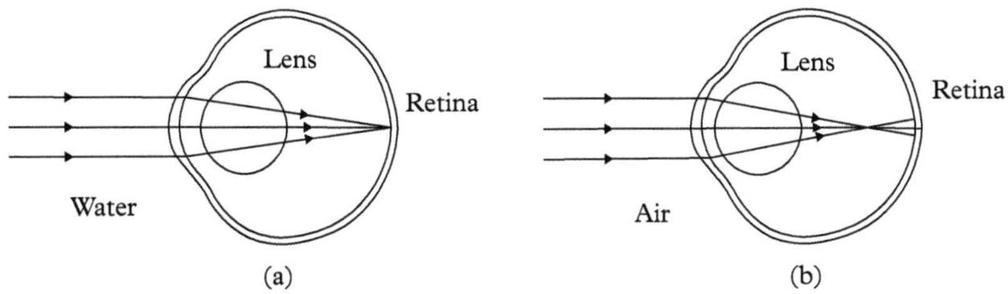

Figure A.2 *The eye of the marine animal (a) in water and (b) in air.*

or air. These mechanisms will necessarily be related to one of the three primary optical elements of the eye: the cornea, the pupil and the lens.

As for the cornea, it is possible that there will be a decrease in curvature or even a flattening in part of its area, thus decreasing its focusing power. This has already been confirmed in the case of some sea lions.

Regarding the pupil, it should be noted the great variation that its area can experience in these animals, reflecting the need they have to adapt to very different levels of light. While the pupil area in the human eye can vary between 7 and 38 mm^2, in the case of some dolphins, for example, this variation can be between 5 and 70 mm^2. In some types of seals the variation in the pupil area can be even much more significant: between 0.9 and 422 mm^2.

Finally, as for the role of the lens in the process of focusing on the retina, it can be performed through two processes, commonly called *accommodation*. The first has to do with the alteration of the shape of the lens by the action of the ciliary muscles, as happens in man and many other mammals. The second form of accommodation involves adjusting the distance between the lens and the retina. This adjustment is carried out in several species of seals through some well-developed ciliary muscles. Some types of whales and dolphins do not have this kind of muscles, and it is assumed that such an effect is achieved in these cases by the action of well-developed extraocular muscles. These muscles can be used to conveniently change the diameter of the eye, thus varying the distance between the retina and the lens.

A.4 Compound Eyes

More than half of animal species have so-called *compound eyes*. These eyes owe their name to the fact that they consist not of a single optical system, but of multiple optical units arranged together on a spherical surface. In the simplest cases—corresponding to the so-called *juxtaposition eyes*—these units act independently of each other, each having its own lens and its own receiver, and each covering a relatively limited field of vision (Fig. A.3).

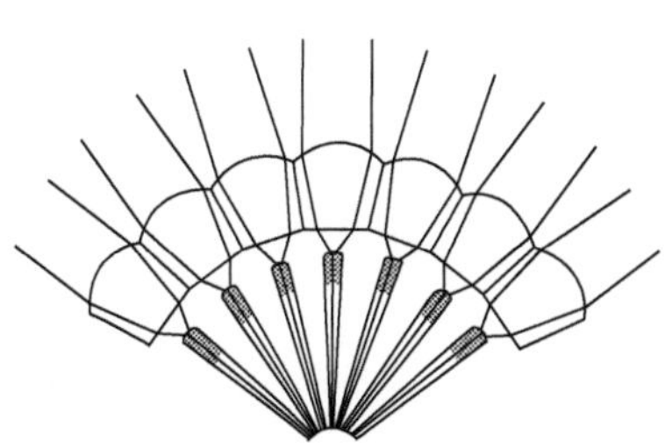

Figure A.3 *Representation of a juxtaposition compound eye.*

In juxtaposition-compound eyes, light reaches each of the receptors through a cylindrical fibre with a diameter of only a few wavelengths. The propagation of light in these fibres cannot be described on the basis of geometrical optics, but only using wave optics. It can be said that nature anticipated in this case the development of today's single-mode optical fibres. In some insects—notably bees—the propagation in these fibres is sensitive to the polarization of light, a property that these insects use to properly guide their movements.

In addition to juxtaposition eyes, which are characteristic of diurnal insects such as flies and bees, there are also so-called *overlapping eyes*, in which several lenses are connected to a single receptor. This type of compound eye is characteristic of some nocturnal insects.

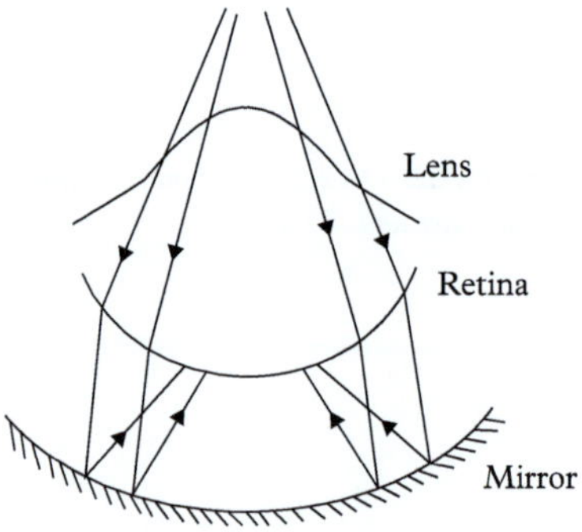

Figure A.4 *Illustration of the scallop eye incorporating a concave mirror.*

A completely different type of eye can be found in some crustaceans and molluscs, including the scallop, and is depicted in Fig. A.4. This eye uses a concave mirror to form an image on the retina, in a geometry similar to that of the Newtonian telescope illustrated in Fig. 3.6. An obvious disadvantage of this type of eye is that the light hits the retina from both sides, even though only the light reflected in the concave mirror is focused on it.

A.5 Reflector Films

The concave mirror present in the eye of the scallop is an example of a good biological reflector. The shiny scales of fish are another example. The high reflectivity of these elements has to do, in general, with their structure, composed of multiple layers of materials with alternately high and low refractive indices. These layers are thin, and their reflectivity depends on the constructive interference effect provided by the dielectric films, as analysed in Chapter 8.

The reflectivity of the aforementioned layers depends on the wavelength of the light, and both the colour and irradiance of the light reflected by them vary with the angle of incidence. When these layers are present in high numbers, the reflectivity can exceed 90% over a certain range of wavelengths. The existence of layers with different thicknesses provides reflections with different colours, as can be seen with the scales of fish.

As far as insects are concerned, they can be considered true experts in materials science and optical engineering. In fact, their very existence is only possible due to the existence of a protective external film, consisting of a material that is both light, resistant and very versatile. On the other hand, insects know not only how to colour the film with various pigments but also how to cut it with a wide variety of microstructures, able to provide a dazzling range of so-called *structural* colours.

The multiple colours usually exhibited by butterfly wings, for example, are determined by the structure of the scales on their surfaces. These scales have a periodic network of grooves, resulting in a structure with multiple layers of chitin and air, in which the refractive index alternates between 1.54 and 1.0. The maximum reflection, in these cases, can correspond to colours for which there are no natural pigments. The fact that these structural colours are due to the interference of light reflected in the various films can be easily demonstrated by placing a drop of some liquid on the wing of the butterfly. This liquid will replace the air in some of these films, and the colour displayed by the wing is consequently altered. The range of optical effects can be extended when two or more scales act together, and the characteristics of each of these scales are then mutually affected.

In contrast to the cases mentioned earlier, some types of butterflies—called 'light-winged'—do not show any significant reflection of light on their wings. In fact, it has been observed that the wings of these butterflies are covered with microscopic protuberances, forming a very effective anti-reflective coating. This type of structure was first found on the surface of the cornea of the eyes of some insects and is present whenever transparency is a property to be privileged.

There are still some mysteries about how some animals—and particularly insects—manage to make some of their structures. In fact, it is admirable the intelligent way in which evolution has produced, in some cases, materials and structures with such extraordinary characteristics. The admiration becomes greater when one realizes that many of these structures are multifunctional. Each scale of a butterfly's wing, for example, is made up of a single cell, which somehow seems to know the position it occupies in that wing, as well as the specific characteristics it must have at a physical, chemical and architectural level. In some types of multicoloured scales, on the other hand, the thin-film stacks also serve to control transmissivity in the infrared, thereby influencing the heat of the body protected by them.

A.6 The Aurora

Of all the optical phenomena that nature offers us, the aurora must be the most spectacular, namely due to its splendour and the great variety of aspects it usually assumes. In fact, the colours, shape, and movements of the aurora can vary, sometimes slowly, sometimes rapidly. On some nights, the aurora can take on the appearance of a faint glow on the horizon, while on other nights it can swirl and dance over our heads, as if it were a ballet. This phenomenon is usually observed around midnight and occurs mainly in high-latitude regions, both in the Northern Hemisphere (where it is called the Aurora Borealis) and in the Southern Hemisphere (the Aurora Australis).

The Aurora Borealis and the Aurora Australis occur simultaneously, like images in a mirror. The so-called solar wind, a plasma made up of protons and electrons, flows from the sun at a speed of about 1000 km/s. Being made up of electric charges, this solar wind is deflected by the earth's magnetic field, so that it ends up distributing itself in a long cylindrical region around the earth, called the magnetosphere. The interaction of the solar wind with the geomagnetic field causes the flow of electric currents in a vast region around the earth. These currents intersect the ionosphere in a roughly oval-shaped region centred at the geomagnetic poles.

The charged particles first collide with oxygen and nitrogen atoms or molecules in the upper atmosphere, causing bursts of light. The colours resulting from these explosions are determined by the energy emitted in each collision, which depends on the type of gas involved. The colour patterns thus generated can be relatively complex and vary continuously due to the influence of the geomagnetic field.

The most common colour at dawn is light green, due to emission by oxygen atoms located about 100 km above the earth's surface. The pinkish-red colour due to emission by nitrogen molecules can also be seen underneath the green colour. In the case of the most active auroras, bright red regions can also be seen above the green emission. This colour is due to emission by oxygen atoms located in the atmosphere above the atoms responsible for emitting the green colour. Due to the time at which it occurs and the intensity of the magnetic storms that determine it, this bright red colour—which is relatively rare—can be observed in auroras visible at intermediate latitudes.

A.7 The Rainbow

The rainbow has its origin in the phenomenon of dispersive refraction produced by water droplets in the air. When falling on one of these droplets, sunlight is refracted on its front surface, then reflected on the back surface, and refracted again when exiting the droplet. Dispersion occurs when light propagates through the water droplet. When this process occurs in a large number of water droplets, the primary rainbow appears, which makes an angle of 42° with the direction of the antisolar point (Fig. A.5). When the observer moves, the position of the rainbow also changes. In this way, it can be said that each person has his or her own rainbow.

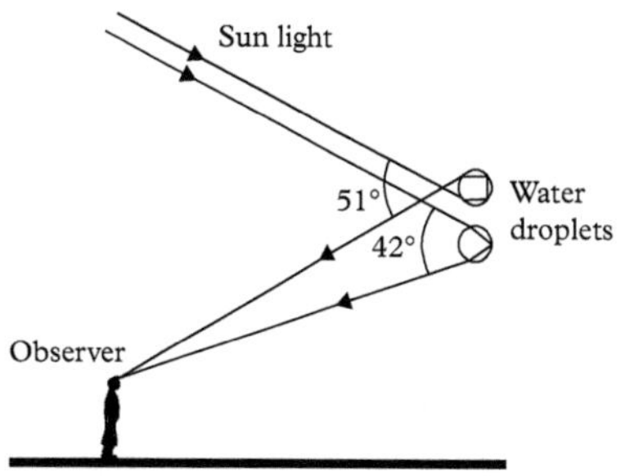

Figure A.5 *Formation of primary and secondary rainbows.*

In general, a secondary rainbow can also be observed, originating from light experiencing two total internal reflections in the water droplets. This second rainbow is formed at an angle of 51° to the direction of the antisolar point (Fig. A.5). Since the refractive index for red is lower than for violet, the primary rainbow is red on the outside, while in the secondary rainbow the colour sequence is reversed. It can also be noticed that the space between the two rainbows is dark. This is due to the fact that each rainbow corresponds to a limit value for the angular deviation of light, allowing the distribution of light inside the primary and outside the secondary rainbow, but not between them.

Rainbows of a higher order than the second are also theoretically possible, although humans cannot observe them in nature. However, some species of animals, such as eagles and lynxes, should have the ability to see the tertiary rainbow.

Example A.1

A ray of light enters a drop of water with refractive index n, as shown in Fig. A.6.

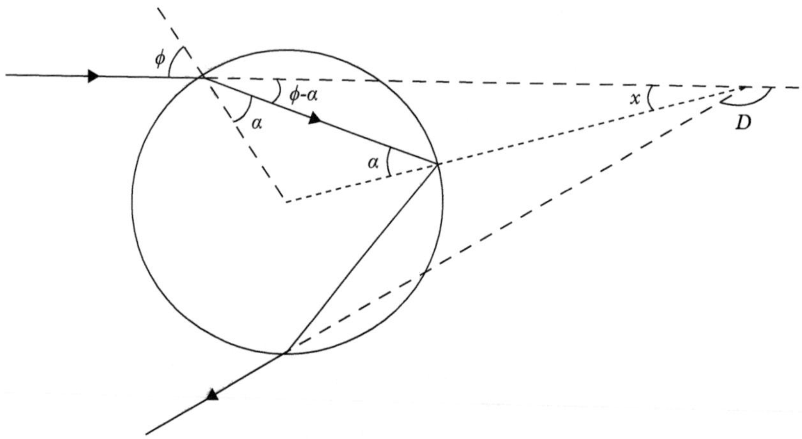

Figure A.6 *Geometry for calculating the angle of deviation in the primary rainbow.*

a) Obtain the angle of incidence on the back surface of the droplet and check whether the ray is fully or partially reflected there.

b) Derive an expression for the deviation angle D.

c) Obtain an expression for the angle of incidence ϕ that produces a minimum deviation angle D. Assuming that $n = 1.33$ for red, show that in this case the incident and emergent rays make an angle of 42.1°.

Solution

a) The angle α in Fig. A.6 is given by Snell's law:

$$n \sin\alpha = \sin\phi \tag{A.1}$$

From Eq. (A.1) one has

$$\sin\alpha = \frac{1}{n}\sin\phi < \frac{1}{n} \tag{A.2}$$

Thus, it can be seen that the angle of incidence at the back of the droplet, α, is less than the critical angle for total reflection, given by $\sin^{-1}\left(\frac{1}{n}\right)$, so there is a partial reflection in this case.

continued

Example A.1 continued

b) Considering the geometry of Fig. A.6, one has $\alpha = (\phi - \alpha) + x$, that is $x = 2\alpha - \phi$. The deviation angle D is then given by

$$D = \pi - 2x = \pi - 4\alpha + 2\phi \tag{A.3}$$

c) Considering the above result for deviation D, one has

$$\frac{dD}{d\phi} = 2\left(1 - 2\frac{d\alpha}{d\phi}\right) \quad \text{and} \quad \frac{d^2D}{d\phi^2} = -4\frac{d^2\alpha}{d\phi^2} \tag{A.4}$$

On the other hand, differentiating Eq. (A.1), one has

$$n \cos \alpha d\alpha = \cos \phi d\phi \tag{A.5}$$

Using Eq. (A.5) in Eq. (A.4) provides the following results:

$$\frac{dD}{d\phi} = 2\left(1 - \frac{2\cos\phi}{n\cos\alpha}\right) \quad \text{and} \quad \frac{d^2D}{d\phi^2} = \frac{2}{n}\frac{\sin(\phi - \alpha)}{\cos^2\alpha} \tag{A.6}$$

Thus, it can be seen that for

$$\frac{\cos\varphi}{\cos\alpha} = \frac{n}{2} \tag{A.7}$$

The deviation D is minimal, since $\frac{d^2D}{d\phi^2}$ is positive in that case. By squaring Eq. (A.7) and using Eq. (A.1) one obtains the following result:

$$\sin^2\phi = \frac{4 - n^2}{3} \tag{A.8}$$

For $n = 1.33$, one has $\sin\phi = 0.861$ and $\sin\alpha = 0.646$. The angle between the incident and emergent rays is then given by

$$2x = 4\alpha - 2\phi = 42.1° \tag{A.9}$$

A.8 Atmospheric Refraction

Due to the earth's rotation around its axis, the stars—and in particular the sun—appear to move uniformly across the sky in circular orbits, which makes it possible to predict their positions at successive instants. However, when a given star approaches the horizon, there is a discrepancy between these predictions and the direct observation data. In fact, the observed height is higher than expected. This discrepancy becomes more significant when the star is closer to the horizon.

The explanation for such discrepancy lies in the refraction of light coming from the star as it passes through the earth's atmosphere. The light trajectory thus satisfies Fermat's principle of minimum time, referred to in Chapter 1. As the speed of light is higher in the less dense layers of the atmosphere, its travel time decreases by choosing a trajectory that—although longer—lies as much as possible in these layers. As a result, it may happen that an observer at a certain point on the earth's surface sees a star above the horizon when in fact it is already below it (Fig. A.7).

When the sun is close to the horizon, the atmospheric refraction referred to above raises the apparent position of its upper edge. However, the apparent position of its lower edge

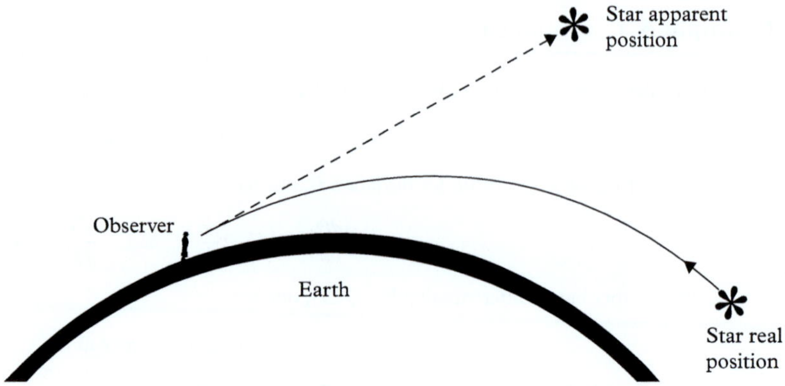

Figure A.7 *Refraction of light from a near-horizon star.*

is even more affected by this effect. This results in the flattened shape that the sun usually exhibits at such times.

A.9 Mirages

The atmospheric refraction discussed in the previous section is caused by the variation in the density of the air, resulting mainly from the variation of pressure with altitude. However, when considering a relatively limited range of altitudes in the lower atmosphere, the decisive factor in the variation in air density is the variation in temperature. Warming causes the air to expand, so under the same pressure warm air is less dense than cold air. The refraction caused by this variation in density is at the origin of the so-called *desert mirage*, as well as the *road mirage*.

During the day, the desert sand becomes hotter than the air, raising the temperature of the layer of air adjacent to it to higher values than that of the layers above. The path of a light ray that propagates near the sand is curved upwards towards the cooler air layers, as shown in Fig. A.8. The dashed lines in this figure indicate the points from which the lower rays appear to come, corresponding to an inverted image of the object, while the upper rays give a right view of the object.

The only environment in nature that provides a right image of an object, simultaneously with its inverted image, is a water mirror. Thus, when observing a mirage in the desert, the observer is led to interpret it as a reflection on a water surface. However, to the hiker's frustration, there is no water, nor is it a reflection—rather it is a simple phenomenon of atmospheric refraction.

A phenomenon similar to that of the desert mirage can be observed with some frequency along roads on a sufficiently hot day. In fact, looking at a distant position on this road, it seems that it is covered by a sheet of water due to the apparent reflection of the sky and the various objects that are on it. Obviously, as one moves forward, this scenario also moves, as with the mirage of the desert. In both cases, the flickering appearance of the mirage is due to the fact that the layers of air are constantly moving relative to each other, thus affecting the value of their refractive indices. The mirages discussed earlier are known as *lower mirages* and form when the land surface is at a higher temperature than the adjacent air layer. However, a different type of mirage—known as a *superior mirage*—can occur when there is warm air over the sea or over a lake whose water is colder. In this case, the layer of air

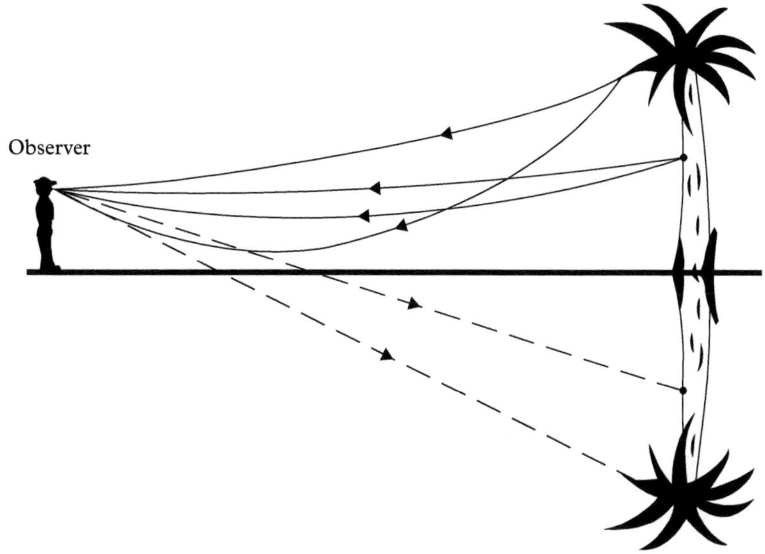

Figure A.8 *Origin of the upright and inverted images in the desert mirage.*

adjacent to the water is cooled, which causes a bending of the path of the light rays towards the earth. As a result, the apparent position of an object observed in these circumstances is above its actual position. This type of mirage can allow, for example, one to see a distant ship when it is, in fact, geometrically below the horizon.

A.10 The Green Flash

The designation 'green flash' refers to different phenomena that can be observed at sunrise or sunset. The colours displayed at these times—most often green—are due to a combination of differential refraction, which cuts out the red part of the spectrum, along with Rayleigh scattering and aerosol scattering, which attenuate the blue and violet colours. On the other hand, orange and yellow colours are attenuated by ozone and water vapour.

Some scientists have sought in the past to explain the *green flash* as an optical illusion. However, the photographic record of this phenomenon, as well as some simulations already carried out, has confirmed its objective existence. Currently, the *green flash* is considered to be a by-product of some kind of mirage. Like all the phenomena of this kind, its characteristics vary greatly with the height at which the observer is, and it is generally observed when the horizon line is at a lower level. The horizon provided by the sea is, in general, the most suitable for this purpose. In fact, the phenomenon of the *green flash* was first observed and sometimes described in detail by sailors, long before scientists themselves became interested in it. However, this phenomenon can also be observed from an airplane, from the top of a mountain, or even from a sufficiently high building.

The *green flash* is typically elliptical in shape, with the longest axis in the horizontal direction. The more intense the mirage, the larger the *flash*, although its duration does not vary significantly. For moderate latitudes, this duration is only 2 s.

A.11 The Twinkling of the Stars

The earth's atmosphere is constantly moving and has local temperature fluctuations, which determine corresponding fluctuations in its refractive index. As a result of these fluctuations, the path of light rays coming from the stars is randomly affected, and it then appears that it is the star itself that is moving.

On the other hand, an air mass between an observer and a given star can act as a converging lens at a given moment, bringing rays of light in slightly different directions into that observer's view. This effect causes an instantaneous increase in both the apparent size and luminosity of that star. Another air mass, which acts as a divergent lens, may cause, at a later instant, an opposite effect in the view of the same observer. These fluctuations in the brightness and even the apparent size of the star are a second factor that explains its flickering.

In addition to causing a deviation in the direction of light rays, the air masses mentioned earlier can also act as a prism, determining the dispersion of the various spectral components of the light coming from the stars. This scattering effect can, for example, mean that at a given instant only the blue light of the respective spectrum is received in the observer's view. A slight change in this prism of air may cause the light that arrives at the same observer at a later instant to correspond to a different colour. This rapid change in colour is one of the characteristic features of the twinkling of stars.

A.12 The Blue Sky and the White Clouds

When, during the day, we look at the sky in a direction away from the sun and contemplate its blue colour, it means that there is (blue) light coming in that direction from the sky to our eyes. Light originates from the sun, but its direction of propagation is changed somewhere in the sky. In fact, when a ray of light passes through the earth's atmosphere, some of its energy is scattered in all directions. If there were no atmosphere, this dispersion would not happen and we would see the dark sky during the day, out of the direction of the sun. This is what is observed, for example, on the moon.

Air causes light to scatter because it is not a continuous fluid, but is made up of discrete molecules. These molecules are sometimes temporarily grouped together, which causes statistical fluctuations in air density. The typical dimensions of these clusters are about a hundred times smaller than the wavelengths of visible light. For this reason, the scattering phenomenon is more significant for light with short wavelengths than for light with long wavelengths. That is, blue light is more effectively dispersed than light with longer wavelengths, which is why the sky appears blue during the day.

As for the clouds that we also usually see in the sky, they are made up of small transparent and colourless water droplets or ice crystals. The dimensions of these droplets are a few tens of times greater than the wavelengths of visible light. Consequently, all components of the solar spectrum are dispersed in a similar way. The overlapping of all these components explains the white colour usually exhibited by clouds (Fig. A.9).

In conclusion, it must be recognized that both the blue colour observed in the sky and the white colour of the clouds have nothing to do with the colours we observe more often in everyday life. These colours are usually determined by the existence of various pigments, which absorb certain wavelengths of incident light. There is no such material in the sky, but only particles. In fact, it turns out that neither the white clouds have anything that is white, nor does the blue sky have anything that is actually blue.

Figure A.9 *A picture illustrating a blue sky with white clouds, as well as their mirror images on a lagoon surface.*

A.13 The Sunset

As the sun descends on the horizon, the distance travelled by the sun's rays through the atmosphere to reach the observer's eyes increases. Along this path, the spectral components with shorter wavelengths (corresponding to the blue colour) are selectively removed, so that the colour of the sunlight that passes through the atmosphere and actually reaches the observer—corresponding to the non-scattered spectral components—varies. Thus, it is understandable that in the course of an afternoon the sunlight which passes through the atmosphere and reaches us exhibits first a yellowish colour, then an orange colour, and finally, towards the end of the afternoon, a reddish colour (Fig. A.10). In reality, blue skies and reddish sunsets are complementary aspects of the same phenomenon.

The reddish colour of the sunset becomes more intense when the atmosphere contains an excess of microscopic particles, which contribute to the scattering of light at longer wavelengths. This phenomenon can be seen, for example, when volcanic eruptions or large forest fires occur, releasing large amounts of these microscopic particles into the atmosphere. Under these circumstances, both clouds and the atmosphere near the direction of the sun are illuminated with reddish light, so the light scattered by these elements exhibits the same colour.

In the most general case, the colour displayed by clouds during a sunset is the result of a complex combination of effects. On the one hand, there is the colour of the light that illuminates them, which is determined by its previous dispersive history. On the other hand, there are the dispersive characteristics of the clouds themselves, which are determined by the dimensions of the droplets that constitute them. The appearance of the clouds is also affected by the colour of the background against which they are observed. The combination of these effects can generate the most varied pictures of the sunset.

Figure A.10 *A picture illustrating the reddish colours of the sunset.*

A.14 Further Reading

Aden and M. Meinel, *Sunsets, Twilights, and Evening Skies*, Cambridge University Press, 1983.

Greenler, R., *Rainbows, Halos, and Glories*, Cambridge University Press, 1980.

Land, M. F., The Optics of Animal Eyes, Contemp. Phys., Vol. 29, 435, 1988.

Lynch, D. K., W. Livingstone, *Color and Light in Nature*, Cambridge University Press, 1995.

Index